# Table of Atomic Weights

| Element | Symbol | Atomic Number | Atomic Weight | Element | Symbol | Atomic Number | Atomic Weight |
|---|---|---|---|---|---|---|---|
| Actinium | Ac | 89 | (227)[a] | Manganese | Mn | 25 | 54.9380 |
| Aluminum | Al | 13 | 26.98154 | Mendelevium | Md | 101 | (258) |
| Americium | Am | 95 | (243) | Mercury | Hg | 80 | 200.59 |
| Antimony | Sb | 51 | 121.75 | Molybdenum | Mo | 42 | 95.94 |
| Argon | Ar | 18 | 39.948 | Neodymium | Nd | 60 | 144.24 |
| Arsenic | As | 33 | 74.9216 | Neon | Ne | 10 | 20.179 |
| Astatine | At | 85 | (210) | Neptunium | Np | 93 | 237.0482 |
| Barium | Ba | 56 | 137.34 | Nickel | Ni | 28 | 58.71 |
| Berkelium | Bk | 97 | (247) | Niobium | Nb | 41 | 92.9064 |
| Beryllium | Be | 4 | 9.01218 | Nitrogen | N | 7 | 14.0067 |
| Bismuth | Bi | 83 | 208.9804 | Nobelium | No | 102 | (255) |
| Boron | B | 5 | 10.81 | Osmium | Os | 76 | 190.2 |
| Bromine | Br | 35 | 79.904 | Oxygen | O | 8 | 15.9994 |
| Cadmium | Cd | 48 | 112.40 | Palladium | Pd | 46 | 106.4 |
| Calcium | Ca | 20 | 40.08 | Phosphorus | P | 15 | 30.97376 |
| Californium | Cf | 98 | (251) | Platinum | Pt | 78 | 195.09 |
| Carbon | C | 6 | 12.01115 | Plutonium | Pu | 94 | (244) |
| Cerium | Ce | 58 | 140.12 | Polonium | Po | 84 | (210) |
| Cesium | Cs | 55 | 132.9054 | Potassium | K | 19 | 39.098 |
| Chlorine | Cl | 17 | 35.453 | Praseodymium | Pr | 59 | 140.9077 |
| Chromium | Cr | 24 | 51.996 | Promethium | Pm | 61 | (147) |
| Cobalt | Co | 27 | 58.9332 | Protactinium | Pa | 91 | 231.0359 |
| Copper | Cu | 29 | 63.546 | Radium | Ra | 88 | 226.0254 |
| Curium | Cm | 96 | (247) | Radon | Rn | 86 | (222) |
| Dysprosium | Dy | 66 | 162.50 | Rhenium | Re | 75 | 186.2 |
| Einsteinium | Es | 99 | (254) | Rhodium | Rh | 45 | 102.9055 |
| Erbium | Er | 68 | 167.26 | Rubidium | Rb | 37 | 85.4678 |
| Europium | Eu | 63 | 151.96 | Ruthenium | Ru | 44 | 101.07 |
| Fermium | Fm | 100 | (257) | Samarium | Sm | 62 | 150.4 |
| Fluorine | F | 9 | 18.99840 | Scandium | Sc | 21 | 44.9559 |
| Francium | Fr | 87 | (223) | Selenium | Se | 34 | 78.96 |
| Gadolinium | Gd | 64 | 157.25 | Silicon | Si | 14 | 28.086 |
| Gallium | Ga | 31 | 69.72 | Silver | Ag | 47 | 107.868 |
| Germanium | Ge | 32 | 72.59 | Sodium | Na | 11 | 22.98977 |
| Gold | Au | 79 | 196.9665 | Strontium | Sr | 38 | 87.62 |
| Hafnium | Hf | 72 | 178.49 | Sulfur | S | 16 | 32.06 |
| Hahnium | Ha | 105 | (260)[b] | Tantalum | Ta | 73 | 180.9479 |
| Helium | He | 2 | 4.00260 | Technetium | Tc | 43 | 98.9062 |
| Holmium | Ho | 67 | 164.9304 | Tellurium | Te | 52 | 127.60 |
| Hydrogen | H | 1 | 1.00797 | Terbium | Tb | 65 | 158.9254 |
| Indium | In | 49 | 114.82 | Thallium | Tl | 81 | 204.37 |
| Iodine | I | 53 | 126.9045 | Thorium | Th | 90 | 232.0381 |
| Iridium | Ir | 77 | 192.22 | Thulium | Tm | 69 | 168.9342 |
| Iron | Fe | 26 | 55.847 | Tin | Sn | 50 | 118.69 |
| Krypton | Kr | 36 | 83.80 | Titanium | Ti | 22 | 47.90 |
| Kurchatovium | Ku | 104 | (260)[c] | Tungsten | W | 74 | 183.85 |
| Lanthanum | La | 57 | 138.9055 | Uranium | U | 92 | 238.029 |
| Lawrencium | Lr | 103 | (256) | Vanadium | V | 23 | 50.9414 |
| Lead | Pb | 82 | 207.19 | Xenon | Xe | 54 | 131.30 |
| Lithium | Li | 3 | 6.941 | Ytterbium | Yb | 70 | 173.04 |
| Lutetium | Lu | 71 | 174.97 | Yttrium | Y | 39 | 88.9059 |
| Magnesium | Mg | 12 | 24.305 | Zinc | Zn | 30 | 65.38 |
| | | | | Zirconium | Zr | 40 | 91.22 |

[a] Value in parentheses is the mass number of the most stable or best-known isotope.

[b] Suggested by American workers but not yet accepted internationally.

[c] Suggested by Russian workers. American workers have suggested the name Rutherfordium.

# Basic Concepts of Chemistry

## To the Student

A Study Guide for the textbook is available through your college bookstore under the title Study Guide to accompany BASIC CONCEPTS OF CHEMISTRY, Second Edition by Leo J. Malone. The Study Guide can help you with course material by acting as a tutorial, review and study aid. If the Study Guide is not in stock, ask the bookstore manager to order a copy for you.

# Basic Concepts of Chemistry

## Second Edition

**Leo J. Malone**
*Saint Louis University*
*St. Louis, Missouri*

**John Wiley & Sons**
New York • Chichester • Brisbane • Toronto • Singapore

Production supervised by Linda R. Indig
Interior designed by Laura C. Ierardi
Cover designed by Lisa A. Krongold
Copy editing supervised by Debbie Herbert
Photo researched by Linda Gutierrez
Cover photo by Phillip A. Harrington/The Image Bank



***Library of Congress Cataloging in Publication Data:***

Malone, Leo J., 1938–
Basic concepts of chemistry.

Includes index.
1. Chemistry. I. Title.
QD31.2.M344 1985 540 84-19476
ISBN 0-471-88600-9

Printed in the United States of America

10 9 8 7 6 5 4

# Preface

*Basic Concepts of Chemistry* is not only an introduction to chemical concepts but is also an introduction to the way chemistry is studied and mastered. The level, content, and sequence of topics have been chosen with a sensitivity to students who have had little or no background in chemistry or who have had a significant interruption in their studies of chemistry. The text is primarily for students who wish to obtain the background and confidence needed to pursue a main sequence chemistry course. It can also be used in the general chemistry part of a course for the allied health professions or for a one-time course in chemistry.

The style is conversational but concise. The traditional sequence of topics provides a step-by-step construction of the science, with one topic building logically on the previous one. It is understood that many, if not most, students using this book will require some mathematical preparation or review in the extensive quantitative concepts applied to chemical systems. The text is written with this very important fact in mind. The mathematical tools of measurement and conversions are presented in Chapter 2 and are supplemented by detailed reviews in the appendixes. The next five chapters introduce many basic chemical concepts that are not primarily mathematical in nature. However, in Chapter 8 the quantitative nature of chemistry becomes apparent with the introduction of the mole. In between Chapters 2 and 8, the students, with the direction of the instructor, are given an opportunity to improve their mathematical, algebraic, and problem-solving skills. There are separate appendixes on basic math (Appendix A), basic algebra (Appendix B), scientific notation (Appendix C), logarithms (Appendix E), and graphs (Appendix F). The appendixes include discussions, worked-out examples, and drill problems (with the answers provided). They are specially designed for application to the types of problems encountered in chemistry. Appendixes A, B, and C also include self-diagnostic tests so that students can easily determine the extent of review they need.

Almost all introductory and general chemistry textbooks now use the unit-factor (dimensional analysis) method of problem solving. I believe, however, that the typical student, who is not familiar with this method, needs more than a one- or two-page introduction to apply this tool consistently and confidently. In Chapter 2, there is a detailed introduction to the topic, *supplemented* by an extensive appendix (Appendix D) that develops the unit-factor method step by step from the construction of conversion factors, to simple one-step conversions, to more complex multiple-step conversions. Solved example and exercise problems (with answers) are provided. Diligent students should become comfortable with this problem-solving method by the time it is applied extensively to chemical systems.

The book is designed with considerable flexibility in the sequence of topics and chapters. Many may prefer to cover moles and stoichiometry as early as possible. Hence, the instructor may proceed directly from Chapter 3 (after Section 3-5) to Chapter 8 and then to Section 9-3 on stoichiometry. I feel that it is preferable to present the topic of moles in a thorough manner, as is done in Chapter 8, instead of superficially introducing this topic in an early chapter without detailed, supporting explanations and examples.

There is obviously more material in this book than can be easily covered in a one-semester course. Generally, Chapters 1 to 10 would be covered, but there will be much variation in later chapters depending on the topics that the instructor wishes to emphasize. The nuclear discussion in Chapter 3 is completely optional: discussion and problems are separated from the rest of the chapter so they can be included at any time. This topic is included here because it follows logically from the discussion of the nature of the nucleus and is effective in building early interest regarding the popular concerns in this area. It can be omitted or included later. The discussions of orbitals and box diagrams in Chapter 4 are also optional and may be deleted without prejudice in later discussions.

Many other features of this book are designed to help students understand and organize the sequence of topics. The following is a list of these special features retained from the first edition of this text.

1 Simple analogies are used that relate the concrete to the abstract. Analogies that are easily understood themselves can be helpful in making new concepts understandable.

2 Introductions to each chapter discuss the overall objectives of the chapter, how it follows from previous discussions and, most important, what specific topics previously discussed are relevant and should be reviewed. This approach emphasizes the continuity of chemistry as a subject where topics build one upon another.

3 Numerous example problems are worked out in the text, step by step. There are usually two or three examples of each type of problem with careful explanations of procedures.

4 End-of-chapter problems are assigned in the margins of the text after a particular topic has been discussed and examples are shown. This is designed to give students direction for the immediate reinforcement of a concept without affecting the continuity of the discussion.

5 End-of-chapter problems are numerous and of varying difficulty. They are categorized by topic. Over one half of the answers are provided in Appendix H (many of the quantitative problems include solutions).

6 New terms are introduced in boldface type. The definitions of the terms are in italics.

7 A comprehensive glossary of terms provides easy reference to the definitions used throughout the text.

The following are additions and other improvements that have been included in the second edition of the book.

1 The number of end-of-chapter exercises are nearly tripled.

2 A chapter purpose and a list of chapter objectives are included at the beginning of each chapter.

3 Chapter summaries are added that give a synopsis of the chapter by using tables, diagrams, and flow charts where appropriate.

4 Comprehensive review tests are added after Chapters 3, 7, 10, and 14. These are designed to integrate the material in the intervening chapters.

5 Chapters are extensively rewritten, and topics within the chapters have been reorganized for improved flow and clarity. Information of current interest is updated.

6 A chapter (Chapter 11) is added on the liquid and solid states of water and changes between states. Some other topics now covered in this text are: colligative properties (Chapter 12), specific heat (Chapter 2), and activation energy (Chapter 15). Two chapters on acid-base chemistry in the first edition are combined into one chapter (Chapter 13) in the second edition. All or parts of this chapter may be included depending on the depth of coverage desired.

7 Several additional study aid materials are now available. The Study Guide to accompany this text is rewritten to provide closer support. In the Study Guide, related sections within a chapter have been grouped for discussion and self-testing rather than the chapter as a whole. Some short stories of current interest are also included in the Study Guide. They stress how chemistry relates to the discoveries and progress in many other disciplines. An innovative, new laboratory manual, written

by Professors Steven Murov and Brian Stedjee, accompanies this text. Finally, excellent computer-assisted instructional software, prepared by Professor Frank P. Rinehart, is being made available to instructors of this course. It will help make the challenge of problem solving fun as well as satisfying. This software will supplement Chapter 9 Chemical Reactions and is one of a forthcoming series for Chemical Education.

I hope you find the study or the teaching of this course rewarding and that you sense the author's enthusiasm for this fascinating discipline.

Leo J . Malone

# Acknowledgments

Writing a book like this is never an individual project. I thank my colleagues in the Chemistry Department of the Saint Louis University for their many helpful comments, especially during the preparation of the first edition. My special thanks are owed to Dr. Judith Durham, who was particularly helpful. I am grateful to my family for their assistance in proofreading, especially my wife, Sharon, and my daughter, Mary. I thank my editors at Wiley, Cliff Mills and then Dennis Sawicki, for their help, humor, and encouragement. Finally, the following people reviewed the manuscript and offered many useful comments and suggestions:

Dr. Howard De Voe
*Department of Chemistry*
*University of Maryland*
*College Park, Maryland 20742*

Prof. Clarence Cunningham
*Department of Chemistry*
*Oklahoma State University*
*Stillwater, Oklahoma 74078*

Prof. James Weber
*Department of Chemistry*
*University of New Hampshire*
*Durham, New Hampshire 03824*

Dr. Forrest C. Hentz
*Department of Chemistry*
*North Carolina State University*
*Raleigh, North Carolina 27650*

Prof. Robert O'Malley
*Department of Chemistry*
*Boston College*
*Chestnut Hill, Massachusetts 02167*

Prof. R. Dyal
*Department of Chemistry*
*University of Illinois*
*Urbana, Illinois 61801*

James E. Hunter
*Department of Chemistry*
*University of Illinois*
*Urbana, Illinois 61801*

Prof. Elizabeth P. Rogers
*University of Illinois*
*Urbana, Illinois 61801*

Prof. Timothy A. Kling
*Department of Chemistry*
*Lakeland Community College*
*1-90 and Route 306*
*Mentor, Ohio 44060*

Prof. Moheb M. Seif El-Nasr
*Department of Chemistry*
*The Lindenwood Colleges*
*Saint Charles, Missouri 63301*

Prof. K. Thomas Finley
*Department of Chemistry*
*State University College– Brockport*
*Brockport, New York 14420*

Prof. Martha J. Gilleland
*Department of Chemistry*
*California State College*
*Bakersfield, California 93309*

Prof. Roger Penn
*Department of Chemistry*
*Sinclair Community College*
*444 West third Street*
*Dayton, Ohio 45402*

Prof. Richard Monson
*Department of Chemistry*
*California State University Hayward*
*Hayward, California 94542*

Prof. Sol Shulman
*Department of Chemistry*
*Illinois State University*
*Normal, Illinois 61761*

Prof. Steven Murov
*3912 Midcrest Ct.*
*Modesto, California 95355*

# Contents

## Chapter 3
## The Atom, the Structure of Matter, and Nuclear Reactions 57



## Chapter 4
## The Arrangement of the Electrons in the Atom 91



## Chapter 5
## The Periodic Nature of the Elements 115

## Chapter 6
## The Nature of the Chemical Bond 137



## Chapter 7
## The Naming of Compounds 167



## Chapter 8
## Quantitative Relationships: The Mole 187

## Appendix A
## Basic Mathematics 474



## Appendix B
## Basic Algebra 488



## Appendix C
## Scientific Notation 504



## Appendix D
## Problem Solving by the Unit-Factor Method (Dimensional Analysis) 513

# Basic Concepts of Chemistry

The emergence of the human race from among the other animals can be traced to the use of fire.

# Chapter 1
# Chemistry: Matter, Changes, and Energy

## Purposes of Chapter 1
In Chapter 1 we introduce the discipline of chemistry, distinguish and classify various forms of matter, and discuss the physical, chemical, and energy changes involved in chemistry.

## Objectives for Chapter 1
(Each objective is followed by the appropriate section in the text.) After completion of this chapter, you should be able to:

**1** Give the definitions of chemistry and matter. (1-2)
**2** Describe the three states of matter and give examples of each. (1-2)
**3** Classify a sample of matter as either heterogeneous or homogeneous. (1-2)
**4** Distinguish between the properties of solutions and pure substances. (1-3)
**5** Classify a list of pure substances as either elements or compounds. (1-4)
**6** List the names and symbols of the common elements. (1-5)
**7** Give examples of physical and chemical properties and physical and chemical changes. (1-6,7)
**8** Apply the law of conservation of mass to explain observed chemical changes. (1-8)
**9** Classify a physical or chemical change as either endothermic or exothermic. (1-9)
**10** Name and describe the various forms of energy. (1-9)
**11** Distinguish between potential and kinetic energy. (1-9)
**12** Describe the relationship between matter and energy. (1-10)

In Greek mythology there existed a god named Prometheus. Prometheus supposedly gave to animals the special tools they needed to survive in a hostile world. In the eyes of the other gods, however, he went too far when he got to humans, because he gave them a tool reserved for the gods themselves. This tool was fire.

In reality, the use of fire by our ancient ancestors is thought to be one of the most significant events that marked the emergence of our species as "human" and thus truly superior to all other animal species. It appears that for at least 400,000 years humans have used fire to bring light to the darkness, to give heat during the long, cold winters, and to tenderize food by cooking. Other species of animals can communicate with each other, and some can even use simple stone tools. Only humans have dared to use fire, the tool of the gods.

Fire is still fascinating. One can watch the flames in a log fire leap and toss about for hours at a time. It is no wonder that it was once thought to be the source of so much magic and mystery. The ancient Greeks considered fire, along with earth, air, and water, as the basic elements of nature. Fire was the transforming element, the one that caused one substance to change into another. We now realize that fire itself is a result of the hot gases and energy associated with a transformation. As we will study later in this chapter, the transformation of one substance into another is called a **chemical reaction.** It is thus logical to state that our species became "human" thousands of years ago when it first put to use the science of chemistry in the form of fire.

Since the taming of fire, many other significant advances have been based on the observance and use of a chemical phenomenon. At first, people shaped tools from stone, but about 10,000 years ago a different naturally occurring substance gained value because it could be pounded into various shapes and sharpened to a much finer edge than ordinary stone. This substance was the metal copper. Still, this metal was relatively rare in its natural state, so it could not be utilized to a large extent. What was probably an accidental discovery opened the way for the age of metals. In about 5000 B.C., someone in ancient Persia discovered that the fire from the glowing coals of an earthen furnace would transform the green stone, malachite, into the red metal, copper. Imagine the commotion that must have occurred as some prehistoric citizen insisted that one could change a common rock into a valuable metal! The conversion of ores to metals is now a well-known chemical process called *metallurgy*. When first discovered, however, it must have been considered an early example of the magic of fire.

The evidence that chemistry has been a major factor in our progress is impressive. In about 3000 B.C. the Egyptians could dye cloth and embalm their dead. They were so good at these chemical processes that we can still tell what caused the death of some of the mummies from these times and even what diseases they may have had during their lifetimes. Despite their use of chemistry, however, the early Egyptians actually had no idea of why any of these procedures occurred.

Around 400 B.C. the Greeks went in the other direction. They thought and argued a great deal about why things occurred in the world around them, but they didn't really put their ideas to practical use or even check out their ideas by experimentation. Still, some of their thoughts have proved to be consistent with modern developments.

The Middle Ages (500–1600 A.D.) are usually referred to as the Dark Ages because of the retreat from the arts and literature, the decline of central governments, and the general loss of so much of the civilization that Egypt, Greece, and Rome had previously built. In the case of chemistry, however, considerable advancement occurred. Some of the chemists of the time, especially those in Europe, were known as alchemists and were thought to possess magical powers. Among other things, alchemists were searching for ways to transform cheaper metals such as lead and zinc into gold, which was thought to be the perfect metal. Obviously, they never accomplished the impossible, and many good alchemists were executed for their failure. However, they may have been pleased to know that they did establish certain important chemical procedures, such as distillation and crystallization. They also discovered and prepared many previously unknown chemicals that we now refer to as elements and compounds.

Modern chemistry had its foundation in the late 1700s when the use of the analytical balance became widespread. At that time chemistry became a quantitative science, and from then on theories used to describe chemistry had to be checked and correlated with the results of experiments performed in the laboratory. From this came the modern atomic theory, first proposed by John Dalton around 1803. This theory, which in a slightly modified form is still accepted today, gave chemistry the solid base from which it could grow to serve humanity on such an impressive scale. Actually, most of our understanding of chemistry has evolved in just the last 100 years. In a way this makes chemistry a very young science. However, if we mark the beginning of "human" behavior as the use of fire, it is also the oldest science.

## 1-1 The Study of Chemistry

Curiosity is defined as the desire to know and to learn about things. We all know how that can get you into trouble, but it is also why science has had such a significant impact on our daily lives. Someone's curiosity has brought us airplanes, automobiles, plastics, synthetic fibers (Orlon, Dacron, and nylon), chemicals that treat cancer and other diseases, microchips for computers, and genetic engineering. The modern advancements of science make 100 years ago look almost like the stone age. At this rate, it is hard to imagine how things will be 100 years from now. In any case, like all other sciences, chemistry attracts the profoundly curious person, one who never outgrew the childhood habit of constantly asking the question "why?" As children, most scientists probably drove their parents crazy with their continual questions about how and why things happen.

Why study chemistry? The obvious answer is because it is required. But maybe you can appreciate *why* it is required. Chemistry is *the* fundamental natural science. Biology, physics, geology, as well as all branches of engineering and medicine, are based on an understanding of the chemical substances of which matter is composed. It is the beginning point in the course of studies that eventually produces all scientists, engineers, and physicians.

Besides scientists and engineers, all intelligent citizens need a foundation of scientific knowledge in order to make intelligent decisions concerning the future of our fragile environment. What is air and water pollution (see Figure 1-1)? How did it get there and how do we get it out? What about our need for energy and its relation to the pollution problem? These questions and others can't be discussed intelligently without some knowledge about the nature of the matter involved. Neither should these questions be left solely to scientists. Everyone has a big stake in the answers. Thus it seems reasonable to expect that a fundamental knowledge of chemistry could be a prerequisite not only for a course of study but for anyone living in confusing and complicated times.

Chemistry can be studied and appreciated by the average person. It need not be feared. Chemistry is orderly, predictable, and entirely reasonable,

**Figure 1-1** SMOG. To fully appreciate the origin and control of pollution, citizens need a basic understanding of chemistry.

which gives the study of the science an advantage over the social sciences. Psychology, for example, deals with the whims of human nature. The nature of humans is constantly changing. On the other hand, once we know the facts about chemicals, we can be assured that the nature of the matter that we have studied will never change.

If you enjoy mastering concepts in your mind or find a sense of satisfaction in solving a problem successfully, then you will find the study of chemistry challenging. Keep an open mind; it may be fun. The study of chemistry is much like the study of a language. As such, it requires special study habits and academic self-discipline. What follows is a checklist of questions dealing with the academic self-discipline needed in the study of this science.

**1** *Can you budget time on a regular basis to study and work problems?*

Like playing basketball or the piano, chemistry requires practice to master. The practice should be done very soon after the topic is covered and not right before the test. One wouldn't wait until the night before the big game to first practice jump shots or the night before the recital to first practice the sonata. This text uses the margins to assign problems that emphasize a certain concept. It is wise to stop at that point and practice before you proceed. Each chapter and each topic within a chapter is like constructing a building. You can't put up the second floor until you have built a firm foundation with the first. The night before a test should be reserved for *review* and a good night's sleep—it is not time to break new ground.

**2** *Are you willing to read ahead?*

If you were captain of a football team you would probably want to come out onto the field early just to get a look at the other team warming up. It may give you a feeling of what lies ahead. Likewise, reading ahead makes you aware of some of the concepts that you will be discussing. You may not completely understand the concepts, but that just makes you listen more intently when it's covered in lecture. Reading ahead can save time in note taking since you'll know that certain tables and definitions are available in the book.

**3** *Will you attend lectures regularly and take good notes?*

You will notice (if you haven't already) that the successful students in college miss very few classes. It is almost impossible to stay up with the lecture unless you're there to hear it. A text such as this is meant to help you understand—it is not meant to replace your instructor.

Once you are there at the lecture you need to take orderly notes. Remember, you will need these notes to know what your instructor emphasizes and to help organize your review. When a problem is worked out on the board, copy the problem so that you can cover it up later and rework it yourself.

**4** *Can you ask questions in class or afterwards?*

You can't let a basic concept go by without owning it. If the class is such that asking questions during class is impossible (or embarrassing), ask later. Your instructor is available, and you owe yourself the answers. Step forward; this is your life, and your question is not "dumb."

**5** *Can you keep at it even if you are disappointed on the first test?*

An "A" is a great way to start, but what if that doesn't happen? In that case you'll have to reanalyze your study habits and try again. Don't be afraid to ask your instructor for advice in this matter. For most people, mastering chemistry takes perseverance. Give the course a chance with your best shot. You will know eventually whether you should "haul up the flag."

**6** *Can you memorize?*

Some definitions, names, and formulas have to be memorized. Some people will say, "I don't want to memorize, I want to understand." To be sure, understanding is what you are after, but memorization is often necessary before understanding develops. In this respect studying chemistry is like studying a foreign language. Before you can learn to speak or write a new language, such things as vocabulary and verb declensions must be committed to memory.

**7** *Can you do problems systematically and neatly?*

It is amazing how the good performances in chemistry courses usually correlate with neat and orderly notes, papers, and tests. Order on paper shows an orderly mind. That helps in chemistry.

If you are able to show by your habits positive answers to these questions, you just may surprise yourself with how well this material comes to you and, as a result, just how fascinating chemistry can be.

## 1-2 Chemistry and the Nature of Matter

The definition of chemistry at first seems simple. **Chemistry** *is the branch of science that deals with the nature and composition of matter and the changes that matter undergoes.* Actually, there are two key words in this definition that require further discussion. These words are "matter" and "changes." First, we will discuss **matter,** which is defined as *anything that has mass and occupies space*. After this we will discuss changes.

All matter exists in one of three distinct physical states. *The* **solid state** *has a fixed shape and a fixed volume. The* **liquid state** *has a fixed volume but not a fixed shape*. Thus liquids take the shape of the lower part of the container. Finally, *the* **gaseous state** *has neither a fixed shape nor a fixed volume*. Thus a gas fills a container uniformly. (See Figure 1-2.)

We are already familiar with many examples of all three physical states. Ice, rock, salt, and steel are substances that exist as solids; water, gasoline, and alcohol exist as liquids; ammonia, natural gas, and air are present in nature as gases. As we will mention later, temperature and the nature of the substance determine the physical state of the particular substance. For example, water is a solid (ice) at low temperatures and rock is a liquid (lava from volcanos) at high temperatures.

Matter can also be classified (arranged systematically) as to the uniformity of its properties. *The* **properties** *of a substance describe its unique, observable characteristics or traits. If the properties of a sample of matter are the same throughout and it contains but one phase, the sample is said to be* **homogeneous.** *A* **phase** *is one physical state (solid, liquid, or gas) with uniform properties*. Sugar, air, an alcohol-water mixture (see Figure 1-3*a*), and copper pennies all exist in one phase, are uniform throughout, and are thus homogeneous. On the other hand, **heterogeneous matter** *is a nonuniform mixture* and thus is the more complex form. *Heterogeneous matter is a mixture of two or more phases with definite boundaries between the phases*. Examples of heterogeneous matter that are obvious to the naked

(*a*)

(*b*)

(*c*)

**Figure 1-2** THE THREE PHYSICAL STATES OF MATTER.
(*a*) Solids have a definite shape and a definite volume.
(*b*) Liquids have a definite volume but an indefinite shape.
(*c*) Gases have an indefinite volume and an indefinite shape.

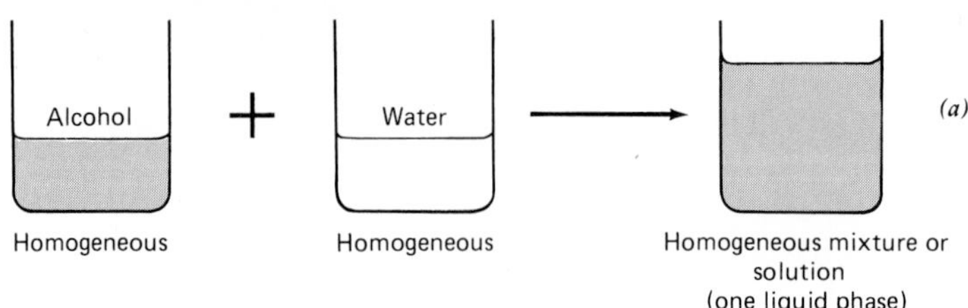


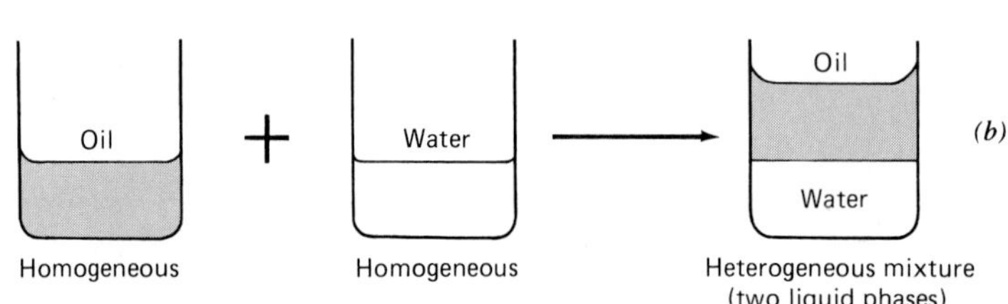


**Figure 1-3** MIXTURES. Water and alcohol form a homogeneous mixture (solution) (*a*) while water and oil form a heterogeneous mixture (*b*).

eye are oil and water (two liquid phases, see Figure 1-3*b*), salt and sand (two solid phases), shaving cream foam (liquid and gas phase), and ice water (solid and liquid phases). Sometimes it is not easy to tell whether matter is homogeneous or heterogeneous. Creamy salad dressing and fog both appear uniform and thus homogeneous to the naked eye. If we put a sample of each under a small microscope, however, the truth would become apparent. The salad dressing has little droplets of oil suspended in the vinegar (two liquid phases), and the fog has droplets of water suspended in the air (liquid and gas phases).

Heterogeneous matter can be separated into two or more homogeneous components. In Figure 1-4 we have started with a heterogeneous mixture containing a liquid phase (ordinary table salt dissolved in water) and a solid phase that has settled to the bottom (copper powder). By a simple laboratory procedure called filtration, we can separate the heterogeneous mixture into its two homogeneous phases, salt water and copper (Figure 1-5).

**See Problems 1-1 through 1-6.**

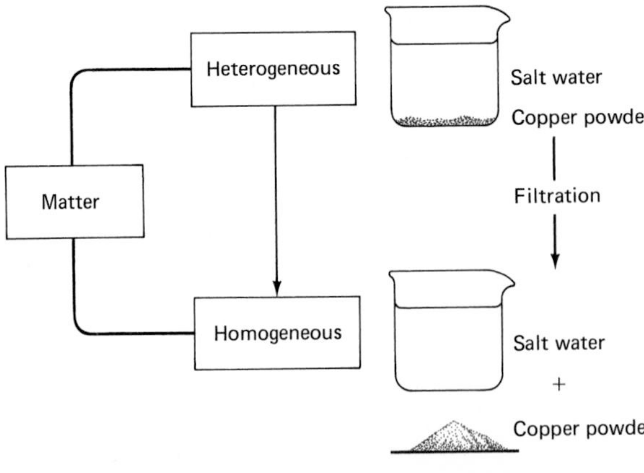


**Figure 1-4** CLASSIFICATION OF MATTER. A heterogeneous mixture can be separated into homogeneous substances.

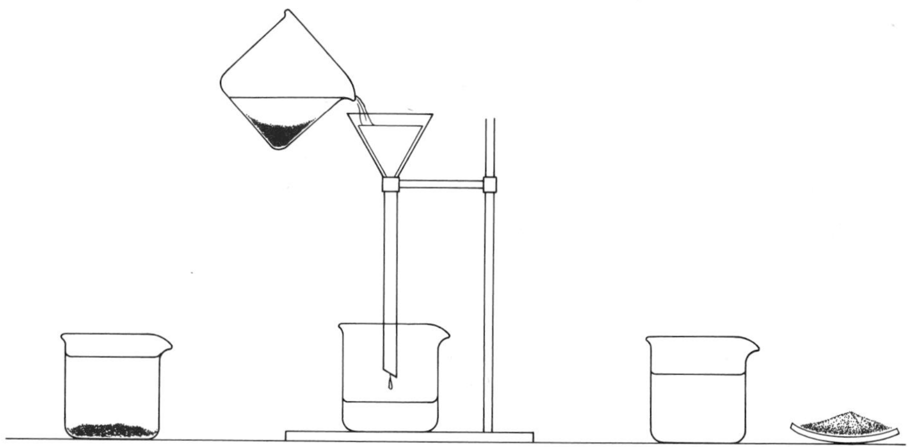

**Figure 1-5** FILTRATION. This laboratory procedure is used to separate a solid phase from a liquid phase.

## 1-3 Classification of Homogeneous Matter

We are now ready to examine homogeneous matter. Homogeneous matter can be also classified into two categories: solutions and pure substances. **Pure substances** *have definite compositions with definite, unchanging, and unique properties.* A **solution** is the more complex, as it *is composed of a mixture of two or more pure substances.* Unlike the heterogeneous mixtures discussed earlier, the components of a homogeneous solution are thoroughly and intimately mixed. (See Figure 1-3*a*.) Even the most high-powered microscope would not reveal evidence of salt crystals suspended in water in the salt water solution of our example. Many solutions can be separated into their pure components by laboratory procedures such as distillation, as shown in Figure 1-6.

In our example, the salt water can be separated into two pure components, salt and water. The copper filtered off earlier is also a pure substance. (See Figure 1-7.) We can identify these three materials as pure substances by their unique properties, which serve as their fingerprints. For example, pure water freezes or conversely, ice melts sharply at 0 °C (32 °F), no matter where on earth it is found. Pure table salt, whose chemical name is sodium chloride, melts at 804 °C. Copper metal can be identified by its reddish color and melting point of 1083 °C. Solutions, on the other hand, have properties that vary according to the proportions of the components. Although salt and water alone have definite melting or freezing points, a salt water solution has a freezing point that varies from about −18 °C to just under 0 °C depending on the amount of salt dissolved.

A mixture not only melts and boils at temperatures different from its components, but it does so over a range of temperatures. For example, if water is boiled until it is all gone, we would notice that the temperature remains at exactly 100 °C (212 °F) from start to finish. When a solution

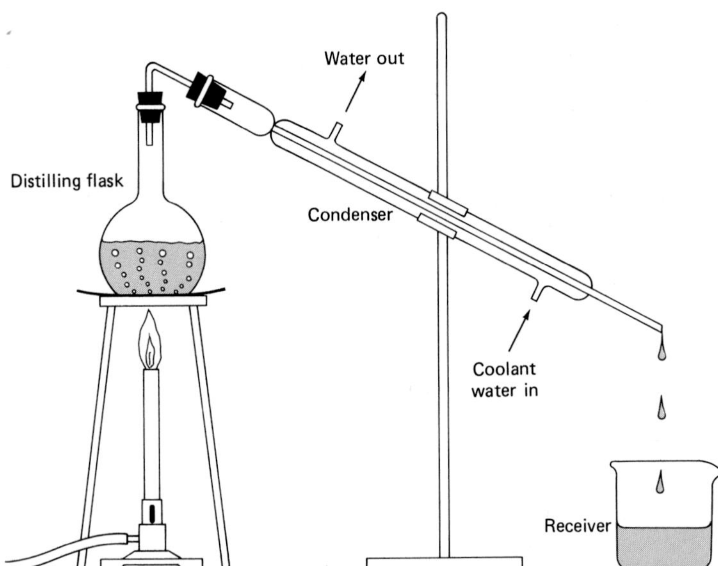


**Figure 1-6** DISTILLATION. This laboratory procedure is used to separate a solution into pure substances.

boils, however, in most cases the temperature of the boiling solution changes as the liquid boils away.

Another variable property of solutions that we can appreciate is the taste of sweetened coffee or tea. The more sugar that is dissolved, the sweeter it tastes, although there is a limit to how much sugar can be dissolved in a given amount of water.

A solution is much like a mixture of colored paints. If you mix red and blue you get violet. The shade of violet depends on the proportions of the two components. The violet color produced is simply a blend of the two and retains characteristics of the two component colors.

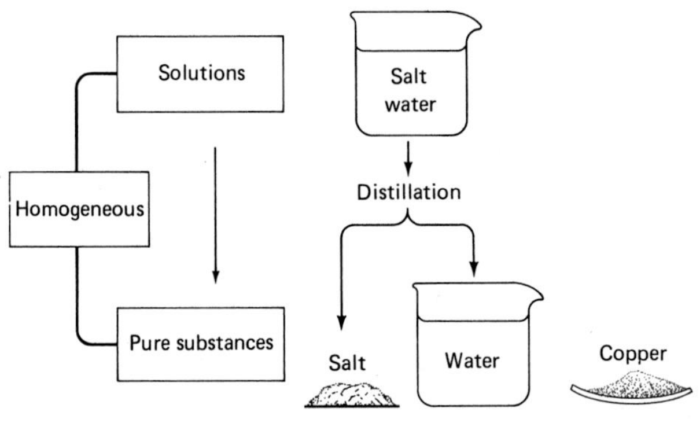


**Figure 1-7** CLASSIFICATION OF HOMOGENEOUS MATTER. A solution can be separated into pure substances.

## 1-4 Classification of Pure Substances

Table salt, water, and copper are all pure substances with distinct and unchanging composition and properties. However, pure substances can also be classified into two categories: elements and compounds. **Elements** *are the most fundamental forms of matter that can exist under normal laboratory conditions.* They cannot be broken down further by chemical means. **Compounds** are the more complex and *are composed of two or more elements combined in fixed proportions.* In our example, the salt and water are compounds and can thus be broken down into elements. The chemical decomposition of water produces the elements hydrogen and oxygen, and the decomposition of salt (sodium chloride) yields the elements sodium and chlorine. The copper powder cannot be separated further, so it is also classified as an element. (See Figure 1-8.)

Compounds are formed from elements or other compounds in what is called a **chemical reaction.** When a chemical reaction occurs, there is a fundamental change in the substances involved. For example, in Figure 1-9 two elements, zinc and sulfur, can be mixed as powders to form an obviously heterogeneous mixture in which both solid phases can be easily seen. Zinc is a shiny, bluish-white metal that melts at 419 °C. Sulfur is a soft, yellow solid that melts at 113 °C. If this intimate mixture of elements is ignited by a hot flame, a chemical reaction occurs forming a compound called zinc sulfide. If the proportions are just right, there is no zinc or sulfur remaining. The zinc sulfide does not have properties similar to either the zinc or sulfur from which it was formed. It has distinct, definite, and uniform properties of its own. Zinc sulfide is a white solid with a melting point of 1850 °C.

Formation of a compound from the elements is much like baking a cake. We mix eggs, flour, and other ingredients and allow the mixture to heat. What comes out is a unique substance that no longer looks, tastes, or actually

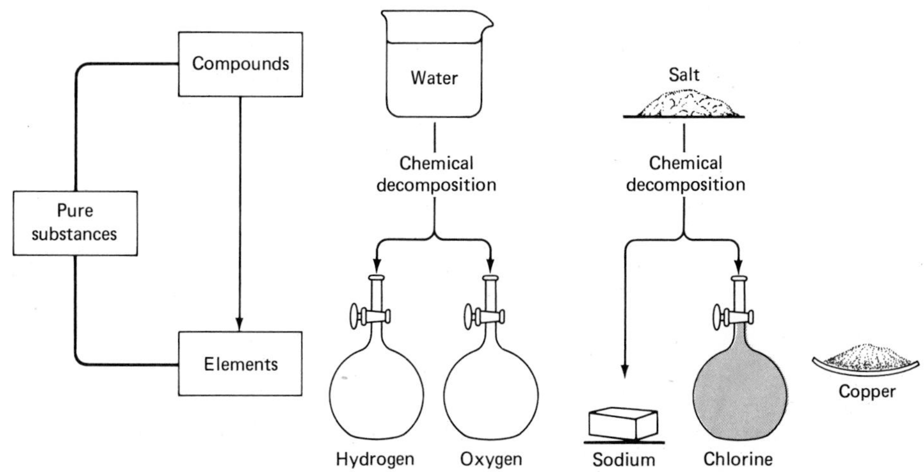


**Figure 1-8** CLASSIFICATION OF PURE SUBSTANCES. A compound can be separated into elements.

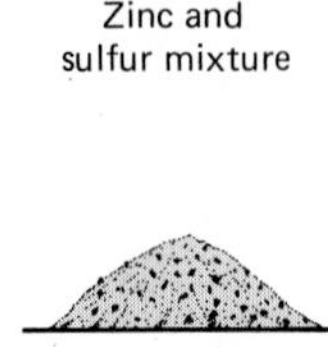


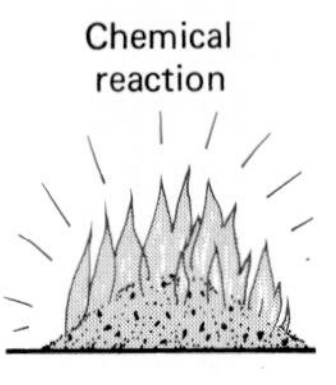


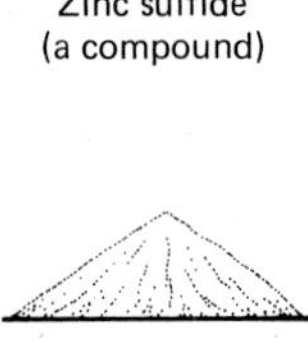


**Figure 1-9** FORMATION OF ZINC SULFIDE. In this chemical reaction a mixture of elements is changed into a compound.

is the mixture we put in the oven. The mix of ingredients has been transformed into a new homogeneous substance (we hope).

## 1-5 The Names and Symbols of the Elements

An element is designated by its name and symbol. *In most cases the first one or two letters of the name are used as the element's* **symbol.** When elements have a two-letter symbol, the first is capitalized but the second is not. The names and symbols of all of the 105 named elements are shown in the inside front cover (elements 106, 107, and 109 have been reported but not yet named). In a chart of the elements called the **periodic table** (which is shown inside the back cover), the symbols of the elements are listed in addition to other important information that will be discussed in Chapter 3. Some common elements and their symbols are shown in Table 1-1.

In the table you probably noticed that some symbols do not seem to relate to their names. Many of these elements, however, derive their symbols from their original Latin or Greek names. For example, the Latin name for sodium is *natrium* (Na); for potassium, *kalium* (K); for gold, *aurum* (Au); and for lead, *plumbum* (Pb). In the chapters that follow more and more

**Table 1-1** Some Common Elements[a]

| Element | Symbol | Element | Symbol |
|---|---|---|---|
| Aluminum | Al | Lead | Pb |
| Bromine | Br | Magnesium | Mg |
| Calcium | Ca | Nickel | Ni |
| Carbon | C | Nitrogen | N |
| Chlorine | Cl | Oxygen | O |
| Chromium | Cr | Phosphorus | P |
| Copper | Cu | Potassium | K |
| Fluorine | F | Silicon | Si |
| Gold | Au | Silver | Ag |
| Helium | He | Sodium | Na |
| Hydrogen | H | Sulfur | S |
| Iodine | I | Tin | Sn |
| Iron | Fe | Zinc | Zn |

[a]An element that isn't particularly common is promethium (Pm), named after the giver of fire who was mentioned earlier.

elements are used in the discussions. Eventually, you will recognize the names and symbols of a large number of elements.

The names of compounds are usually derived from the names of the elements from which they are made. The names of many of the compounds that are referred to in this text contain two words where the second word ends in *ide, ite,* or *ate.* Thus names such as carbon dioxide, zinc sulfate, and iron(III) nitrite all refer to specific compounds. A large number of compounds also have common names such as water, ammonia, lye, and methane. We will talk more about the names of compounds in Chapter 7.

**See Problems 1-7 through 1-15.**

## 1-6 Physical and Chemical Properties

In the discussion in Section 1-4 we discussed the properties of three pure substances, two elements (zinc and sulfur), and one compound (zinc sulfide). There are two kinds of properties of matter, physical and chemical. **Physical properties** *can be observed without changing the substance into another substance.* Physical properties of substances are properties such as odor, color, state of matter (solid, liquid, or gas), melting point, and boiling point. Specific gravity, which compares the weight of a specified volume of the substance to that of water, is also a common physical property (see Figure 1-10). Your physical properties may be listed on your driver's license, such as height, weight, sex, and the color of your eyes and hair.

Just as there's more to falling in love with a person than just noticing his or her physical properties (theoretically, anyway, there should be some "chemistry"), there's more to describing a substance than noting physical properties. A substance also has **chemical properties,** *which relate to its*

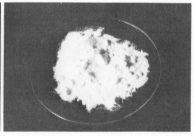

*Zinc*

1. Solid
2. Melts at 419 °C
3. Boils at 907 °C
4. Bluish-white
5. Metallic taste
6. Specific gravity 7.1

*Sulfur*

1. Solid
2. Melts at 113 °C
3. Boils at 444 °C
4. Yellow
5. Pungent smell
6. Specific gravity 2.1

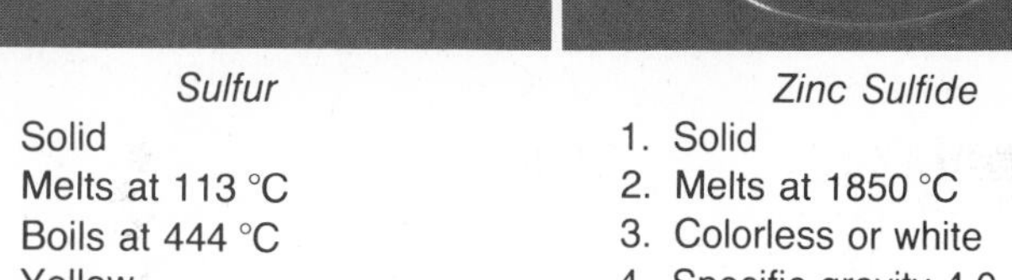

*Zinc Sulfide*

1. Solid
2. Melts at 1850 °C
3. Colorless or white
4. Specific gravity 4.0

**Figure 1-10** PHYSICAL PROPERTIES. Zinc, sulfur, and zinc sulfide all have distinct and different physical properties.

*ability or tendency to change into other substances by a chemical reaction.* Chemical properties are like a personality. You have to see how the substance acts in the presence of other substances to see what its "chemistry" is all about. Since chemical properties describe chemical changes, examples are given in the next section.

## 1-7 Physical and Chemical Changes

A **physical change** *in a substance does not involve a change in the composition of the substance but is simply a change in physical state or dimensions.* When water freezes to form ice or boils to form steam, the liquid state has simply undergone a physical change to another phase. Ice and steam are both forms of water just as much as is the more familiar liquid state. On the other hand, a **chemical change** *involves a change of substances into other substances by means of a chemical reaction.* (See Figure 1-11.) As discussed previously, zinc sulfide is fundamentally different than the elements from which it came. The decay of vegetation, burning of coal, and growth of a tree all involve chemical changes.

**See Problems 1-16 through 1-19.**

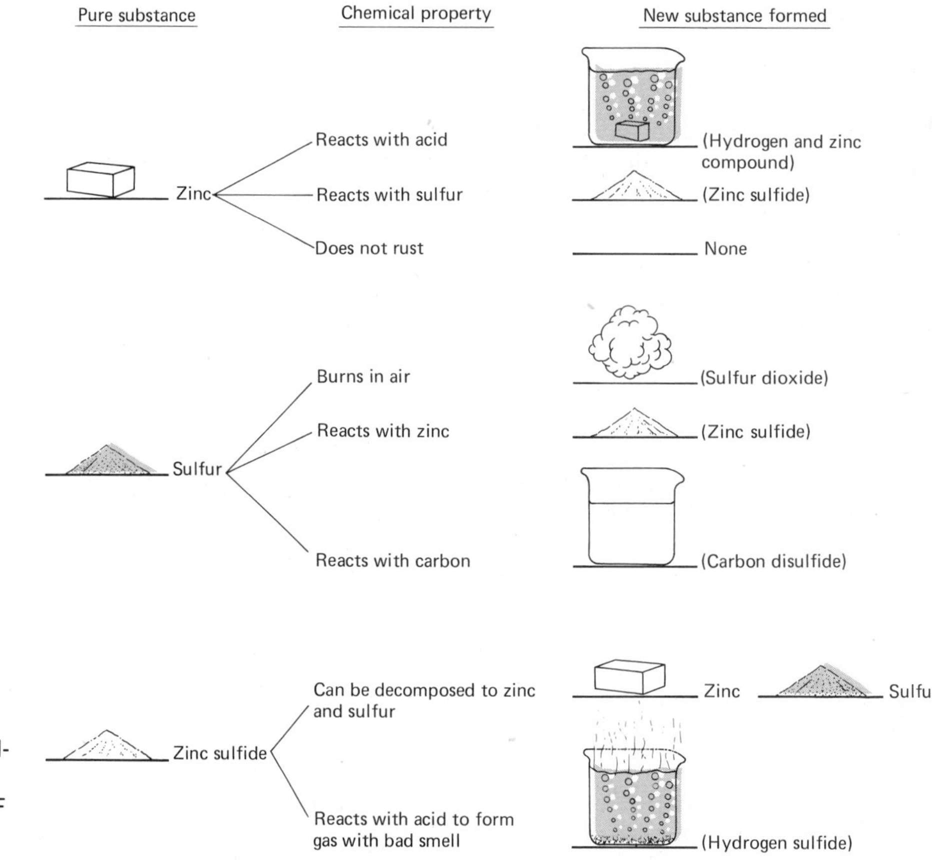


**Figure 1-11** CHEMICAL PROPERTIES AND CHANGES OF SUBSTANCES.

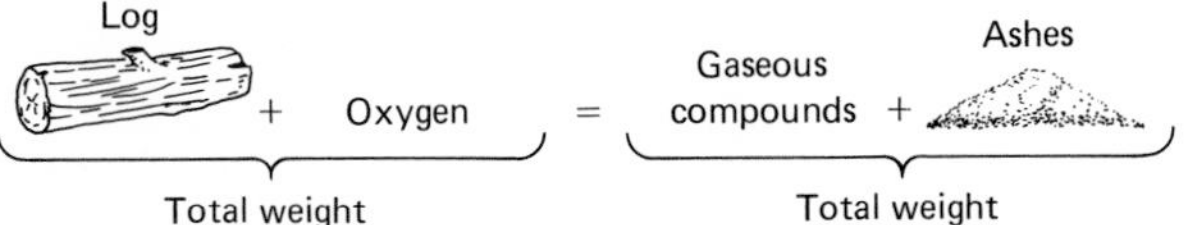


**Figure 1-12** THE COMBUSTION OF WOOD. In a chemical reaction mass is conserved.

## 1-8 The Conservation of Mass

In Figure 1-9, the zinc sulfide weighs exactly the same as the zinc and sulfur mixture from which it was formed. Not more than 200 years ago, however, scientists were still puzzled when wood burned and most of the weight seemed to disappear. This was explained by various theories, but our understanding of chemical changes is now based on the **law of conservation of mass,** which states that *matter is neither created nor destroyed in a chemical reaction.* Two hundred years ago the involvement of gases in chemical reactions was not fully understood. It is now known that the weight of the wood plus the weight of the oxygen in the air used in combustion equals the weight of the gases formed plus the weight of the ashes (see Figure 1-12). The gases involved are all invisible but still have mass.

In Chapter 8, chemical reactions from a quantitative point of view will be introduced. At that time you will notice that many of the calculations are based on the law of conservation of mass.

## 1-9 Energy Changes in Chemical Reactions

When a log burns, it is obvious that more than just gas and ashes are formed. Large amounts of heat are also released by the combustion process. *Chemical reactions that release heat are called* **exothermic,** *and those that absorb heat are called* **endothermic.** An example of an endothermic process involves an "instant cold-pack." When the compounds ammonium nitrate (a solid) and water are brought together in a plastic bag, a solution is formed. The endothermic solution process causes enough cooling to make an ice pack useful for treating sprains and minor aches for athletes. (See Figure 1-13.)

**Figure 1-13** INSTANT COLD-PACK. When capsules of ammonium nitrate are broken and mixed with water, a cooling effect results.

Heat is but one form of energy. But what is energy? When we have no energy left at the end of the day, the implication is clear: We have no capacity to do work. That is exactly how energy is defined. **Energy** *is the capacity or the ability to do work.* Energy exists in several forms, and chemistry is vitally concerned with any changes in energy involving chemical reactions. Almost all of our energy on earth originates from the sun in the form of *radiant* or *light energy.* (Light energy will be discussed in Chapter 4.) By a process called photosynthesis, a living tree can transform the solar, radiant energy into *chemical energy.* In this process, energy-poor compounds in the environment are transformed into energy-rich compounds in the tree. When logs from the tree burn, the chemical energy is transformed into *heat energy,* which we observed earlier. In this process, energy-poor compounds are also formed and returned to the environment. If the heat energy is used to produce steam that is used to turn a turbine, the energy is converted into

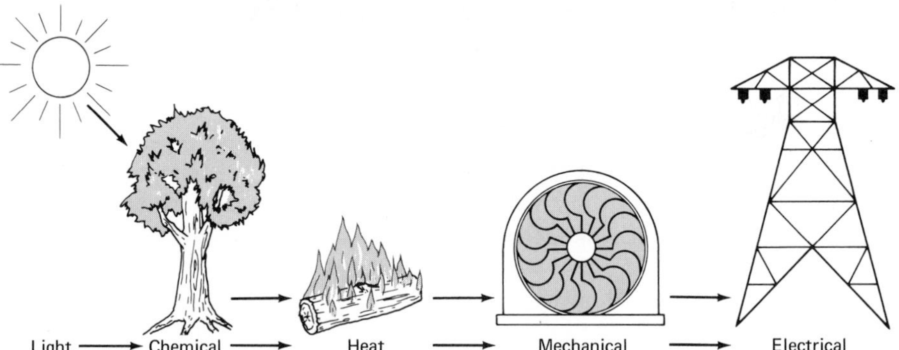


**Figure 1-14** ENERGY. Energy is neither created nor destroyed but can be transformed.

*mechanical energy,* which is then converted into *electrical energy*. The chemical reaction that occurs in a car battery, however, is used to convert chemical energy directly into electrical energy. In all of these energy transformations that have been discussed, energy is not gained or lost. It is simply transformed. (See Figure 1-14.) **The law of conservation of energy** *states that energy cannot be created or destroyed.*

In addition to the forms of energy there are two *types* of energy depending on whether the energy is available but not being used or actually in use. For example, a weight suspended above the ground has energy available because of its suspended position and the attraction of gravity for the weight. *Energy that is available because of position or composition is known as* **potential energy.** Other examples are water stored behind a dam, and a coiled spring. In our example, when the weight is released, the energy can be put to use by the downward motion. *Energy due to motion is called* **kinetic energy.** In order to move the weight back into position, energy must be supplied. (See Figure 1-15.) In future chapters, potential and kinetic energy will play a part in discussions of some chemical systems.

**See Problems 1-20 through 1-25.**

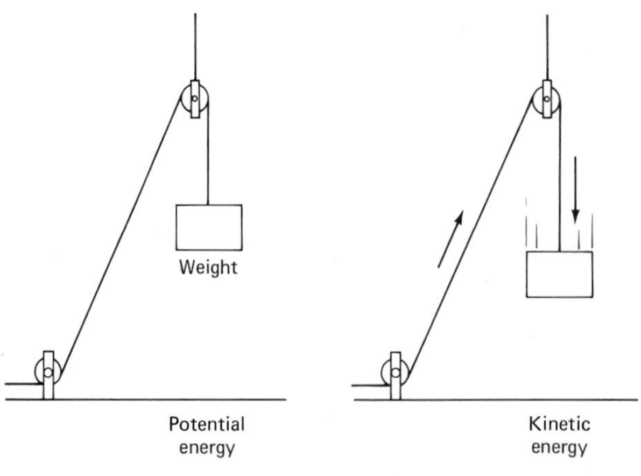


**Figure 1-15** POTENTIAL AND KINETIC ENERGY. Potential energy is energy of position, and kinetic energy is energy of motion.

## 1-10 The Relationship of Matter and Energy

The laws of conservation of mass and energy turn out to be less exact than was originally thought. In 1905 Albert Einstein proposed the now well-known relationship between mass and energy,

$$E = mc^2$$

$E$ = energy  $m$ = mass  $c$ = the speed of light

This amazing theory tells us that, in fact, energy results from a conversion of mass. Since a small amount of mass produces a tremendous amount of energy, the mass loss or gain in a chemical reaction is far too small to be detectable. For our purposes in chemistry, the laws of conservation of mass and energy remain valid and can be applied to the understanding of the nature of matter and its changes. In nuclear reactions, however, there is a significant mass loss, which results in a vast amount of energy. *Nuclear energy* is another form of energy and is the source of the energy of the sun. Nuclear energy is discussed in Chapter 3.

## 1-11 Chapter Summary

Chemistry is the oldest and yet, in a way, one of the youngest sciences. Chemistry is the branch of science that deals with matter and the changes it undergoes. Matter was discussed first, and we found that there are four types of matter depending on the complexity of its composition. These forms, in order of increasing complexity, are:

| | |
|---|---|
| 1 Elements | The most fundamental forms of pure substances. |
| 2 Compounds | Pure substances made up of two or more elements chemically combined in fixed proportions. |
| 3 Homogeneous mixtures or solutions | Intimate and uniform mixtures of elements and/or compounds. A solution exists in one phase. |
| 4 Heterogeneous mixtures | Nonuniform mixtures of elements and/or compounds that exist in two or more distinct phases. |

The two forms of pure substances have definite and distinct properties. Mixtures, on the other hand, have properties that vary depending on the proportions of substances in the mixture. The elements are identified by name and symbol. The chemical names of many compounds end in *ide, ite,* or *ate.*

Substances have both physical and chemical properties and undergo both physical and chemical changes as summarized below.

| Property | Definition | Change |
|---|---|---|
| Physical | Describes properties that do not result in a change into another substance | Changes in dimensions or physical state |
| Chemical | Describes types of chemical reactions the substance undergoes | Changes into other pure substances |

All changes, either chemical or physical, are governed by the laws of conservation of mass and energy. The energy involved in these processes can be classified as to form and type as follows.

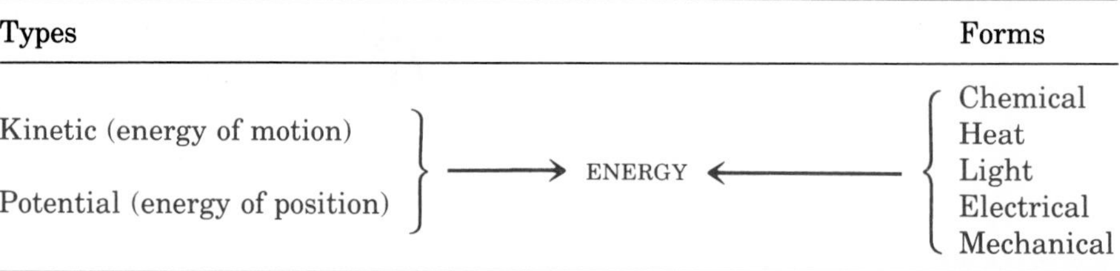


One form of energy can be converted into another. An important conversion in chemistry is between chemical and heat energies. When chemical energy is converted into heat energy, as in the combustion of a log, the reaction is said to be exothermic. When heat energy is converted into chemical energy, the reaction is said to be endothermic. Finally, we found that despite a relationship between mass and energy, there is a negligible mass loss in a chemical reaction.

## Exercises

*Throughout the text, answers to all problems except those with the problem number in color are given in Appendix H.*

### The Nature of Matter

**1-1** Which of the following describes the liquid phase?
(a) It has a fixed shape and a fixed volume.
(b) It has a fixed shape but not a fixed volume.
(c) It has a fixed volume but not a fixed shape.
(d) It has neither a fixed shape nor volume.

**1-2** When a teaspoon of solid sugar is dissolved in a glass of liquid water, what phase or phases are present after mixing?
(a) liquid only
(b) still solid and liquid
(c) solid only

**1-3** Identify the following as homogeneous or heterogeneous matter.
(a) gasoline (b) dirt (c) smog (d) alcohol (e) a new nail (f) vinegar (g) ice cubes in water (h) aerosol spray (i) air

**1-4** Identify the following as homogeneous or heterogeneous matter.
(a) a cloud (b) dry ice (c) whipped cream (d) bourbon (e) natural gas (f) a grapefruit

**1-5** What is the physical state or states present in each of the substances listed in Problem 1-3?

**1-6** What is the physical state or states present in each of the substances listed in Problem 1-4?

### Elements and Compounds

**1-7** Carbon dioxide is not a mixture of carbon and oxygen. Explain.

**1-8** Identify the following pure substances as either elements or compounds.
(a) carbon monoxide (b) hydrogen (c) iron (d) titanium dioxide (e) potash (f) sodium bicarbonate

**1-9** Identify the following pure substances as either elements or compounds.
(a) ammonia (b) hydrogen peroxide (c) aspirin (d) mercury (e) stannous fluoride (f) uranium

**1-10** Tell whether each of the following properties describes a heterogeneous mixture, a solution (homogeneous mixture), a compound, or an element.
(a) A homogeneous liquid that when boiled away leaves a solid residue.
(b) A cloudy liquid that after a certain time seems more cloudy toward the bottom.
(c) A uniform red solid with a definite and sharp melting point that cannot be decomposed into simpler substances.
(d) A colorless liquid that boils at one unchanging temperature and can be decomposed into simpler substances.
(e) A liquid that at first boils at one temperature but as the boiling continues, the temperature of the boiling liquid slowly rises. There is only one liquid phase.

**1-11** Tell whether each of the following properties describes a heterogeneous mixture, a solution (a homogeneous mixture), a compound, or an element.
(a) A nonuniform powder that when heated first turns mushy and continues to melt as the temperature rises.
(b) A colored gas that can be decomposed into a solid and another gas. The entire sample of gas seems to have the same chemical properties.
(c) A sample of colorless gas, only part of which reacts with hot copper.

**1-12** Without reference to a table of elements, give the symbol of the following elements.
(a) carbon (b) helium (c) chlorine (d) tin (e) sodium (f) oxygen

**1-13** Using the table of elements in the front of the book, give the symbol for the following elements.
(a) barium (b) neon (c) cesium (d) platinum (e) manganese (f) tungsten

**1-14** Without reference to a table of elements, give the name corresponding to the following symbols.
(a) S (b) K (c) Fe (d) N (e) Mg (f) Al

**1-15** Using the table of elements in the front of the book, give the name corresponding to the following symbols.
(a) B (b) Bi (c) Ge (d) U (e) Co (f) Hg (g) Be (h) As

### Physical and Chemical Properties and Changes

**1-16** Identify the following as either a physical or chemical property.
(a) Diamond is the hardest known substance.
(b) Carbon monoxide is a poisonous gas.
(c) Soap is slippery.
(d) Silver tarnishes.
(e) Gold does not rust.
(f) Carbon dioxide freezes as −78 °C.
(g) Tin is a shiny, gray metal.
(h) Sulfur burns in air.
(i) Aluminum has a low density.

**1-17** Identify the following as either a physical or chemical property.
(a) Sodium burns in the presence of chlorine gas.
(b) Mercury is a liquid at room temperature.
(c) Water boils at 100 °C at sea level.
(d) Limestone gives off carbon dioxide when heated.
(e) Hydrogen sulfide has a pungent odor.

**1-18** Identify the following as a physical or chemical change.
(a) the frying of an egg
(b) the vaporization of dry ice
(c) the boiling of water
(d) the burning of gasoline
(e) the breaking of glass

**1-19** Identify the following as a physical or chemical change.
(a) the tanning of leather
(b) the fermentation of apple cider
(c) the compression of a spring
(d) the grinding of a stone

### Mass and Energy

**1-20** When a log burns, the ashes weigh less than the log. When zinc reacts with sulfur, the zinc sulfide weighs the same as the zinc and sulfur. When iron burns in air, however, the compound formed weighs more than the original iron. Explain in terms of the law of conservation of mass.

**1-21** From your own experiences tell whether the following processes are exothermic or endothermic.
(a) the decay of grass clippings
(b) the melting of ice
(c) the change in an egg when it is fried
(d) the condensation of steam
(e) the curing of freshly poured cement

**1-22** A car battery can be recharged after the engine starts. Trace the different energy conversions from gasoline to the battery.

**1-23** Windmills are used to generate electricity. What are all of the different forms of energy involved with generation of electricity by this method?

**1-24** Identify the principal type of energy (kinetic or potential) that is exhibited by each of the following.
(a) a car parked on a hill
(b) a train traveling 60 miles/hr
(c) chemical energy
(d) an uncoiling spring
(e) a falling brick

**1-25** Identify the following as having either potential or kinetic energy or both.
(a) an arrow in a fully extended bow
(b) a baseball traveling high in the air
(c) two magnets held apart
(d) a chair on the fourth floor of a building

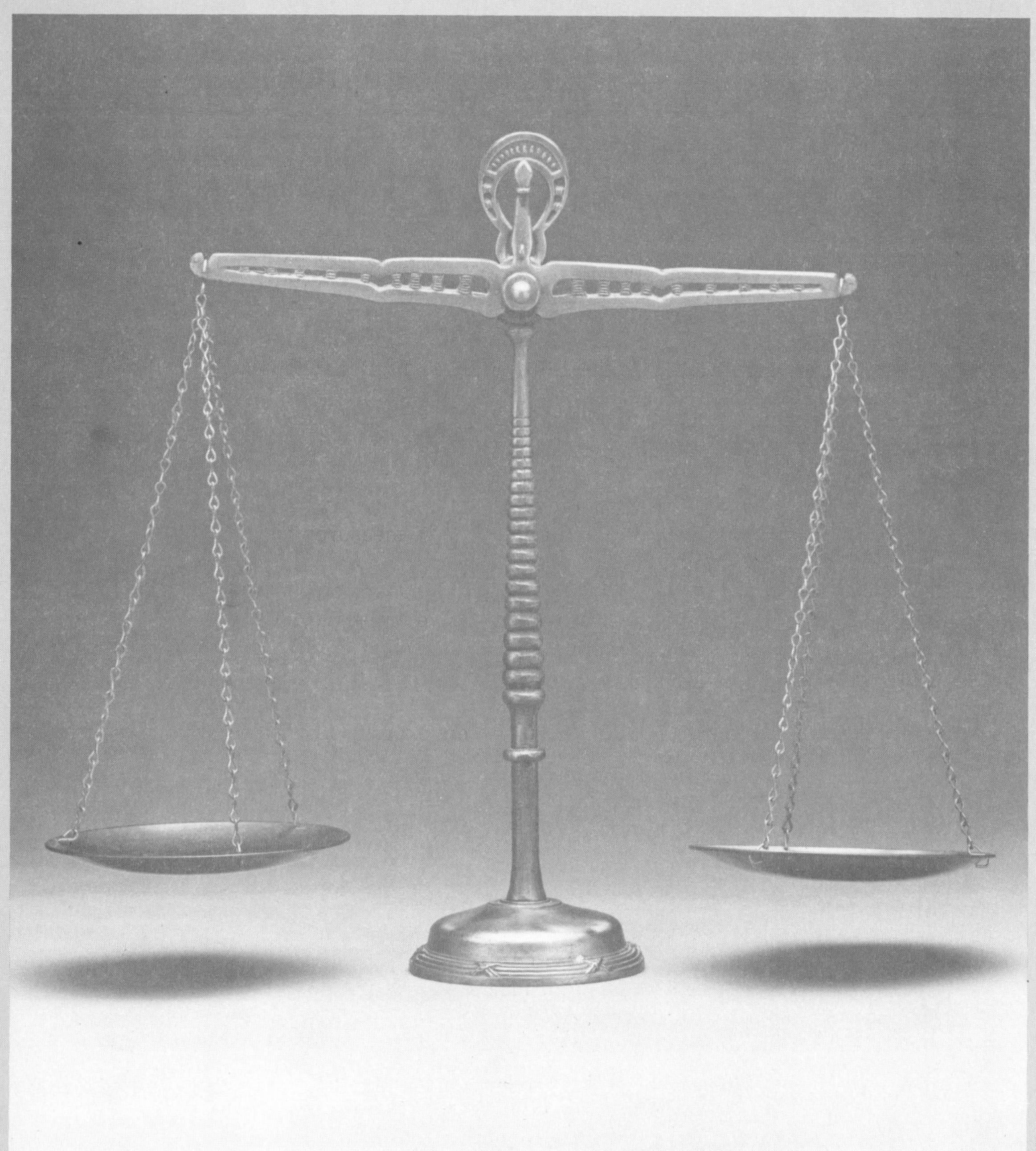

The science of chemistry was changed dramatically when the analytical balance was put to use.

# Chapter 2
# Measurements in Chemistry

## Purpose of Chapter 2

In Chapter 2 we develop an understanding of the importance of measurements in chemistry including the various units used and the tools to manipulate both the measurements and their units.

## Objectives for Chapter 2

After completion of this chapter, you should be able to:

1. Distinguish between the precision and accuracy of a measurement. (2-1)
2. Determine the number of significant figures in a measurement. (2-1)
3. Express a result of a mathematical operation to the proper decimal place (addition and subtraction) or to the proper number of significant figures (multiplication and division). (2-2)
4. Write a number in scientific notation. (2-3)
5. Carry out mathematical operations with numbers expressed in scientific notation. (2-3, Appendix C)
6. Name the basic metric units and their symbols. (2-4)
7. Write the relationships of the metric prefixes *kilo, centi,* and *milli* to the basic unit. (2-4)
8. Distinguish between mass and weight. (2-4)
9. Make conversions between English and metric units using the unit-factor method (dimensional analysis). (2-5)
10. Distinguish between density, buoyancy, and specific gravity. (2-6)
11. Calculate density given appropriate experimental data and use it as a conversion factor between mass and volume. (2-6)
12. Convert between the Celsius, Fahrenheit, and Kelvin scales. (2-7)
13. Calculate specific heat given appropriate experimental data and use it as a conversion factor between heat, temperature, and mass. (2-8)

In the early 1980s there was a popular television series about an alien from another planet who used an undefined unit of time measurement called a "bleem." Actually, modern science on this planet has plenty of legitimate units that may sound even stranger. For example, there are rads, rems, roentgens, and barns; ergs, oersteds, and henrys; ohms, gauss, and newtons. All of these units are useful, however, because they answer in some particular way the quantitative question: "How much?"

Quantitative questions concern measurements. **Measurements** *determine quantity, dimensions, or extent of something, usually in comparison to some standard.* For example, if you are six feet tall, you are six times the length of a long-dead English king's foot, which was the original standard. We will see that modern units of measurement have less arbitrary standards than somebody's foot.

The purpose of this chapter is to examine the language the chemist uses to describe measurements. We will look into weight, length, volume, and temperature. In doing this we convert from our familiar English system to the metric system by the unit-factor method, which is introduced in this chapter and used throughout the text.

It is hard to imagine how a mechanic could repair an automobile engine without being able to use tools such as wrenches and screwdrivers. It is also hard to picture how one could grasp the meaning of measurements without the ability to use mathematics. "Why is there so much math in chemistry?" is a question often heard from the student new to this science. The answer is that chemistry, like physics and astronomy, is a **physical science,** that is, *a science concerned with the natural laws of matter other than those laws concerned with life (biology, botany) and the earth (geology).* As we will see throughout this text, most of these natural laws are based on quantitative, reproducible measurements.

The math involved in this course and in most general chemistry courses at the college or university level is not at all awesome. In fact, no more than a year of high school algebra is needed. Still, many students need a review of mathematics. As you know, being in good physical shape allows physical exertion in sports to be relaxing, challenging, and just downright fun. If you're not in shape, exertion is not only painful but unsatisfying. Being in good mathematical shape has a similar effect on the study of chemistry. The normal sequence of chapters in this text will allow you time to review and strengthen your math background before the quantitative aspects of chemistry are introduced in force in Chapter 8. Hopefully, students with specific needs will take full advantage of this time to "get in shape."

**Refer to Appendixes A and B for a review of basic math and algebra.**

One of the unique characteristics of this text is the reviews and help provided for those with some rusty spots in math, algebra, and problem solving. To this end, Appendixes A and B are provided for a review of basic math and algebra. To see if you need the review, it is advised that you take the self-tests and then evaluate your needs based on the results. *A word of caution:* Where units are used in Appendixes A and B, answers are rounded off to the proper number of significant figures. This topic is covered in the next two sections.

## 2-1 Significant Figures

In most large cities there is one official who is an expert at estimating the size of a crowd. Let's say that this official estimates the size of a certain gathering at 9000 people, and then notices that eight people leave. Should the new estimate be changed to 8992? No, of course not. The original number wasn't that good. The three zeros in the estimate of 9000 were not really known numbers (as opposed to the zero in 108), but were there merely as fillers to indicate the magnitude of the number or, in other words, to locate the decimal point. In this example, the crowd would have to diminish by at least 500 before the official might feel justified in changing the estimate of the crowd to 8000. In the number 9000 only one of the numbers, the 9, is "significant." *The number of* **significant figures** *in a measurement is simply the number of valid numbers and refers to the precision of the measurement.* **Precision** *refers to how close repeated measurements come to each other.* For example, repeated estimates of the crowd of 9000 may range from 8000 to 10,000, indicating an uncertainty of ±1000. If the same crowd were seated in the bleachers of a stadium instead of milling about, a more *precise* estimate is possible. In this case, repeated estimates range from 9400 to 9600 with an average of 9500. Uncertainty has been reduced to ±100. If the crowd went through a turnstile, an even more precise number could be given. In any case, in a measurement the significant figure farthest to the right is considered to be estimated.

In another example, the precision of a measurement is illustrated in Figure 2-1. The clock in the tower measures time to the nearest minute. Notice that even so the actual minute must be estimated. The nondigital wrist watch measures time more *precisely,* since the seconds as well as the minute can be read. Again, the last digit of the second hand is estimated.

Big Ben
Time 1:21

Wrist watch
Time 1:22:47

Stopwatch
Time 1:22:45.6

**Figure 2-1** PRECISION. Precision refers to the "exactness" of a measurement.

Finally, the stopwatch used to time track events is even more precise, since time can be measured with it to tenths of a second.

In Figure 2-1, the actual time can be compared with the standard, known as Greenwich time, which is established at an observatory in Greenwich, England. In the example the Greenwich time is 1:22:45.83. Notice that the stopwatch time is not only the most precise, since its value has the largest number of significant figures, but it is also the most accurate. **Accuracy** *refers to how close the value of the measurement is to the true value.* In scientific measurements the most precise number is generally (but certainly not always) the most accurate.

Since the advent of the inexpensive hand calculator, it has become even more important for the student of chemistry to understand the precision of the answers to quantitative problems. For example, 7.8 divided by 2.3 reads 3.33913043 on a calculator. But if the original numbers represent measurements (e.g., 7.8 in. and 2.3 in.) the calculator's answer is much too precise. The calculator makes no judgment about the precision of the numbers punched into it. Therefore, *we* must know how to report an answer properly despite what the calculator reads. The answer must be expressed to the correct number of significant figures.

As we've mentioned, the number of significant figures in a measurement is simply the number of valid numbers. For example, 76.3 has three significant figures, and 4562 has four significant figures. That would be all there is to it except for the number zero. Zero serves two functions: It can be a legitimate number indicating "none," or simply a filler to locate a decimal point (like the zeros in the crowd of 9000 we had earlier). It is unfortunate that centuries ago when numbers were invented no one invented a symbol that could be used in place of a zero that is not significant. Since we don't have a separate symbol, we have some rules that tell us whether a zero is a significant figure or not.

1 When a zero is between other digits, it is significant (e.g., $\underline{709}$ has three significant figures).

2 Zeros to the right of a digit and to the right of the decimal point are significant (e.g., $\underline{8.0}$ has two significant figures as does $\underline{7.9}$ and $\underline{8.1}$; $\underline{5.700}$ has four significant figures).

3 Zeros to the left of the first digit are not significant. They are used to locate the decimal point (e.g., $0.00\underline{78}$ has two significant figures, $0.0\underline{4060}$ has four significant figures).

4 Zeros to the left of a decimal point when no decimal point is shown may or may not be significant. It is usually assumed that they are not (e.g., $\underline{9}000$ has one signficant figure and $\underline{66}00$ has two). What if a number such as 890 is actually known to three significant figures? That is, the zero is really a zero. This is a tough question. Some texts use a line over the

zero to indicate that it is significant (i.e., $89\bar{0}$). As we will see shortly, scientific notation gives us a way to solve the problem. In most calculations in this text, measurements are expressed with three significant figures. Therefore, in calculations where other numbers have three significant figures we will assume that numbers such as 890 also have three significant figures.

**Example 2-1**

How many significant figures are in the following measurements?

(a) 1508 cm (b) 300.0 ft (c) 20.003 lb (d) 0.00705 gal (e) 20,000 in.$^2$

**See Problem 2-1 through 2-5.**

**Answers**

(a) four (rule 1) (b) four (rule 2) (*Note:* Since the zero to the right of the decimal point is significant, the other two zeros are also significant because they lie between significant figures.) (c) five (rule 1) (d) three (rules 1 and 3) (e) one (rule 4)

## 2-2 Significant Figures in Mathematical Operations

When we use measurements in any mathematical operation (addition, multiplication, etc.), it is important to express the resulting number or answer to the proper number of significant figures. There is one rule for addition and subtraction and another for multiplication and division. First, we will consider addition and subtraction. *When numbers are added or subtracted, the answer should be rounded off to the place farthest to the right of the decimal that is common to all numbers.* This rule is illustrated by the following summation:

$$\begin{array}{r} 10.68\phantom{3} \\ 0.473 \\ 1.32\phantom{3} \\ \hline 12.473 \end{array} = 12.47$$

Notice that the "3" in the addition cannot be expressed in the answer, since two numbers are not known to that many decimal places. In fact, since the last significant figure in a measured number is estimated, it is assumed that there is an uncertainty of $\pm 1$ in that number. Therefore, the number above is expressed as 12.47 although it may actually range from 12.46 to 12.48.

In addition and subtraction we must also be careful with manipulations involving numbers containing nonsignificant zeros. The final answer must be **rounded off** to show the same number of nonsignificant zeros as in the original number. This can be illustrated with the crowd estimated at 9000 people discussed earlier.

$$\begin{array}{r} 9000 \\ -\ \ 8 \\ \hline 8992 \end{array} = \underline{\underline{9000}} \qquad \begin{array}{r} 9000 \\ -80 \\ \hline 8920 \end{array} = \underline{\underline{9000}} \qquad \begin{array}{r} 9000 \\ -800 \\ \hline 8200 \end{array} = \underline{\underline{8000}}$$

The rules for rounding off are as follows (the examples are all rounded off to three significant figures):

1 If the digit to be dropped is less than 5, simply drop that digit (e.g., 12.44 rounds off to 12.4).

2 If the digit to be dropped is 5 or greater, increase the preceding digit by one (e.g., 0.3568 rounds off to 0.357 and 13.75 rounds off to 13.8).

**Example 2-2**
Carry out the following calculations, rounding off the answer to the proper decimal place.

$$\begin{array}{rr} & 7.56 \\ + & 0.375 \\ + & 14.2203 \\ \hline & 22.1553 \end{array} = \underline{\underline{22.16}} \qquad \begin{array}{rr} & 14{,}000 \\ + & 580 \\ + & 75 \\ \hline & 14{,}655 \end{array} = \underline{\underline{15{,}000}} \qquad \begin{array}{rr} & 0.0327 \\ - & 0.00068 \\ \hline & 0.03202 \end{array} = \underline{\underline{0.0320}}$$

In multiplication and division the concern does not center on the location of the decimal point. Instead, *the answer must be expressed with the same number of significant figures as the multiplier, dividend, or divisor with the least number of significant figures.* In other words, the answer is only as precise as the least precise part of the problem (i.e., the chain is only as strong as its weakest link).

**Example 2-3**
Carry out the following calculations, rounding off the answer to the proper number of significant figures.

(a) 2.34 in. × 3.225 in.
The answer on the calculator reads 7.5465. Since the first multiplier has three significant figures and the second has four, the answer should be expressed to three figures. The answer is rounded off to $\underline{\underline{7.55\ \text{in.}^2}}$

(b) 11.688 ft/4.0 sec

The answer, 2.922, should be rounded off to two significant figures. The answer is 2.9 ft/sec.

(c) 148.6 cm × 0.224 cm = 33.2864 $cm^2$ = 33.3 $cm^2$

(d) $\dfrac{875.6 \text{ cm}^2}{62 \text{ cm}} = 14.122 \text{ cm} = 14 \text{ cm}$

Before leaving the topic of significant figures, we should mention the effect of *exact* numbers on a calculation. An exact number is one that is either defined exactly or originates from a count. For example, 1 gal = 4 qt is an exactly defined relationship. As such it has unlimited significant figures (i.e., 1.00 . . . etc. gal = 4.00 . . . etc. qt). Counted numbers are also exact and have unlimited significant figures. For example, if there are 32 students in a certain classroom, the number 32 is considered exact (i.e., 32.00 . . . etc.). As we will see in this chapter, an exact relationship does not limit or affect the number of significant figures in a calculation. Relationships that originate from a measurement are not exact and limit the number of significant figures in a calculation as described. You will become more familiar with exact numbers as we proceed.

**See Problems 2-6 through 2-11.**

## 2-3 Scientific Notation

Sciences require the use of some very large and some very small numbers (see Figure 2-2). In chemistry, for example, numbers such as

602,000,000,000,000,000,000,000

are used in Chapter 8. Obviously, this number as written is extremely awkward. A much more convenient way of both writing and reading the above number is

$6.02 \times 10^{23}$ ("six point oh two times ten to the twenty-third")

The latter number is expressed in scientific notation. **Scientific notation** *expresses zeros used to locate the decimal point in powers of 10.* A number expressed in scientific notation has two parts: *the* **coefficient** *is the number that is multiplied by 10 raised to a power, and the* **exponent** *is the power to which 10 is raised.*

$$\underbrace{6.02}_{\text{Coefficient}} \times 10^{\overset{\text{Exponent}}{23}}$$

**Figure 2-2** POWERS OF TEN. In astronomy extremely large numbers are encountered. In cell biology, very small numbers are used.

The planet Saturn is $10^{12}$ meters away at its closest approach to Earth.

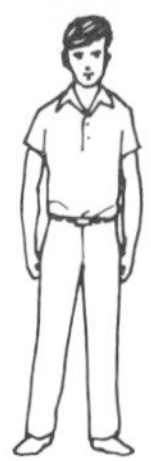

2 meters tall

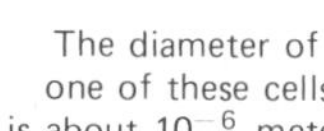

The diameter of one of these cells is about $10^{-6}$ meters

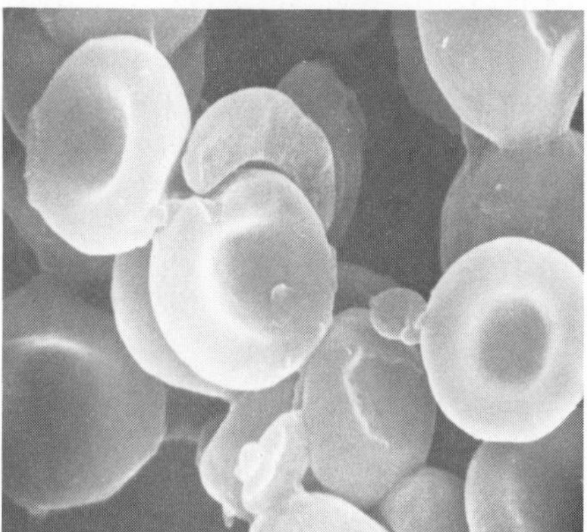

The following reviews some powers of 10 and their equivalent numbers:

| | |
|---|---|
| $10^0 = 1$ | $10^{-1} = \frac{1}{10^1} = 0.1$ |
| $10^1 = 10$ | $10^{-2} = \frac{1}{10^2} = 0.01$ |
| $10^2 = 10 \times 10 = 100$ | $10^{-3} = \frac{1}{10^3} = 0.001$ |
| $10^3 = 10 \times 10 \times 10 = 1000$ | $10^{-4} = \frac{1}{10^4} = 0.0001$ |
| $10^4 = 10 \times 10 \times 10 \times 10 = 10{,}000$ | etc. |
| etc. | |

The *standard* method of expressing numbers in scientific notation is with one digit to the left of the decimal point (e.g., $4.56 \times 10^6$ rather than $45.6 \times 10^5$).

**Example 2-4**

Express each of the following numbers in scientific notation with one digit to the left of the decimal point.

(a) 47,500 (b) 5,030,000 (c) 0.0023 (d) 0.0000470

**Solution**

(a) The number 47,500 can be factored as $4.75 \times 10{,}000$. Since $10{,}000 = 10^4$, the number can be expressed as

$$\underline{\underline{4.75 \times 10^4}}$$

A more practical way to transform this number to scientific notation is to count to the left from the *old* decimal point to where you wish to put the new decimal point. The number of places counted to the left will be the positive exponent of 10.

$$\overset{\phantom{4}}{4}\,\overset{4}{7}\,\overset{3}{5}\,\overset{2}{0}\,\overset{1}{0} = 4.75 \times 10^4$$

(b)

$$5,\overset{6}{0}\,\overset{5}{3}\,\overset{4}{0},\overset{3}{0}\,\overset{2}{0}\,\overset{1}{0} = 5.03 \times 10^6$$

(c) 0.0023 can be factored into $2.3 \times 0.001$. Since $0.001 = 10^{-3}$, the number can be expressed as

$$\underline{\underline{2.3 \times 10^{-3}}}$$

A more practical way is to count to the right from the *old* decimal point to where you wish the new decimal point. The number of places counted to the *right* will be the negative exponent of 10.

$$0.\overset{1}{0}\,\overset{2}{0}\,\overset{3}{2}\,3 = 2.3 \times 10^{-3}$$

(d)

$$0.\overset{1}{0}\,\overset{2}{0}\,\overset{3}{0}\,\overset{4}{0}\,\overset{5}{4}\,7\;0 = \underline{\underline{4.70 \times 10^{-5}}}$$

The use of scientific notation can help remove the ambiguity of numbers containing zero that may or may not be significant. For example, the number 9000 can have from one to four significant figures as written. The following expressions, however, are clear as to the number of significant figures:

$9 \times 10^3$ has one significant figure

$9.0 \times 10^3$ has two significant figures

$9.00 \times 10^3$ has three significant figures

$9.000 \times 10^3$ has four significant figures

**Refer to Appendix C for further review.**

The preceding discussion is supplemented in Appendix C. This appendix includes a diagnostic test and a disucssion of the mathematical manipulation of numbers expressed in scientific notation. You are urged to take the short test at this time and work through the appendix if practice is indicated by the results of the test.

**See Problems 2-12 through 2-21.**

## 2-4 Length, Volume, and Mass in the Metric System

If you were skillful (or lucky) enough to catch a trophy fish, you would certainly describe this fish by means of its length and weight. Even more detail could be given by including its girth (circumference), which would give the listener an idea of its volume. In the United States, you would most likely use the English system and report numbers such as 2 ft 4 in. and 6 lb 3 oz for the fish. Notice that we often use two units (e.g., ft and in.) for one measurement. This awkward situation arises because of a lack

(*a*)

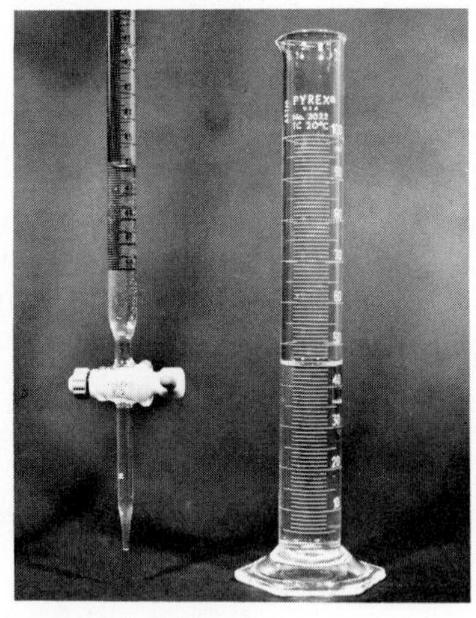


(*b*)

(*c*)

**Figure 2-3** LENGTH, VOLUME, MASS. These properties of a quantity of matter are measured with common laboratory equipment. (*a*) Length. Metric rulers. (*b*) Volume. A buret and a graduate cylinder. (*c*) Mass. A pan balance and an electric balance.

**Table 2-1** Metric or SI Units

| Measurement | Unit | Symbol |
|---|---|---|
| Mass | gram | g |
| Length | meter | m |
| Volume | liter | L |
| Time | second | s |
| Temperature | Celsius | °C |
| | kelvin | K |
| Quantity | mole | mol |
| Energy | joule | J |
| Pressure | pascal | Pa |

of systematic relationships between units. That is, units are not related by the same number. For example, we have relationships of length of 12 in. = 1 ft, 3 ft = 1 yd, and 1750 yd = 1 mile. Our monetary system is an exception because it is based on the decimal system. Therefore, only one unit (dollars) is needed to show a typical student's dismal financial condition (e.g., $11.98). The English system that the United States almost adopted required the use of two units (e.g., 2 pounds, 3 shillings).

The rest of the world and the sciences use the metric system of measurement for length, volume, and mass (see Figure 2-3). This system, like our monetary system, uses decimal fractions. Units in the metric system are conveniently related by multiples of 10. The United States still plans to adopt the metric system, but many years will pass before a complete switch is made. In slightly modified form the metric system is also known as the SI system after the French words for International System. The basic SI or metric units that will concern us are listed in Table 2-1. There are many other SI units that designate quantities that are not used in this text. Most SI units have very precisely defined standards based on certain precisely known properties of matter and light. Although English units were

**Table 2-2** Prefixes Used in the Metric System

| Prefix | Symbol | Relation to Basic Unit | Prefix | Symbol | Relation to Basic Unit |
|---|---|---|---|---|---|
| tera- | T | $10^{12}$ | deci- | d | $10^{-1}$ |
| giga- | G | $10^{9}$ | centi- | c | $10^{-2}$ |
| mega- | M | $10^{6}$ | milli- | m | $10^{-3}$ |
| kilo- | k | $10^{3}$ | micro- | $\mu$[a] | $10^{-6}$ |
| hecto- | h | $10^{2}$ | nano- | n | $10^{-9}$ |
| deca- | da | $10^{1}$ | pico- | p | $10^{-12}$ |

[a]A Greek letter, mu.

**Table 2-3** Relationships among Metric Units Using Common Prefixes

| Mass Unit | Symbol | Relation to Basic Unit | Volume Unit | Symbol | Relation to Basic Unit | Length Unit | Symbol | Relation to Basic Unit |
|---|---|---|---|---|---|---|---|---|
| kilogram | kg | $10^3$ g | kiloliter | kL | $10^3$ L | kilometer | km | $10^3$ m |
| decigram | dg | $10^{-1}$ g | deciliter | dL | $10^{-1}$ L | decimeter | dm | $10^{-1}$ m |
| centigram | cg | $10^{-2}$ g | centiliter | cL | $10^{-2}$ L | centimeter | cm | $10^{-2}$ m |
| milligram | mg | $10^{-3}$ g | milliliter | mL | $10^{-3}$ L | millimeter | mm | $10^{-3}$ m |
| microgram | $\mu$g | $10^{-6}$ g | | | | | | |

formerly based on a king's anatomy, in modern times many have been redefined more precisely based on a corresponding metric unit.

Use of SI or metric units is simplified by their exact relationships by powers of 10. This is illustrated in Table 2-3 by use of the more common prefixes listed in Table 2-2.

In the metric system there is also an exact relationship between length and volume. Thus one liter is defined as the volume occupied by one cubic decimeter (i.e., 1 L = 1 $dm^3$). One milliliter is the volume occupied by one cubic centimeter (i.e., 1 mL = 1 $cm^3$ = 1 cc). The units milliliter (mL) and cubic centimeter ($cm^3$ or cc) can be used interchangeably when expressing volume. (See Figure 2-4.)

There is often confusion between the terms "mass" and "weight". **Mass** *is the quantity of matter that a sample contains.* It is the same for the sample anywhere in the universe. **Weight** *is a measure of the attraction of gravity for the sample.* An astronaut has the same mass on the moon as on the earth. A 170-lb astronaut on earth, however, weighs only about 29 lb on

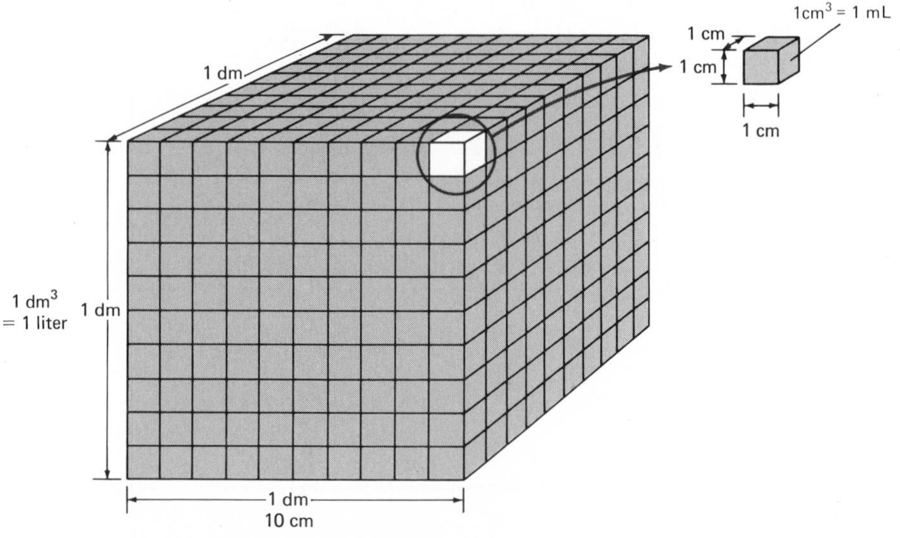


**Figure 2-4** VOLUME AND LENGTH. One milliliter is the volume occupied by one cubic centimeter.

**Table 2-4** The Relationship between English and Metric Units

| | English | Metric Equivalent |
|---|---|---|
| Length | 1.00 in. | 2.54 cm |
| | 1.00 mile | 1.61 km |
| Mass | 1.00 lb | 454 g |
| | 2.20 lb | 1.00 kg |
| Volume | 1.06 qt | 1.00 L |
| | 1.00 gal | 3.78 L |
| Time | 1.00 sec | 1.00 sec |

the moon. In earth orbit, where there is no gravity, the astronaut becomes "weightless." Obviously, the astronaut lost weight on the moon, but matter (mass) was not lost. Since the attraction of gravity is essentially the same anywhere on the face of the earth, we use weight as a measure of mass. The terms "mass" and "weight" are thus used interchangeably.

Several relationships between the metric and English systems are listed in Table 2-4. (See also Figure 2-5.) The relationships *within* systems (e.g., 12 in. = 1 ft, $10^3$ m = 1 km) are always defined and exact. The relationships *between* systems, however, can be expressed to various degrees of precision. All of the relationships shown are known to more than the three significant figures given. Although there are exact, defined relationships between the two systems, the definitions usually include many more significant figures than our needs in chemistry require. For example, one yard is defined by the U.S. Bureau of Standards as exactly equal to 0.91440183 meters.

**Figure 2-5** METRIC UNITS. The use of metric units in this country is becoming increasingly evident.

## 2-5 Conversion of Units by the Unit-Factor Method

A large number of mathematical problems that you will encounter in this text are basically of the same type. They simply involve the conversion of one unit of measurement to another. To do this we use the unit-factor method, which is also known as dimensional analysis or the factor-label method. A brief introduction is included in this chapter, with supplemental explanations and exercises presented in Appendix D.

*The* **unit-factor method** *converts from one unit to another by use of conversion factors. A* **conversion factor** *is an equality or equivalency relationship between units or quantities expressed in fractional form.* For example, in the metric system we have the exact relationship

$$10^3\ \text{m} = 1\ \text{km}$$

This can be expressed in fractional or factor form as

$$\frac{1\ \text{km}}{10^3\ \text{m}} \quad \text{or the reciprocal} \quad \frac{10^3\ \text{m}}{1\ \text{km}} \quad \left(\text{or simply } \frac{10^3\ \text{m}}{\text{km}}\right)$$

This is read as "one kilometer per $10^3$ meters" or "$10^3$ meters per one kilometer." The latter factor is usually simplified to "$10^3$ meters per kilometer." When just a unit (no number) is read or written in the denominator, the number is assumed to be "one."

Conversion factors are also constructed from the equalities between the English and metric units. For example, the equality

$$1.00\ \text{in.} = 2.54\ \text{cm}$$

can be expressed in factor form as

$$\frac{1.00\ \text{in.}}{2.54\ \text{cm}} \quad \text{or} \quad \frac{2.54\ \text{cm}}{1.00\ \text{in.}}$$

Factors such as the examples above are sometimes referred to as **unity factors,** since *they relate a quantity to "one" of another.* As such, the factors are often written in a somewhat simplified form as follows:

$$\frac{1\ \text{in.}}{2.54\ \text{cm}} \quad \text{or} \quad \frac{2.54\ \text{cm}}{\text{in.}}$$

Unity factors are written as such in this text. *It should be understood, however, that "one" (as written in the numerator or implied in the denominator) is known exactly.*

These factors can be used to convert a measurement in one unit to the other. In the unit-factor method all units of the numbers are maintained in the calculation. Like the numbers themselves, the units are multiplied, divided, and canceled in the course of the calculation. This adds a little time to the calculation, but the payoff is *correct* answers. Students who consistently use this method are often amazed at the orderliness and logic of chemistry. In more complex conversions, where several conversion factors are required, the student can plan the calculation in a stepwise and orderly manner from what is given to what is requested.

The general procedure of a one-step conversion by the unit-factor method is as follows:

$$(\text{What's given}) \times (\text{conversion factor}) = (\text{what's requested})$$

In most calculations, the proper conversion factor that converts "given" to "requested" has the unit of what is given in the denominator (the old unit) and the unit of what is requested in the numerator (the new unit). In this way, the old unit *cancels* as would an identical numerical quantity. This leaves the new unit in the numerator.

$$\begin{pmatrix}\text{Given}\\ \text{quantity}\end{pmatrix} \cancel{\text{old unit}} \times \begin{pmatrix}\text{conversion}\\ \text{factor}\\ \text{quantity}\end{pmatrix} \frac{\text{new unit}}{\cancel{\text{old unit}}} = \begin{pmatrix}\text{requested}\\ \text{quantity}\end{pmatrix} \text{new unit}$$

The key, then, is to select the conversion factor that does the right job. As we will see, many conversions require several steps and thus need more than one conversion factor to convert from what's given to what's requested. Proper organization and planning help avoid difficulties for these problems. The following examples illustrate the use of the unit-factor method in one-step conversion between units.

**Example 2-5**

Convert 0.468 m to (a) kilometers and (b) millimeters.

(a)

1 Given: 0.468 m. Requested: ___?___ km.

2 Procedure: Use a conversion factor that cancels m and leaves km in the numerator. In shorthand our procedure is

$$\text{m} \rightarrow \text{km}$$

3 Relationship: $10^3$ m = 1 km (from Table 2-3).

4 Conversion factor: Of the two possible conversion factors that originate from the above relationship, we choose the one with km (requested) in the numerator and m (given) in the denominator. This is 1 km/$10^3$ m.

5 Solution:

$$0.468\ \cancel{\text{m}} \times \frac{1\ \text{km}}{10^3\ \cancel{\text{m}}} = 0.468 \times 10^{-3}\ \text{km} = \underline{\underline{4.68 \times 10^{-4}\ \text{km}}}$$

(b)

1 Given: 0.468 m. Requested: ___?___ mm.

2 Procedure: m → mm.

3 Relationship: 1 mm = $10^{-3}$ m.

4 Conversion factor: 1 mm/$10^{-3}$ m.

5 Solution:

$$0.468\ \cancel{\text{m}} \times \frac{1\ \text{mm}}{10^{-3}\ \cancel{\text{m}}} = 0.468 \times 10^3\ \text{mm} = \underline{\underline{468\ \text{mm}}}$$

**Example 2-6**
Convert 825 cm to inches.

1 Given: 825 cm. Requested: ___?___ in.

2 Procedure: cm → in.

3 Relationship: 1.00 in. = 2.54 cm (from Table 2-4).

4 Conversion factor: 1 in./2.54 cm.

5 Solution:

$$825\ \cancel{\text{cm}} \times \frac{1\ \text{in.}}{2.54\ \cancel{\text{cm}}} = \underline{\underline{325\ \text{in.}}}$$

**Example 2-7**
Convert 6.85 qt to liters.

1 Given: 6.85 qt. Requested: ___?___ L.

2 Procedure: qt → L.

3 Relationship: 1.00 L = 1.06 qt.

4 Factor: 1 L/1.06 qt.

5 Solution:

$$6.85\ \cancel{\text{qt}} \times \frac{1\ \text{L}}{1.06\ \cancel{\text{qt}}} = \underline{\underline{6.46\ \text{L}}}$$

What would have happened if we had multiplied instead of divided? In that case, the units would have served as a red flag indicating that a mistake had been made. The units of the answer would have been $qt^2/L$, which is obviously not correct.

$$6.85\ \text{qt} \times \frac{1.06\ \text{qt}}{1\ \text{L}} = 7.26\ \frac{\text{qt}^2}{\text{L}} \qquad ??????$$

The one-step conversions that have been worked so far are analogous to a direct, nonstop airline flight between your home city and your destination. The problems that follow are analogous to the situation in which a nonstop flight is not available so you have to make the flight in several steps before reaching your destination. Each step in the flight is a separate journey, but it gets you closer to your ultimate destination. For multistep conversions you must plan a step-by-step path from your origin (what's given) to your destination (what's requested). Each step along the path toward the answer requires one conversion factor.

**Example 2-8**
Convert 4978 mg to kilograms.

**1** Given 4978 mg. Requested: ___?___ kg.

**2** Procedure: In Table 2-3, notice that one relationship between milligrams and kilograms is not directly available. However, we can take a two-step journey by (a) converting milligrams to grams and then (b) converting grams to kilograms. In shorthand the procedure is

$$\text{mg} \xrightarrow{(a)} \text{g} \xrightarrow{(b)} \text{kg}$$

**3** Relationships: 1 mg = $10^{-3}$ g; 1 kg = $10^3$ g.

**4** Conversion factors: Treat each step as a distinct operation or problem. In step (a) we need g in the numerator and mg in the denominator. In step (b) we need kg in the numerator and g in the denominator.

(a) $10^{-3}$ g/mg (b) 1 kg/$10^3$ g

**5** Solution

$$4978\ \cancel{\text{mg}} \times \overset{(a)}{\frac{10^{-3}\ \cancel{\text{g}}}{\cancel{\text{mg}}}} \times \overset{(b)}{\frac{1\ \text{kg}}{10^3\ \cancel{\text{g}}}} = 4978 \times 10^{-6}\ \text{kg} = \underline{\underline{4.978 \times 10^{-3}\ \text{kg}}}$$

**Example 2-9**
Convert 9.85 L to gallons.

**1** Given: 9.85 L. Requested: ___?___ gal.

**2** Procedure

$$L \xrightarrow{(a)} qt \xrightarrow{(b)} gal$$

**3** Relationships: 1.06 qt = 1.00 L; 4 qt = 1 gal.

**4** Factors

$$(a)\ \frac{1.06\ qt}{L} \qquad (b)\ \frac{1\ gal}{4\ qt}$$

**5** Solution

$$9.85\ \cancel{L} \times \overset{(a)}{\frac{1.06\ \cancel{qt}}{\cancel{L}}} \times \overset{(b)}{\frac{1\ gal^*}{4\ \cancel{qt}}} = \underline{\underline{2.61\ gal}}$$

(* An exact number does not limit the number of significant figures in the answer.)

---

**Example 2-10**
Convert 55 miles/hr to meters per minute.

**1** Given: 55 miles/hr. Requested: __?__ m/min.

**2** Procedure: Notice in this problem that both the numerator and denominator must be converted to other units. The first two steps convert the numerator and the third the denominator.

$$\text{Numerator:} \quad \text{miles} \xrightarrow{(a)} \text{km} \xrightarrow{(b)} \text{m}$$

$$\text{Denominator:} \quad \text{hr} \xrightarrow{(c)} \text{min}$$

**3** Relationships: 1.00 mile = 1.61 km; $10^3$ m = 1 km*;
60 min = 1 hr*
(*Exact numbers.)

**4** Conversion factors

$$(a)\ \frac{1.61\ km}{mile} \qquad (b)\ \frac{10^3 m}{km}$$

To convert a denominator to another unit, notice that the conversion factor must be set up with what's given in the *numerator* and what's requested in the *denominator*.

$$(c)\ \frac{1\ hr}{60\ min}$$

**5 Solution**

$$55\ \frac{\cancel{\text{miles}}}{\cancel{\text{hr}}} \times \overset{\text{(a)}}{\frac{1.61\ \cancel{\text{km}}}{\cancel{\text{mile}}}} \times \overset{\text{(b)}}{\frac{10^3\text{m}}{\cancel{\text{km}}}} \times \overset{\text{(c)}}{\frac{1\ \cancel{\text{hr}}}{60\ \text{min}}} = \underline{\underline{1.5 \times 10^3\ \text{m/min}}}$$

The general procedure for working conversion problems can be summarized as follows:

1 Write down what is given and what is requested.

2 Outline a procedure in shorthand for a path from given to requested. Determine whether one or more steps are required to solve the problem.

3 Write down the relationship for each step in the conversion. Consult the appropriate tables if necessary.

4 Determine the appropriate conversion factor from the relationsips in step 3. *For each step,* the unit of what's wanted should be in the numerator and that of what's given in the denominator. (An exception to this is where a problem requires a change in a denominator such as miles per hour to miles per minute.)

5 Put the equation together. Make sure that the proper units cancel and that you are left with only the unit or units requested. *Carefully* do the math.

**See Problems 2-22 through 2-47.**

If these examples and the problems at the end of this chapter cause you any difficulties, you are strongly urged to work through Appendix D. There you will find an extensive supplement including solved problems and exercises employing familiar English units, imaginary units, and actual chemical units. It is designed to increase your proficiency in the mechanics of the unit-factor method so that it can serve as your "backup system" for obtaining correct answers. (Your primary system should eventually be a "feeling" for the type and general magnitude of the answer you are seeking.)

**Refer to Appendix D for additional discussion on problem solving by the unit-factor method.**

## 2-6 Density

Styrofoam is considered "light" and lead "heavy." However, since a truckload of styrofoam is certainly "heavy," we must compare the masses of equal volumes if our terms are to have any meaning. (See Figure 2-6.) What we wish to compare is the densities of the two substances. *The* **density** *of a substance is the ratio of the mass of a substance to the volume.* The mass is usually expressed in grams and the volume in milliliters or liters. Both mass and volume are variable properties that depend on the amount of substance

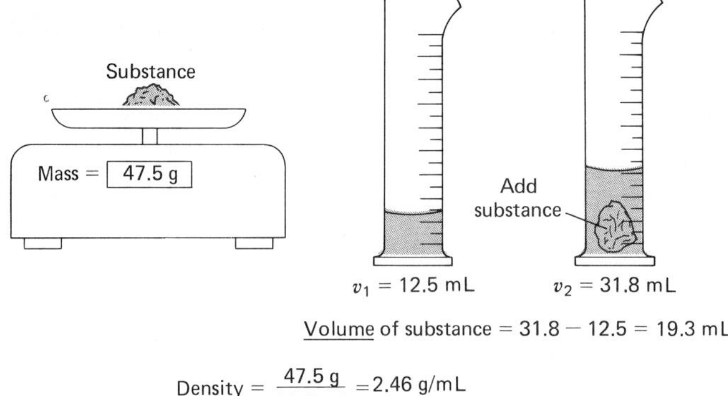


**Figure 2-6** DENSITY. To determine the density, both the mass and the volume of a substance are measured.

present. The density (the ratio of the two) is the *same* for any amount or portion of a homogeneous substance *under the same conditions of pressure and temperature.* Since pure substances (elements and compounds) are homogeneous, densities can be used as an identifying characteristic. The densities of several liquids and solids are listed in Table 2-5. The densities of gases are discussed in Chapter 10.

The following examples illustrate how density is calculated and how it is used to identify pure substances.

**Example 2-11**

A young woman was interested in purchasing a sample of pure gold weighing 8.99 g. Being wise, she wished to confirm that it was actually gold before she paid for it. With a quick test using a graduated cylinder like that shown in Figure 2-6, she found that the "gold" had a volume of 0.796 mL. Was the substance gold?

**Solution**

From the volume and the mass, the density can be calculated and compared with that of pure gold.

$$\text{Density} = \frac{8.99 \text{ g}}{0.796 \text{ mL}} = 11.3 \text{ g/mL}$$

The sample was *not* gold. It apparently was lead that was dipped in gold paint.

**Example 2-12**

A sample of a pure substance was weighed and found to have a mass of 47.5 g. As shown in Figure 2-6, a quantity of water has a volume of 12.5 mL. When the substance is added to the water the volume

**Table 2-5** Density (at 20 °C)

| Substance (Liquid) | Density (g/mL) | Substance (Solid) | Density (g/mL) |
|---|---|---|---|
| Ethyl alcohol | 0.790 | Aluminum | 2.70 |
| Gasoline (a mixture) | ~0.67 (variable) | Gold | 19.3 |
| Carbon tetrachloride | 1.60 | Ice | 0.92 (0 °C) |
| Kerosene | 0.82 | Lead | 11.3 |
| Water | 1.00 | Lithium | 0.53 |
| Mercury | 13.6 | Magnesium | 1.74 |
| | | Table salt | 2.16 |

reads 31.8 mL. The difference in volume is the volume of the substance. What is the density?

**Solution**

Refer to Figure 2-6.

In the previous examples, density was calculated from a given weight and volume. Density itself is used as a conversion factor to:

1 Convert a given weight to an equivalent volume.

2 Convert a given volume to an equivalent weight.

**Example 2-13**

Using Table 2-5, determine the volume occupied by 485 g of table salt.

1 Given: 485 g of salt. Requested: ___?___ mL of salt.

2 Procedure: g → mL.

3 Relationship: 1.00 mL = 2.16 g.

4 Conversion factor: 1 mL/2.16 g.

5 Solution

$$485 \not{g} \times \frac{1 \text{ mL}}{2.16 \not{g}} = \underline{\underline{225 \text{ mL}}}$$

**Example 2-14**

What is the weight of 1.52 L of kerosene?

1 Given: 1.52 L. Requested: ___?___ g of kerosene.

**2** Procedure

$$L \xrightarrow{(a)} mL \xrightarrow{(b)} g$$

**3** Relationships: 1 mL = $10^{-3}$ L, 0.82 g = 1.0 mL.

**4** Conversion factors: (a) 1 mL/$10^{-3}$ L; (b) 0.82 g/mL.

**5** Solution

$$1.52\ \cancel{L} \times \frac{1\ \cancel{mL}}{10^{-3}\ \cancel{L}} \times \frac{0.82\ g}{\cancel{mL}} = \underline{\underline{1.2 \times 10^{3}\ g}}$$

Assuming that there is no reaction or mixing, *a substance with a density lower than a certain liquid floats or is* **buoyant** *in that liquid.* In the case of water, anything with a density less than 1.00 g/mL floats. Notice that gasoline and ice float on water, but most solids and carbon tetrachloride sink.

Liquids can be mixed in such proportions to provide a range of densities. Gemologists use this principle to determine the authenticity of gemstones. For example, liquids can be mixed, whereby an emerald (density 2.70 g/mL) floats in one mixture but sinks in another. Fakes do not have the same density as the authentic stone and can be discovered as shown in Figure 2.7.

*The density of a liquid can be determined by a device called a* **hydrometer.** The hydrometer tube, as shown in Figure 2-8, is exactly balanced with weights in the bottom so that its level in pure water is exactly at the 1.00 mark. When immersed in liquids of other densities, it becomes more or less buoyant. The scale is calibrated (the divisions previously determined and checked) in such a manner that the density of the liquid is read directly from the scale. A hydrometer is used to measure the density of the acid in a car battery because this indicates its charge condition.

"Specific gravity" is a term sometimes used in place of density. **Specific gravity** *is a comparison of the density of a substance to that of water.* Since

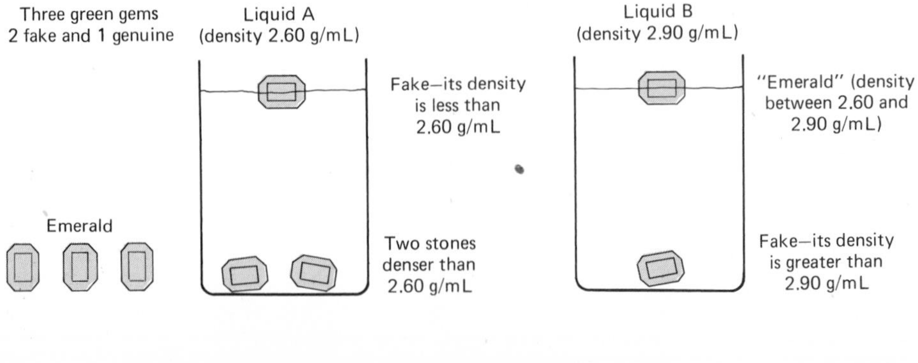

**Figure 2-7** BUOYANCY OF AN EMERALD. A stone sinks in a liquid that is less dense than the stone but floats in a more dense liquid.

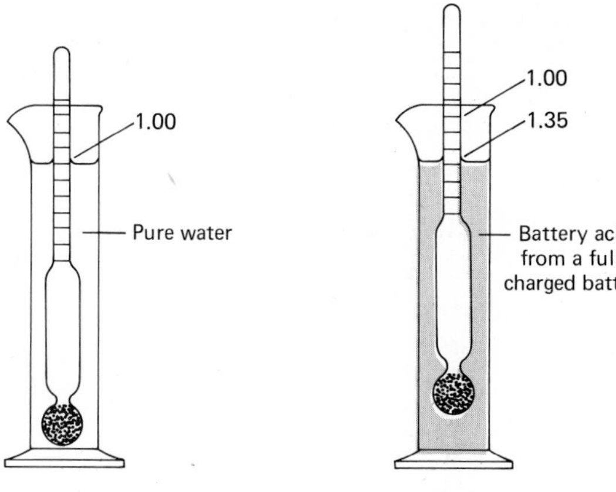


**Figure 2-8** THE HYDROMETER. This device is used to measure the density of a liquid.

the density of water is 1.00 g/mL, specific gravity is simply the density expressed without units.

$$\text{Density of Hg} = 13.6 \text{ g/mL}$$

$$\text{Density of water} = 1.00 \text{ g/mL}$$

$$\text{Specific gravity} = \frac{\text{density of mercury}}{\text{density of water}} = \frac{13.6 \cancel{\text{g/mL}}}{1.00 \cancel{\text{g/mL}}} = \underline{\underline{13.6}}$$

**See Problems 2-48 through 2-62.**

## 2-7 Temperature

Another important property of a substance is its temperature. **Temperature** *is a measure of the intensity of heat of a substance.* A **thermometer** *is a device that measures temperature.* The thermometer scale with which we are most familiar is probably the Fahrenheit scale (°F), but the Celsius scale (°C) is used in most of the rest of the world and in science. As you've probably noticed on the TV news, most weather reports now give the temperature readings in both scales as this country slowly switches to use of the Celsius scale (see Figure 2-9).

We learned in Chapter 1 that the boiling and melting points of pure water (at sea level) are constant and definite properties. We can take advantage of this fact to compare the two temperature scales and establish a relationship between them. In Figure 2-10 the temperature of an ice and water mixture is shown to be exactly 0 °C. This temperature was originally established by definition. This corresponds to exactly 32 °F on the Fahrenheit thermometer. The boiling point of pure water is exactly 100 °C, which corresponds to 212 °F.

On the Celsius scale there are 100 equal divisions between these two temperatures, whereas on the Fahrenheit scale there are 212 − 32 = 180

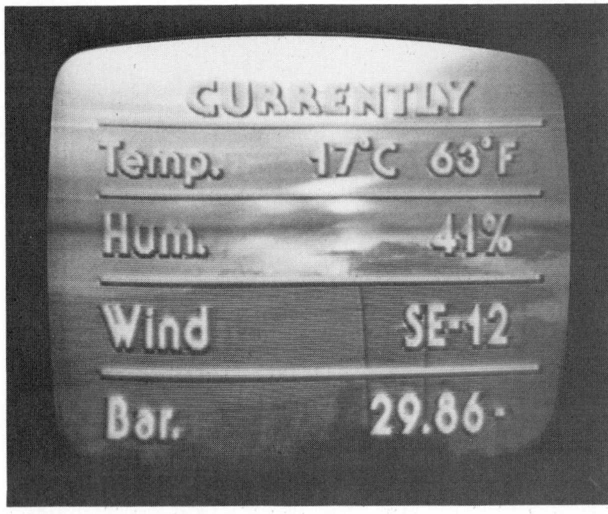


**Figure 2-9** CELSIUS TEMPERATURE. Thanks to the evening news, Americans are becoming more familiar with the Celsius scale.

equal divisions between the two temperatures. Thus we have the following relation between scale divisions:

$$100 \text{ C div.} = 180 \text{ F div.}$$

This relationship can be used to construct conversion factors between an equivalent number of Celsius and Fahrenheit degrees.

$$\frac{1.8\ °\text{F}}{1\ °\text{C}} \quad \text{and} \quad \frac{1\ °\text{C}}{1.8\ °\text{F}}$$

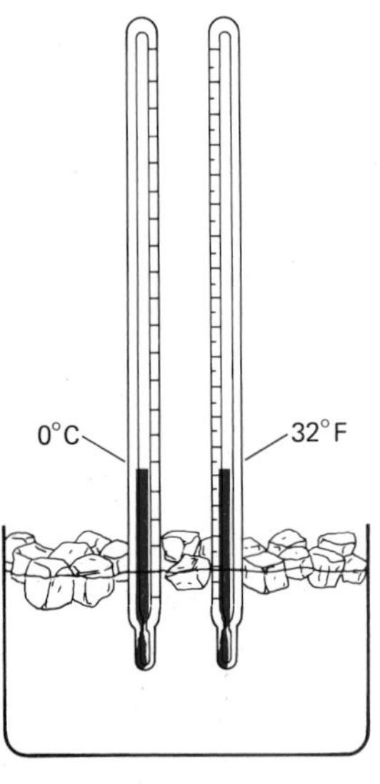


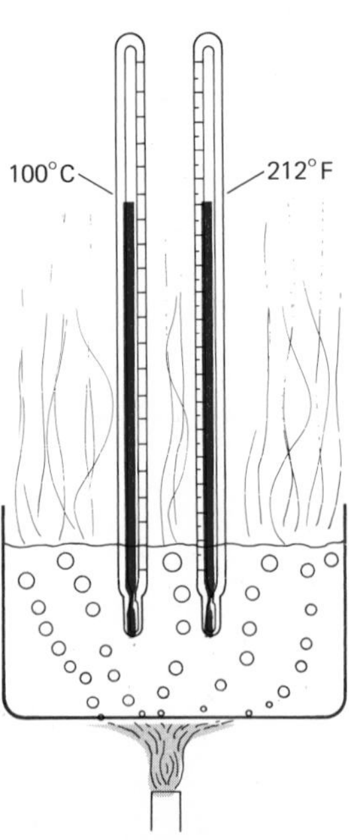


**Figure 2-10** THE TEMPERATURE SCALES. The freezing and boiling points of water are used to calibrate the temperature scales.

To convert the Celsius temperature [$t$(C)] to the Fahrenheit temperature [$t$(F)]:

1 Multiply the Celsius temperature by the proper conversion factor (1.8 °F/1 °C) to convert to the equivalent number of Fahrenheit degrees.

2 Add 32 °F to this number so that both scales start at the same point (the freezing point of water).

$$t(\text{F}) = \left[t(\text{C}) \times \frac{1.8\ ^\circ\text{F}}{1\ ^\circ\text{C}}\right] + 32\ ^\circ\text{F} = [t(\text{C}) \times 1.8] + 32$$

To covert the Fahrenheit temperature to the Celsius temperature:

1 Subtract 32 °F from the Fahrenheit temperature so that both scales start at zero.

2 Multiply that number by the proper conversion factor (1 °C/1.8 °F) to convert to the equivalent number of Celsius degrees.

$$t(\text{C}) = [t(\text{F}) - 32\ ^\circ\text{F}] \times \frac{1\ ^\circ\text{C}}{1.8\ ^\circ\text{F}} = \frac{[t(\text{F}) - 32]}{1.8}$$

**Example 2-15**
A person with a cold has a fever of 102 °F. What would be the reading on a Celsius thermometer?

$$t(\text{C}) = \frac{[\text{t(F)} - 32]}{1.8} = \frac{(102 - 32)}{1.8} = \underline{\underline{38.9\ ^\circ\text{C}}}$$

**Example 2-16**
On a cold winter day the temperature is −10.0 °C. What is the reading on the Fahrenheit scale?

$$t(\text{F}) = [\text{t(C)} \times 1.8] + 32$$

$$t(\text{F}) = (1.8 \times -10.0) + 32.0 = -18.0 + 32.0 = \underline{\underline{14.0\ ^\circ\text{F}}}$$

The SI temperature unit is called **the kelvin** (K). The zero on the Kelvin scale is theoretically the lowest possible temperature (the temperature at which the heat energy is zero). This corresponds to −273 °C (or more precisely −273.15 °C). Since the magnitude of a kelvin and a Celsius degree unit is the same, we have the following simple relationship between the two scales, where $T$ represents the number of kelvins and $t$(C) represents the Celsius degrees.

$$T = [t(\text{C}) + 273]$$

**See Problems 2-63 through 2-71.**

Thus the freezing point of water is 0 °C or 273 K and the boiling point is 100 °C or 373 K. We will use the Kelvin scale more in later chapters.

## 2-8 Specific Heat

Changes in temperature arise from the increase or decrease in heat energy of the surroundings. The *amount* of heat energy is reported in either of two units used extensively in science: the calorie and the joule. The *calorie* was originally defined as the amount of heat required to raise one gram of water from 14.5 to 15.5 °C. *The calorie is now defined in terms of its equivalent SI unit, the joule.* The modern definition of the calorie is thus:

$$1 \text{ cal} = 4.184 \text{ joules (J)} \qquad \text{(exactly)}$$

The nutritional calorie (1 Cal) is actually one kilocalorie ($10^3$ cal). It is the amount of heat energy a certain substance (like a piece of cake) would yield if it was burned. Although both calories and joules are used in chemistry, the joule is slowly becoming the unit of choice.

A property of matter similar to density is called specific heat. **Specific heat** *is defined as the heat required to raise one gram of the substance one degree Celsius.* The specific heat of water is therefore 1.00 cal/g · °C (by definition) or 4.184 J/g · °C. (The °C in the denominator represents a temperature change and not an actual temperature reading.) Specific heats of several pure substances are listed in Table 2-6.

Notice that the specific heat of water is comparatively quite large. This fortunate property of water helps even out the temperature of our planet. For example, the Gulf Stream is a river of warm water within the Atlantic Ocean that transports a large amount of heat energy from the Gulf of Mexico to England and Scandinavia. If it wasn't for this warm current and the ability of water to hold heat energy, these countries would be covered by ice, as is Greenland. Climate is more moderate near any ocean or large lake. The water can slowly absorb large amounts of heat energy in the summer, thus keeping the surroundings comparatively cool. In the winter the water releases the stored heat energy and keeps the surroundings comparatively warm.

**Table 2-6** Specific Heats

| Substance | Specific Heat (cal/g · °C) | Specific Heat (J/g · °C) |
|---|---|---|
| Water | 1.00 | 4.18 |
| Ice | 0.492 | 2.06 |
| Aluminum (Al) | 0.214 | 0.895 |
| Gold (Au) | 0.031 | 0.13 |
| Copper (Cu) | 0.092 | 0.38 |
| Zinc (Zn) | 0.093 | 0.39 |

**Example 2-17**
If 150 cal of heat energy is added to 50.0 g of water at 25 °C, what is the final temperature of the water?

$$\frac{150\ \text{cal}}{50.0\ \cancel{\text{g}}} \times \frac{1\ \cancel{\text{g}} \cdot {}^{\circ}\text{C}}{1.00\ \text{cal}} = 3\ {}^{\circ}\text{C rise} \qquad t(\text{C}) = 25 + 3 = \underline{\underline{28\ {}^{\circ}\text{C}}}$$

**Example 2-18**
It takes 15.0 cal to raise a 125-g quantity of silver 0.714 °C. What is the heat capacity of silver?

$$\frac{15.0\ \text{cal}}{125\ \text{g} \cdot 0.714\ {}^{\circ}\text{C}} = \underline{\underline{0.168\ \text{cal/g} \cdot {}^{\circ}\text{C}}}$$

**Example 2-19**
A 255-g quantity of gold is heated from 28 °C to 100 °C. How many joules were added to the sample?

Rise in temperature = 100 − 28 = 72 °C

$$225\ \cancel{\text{g}} \times 72\ {}^{\circ}\text{C} \times \frac{0.13\ \text{J}}{\cancel{\text{g}} \cdot \cancel{{}^{\circ}\text{C}}} = \underline{\underline{2400\ \text{J}}}$$

**See Problems 2-72 through 2-82.**

## 2-9 Chapter Summary

Since chemistry is a physical science it is vitally concerned with measurements. Two qualities of a measurement are important: its precision and its accuracy.

Measurements have
- Precision — Refers to the degree of reproducibility or the number of significant figures in a measurement
- Accuracy — Refers to how close a measurement is to the true value

Rules were given for handling zero as a significant figure and for rounding off. We then discussed how to express the answers to mathematical operations as follows:

Calculations
- Addition or subtraction — Answer expressed to the proper decimal place
- Multiplication or division — Answer expressed to the proper number of significant figures

To handle the very large and the very small numbers used in chemistry the use of scientific notation was discussed.

The metric or SI system of measurement was introduced specifically with regard to length, volume, and mass. The decimal relationships between units of the metric system are convenient compared with the unsystematic English system. Relationships within a system are exact numbers, while relationships between systems can be expressed to various degrees of precision.

The unit-factor method was introduced as a convenient tool to convert a measurement to other units within the same system or to another system. The relationships listed in Tables 2-3 and 2-4 were used as conversion factors to convert what is "given" to what is "requested." The careful use of units helps guide us to the proper mathematical manipulation.

Several other measurable properties of matter were then discussed. Density is a property of homogeneous matter that relates mass to a specific volume. Substances that are less dense than a liquid are buoyant, or float in the liquid. A hydrometer uses the principle of buoyancy to determine the density of specific gravity of a liquid. Specific gravity is simply the density of the substance compared with that of water.

Temperature is a property of matter that measures heat intensity. Thermometers measure temperature. Three temperature scales are common: Fahrenheit (°F), principally used in the United States and Britain; Celsius (°C), used in most of the world and science; and Kelvin (K) used almost entirely in science as the SI unit. Relationships between scales were given. Specific heat was discussed as a property of matter that relates the amount of heat in calories or joules to a mass and a change of temperature.

The six properties of matter that were discussed can be summarized as follows.

| Property | Principal Unit | Comment |
|---|---|---|
| Length | meter (m) | $m^2$ refers to area. |
| Volume | liter (L) | $m^3$ refers to volume, $1\ dm^3 = 1$ L. |
| Mass | gram (g) | In the metric system 1 mL of $H_2O$ weighs 1 g. |
| Density | g/mL | This property can be used to identify pure substances and as a conversion factor between volume and mass. |
| Temperature | °C | $T = [t(C) + 273]$ K. Zero on the Kelvin scale is the lowest possible temperature |

| Property | Principal Unit | Comment |
|---|---|---|
| Specific heat | cal/g · °C | This property can be used to identify pure substances and as a conversion factor between heat, mass, and temperature change. |

## Exercises

*More difficult problems are marked by an asterisk.*

### Significant Figures

**2-1** Which of the following measurements is the most precise?
(a) 75.2 gal (b) 74.212 gal (c) 75.22 gal (d) 75 gal

**2-2** How can a measurement be precise but not accurate?

2-3 The actual length of a certain plank is 26.782 in. Which of the following measurements is the most precise and which is the most accurate?
(a) 26.5 in. (b) 26.8 in. (c) 26.202 in. (d) 26.98 in.

**2-4** How many significant figures are in each of the following measurements?
(a) 7030 g (b) 4.0 kg (c) 4.01 lb (d) 0.01 ft (e) 4002 m (f) 0.060 hr (g) 8200 km (h) 0.00705 ton

2-5 How many significant figures are in each of the following measurements?
(a) 10.070 in. (b) 0.0023 mm (c) 0.606 mL (d) 0.300 sec (e) 3000 lb (f) 100.0 cm (g) 1.002 kL (h) 2060 yr

### Significant Figures in Mathematical Operations

**2-6** Round off each of the following numbers to three significant figures.
(a) 15.9994 (b) 1.0080 (c) 0.6654 (d) 4890 (e) 87,550 (f) 0.027225 (g) 301.4

2-7 Round off each of the following numbers to two significant figures.
(a) 18.998 (b) 10.81 (c) 3,650,000 (d) 0.07482 (e) 444 (f) 9750 (g) 6.02 (h) 9500

**2-8** Carry out each of the following operations. Assume that the numbers represent measurements so that the answer should be expressed to the proper decimal place.
(a) $14.72 + 0.611 + 173$
(b) $0.062 + 11.38 + 1.4578$
(c) $1600 - 4 + 700$
(d) $47 + 0.91 - 0.286$
(e) $0.125 + 0.71$

2-9 Carry out each of the following operations. Assume that the numbers represent measurements so that the answer should be expressed to the proper decimal place.
(a) $0.0128 + 0.00102 + 0.00416$
(b) $173 + 150 + 122$
(c) $15.3 - 1.12 + 3.377$
(d) $9300 + 1100 - 188$

**2-10** Carry out the following calculations. Express your answer to the proper number of significant figures.
(a) 40.0 cm × 3.0 cm
(b) 179 ft × 2.20 lb
(c) 4.386 cm$^2$ ÷ 2 cm
(d) (14.65 in. × 0.32 in.) ÷ 2.00 in.

2-11 Carry out the following calculations. Express your answer to the proper number of significant figures.

(a) 1400 m ÷ 0.6 m
(b) 0.565 $mm^3$ ÷ (1.62 mm × 0.50 mm)
(c) 106.0 ft × 3.0 ft
(d) (14.72 $cm^2$ × 2 cm) ÷ 5.678 cm

### Scientific Notation

**2-12** Express the following numbers in scientific notation with one digit to the left of the decimal point in the coefficient.

| | |
|---|---|
| (a) 157 | (e) 0.0349 |
| (b) 0.157 | (f) 32,000 |
| (c) 0.0300 | (g) 32 billion |
| (d) 40,000,000 (two significant figures) | (h) 0.000771 |
| | (i) 2340 |

**2-13** Express the following as numbers without powers of 10.

| | |
|---|---|
| (a) $4.76 \times 10^{-4}$ | (d) $0.489 \times 10^{5}$ |
| (b) $6.55 \times 10^{3}$ | (e) $475 \times 10^{-2}$ |
| (c) $788 \times 10^{-5}$ | (f) $0.0034 \times 10^{-3}$ |

**2-14** Change the following numbers to scientific notation with one digit to the left of the decimal point in the coefficient.

| | |
|---|---|
| (a) $489 \times 10^{-6}$ | (d) $571 \times 10^{-4}$ |
| (b) $0.456 \times 10^{-4}$ | (e) $4975 \times 10^{5}$ |
| (c) $0.0078 \times 10^{6}$ | (f) $0.030 \times 10^{-2}$ |

**2-15** Change the following numbers to standard scientific notation (one digit to the left of the decimal point).

| | |
|---|---|
| (a) $0.99 \times 10^{9}$ | (d) $8900 \times 10^{22}$ |
| (b) $787 \times 10^{-6}$ | (e) $18.08 \times 10^{-13}$ |
| (c) $0.00560 \times 10^{10}$ | (f) $0.0000205 \times 10^{-15}$ |

**2-16** Carry out the following operations. Assume that the numbers represent measurements so that the answer should be expressed to the proper decimal place.
(a) $(1.82 \times 10^{-4}) + (0.037 \times 10^{-4}) + (14.11 \times 10^{-4})$
(b) $(13.7 \times 10^{6}) - (2.31 \times 10^{6}) + (116.28 \times 10^{5})$
(c) $(0.61 \times 10^{-6}) + (0.11 \times 10^{-4}) + (0.0232 \times 10^{-3})$
(d) $(372 \times 10^{12}) + (1200 \times 10^{10}) - (0.18 \times 10^{15})$

**2-17** Carry out the following operations. Assume that the numbers represent measurements so that the answer should be expressed to the proper decimal place.
(a) $(1.42 \times 10^{-10}) + (0.17 \times 10^{-10}) - (0.009 \times 10^{-10})$
(b) $(146 \times 10^{8}) + (0.723 \times 10^{10}) + (11 \times 10^{8})$
(c) $(1.48 \times 10^{-7}) + (2911 \times 10^{-9}) + (0.6318 \times 10^{-6})$
(d) $(299 \times 10^{10}) + (823 \times 10^{8}) + (0.75 \times 10^{11})$

**2-18** Carry out the following operations. Assume that the numbers represent measurements so that the answer should be expressed to the proper number of significant figures.
(a) $(149 \times 10^{6}) \times (0.21 \times 10^{3})$
(b) $(0.371 \times 10^{14}) \div (2 \times 10^{4})$
(c) $(6 \times 10^{6}) \times (6 \times 10^{6})$
(d) $(0.1186 \times 10^{6}) \times (12 \times 10^{-5})$
(e) $(18.21 \times 10^{-10}) \div (0.0712 \times 10^{6})$

**2-19** Carry out the following operations. Assume that the numbers represent measurements so that the answer should be expressed to the proper number of significant figures.
(a) $(0.88 \times 10^{5}) \times (1.62 \times 10^{4})$
(b) $(448 \times 10^{10}) \div (4.4 \times 10^{10})$
(c) $(7637 \times 10^{12}) \div (0.028 \times 10^{-4})$
(d) $(4 \times 10^{-6}) \times (4.0 \times 10^{-6})$
(e) $(0.04772 \times 10^{-8}) \div (148 \times 10^{6})$

**2-20** Using scientific notation, express the number 87,000,000 to (a) one significant figure, (b) two significant figures, and (c) three significant figures.

**2-21** Using scientific notation, express the number 12,600 to (a) one significant figure, (b) two significant figures, (c) three significant figures, and (d) four significant figures.

### Length, Volume, and Mass in the Metric System

**2-22** Complete the following table:

| | mm | cm | m | km |
|---|---|---|---|---|
| (Example) | 108 | 10.8 | 0.108 | $1.08 \times 10^{-4}$ |
| (a) | $7.2 \times 10^{3}$ | ____ | ____ | ____ |
| (b) | ____ | ____ | 56.4 | ____ |
| (c) | ____ | ____ | ____ | 0.250 |

**2-23** Complete the following table:

| | mg | g | kg |
|---|---|---|---|
| (a) | $8.9 \times 10^3$ | ___ | ___ |
| (b) | ___ | 25.7 | ___ |
| (c) | ___ | ___ | 1.25 |

**2-24** Complete the following table:

| | mL | L | kL |
|---|---|---|---|
| (a) | ___ | ___ | 0.0976 |
| (b) | ___ | 432 | ___ |
| (c) | $4.58 \times 10^4$ | ___ | ___ |

### Conversions between the English and Metric Systems

**2-25** Which of the following are "exact" relationships?
(a) 12 = 1 dozen  (d) 1.06 qt = 1 L
(b) 1 gal = 3.78 L  (e) $10^3$ m = 1 km
(c) 3 ft = 1 yd  (f) 454 g = 1 lb

**2-26** Complete the following table:

| | miles | ft | m | km |
|---|---|---|---|---|
| (a) | ___ | ___ | $7.8 \times 10^3$ | ___ |
| (b) | 0.450 | ___ | ___ | ___ |
| (c) | ___ | $8.98 \times 10^3$ | ___ | ___ |
| (d) | ___ | ___ | ___ | 6.78 |

**2-27** Complete the following table.

| | gallons | quarts | liters |
|---|---|---|---|
| (a) | 6.78 | ___ | ___ |
| (b) | ___ | 670 | ___ |
| (c) | ___ | ___ | $7.68 \times 10^3$ |

**2-28** Complete the following table.

| | lb | g | kg |
|---|---|---|---|
| (a) | ___ | $1.98 \times 10^3$ | ___ |
| (b) | 178 | ___ | ___ |
| (c) | ___ | ___ | 4.40 |

**2-29** If a person weighs 122 lb, what is her weight in kilograms?

**2-30** A punter on a professional football team averaged 28.0 m per kick. What is his average in yards? Should he be kept on the team?

**2-31** If a student drinks a 12-oz (0.375 qt) can of soda, what volume did she drink in liters?

**2-32** A prospective basketball player is 2.11 m tall and weighs 98.8 kg. What are his statistics in feet and pounds?

**2-33** Gasoline is sold by the liter in Europe. How many gallons does a 55.0-L gas tank hold?

**2-34** If the length of a football field is changed to 100 m from 100 yd, will the field be longer or shorter than the current field? How many yards would a first and ten be on the metric field?

**2-35** Bourbon used to be sold by the "fifth" (one fifth of a gallon). A bottle now contains 750 mL. Which is greater?

**2-36** The speed limit is 55 miles/hr. What is the speed limit in kilometers per hour?

**2-37** If gold costs $505/oz, what is the cost of 1.00 kg of gold? (Assume 16 oz per pound.)

**2-38** It is 245 miles from Kansas City to St. Louis. How far is this in kilometers?

**2-39** Gasoline sold for $1.12/gal in 1984. What was the cost per liter? What did it cost to fill an 80.0-L tank?

**2-40** Using the price of gas from Problem 2-39, how much does it cost to drive 551 miles if your car averages 21.0 miles/gal? How much does it cost to drive 482 km?

**2-41** Using the price of gas from Problem 2-39, how many km can one drive for $45.00? The car averages 21.0 miles/gal.

**2-42** An aspirin contains 0.324 g (5.00 grains) of aspirin. How many pounds of aspirin are in a 500-aspirin bottle?

**2-43** A certain type of nail costs $0.95/lb. If there are 145 nails per pound, how many nails can you purchase for $2.50?

**2-44** Another type of nail costs $0.92/lb, and

there are 185 nails per pound. What is the cost of 5670 nails?

* **2-45** At a speed of 35 miles/hr, how many centimeters will you travel per second?

* **2-46** The planet Jupiter is about $4.0 \times 10^8$ miles from Earth. If radio signals travel at the speed of light, which is $3.0 \times 10^{10}$ cm/sec, how long does it take a radio command from Earth to reach a Voyager spacecraft passing Jupiter?

**2-47** If grapes sell for $1.15/lb and there are 255 grapes per pound, how many grapes can you buy for $5.15?

### Density

**2-48** A handful of sand weighs 208 g and displaces a volume of 80.0 mL. What is its density?

**2-49** A 45.5-g quantity of iron has a volume of 5.76 mL. What is its density?

**2-50** What is the volume in milliliters occupied by 285 g of mercury? (See Table 2-5.)

**2-51** What is the weight of 671 mL of table salt? (See Table 2-5.)

**2-52** What is the weight of 1.00 L of gasoline? (Refer to Table 2-5.)

**2-53** What is the volume in milliliters occupied by 1.00 kg of carbon tetrachloride?

**2-54** Pumice is a volcanic rock that contains many trapped air bubbles in the rock. A 155-g sample is found to have a volume of 163 mL. What is the density of pumice? What is the volume of 4.56-kg sample? Will pumice float or sink in water? In ethyl alcohol?

**2-55** The density of diamond is 3.51 g/mL. What is the volume of the Hope diamond if it weighs 44.0 carats (1 carat = 0.200 g)?

**2-56** A small box is filled with liquid mercury. The dimensions of the box are 3.00 cm wide, 8.50 cm long, and 6.00 cm high. What is the weight of the mercury in the box? (1.00 mL = 1.00 $cm^3$.)

**2-57** Which has a greater volume, 1 kg of lead or 1 kg of gold?

**2-58** Which has a greater weight, 1 L of gasoline or 1 L of water?

**2-59** What is the weight of 1 gal of gasoline in grams? In pounds?

* **2-60** Calculate the density of water in pounds per cubic foot ($lb/ft^3$)?

* **2-61** In certain stars matter is tremendously compressed. In some cases the density is as high as $2.0 \times 10^7$ g/mL. A tablespoon full of this matter is about 4.5 mL. What is this weight in pounds?

**2-62** What is the difference between density and specific gravity?

### Temperature

**2-63** The temperature of the water around a nuclear reactor core is about 300 °C. What is this temperature in degrees Fahrenheit?

**2-64** The temperature on a comfortable day is 76 °F. What is this temperature in degrees Celsius?

**2-65** The lowest possible temperature is −273 °C and is called absolute zero. What is absolute zero on the Fahrenheit scale?

**2-66** Mercury thermometers cannot be used in cold arctic climates because mercury freezes at −39 °C. What is this temperature in degrees Fahrenheit?

**2-67** The coldest temperature recorded on earth was −110 °F. What is this temperature in degrees Celsius?

**2-68** A hot day in the midwest is 35.0 °C. What is this in degrees Fahrenheit?

**2-69** Convert the following temperatures to the Kelvin scale.

(a) 47 °C (d) −12 °C
(b) 23 °C (e) 65 °F
(c) −73 °C (f) −20 °F

**2-70** Convert the following Kelvin temperatures to degrees Celsius.

(a) 175 K (d) 225 K
(b) 295 K (e) 873 K
(c) 300 K

* **2-71** At what temperature are the Celsius and Fahrenheit scales numerically equal?

### Specific Heat

**2-72** It took 17.5 cal of heat to raise the temperature of 10.0 g of a substance 8.58 °C. What is the specific heat of the substance?

2-73 When 365 g of a certain pure metal cooled from 100 °C to 95 °C, it liberated 56.6 cal. Identify the metal. (Refer to Table 2-6.)

**2-74** If 150 cal of heat energy is added to 50.0 g of copper at 25 °C, what is the final temperature of the copper? Compare this temperature rise with that of an equivalent amount of water in Example 2-17.

**2-75** How many calories are evolved if 43.5 g of aluminum is cooled by 13.0 Celsius degrees?

**2-76** What weight of iron is needed to absorb 16.0 cal if the temperature of the sample rises from 25 °C to 58 °C? The specific heat of iron is 0.106 cal/g·°C

2-77 When 685 g of copper absorbs 260 J of heat, what is the rise in temperature? (Refer to Table 2-6.)

**2-78** A can of diet soda contains 1.00 Cal (1.00 kcal) of heat energy. If this energy was transferred to 50.0 g of water at 25 °C, what would be the final temperature?

**2-79** If 50.0 g of aluminum at 100.0 °C is allowed to cool to 35.0 °C, how many joules are evolved?

* **2-80** In the preceding problem assume that the hot aluminum was added to water originally at 30.0 °C. If all of the heat lost by the aluminum was gained by the water, what weight of water was present if the final temperature was 35.0 °C?

* 2-81 If 50.0 g of water at 75.0 °C is added to 75.0 g of water at 42.0 °C, what is the final temperature?

* **2-82** If 100.0 g of a metal at 100.0 °C is added to 100 g of water at 25.0 °C, the final temperature is 31.3 °C. What is the specific heat of the metal? Identify the metal from Table 2-6.

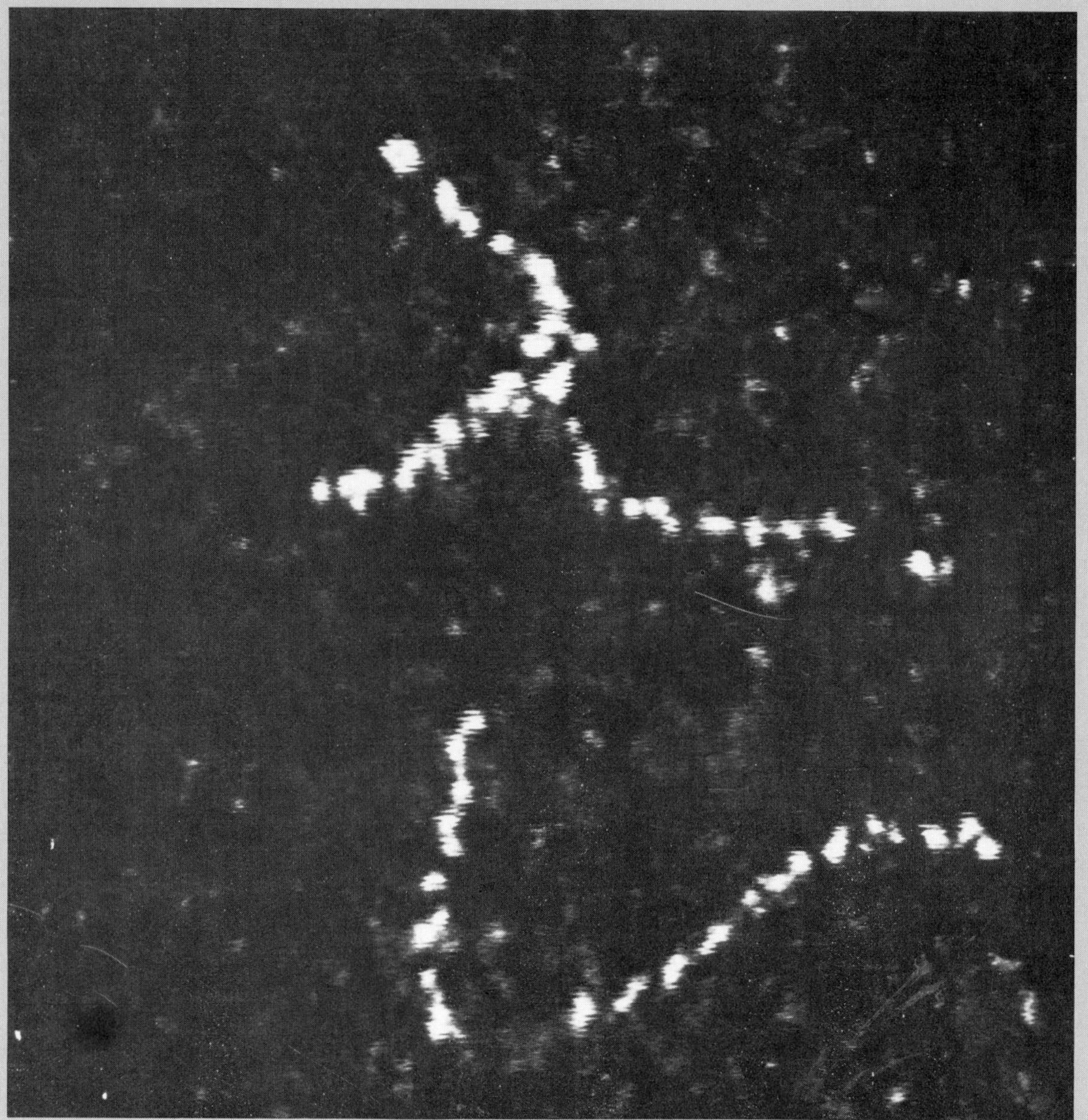

This electron microscope image shows individual thorium atoms.

# Chapter 3
# The Atom, the Structure of Matter, and Nuclear Reactions

## Purposes of Chapter 3

In Chapter 3 we introduce the basic building block of nature—the atom, its composition, and how it exists in elements and compounds. Also in Chapter 3 we describe the changes that the nucleus of the atom undergoes.

## Objectives for Chapter 3

After completion of this chapter, you should be able to:

1. List the three major parts of Dalton's atomic theory. (Intro.)
2. Describe the nuclear atom, including the name, location, mass (in amu), and electrical charge of the three particles in the atom. (3-1)
3. Give the atomic number and mass number of a specified isotope. (3-2)
4. Write the number of protons, neutrons, and electrons from the representation of a specified isotope. (3-2)
5. Define atomic weight and describe how it differs from mass number. (3-2)
6. Calculate the atomic weight of an element given the percent distribution and atomic weights of its isotopes. (3-2)
7. Distinguish between atoms and molecules. (3-3)
8. Describe the function of a covalent bond in a molecule. (3-3)
9. List the elements and the number of atoms of that element in a compound from the formula. (3-4).
10. Write definitions for the terms ion, cation, anion, and electrostatic force. (3-5)
11. Distinguish between molecular and ionic compounds. (3-5)
12. List the number of protons and electrons present in a specified ion. (3-5)

**(Optional: Nuclear Reactions)**

13. Explain what is illustrated by a nuclear equation. (3-6)
14. Define radioactivity and the three common kinds of radiation. (3-6)
15. Complete a nuclear equation given appropriate data. (3-6)

**16** Describe the concept of half-life and radioactive decay series. (3-7)
**17** Describe how the three types of radiation affect nearby matter. (3-8)
**18** Explain how nuclear reactions take place and how elements are artificially synthesized. (3-9)
**19** Describe nuclear fission and how a chain reaction occurs. (3-10)
**20** Explain how fission is controlled in a nuclear reactor. (3-10)
**21** Explain what is meant by nuclear fusion. (3-11)

Two hundred years ago scientists assumed that matter was continuous. That is, if you took a sample of an element like copper it was thought that it could be divided into infinitely smaller pieces without changing the nature of the element. In 1803, however, an English scientist named John Dalton (1766–1844) proposed a theory that had three major parts.

1 Matter is composed of small indivisible particles called atoms.

2 Atoms of the same element are identical and have the same properties.

3 Chemical reactions are merely the rearrangement of atoms into different combinations.

This theory, called the atomic theory, serves as the basis of our modern understanding of the nature of matter. In fact, it was a restatement of a long dormant theory proposed by a Greek philosopher, Democritus, over two thousand years earlier. Dalton's theory, unlike that of Democritus, was soundly based on several quantitative laws including the law of conservation of mass (see p. 15).

In this chapter we will take a close look at the **atom,** *which is the smallest particle of an element that can exist and enter into chemical reactions.* There are three aspects of the atom that will be developed in this chapter. First we will examine the nature of the atom itself in order to determine how the atoms of one element differ from those of another. Then we will examine the nature of the atom as it exists in elements and compounds. Finally (in an optional section), we will discuss how the interior of the atom, known as its nucleus, undergoes changes that are important, yet controversial, in our modern world.

The first thing we must realize is that atoms are extremely small. Since the diameter of an atom is on the order of $10^{-8}$ cm, it would take 100 million atoms in a line to extend for 1 cm (less than half an inch). Even the most powerful microscope now in use can produce only fuzzy images of some of the larger atoms. In early 1983, however, it was announced that a new version of a special type of microscope called an electron microscope was being constructed at the University of Chicago. It is hoped that this new instrument will be able to resolve most individual atoms. Let's assume that this new microscope is already available to us so that we can actually picture individual atoms as small spheres. In what follows we can thus describe how scientists think individual atoms will appear in elements and compounds. First, however, we will assume that we can resolve the image of the atom even further so that we can actually look inside the atom itself.

## 3-1 The Composition of the Atom

From earth with the naked eye, the moon looks like a solid, round sphere. Through a telescope, however, the moon is seen to be complex, with mountains, valleys, and large craters. The new electron microscope will at best only show the atom as a featureless sphere. Although we will never be able to construct an electron microscope that will allow us to actually see inside the atom itself, let's assume that we can picture how the atom should appear. What follows then is a simplification of the currently held view of the atom. How this picture evolved is an intriguing story beginning in the late 1800s and continuing to the present day. Many famous scientists were involved in this story. Scientists by the names of Thomson, Rutherford, Becquerel, Curie, and Roentgen, among others, all had a hand in the development of this picture, known as the nuclear model of the atom.

In a close look into the atom, the first thing to be noticed is that the atom is not solid as it first appeared. In fact, the atom is mostly empty space. Most of this vast space of the atom is the home of small particles called electrons. Electrons were the first particles in the atom to be identified. *In 1897, J. J. Thomson characterized the* **electron** *by noting that it had a negative charge and was common to the atoms of all elements.* In 1911 the English scientist, Lord Ernest Rutherford, proved by his experiments that *these electrons surround a small dense core at the center of the atom called the* **nucleus.** The nucleus is so small compared to the total volume of a typical atom that if it were expanded to the size of a softball, the radius of the atom would extend for about 1 mile. Despite the small size of the nucleus, it contains almost all of the mass of the atom. *The nucleus is composed of particles called* **nucleons.** *There are two types of nucleons,* **protons,** *which have a positive charge, and* **neutrons,** *which do not carry a charge.* (See Figure 3-1.)

The data on the particles in the atom is summarized in Table 3-1. *The unit of weight,* **amu,** *is an abbreviation of* **atomic mass unit** *and is approximately the weight of one proton (or one neutron).* The origin of this unit of weight will be discussed later.

At this point in the discussion it would appear that the atom is composed of just these three particles. In recent years, however, the picture of the atom has become more complex, as physicists are now looking into the basic structures of the three particles themselves. The current view is that these particles are composed of even more fundamental particles. The search for what is considered to be *the* fundamental particles of nature continues with

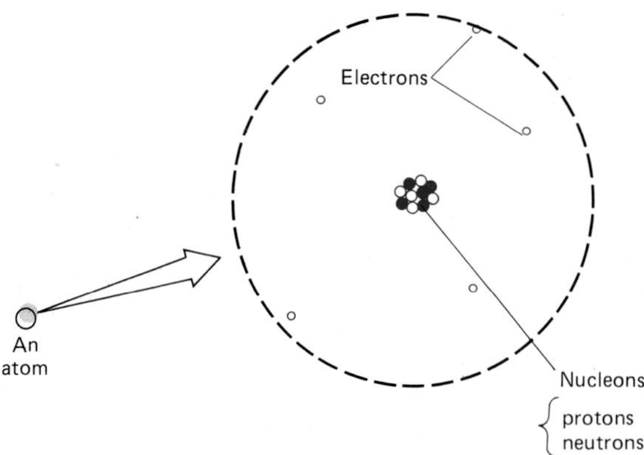


**Figure 3-1** THE COMPOSITION OF THE ATOM. The atom is composed of electrons, neutrons, and protons.

the discovery of nuclear particles with names such as quarks, gluons, and W particles. With each discovery the picture of the nucleus becomes increasingly more complex. Fortunately, the three-particle model of the atom still meets the needs of the chemist.

## 3-2 Atomic Number, Mass Number, and Atomic Weight

A typical atom of the element copper is made up of a nucleus containing a total of 63 nucleons, of which 29 are protons and 34 are neutrons. The neutral atom also contains 29 electrons. *The number of protons in the nucleus* (which is equal to the total positive charge) *is referred to as its* **atomic number.** *The total number of nucleons is called the* **mass number.** Therefore, this particular copper atom has an atomic number of 29 and a mass number of 63. If we were to examine a number of copper atoms, we would soon come upon one that is different from the one we just described. This atom of copper also has an atomic number of 29 but a mass number of 65, so it has 36 neutrons. *Atoms having the same atomic number but different mass numbers are known as* **isotopes.** (Obviously, Dalton wasn't completely right; all of the atoms of an element are not necessarily identical.)

If we looked at the atoms of other elements, we would find the same phenomenon—almost all elements exist in nature as a mixture of isotopes. What we would also notice is that different elements have different atomic

**Table 3-1** Atomic Particles

| Name | Symbol | Electrical Charge | Mass (amu) | Mass (g) |
|---|---|---|---|---|
| Electron | $e$ | $-1$ | 0.000549 | $9.110 \times 10^{-28}$ |
| Proton | $p$ | $+1$ | 1.00728 | $1.673 \times 10^{-24}$ |
| Neutron | $n$ | 0 | 1.00867 | $1.675 \times 10^{-24}$ |

numbers. *Indeed, it is the atomic number that distinguishes one element from another.*

Isotopes of elements are represented by showing the mass number in the upper left-hand corner of the symbol of the element. Often, the atomic number of the element is represented in the lower left-hand corner. This is strictly a convenience, however, since the atomic number determines the identity of the element. For example, if the atomic number is 28, the element is nickel; if it is 30, the element is zinc. The two isotopes of copper are shown below.

The mass number (number of nucleons) (29 protons and 34 neutrons) → $^{63}_{29}Cu$ $^{65}_{29}Cu$ ← (29 protons and 36 neutrons)

The atomic number (number of protons) (29 protons)

---

**Example 3-1**

How many protons, neutrons, and electrons are present in $^{90}_{38}Sr$?

$$\text{Atomic number} = \text{number of protons} = \underline{\underline{38}}$$

$$\text{Number of neutrons} = \text{mass number} - \text{number of protons}$$

$$90 - 38 = \underline{\underline{52}}$$

*In a neutral atom the number of protons and electrons is the same.*

$$\text{Number of electrons} = \underline{\underline{38}}$$

---

**Example 3-2**

How many protons, neutrons, and electrons are in the three isotopes of oxygen, $^{16}_{8}O$, $^{17}_{8}O$, and $^{18}_{8}O$?

$^{16}_{8}O$ Protons = atomic number = $\underline{\underline{8}}$

Neutrons = mass number − atomic number = 16 − 8 = $\underline{\underline{8}}$

Electrons = protons = $\underline{\underline{8}}$

$^{17}_{8}O$ Protons = $\underline{\underline{8}}$

Neutrons = 17 − 8 = $\underline{\underline{9}}$

Electrons = $\underline{\underline{8}}$

$^{18}_{8}O$ Protons = $\underline{\underline{8}}$

Neutrons = 18 − 8 = $\underline{\underline{10}}$

Electrons = $\underline{\underline{8}}$

**See Problems 3-1 through 3-9.**

Notice that the only difference among the isotopes is the number of neutrons in the nucleus.

The mass number of an isotope is a convenient but imprecise measure of its comparative weight. A more precise measure of weight of one isotope relative to another is known as the atomic weight. *The* **atomic weight** *of an isotope is determined by comparison to a standard,* $^{12}C$*, which is defined as having an atomic weight of exactly 12 amu.* By comparison to $^{12}C$, the atomic weight of $^{10}B$ is 10.013 amu and that of $^{11}B$ is 11.009 amu. Since boron, as well as most other naturally occurring elements, are found in nature as a mixture of isotopes, the atomic weight of the element reflects this mixture. Thus the atomic weight of an element as found in nature is the average of the atomic weights of all isotopes present. The average is weighted, however, according to the percent of each isotope present, as illustrated by the following examples.

### Example 3-3

In nature the element boron is present as 19.9% $^{10}_{5}B$, and 80.1% $^{11}_{5}B$. If the atomic weight of $^{10}_{5}B$ is 10.013 amu and of $^{11}_{5}B$ is 11.009 amu, what is the atomic weight of boron?

### Procedure

Find the contribution of each isotope toward the atomic weight by multiplying the percent in decimal fraction form by the weight of the isotope.

$$^{10}_{5}B \quad 0.199 \times 10.013 \text{ amu} = 1.99 \text{ amu}$$

$$^{11}_{5}B \quad 0.801 \times 11.009 \text{ amu} = \underline{8.82 \text{ amu}}$$

$$\underline{\underline{10.81 \text{ amu}}}$$

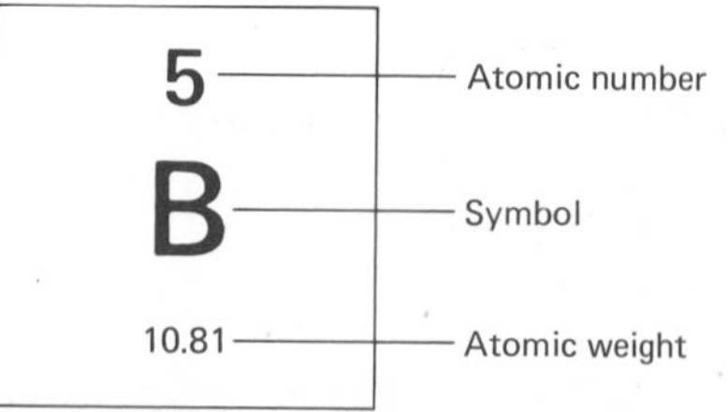


**Figure 3-2** ATOMIC WEIGHT AND ATOMIC NUMBER FROM THE PERIODIC TABLE.

**Example 3-4**

Naturally occurring lead is composed of four isotopes: 1.40% $^{204}_{82}Pb$ (wt. = 203.97 amu), 24.10% $^{206}_{82}Pb$ (wt. = 205.97 amu), 22.10% $^{207}_{82}Pb$ (wt. = 206.98 amu), and 52.40% $^{208}_{82}Pb$ (wt. = 207.98 amu). What is the atomic weight of lead?

$$^{204}_{82}Pb \quad 0.0140 \times 203.97 \text{ amu} = 2.86 \text{ amu}$$

$$^{206}_{82}Pb \quad 0.2410 \times 205.97 \text{ amu} = 49.64 \text{ amu}$$

$$^{207}_{82}Pb \quad 0.2210 \times 206.98 \text{ amu} = 45.74 \text{ amu}$$

$$^{208}_{82}Pb \quad 0.5240 \times 207.98 \text{ amu} = \underline{109.0 \text{ amu}}$$

$$\underline{\underline{207.2 \text{ amu}}}$$

Inside the back cover you will find an arrangement of all of the elements in what is called the periodic table. The symbol of each element is shown in a box. Also included in the box is the atomic number and the atomic weight of the element, which is the weighted average of the isotopes occurring in nature for that element (see Figure 3-2.) Notice that the elements are listed in order of increasing atomic number. The elements with whole-number atomic weights in parentheses are synthetic elements and are not found in nature. These elements undergo nuclear disintegration, as discussed in a later section of this chapter. The atomic weight listed for these elements represents the mass number of the most stable isotope made.

**See Problems 3-10 through 3-19.**

## 3-3 The Basic Structure of the Elements

To the naked eye a wire made from the element copper appears shiny, solid, and continuous. The new electron microscope will show a different picture. The element is indeed composed of small spherical atoms, all of which appear to be identical. This phenomenon is much like the appearance of a brick wall: it looks completely featureless and continuous from a distance, but up

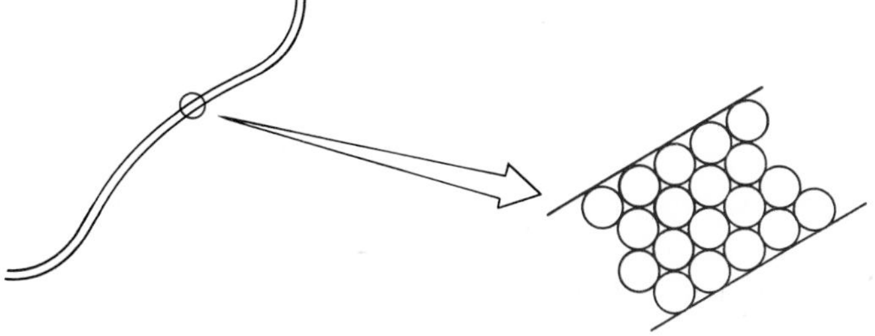

**Figure 3-3** THE ATOMS OF COPPER. Most solid elements exist in nature as closely packed atoms.

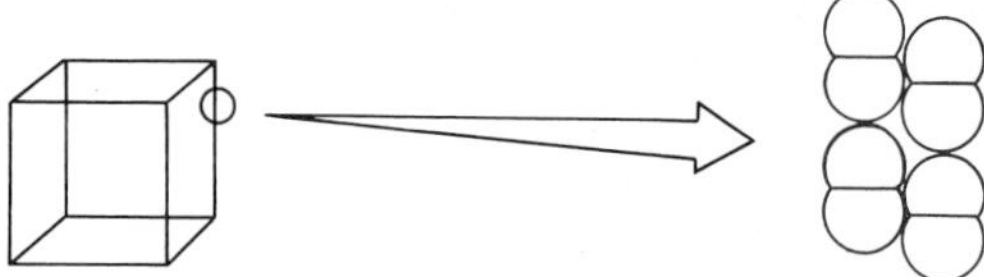

**Figure 3-4** THE MOLECULES OF IODINE. Some elements exist in nature as two or more atoms bound together.

close it is obvious that the wall is constructed of closely packed basic units (bricks). In the wire, the atoms of copper are firmly packed together, as would be any spheres of the same size such as marbles. (See Figure 3-3.)

All elements do not exist in nature as simple congregations of individual atoms, as is the case in copper. In some elements the basic units are composed of molecules. *A* **molecule** *is a group of two or more atoms held together by a force called a* **covalent chemical bond.**

Iodine is an example of an element whose atoms are joined by covalent bonds to form molecules. In Figure 3-4, magnification of a sample of solid iodine will show that the atoms are bonded to each other in pairs. Two spherical atoms in each molecule are firmly joined together and even overlap to a small extent. Atoms of adjoining molecules just touch, however, as did the individual atoms of copper. The nature of the covalent chemical bond will be discussed in more detail in Chapter 6.

Examples of other elements that exist as diatomic (two-atom) molecules under normal conditions such as are found on the surface of the earth are oxygen, nitrogen, hydrogen, fluorine, chlorine, and bromine. A form of phosphorus consists of molecules composed of four atoms, and a form of sulfur consists of molecules composed of eight atoms.

## 3-4 Compounds and Formulas

The molecules of an element are composed of the same type of atoms. Of more significance to the chemist is the fact that atoms of different elements also combine to form molecules. *One type of compound is composed of molecules made up of atoms of two or more different elements.* Water is an example of a molecular compound with molecules composed of two atoms of hydrogen joined to one atom of oxygen by covalent bonds (see Figure 3-5).

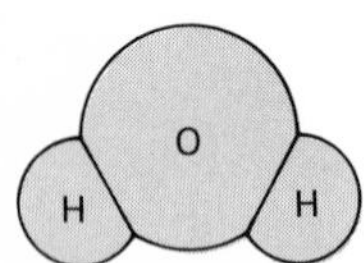


**Figure 3-5** A WATER MOLECULE. A water molecule is composed of two hydrogen atoms bound to one oxygen atom.

A *compound is symbolized by the atoms of which it is composed. This is called the* **formula** *of the compound or element.* The familiar formula for water is therefore

$$H_2O$$

This is, of course, pronounced "H-two-O." You should notice that the "2" is written as a subscript, indicating that the molecule has two hydrogen atoms. When there is only one atom of a certain element present (e.g., oxygen), the subscript of "1" is omitted.

*Each chemical compound has a unique formula or arrangement of atoms in its molecules.* For example, another compound exists whose molecules are composed of hydrogen and oxygen and is called hydrogen peroxide. Since the molecules of hydrogen peroxide are composed of two hydrogen atoms and two oxygen atoms, its formula is $H_2O_2$ and its properties are distinctly different from those of water ($H_2O$). In the previous section, elemental iodine has the formula $I_2$ and phosphorus $P_4$. The formulas of several other well-known molecular compounds are table sugar (sucrose), $C_{12}H_{22}O_{11}$; aspirin, $C_9H_8O_4$; ammonia, $NH_3$; and methane gas, $CH_4$.

It is possible that two compounds may have the same formula. For example, ethyl alcohol and dimethyl ether both have the formula $C_2H_6O$. The two compounds are unique, however, because of the different arrangement of atoms within the molecules, as shown below. (The dashes between atoms represent covalent bonds.)

```
   ethyl alcohol            dimethyl ether
     H  H                    H       H
     |  |                    |       |
  H—C—C—O—H              H—C—O—C—H
     |  |                    |       |
     H  H                    H       H
```

*Formulas such as those above that show the order and arrangement of atoms in a molecule are known as* **structural formulas.**

## 3-5 Ions and Ionic Compounds

Ordinary table salt is primarily composed of the compound sodium chloride, which has the formula NaCl. This type of compound is fundamentally different than the molecular compounds discussed previously. In this case, the sodium and chlorine atoms have acquired a net electrical charge in the compound. *Atoms or groups of atoms that have a net electrical charge are known as* **ions.** *Positive ions are called* **cations** *and negative ions are known as* **anions.** In NaCl, the sodium is present as a cation with a single positive charge and the chlorine is present as an anion with a charge equal and opposite to the sodium cation. This is illustrated with a + or − as a superscript to the right of the symbol of the element.

$$Na^+Cl^-$$

Since the sodium cations and chlorine anions are oppositely charged, the ions are held together by attractive forces. *Forces of attraction between opposite charges and forces of repulsion between like charges are known as* **electrostatic forces.** (See Figure 3-6.)

*Compounds consisting of ions are known as* **ionic compounds.** *The electrostatic forces holding the ions together are known as* **ionic bonds.** In Figure 3-7 an image of the arrangement of the ions in NaCl is shown as they may appear in the new electron microscope. Notice that one cation is

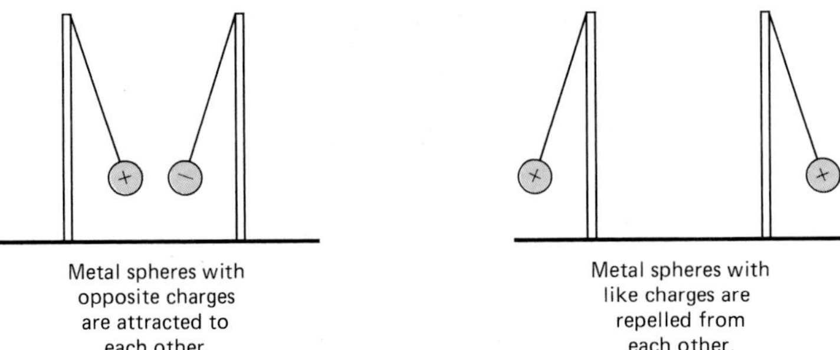

+
−
+
+
Metal spheres with opposite charges are attracted to each other.
Metal spheres with like charges are repelled from each other.


**Figure 3-6** ELECTROSTATIC FORCES. Opposite charges attract, like charges repel.

not attached to one particular anion. In fact, each ion is surrounded by six oppositely charged ions. This is a different situation from the molecular compounds discussed previously where atoms actually overlapped to form discrete groups of molecules. The formula of an ionic compound, such as NaCl, represents the simplest ratio of ions present and is known as a **formula unit.**

Certain formulas of ionic compounds at first may appear more complex than necessary. For example, barium nitrite has the formula

$$Ba(NO_2)_2$$

rather than the simpler formula $BaN_2O_4$. In this case, the ionic compound consists of $Ba^{2+}$ cations and $NO_2^-$ anions. The parentheses around the $NO_2^-$ indicates that there are two $NO_2^-$ ions present for each $Ba^{2+}$ ion. The $NO_2^-$ anion is known as a **polyatomic** (more than one atom) **ion.** The two oxygen atoms and one nitrogen atom in the anion are joined by covalent bonds. An ion (e.g., $NO_2^-$) does not exist alone in nature as does a neutral molecule (e.g., $H_2O$). Anions and cations always exist together in such a ratio that the total positive charge is balanced by the total negative charge. Other familiar examples of ionic compounds are potash, $K_2O$; limestone, $CaCO_3$; baking soda, $NaHCO_3$; and sodium fluoride (in toothpaste), NaF.

An ion has a net electrical charge because it does not have an equal number of protons in its nucleus (or nuclei for polyatomic ions) and electrons.

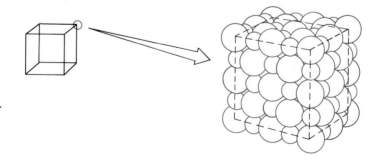

**Figure 3-7** THE IONS IN SODIUM CHLORIDE. An ionic compound exists as an arrangement of charged particles.

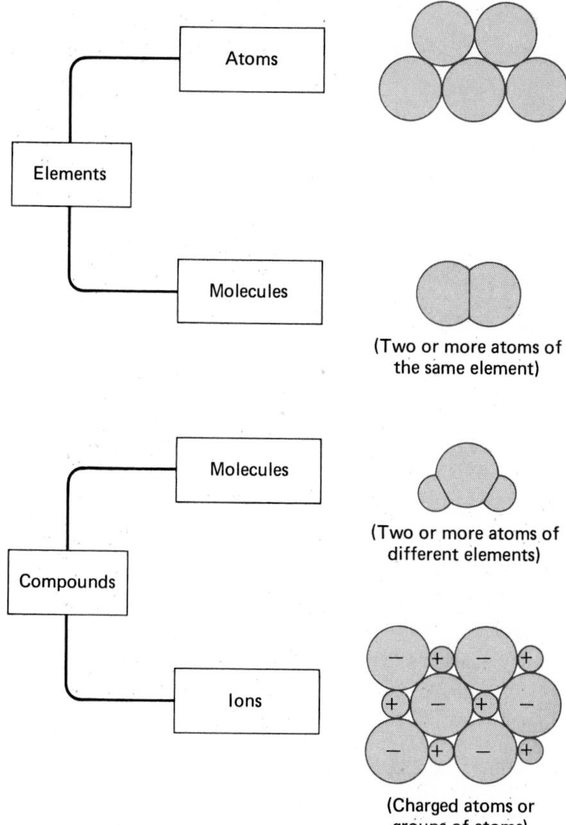


**Figure 3-8** THE BASIC STRUCTURES OF ELEMENTS AND COMPOUNDS.

Cations have fewer electrons than the neutral atom (one less electron for each positive charge) and anions have more electrons than the neutral atom (one extra electron for each negative charge). Thus the $Na^+$ cation has 11 protons in its nucleus but only 10 electrons ($+11 - 10 = +1$) and the $Cl^-$ anion has 17 protons in its nucleus and 18 electrons ($+17 - 18 = -1$).

**See Problems 3-20 through 3-31.**

In Figure 3-8, the discussion in Sections 3-3 through 3-5 has been summarized. Under normal conditions found on the earth, matter is composed of either atoms, molecules, or ions. The formula of a compound represents either a discrete molecular unit in the case of molecules or a ratio of ions in the case of ionic compounds. Eventually, we will discuss why some elements are composed of molecules and why one compound is molecular whereas another is ionic. This is the subject of Chapter 6.

(*Note:* The material in the remainder of this chapter concerns a more in-depth study of the nucleus of the atom. It is not directly relevant to the chapters that follow. It may be covered now as a logical extension of the nature of the nucleus, omitted completely, or covered later as desired.)

## 3-6 Nuclear Decay

In 1896 a French scientist by the name of Henri Becquerel discovered (quite by accident) that one element can spontaneously change into another. This was the process that many an alchemist in the Middle Ages had hoped to promote. The actual process was far different than what the alchemist expected, however. It was eventually discovered that the major isotope of uranium, $^{238}_{92}U$, can spontaneously (without warning or stimulation) emit from its nucleus a helium nucleus. This occurrence can be symbolized by a nuclear equation. *A* **nuclear equation** *shows the change from the original nucleus or nuclei on the left of the arrow to the product nuclei on the right of the arrow.*

$$^{238}_{92}\text{U} \rightarrow {}^{234}_{90}\text{Th} + {}^{4}_{2}\text{He}$$

Notice that the loss of four nucleons from the original (parent) nucleus leaves the remaining (daughter) nucleus with 234 nucleons (238 − 4 = 234); the loss of the two protons leaves 90 protons in the daughter nucleus (92 − 2 = 90). Since a nucleus with 90 protons is now an isotope of thorium, one element has indeed changed into another.

*The helium nucleus, when emitted from another nucleus, is called an* **alpha (α) particle** (see Figure 3-9). *The process of emitting energy or particles from a nucleus is known as* **radioactive decay** *or simply* **radioactivity.** *The particles or "rays" that are emitted are called* **radiation.** *The radioactive isotopes that undergo decay are sometimes referred to as* **radionuclides.**

A second form of radioactive decay involves the emission of an electron from the nucleus. *An electron emitted* from a nucleus *is known as a* **beta (β) particle.** The following equation illustrates beta particle emission:

$$^{131}_{53}\text{I} \rightarrow {}^{131}_{54}\text{Xe} + {}^{0}_{-1}e$$

(See Figure 3-10.) Notice that the loss of a beta particle from a nucleus has the effect of changing a neutron in the nucleus into a proton, which increases the atomic number by one but leaves the mass number (number of nucleons) unchanged.

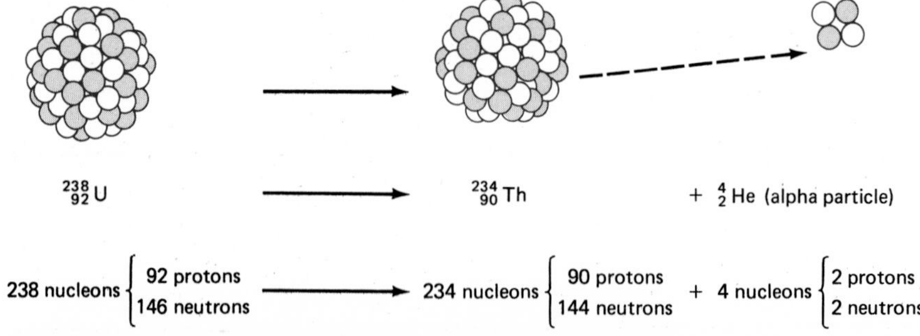


**Figure 3-9** ALPHA RADIATION. An alpha particle is a helium nucleus emitted from a larger nucleus.

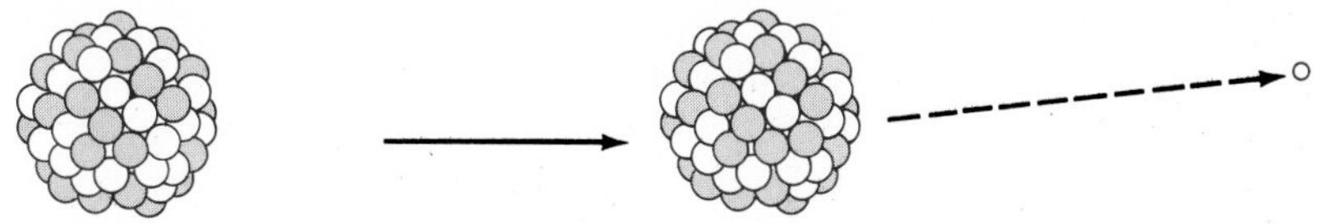

**Figure 3-10** BETA RADIATION. A beta particle is an electron emitted from a nucleus.

*A third type of radioactive decay involves the emission of a high-energy form of light called a* **gamma ($\gamma$) ray.** Like all light, gamma rays travel at $3.0 \times 10^{10}$ cm/sec (186,000 miles/sec). Light as a form of energy will be discussed in the next chapter. This type of radiation may occur alone or in combination with alpha or beta radiation as discussed above. The following equation illustrates gamma ray emission:

$$^{60}_{27}Co^{*} \rightarrow {}^{60}_{27}Co + \gamma$$

(* means nucleus is in a high-energy state)

Notice that gamma radiation by itself does not result in a change in the nucleus. Excess energy contained in a nucleus has simply been released, much as burning coal releases chemical energy.

The three major types of radiation are summarized in Table 3-2.

**Example 3-5**
Complete the following nuclear equations:

(a) $^{218}_{84}Po \rightarrow$ ______ + $^{4}_{2}He$

(b) $^{210}_{81}Tl \rightarrow {}^{210}_{82}Pb$ + ______

**Procedure**
For the missing isotope or particle:

1 Find the number of nucleons; they are equal on both sides of the equation.

**Table 3-2** Types of Radiation

| Radiation | Mass Number | Charge | Identity |
|---|---|---|---|
| Alpha ($\alpha$) | 4 | +2 | Helium nucleus |
| Beta ($\beta$) | 0 | −1 | Electron |
| Gamma ($\gamma$) | 0 | 0 | Light |

2 Find the charge or atomic number; this is also equal on both sides of the equation.

3 If what's missing is the isotope of an element, find the symbol of the element that matches the atomic number from the listing of the elements inside the front cover.

**Solution**

(a) Nucleons: $218 = x + 4$, so $x = 214$.
Atomic number: $84 = y + 2$, so $y = 82$.

From the inside cover, the element having an atomic number of 82 is Pb. The isotope is

$$\underline{\underline{{}^{214}_{82}\text{Pb}}}$$

(b) Nucleons: $210 = 210 + x$, so $x = 0$.
Atomic number: $81 = 82 + y$, so $y = -1$.

An electron or beta particle has negligible mass (compared to a nucleon) and a charge of $-1$.

See Problems 3-32 through 3-35.

$$\underline{\underline{{}^{0}_{-1}e}}$$

## 3-7 Rates of Decay of Radioactive Isotopes

The isotopes of all elements heavier than ${}^{209}_{83}\text{Bi}$ are radioactive. If this is true, how can elements such as ${}^{238}_{92}\text{U}$ occur naturally in the earth? The answer is that many of these isotopes have very long "half-lives." *The* **half-life** (**$t_{1/2}$**) *of a radioactive isotope is the time required for one-half of a given sample to decay.* Each radioactive isotope has a specific and constant half-life. For example, ${}^{238}_{92}\text{U}$ has a half-life of $4.5 \times 10^9$ years. Since that is roughly the age of the earth, about one-half of the ${}^{238}_{92}\text{U}$ originally present when the earth formed from hot gases and dust is still present. Half of what is now present will be gone in another 4.5 billion years. The half-lives and modes of decay of some radioactive isotopes are listed in Table 3-3.

**Table 3-3** Half-lives

| Isotope | $t_{1/2}$ | Mode of Decay | Product |
|---|---|---|---|
| ${}^{238}_{92}\text{U}$ | $4.5 \times 10^9$ years | $\alpha, \gamma$ | ${}^{234}_{90}\text{Th}$ |
| ${}^{234}_{90}\text{Th}$ | 24.1 days | $\beta$ | ${}^{234}_{91}\text{Pa}$ |
| ${}^{226}_{88}\text{Ra}$ | 1620 years | $\alpha$ | ${}^{222}_{86}\text{Rn}$ |
| ${}^{14}_{6}\text{C}$ | 5760 years | $\beta$ | ${}^{14}_{7}\text{N}$ |
| ${}^{131}_{53}\text{I}$ | 8.0 days | $\beta, \gamma$ | ${}^{131}_{54}\text{Xe}$ |
| ${}^{218}_{85}\text{At}$ | 1.3 sec | $\alpha$ | ${}^{214}_{83}\text{Bi}$ |

Notice in Table 3-3 that when $^{238}_{92}U$ decays it forms an isotope ($^{234}_{90}Th$) that also decays (very rapidly compared to $^{238}_{92}U$). The $^{234}_{91}Pa$ formed from the decay of $^{234}_{90}Th$ also decays, and so forth until finally the stable isotope $^{206}_{82}Pb$ is formed. This is known as the $^{238}_{92}U$ radioactive decay series. *A* **radioactive decay series** *starts with a naturally occurring radioactive isotope with a half-life near the age of the earth* (if it was very much shorter, there wouldn't be any left). *The series ends with a stable isotope.* There are two other naturally occurring decay series: the $^{235}_{92}U$ series, which ends with $^{207}_{82}Pb$; and the $^{232}_{90}Th$ series, which ends with $^{208}_{82}Pb$. As a result of the $^{238}_{92}U$ decay series, where uranium is found in rocks we also find other radioactive isotopes as well as lead. In fact, by examining the ratio of $^{238}_{92}U/^{206}_{82}Pb$ in a rock, its age can be determined. For example, a rock from the moon showed about half of the original $^{238}_{92}U$ had decayed to $^{206}_{82}Pb$. This meant that the rock was about $4.5 \times 10^9$ years old.

**Example 3-6**

If we started with 4.0 mg of $^{14}_{6}C$, how long would it take until only 0.50 mg remained?

**Solution**

$$\text{After 5,760 years, } \frac{1}{2} \times 4.0 \text{ mg} = 2.0 \text{ mg remaining}$$

$$\text{After another 5,760 years, } \frac{1}{2} \times 2.0 \text{ mg} = 1.0 \text{ mg remaining}$$

$$\text{After another } \underline{\text{5,760 years,}} \ \frac{1}{2} \times 1.0 \text{ mg} = 0.50 \text{ mg remaining}$$

$$\text{17,280 years}$$

Therefore, in 17,280 years, 0.50 mg remains.

## 3-8 The Effects of Radiation

We mentioned in Chapter 1 that energy-rich compounds undergo chemical reactions to produce energy-poor compounds, releasing the chemical energy as heat. Radioactive isotopes can be considered in a similar manner. When the nucleus of an isotope is radioactive, it is considered to be an energy-rich nucleus. When it decays, the radiation carries away this excess energy from the nucleus by means of a high-energy particle (alpha or beta) or high-energy light (gamma). This nuclear energy carried away by the particles or rays may also be converted to heat energy. In fact, much of the internal heat generated in the interior of the earth and other planets is a direct result of natural radiation from unstable elements. The presence of radioactive elements in wastes from nuclear reactors makes these materials extremely hot so that they must be cooled continuously to avoid melting.

Radiation also affects matter when the high-energy particles or rays penetrate. *The radiation changes neutral molecular compounds into ions in a process known as* **ionization.** For example, if high-energy particles collide with an $H_2O$ molecule, an electron may be removed from the molecule leaving an $H_2O^+$ ion behind. The properties of the ion are distinctly different from those of the neutral molecule. If the molecule in question is a large complex molecule that is part of a cell of a living system, the ionization causes damage or even eventual destruction of the cell. As shown in Figure 3-11, an alpha particle causes the most ionization and is the most destructive along its path. However, these particles do not penetrate matter to any extent and can be stopped even by a piece of paper. The danger of alpha emitters (such as uranium and plutonium) is that they can be ingested through food or inhaled into the lungs where these heavy elements tend to accumulate in bones. There, in intimate contact with the blood-producing cells of the bone marrow, they slowly do damage. Ultimately, the radiation can cause certain cells to change into abnormal cells that reproduce rapidly. These are the dreaded cancer cells such as found with leukemia.

Beta radiation is less ionizing than alpha radiation but is more penetrating. Still, a couple of inches of solid or liquid matter absorbs beta radiation. This type of radiation can cause damage to surface tissue such as skin and eyes but does not reach internal organs unless ingested. The most damaging radiation is gamma radiation. Although much less ionizing along its path than the others, it has tremendous penetrating power. Several feet of concrete or blocks of lead are needed for protection from gamma radiation.

The damage done to cells by gamma radiation is cumulative. That means small doses over a long period of time can be as harmful as one large dose at once. That is why the dosage of radiation absorbed by those who work in radioactive environments such as a nuclear research laboratory must be carefully and continually monitored. If one receives too much radiation over a specified period, that person must be removed from additional exposure for a length of time.

Isotopes that emit gamma radiation (e.g., $^{60}_{27}Co$) can be useful as well as

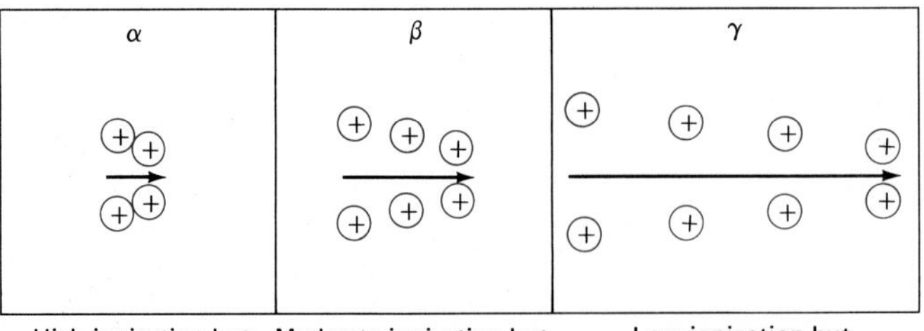


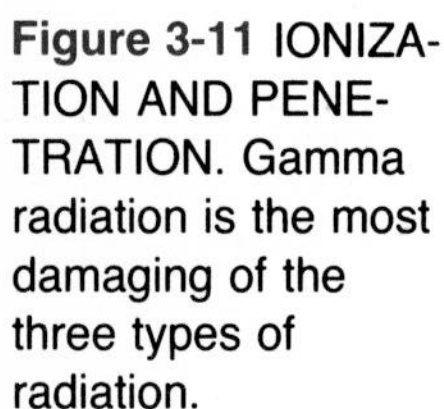
**Figure 3-11** IONIZATION AND PENETRATION. Gamma radiation is the most damaging of the three types of radiation.

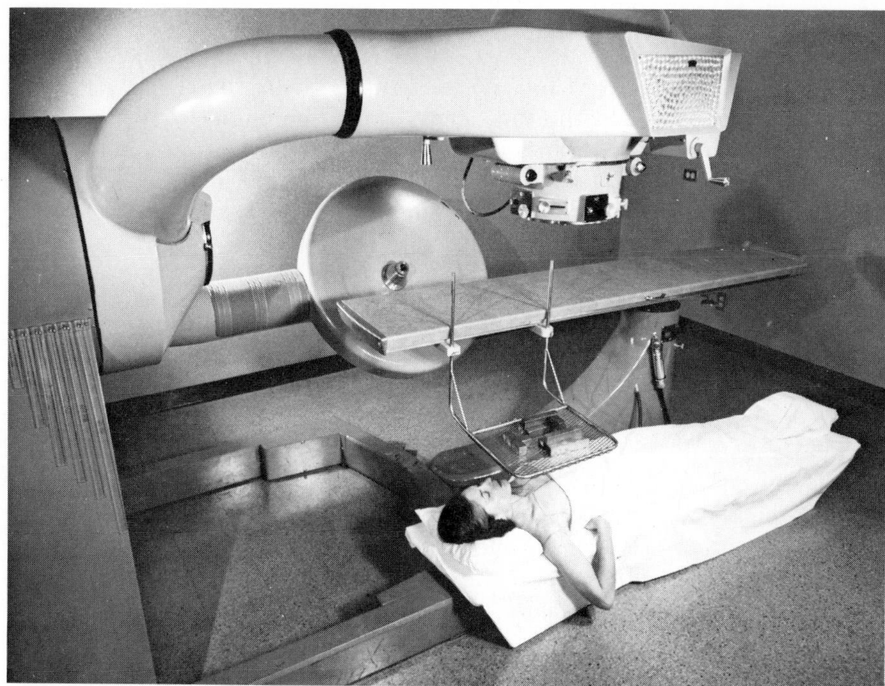

**Figure 3-12** $^{60}Co$ RADIATION TREATMENT. A beam of gamma rays can be used to destroy cancerous tissue.

destructive because of their penetrating rays. The gamma radiation of $^{60}_{27}Co$ can be narrowly focused like a beam of light to destroy cancer cells in localized tumors located deep within the body (see Figure 3-12). Healthy cells are destroyed as well in the process, so this form of cancer treatment has many undesireable side effects. The purpose is to destroy or damage more cancer cells than normal cells. The normal cells can recover faster than the damaged abnormal cells.

**See Problems 3-36 through 3-41.**

## 3-9 Nuclear Reactions

**Transmutation** *of an element is the conversion of that element into another.* We have discussed how this occurs naturally by the spontaneous emission of an alpha or beta particle from a nucleus. Transmutation can also occur by artificial or synthetic means. The first example of transmutation was given by Ernest Rutherford in 1919. (Rutherford had earlier proposed the nuclear model of the atom.) Rutherford found that $^{14}_{7}N$ atoms could be bombarded with alpha particles, causing a nuclear reaction that produced $^{17}_{8}O$ and a proton. This nuclear reaction is illustrated by the nuclear equation

$$^{14}_{7}N + ^{4}_{2}He \rightarrow ^{17}_{8}O + ^{1}_{1}H$$

Notice that the total number of nucleons is conserved by the reaction [14 + 4 (on the left) = 17 + 1 (on the right)]. The total charge is also conserved during the reaction [7 + 2 (on the left) = 8 + 1 (on the right)]. Both the number of nucleons and the total charge must be balanced (the same on both sides of the equation) in a nuclear reaction.

Since the time of Rutherford, thousands of isotopes have been prepared by nuclear reactions. Two other examples are

$$^{209}_{83}\mathrm{Bi} + {}^{2}_{1}\mathrm{H} \rightarrow {}^{210}_{84}\mathrm{Po} + {}^{1}_{0}n$$

$$^{27}_{13}\mathrm{Al} + {}^{4}_{2}\mathrm{He} \rightarrow {}^{30}_{15}\mathrm{P} + {}^{1}_{0}n$$

($^{2}_{1}\mathrm{H}$ is a deuterium nucleus or a deuteron; $^{1}_{0}n$ is a neutron).

Since the nucleus has a positive charge and many of the bombarding nuclei (except neutrons) also have a positive charge (e.g., $^{4}_{2}\mathrm{He}$, $^{1}_{1}\mathrm{H}$, $^{2}_{1}\mathrm{H}$), the particles must have a high energy (velocity) in order to overcome the natural repulsion between these two like charges. From the collision of the nucleus and the particle, a high-energy nucleus is formed. As a result, another particle (usually a neutron) is then emitted by the nucleus to carry away the excess energy from the collision. This is analogous to two cars in a head-on collision where fenders, doors, and bumpers come flying out after impact. This is illustrated in Figure 3-13.

The invention of particle accelerators, which accelerate particles to high velocities and energies, has opened up vast possibilities for artificial nuclear reactions. These accelerators can be large, expensive devices. For example, the circular accelerator at Batavia, Illinois has a diameter of 2 miles, and it cost $245 million when completed in the early 1970s. (See Figure 3-14.) Accelerators are being planned that would have a diameter of 20 miles and cost at least $1.5 billion.

One interesting development from particle accelerators has been the preparation of new, heavy elements. In fact, all elements between neptunium (atomic number 93) and the last element made (atomic number 106, which has not yet been named) are synthetic elements. The following nuclear

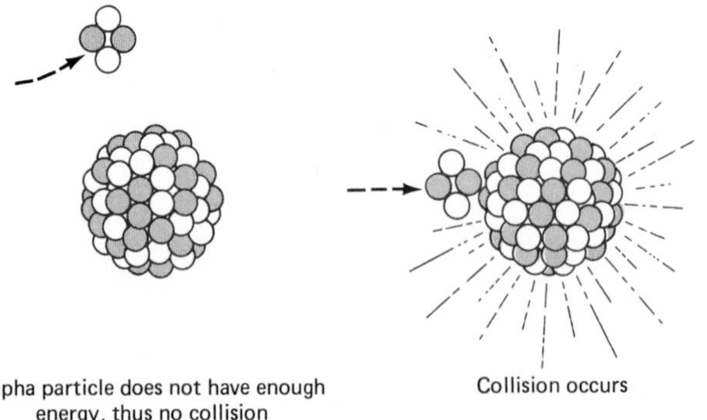
Alpha particle does not have enough energy, thus no collision
Collision occurs

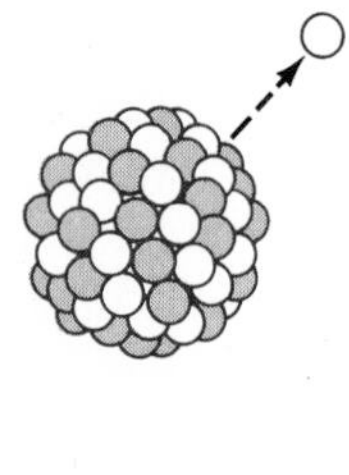
High-energy nucleus gives off particle

**Figure 3-13** COLLISION OF AN $\alpha$ PARTICLE WITH A NUCLEUS. Alpha particles must have sufficient energy to overcome the repulsion of the target nucleus in order for a collision to take place.

**Figure 3-14** The FERMI NATIONAL ACCELERATOR LABORATORY. This aerial view of the accelerator gives an indication of its huge size.

reactions illustrate the formation of heavy elements:

$$^{238}_{92}\text{U} + ^{12}_{6}\text{C} \rightarrow ^{244}_{98}\text{Cf} + 6^{1}_{0}n$$

$$^{238}_{92}\text{U} + ^{16}_{8}\text{O} \rightarrow ^{250}_{100}\text{Fm} + 4^{1}_{0}n$$

**Example 3-7**
Complete the following nuclear equations:

(a) $^{9}_{4}\text{Be} + ^{2}_{1}\text{H} \rightarrow$ ______ $+ ^{1}_{0}n$

(b) $^{252}_{98}\text{Cf} + ^{10}_{5}\text{B} \rightarrow$ ______ $+ 5^{1}_{0}n$

**Procedure**
Same as Example 3-5.
(a) Nucleons: $9 + 2 = x + 1$, so $x = 10$.
Atomic number: $4 + 1 = y + 0$, so $y = 5$.
From a listing of the elements, we find that the element is boron (atomic number 5). The isotope is

$$\underline{\underline{^{10}_{5}\text{B}}}$$

(b) Nucleons: $252 + 10 = x + (5 \times 1)$, so $x = 257$.
Atomic number: $98 + 5 = y + (5 \times 0)$, so $y = 103$.
From the listing of elements, the isotope is

$$\underline{\underline{{}^{257}_{103}\text{Lr}}}$$

Many nuclear reactions form isotopes that are themselves radioactive. For example, $^{60}$Co is prepared as follows:

$$^{59}_{27}\text{Co} + {}^{1}_{0}n \rightarrow {}^{60}_{27}\text{Co}^{*}$$

This type of procedure (called neutron activation) is used to prepare many radioactive isotopes whose naturally occurring isotopes are stable. There has been a great deal of application of these synthetic radioactive isotopes in the field of medicine, mainly in the area of diagnosis of diseases. In fact, in 1976 there were 10 million diagnostic tests involving the use of radioactive isotopes.

## 3-10 Nuclear Fission

In 1938 two German scientists, Otto Hahn and Fritz Strassmann, first explained some strange and unexpected results of a nuclear reaction. When naturally occurring uranium (99.3% $^{238}$U, 0.7% $^{235}$U) was bombarded with neutrons, it was expected that elements with a higher atomic number than uranium could be synthesized as follows:

$$^{238}_{92}\text{U} + {}^{1}_{0}n \rightarrow {}^{239}_{92}\text{U} \rightarrow {}^{239}_{93}\text{X} + {}^{0}_{-1}e$$

Indeed, this reaction seemed to occur, but of more importance was the discovery by these scientists of isotopes present that were about half of the mass of the uranium atoms (e.g., $^{139}_{56}$Ba, $^{90}_{38}$Sr). The German scientists were able to show that the smaller atoms resulted from the reaction of $^{235}$U and not the more abundant $^{238}$U. The following represents a typical nuclear reaction that occurs when $^{235}$U absorbs a neutron:

$$^{235}_{92}\text{U} + {}^{1}_{0}n \rightarrow {}^{139}_{56}\text{Ba} + {}^{94}_{36}\text{Kr} + 3\,{}^{1}_{0}n$$

This type of nuclear reaction is called fission. **Fission** *is the splitting of a large nucleus into two smaller nuclei of similiar size* (see Figure 3-15).

There were two points about this reaction that had monumental consequences for the world. Scientists in Europe and America were quick to grasp the meaning in a world about to go to war.

1 Calculation of the weights of the product nuclei compared with the original nuclei indicated that a significant mass loss occurs in the reaction. According to Einstein's equation (see Section 1-12), this mass loss must

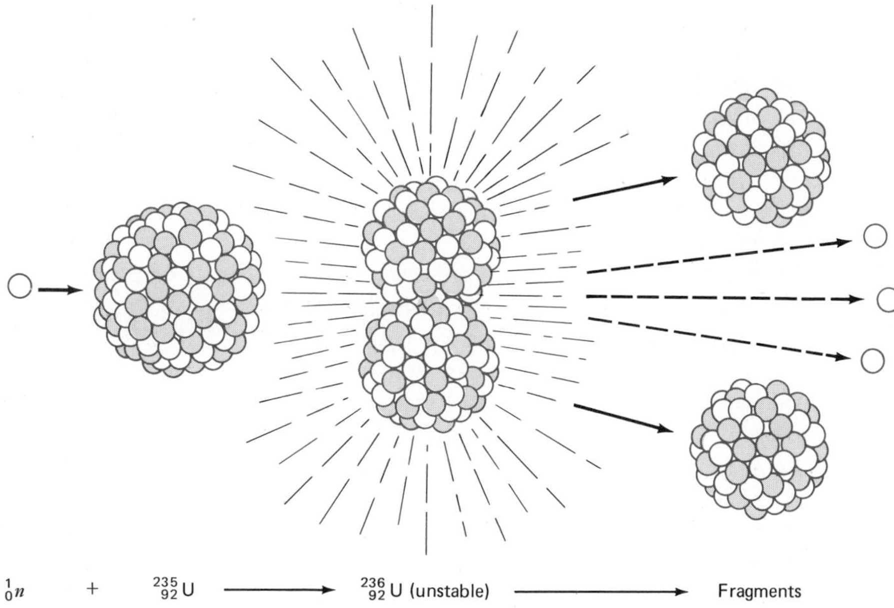


**Figure 3-15** FISSION. In fission, one neutron produces two fragments of the original atom and an average of three neutrons.

be converted to a tremendous amount of energy. Fission of a few kilograms of $^{235}U$ could produce energy equivalent to tens of thousands of tons of the conventional explosive, TNT.

2 What made the rapid fission of a large sample of $^{235}U$ feasible was the potential for a chain reaction. *A* **nuclear chain reaction** *is a reaction*

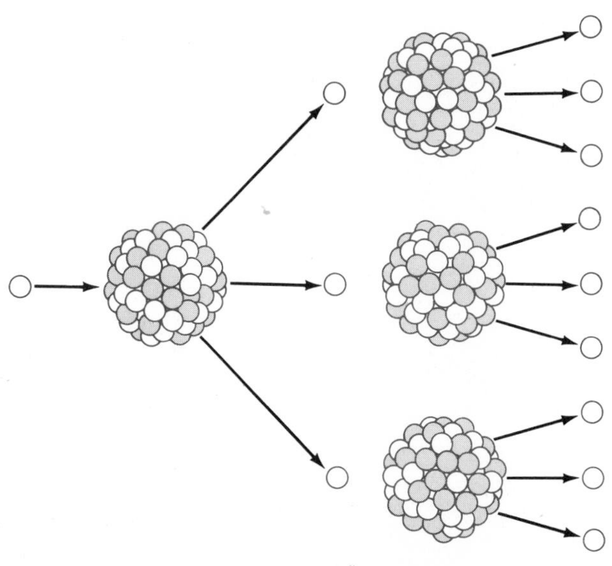

**Figure 3-16** A CHAIN REACTION. The fission of one nucleus produces neutrons, which induces fission in other nuclei.

*that is self-sustaining. The reaction generates the means to trigger additional reactions.* Notice in Figure 3-15 that the reaction of one original neutron caused three to be released. These three neutrons cause the release of nine neutrons, and so forth, as illustrated in Figure 3-16. If a certain densely packed "critical mass" of $^{235}U$ is present, the whole mass of uranium can undergo fission in an instant with a quick release of the energy in the form of radiation.

Unfortunately, the world thus entered the nuclear age in pursuit of a bomb. After a massive but secret effort, the first nuclear bomb was exploded over Alamogardo Flats in New Mexico on July 16, 1945. This bomb and the bomb exploded over Nagasaki, Japan, was made of $^{239}Pu$, which is a synthetic fissionable isotope. The bomb exploded over Hiroshima, Japan, was made of $^{235}U$.

This enormously destructive device, however, can be tamed. The chain reaction can be controlled by absorbing excess neutrons with cadmium bars. A typical nuclear reactor is illustrated in Figure 3-17. In a reactor core the uranium in the form of pellets is encased in long rods called fuel elements. Cadmium bars are raised and lowered among the fuel elements to control the rate of fission by absorbing neutrons. If the cadmium bars are lowered

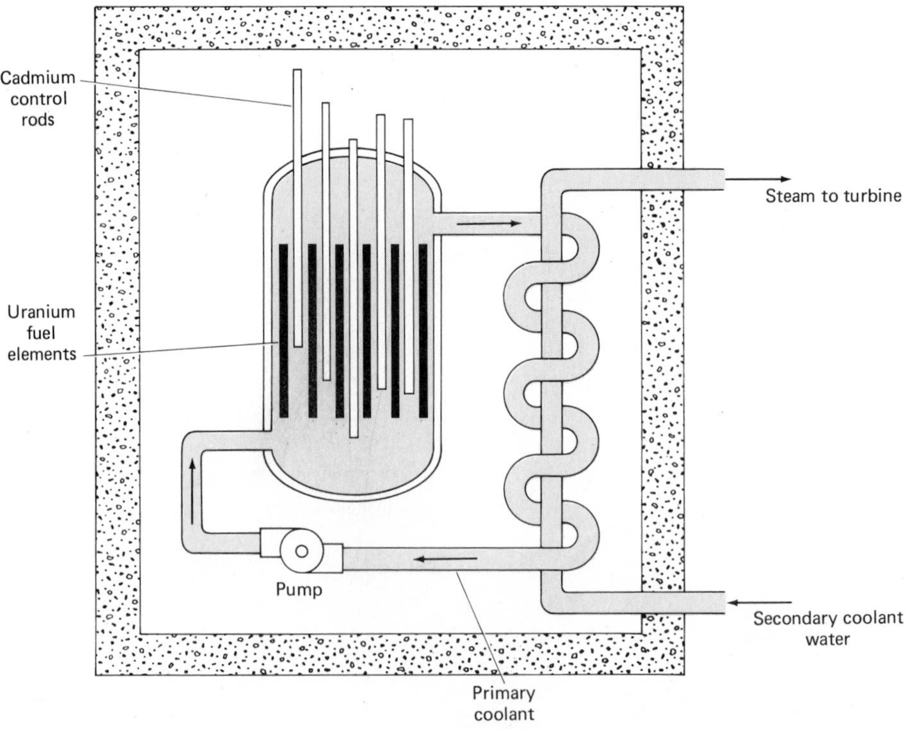


Figure 3-17 A NUCLEAR REACTOR. Nuclear energy is converted to heat energy, the heat energy to mechanical energy and the mechanical energy is converted to electrical energy.

all of the way, the fission process can be halted all together. In normal operation, the bars are raised just enough that the fission reaction occurs at the desired level. Energy released by the fission and the decay of the radioactive products formed from the fission is used to heat water, which circulates among the fuel elements. This water, which is called the primary coolant, becomes very hot (about 300 °C) without boiling because it is under high pressure. The water heats a second source of water, changing it to steam. The steam from the secondary coolant is cycled outside of the containment building, where it is used to turn a turbine that generates electricity.

An advantage of nuclear energy is that it does not deplete the limited supply of fossil fuels (natural gas, coal, and oil). More importantly it does not pollute the air with compounds of sulfur and nitrogen nor does it increase the amount of carbon dioxide in the atmosphere. After years of use there have been few accidents and no loss of life from commercial reactors. In 1981, 13% of the world's electrical power was from nuclear energy. In fact, in France nuclear power is now the primary source of electricity. Many countries are leaning more toward nuclear energy so that they will be less dependent on imports of foreign oil. Proponents of nuclear power feel that adequate safeguards and backup systems are available to prevent a catastrophic accident.

The disadvantage of nuclear reactors received much publicity in the accident at Three Mile Island in Pennsylvania in March, 1979. A series of mechanical failures compounded by human error led to a loss of the primary coolant water, which allowed the nuclear core to overheat, causing an emergency. Although there was some loss of airborne radiation, the main danger and fear of a "melt down" was averted.

A melt down would occur if the temperature of the core exceeded 1130 °C, the melting point of the uranium. Theoretically, the mass of molten uranium, together with all of the highly radioactive decay products, could accumulate on the floor of the reactor, melt through the many feet of protective concrete of the containment building, and eventually reach the ground water. If this happened, vast amounts of deadly radioactive wastes could be released into the environment. In any case, it will take a total of 10 years and over $1 billion to clean up the mess at Three Mile Island. Because of this accident, new regulations for required safeguards have pushed the cost of a nuclear power plant well above that of a conventional plant. As a result, few new nuclear power plants are now being constructed in the United States. Some of those under construction may not be completed because of the costs. (See Figure 3-18.)

Another disadvantage of nuclear reactors involves the used or spent fuel elements and the highly radioactive primary coolant water. Eventually, fuel elements must be replaced, as the buildup of radioactive isotopes from the fission of $^{235}U$ hinders the fission process. These fuel elements will remain highly radioactive for approximately 1000 years. The problem of the processing, disposal, and transportation of wastes has not yet been solved

**Figure 3-18** A NUCLEAR POWER PLANT. In these plants the nuclear energy of uranium is converted into electrical energy.

to the extent necessary to handle the large number of nuclear power plants now in operation.

## 3-11 Nuclear Fusion

Soon after the process of nuclear fission was demonstrated, an even more powerful type of nuclear reaction was shown to be theoretically possible. *This reaction involved the* **fusion** *or bringing together of two small nuclei.* An example of a fusion reaction is

$$^{3}_{1}\text{H} + {}^{2}_{1}\text{H} \rightarrow {}^{4}_{2}\text{He} + {}^{1}_{0}n + \text{energy}$$

($^{3}_{1}$H is called tritium; it is a radioactive isotope of hydrogen.)

As in the fission process, a significant mass loss converts to energy. Fusion energy is the origin of almost all of our energy, since it powers the sun. Millions of tons of matter are converted to energy in the sun every second. Because of its large mass, however, the sun contains enough hydrogen to "burn" for additional billions of years.

The principle of fusion was first demonstrated on this planet with a tremendously destructive device called the hydrogen bomb. This bomb can be more than 1000 times more powerful than the atomic bomb, which uses the fission process.

Fission can be controlled, but what about fusion? This is a big and very important question at present. Research into this area is currently of high priority in much of the industrial world, especially in the Soviet Union and the United States. Technically, controlling fusion for the generation of power is an extremely difficult problem. In order for fusion to take place, temperatures on the order of 100 million degrees Celsius are needed just to start the fusion process. This temperature is hotter than the interior of the sun. No known materials can withstand these temperatures, so alternative containment procedures are being examined. So far, the research efforts of several nations have not reached the "breakeven" point, where as much energy is released from the fusion process as was put into it to get it started.

The advantages of controlled fusion power are impressive.

1 It would be clean. Few radioactive products are formed.

2 Fuel is inexhaustable. The oceans of the world contain enough deuterium, one of the reactants, to provide the world's energy needs for a trillion years. On the other hand, there is a very limited supply of fossil fuels and uranium.

3 There is no possibility of the reaction getting out of control and causing a melt down. Fusion will occur in power plants in short bursts of energy than can be easily stopped in case of mechanical problems.

The disadvantage is, of course, the expense of research and development. Billions of dollars have been spent and billions more will have to be spent before the technical problems are solved. Hopefully, before the end of this century there will be experimental plants in operation that will prove the feasibility of fusion as a source of power. Unfortunately, there is little choice in this matter as there are few alternatives for power in the next century.

**See Problems 3-42 through 3-46.**

## 3-12 Chapter Summary

Since the early 1800s it has been accepted that there is a basic unit of matter called the atom. In this century we have discovered that atoms have structure and are composed of protons, neutrons, and electrons. The charge and approximate mass of these particles (in amu) is summarized in Table 3-1.

Protons and neutrons exist in the small core of the atom called its nucleus. These two particles are thus known as nucleons. The electrons occupy the otherwise empty space surrounding the nucleus. The atoms of an element all have the same number of protons, which is known as its atomic number. The atomic number also represents the number of electrons in a neutral atom. The mass number of an element is equal to the number of nucleons. Atoms of the same element may have different mass numbers (different number of neutrons) and are known as isotopes. This information for a particular isotope is illustrated as follows:

Mass number → $^{X}_{Y}M^{Z}$ ← Charge (if not neutral)
Atomic number → ← Symbol of element

The atomic weight of an isotope is a more exact measure of mass than mass number. This is obtained by comparing the mass of the isotope to the mass of $^{12}C$, which is the defined standard. Finally, the atomic weight of an element as found in nature is the weighted average of all of the naturally occurring isotopes.

Pure matter occurs as either elements or compounds. Elements are composed of either individual atoms or combinations of two or more atoms called molecules. The atoms in a molecule are attached by means of covalent bonds. The molecules of compounds are composed of atoms of two or more different elements. Some compounds are made up of charged atoms or groups of atoms called ions. Ions have an electrical charge because of an imbalance between the number of electrons and protons. Ionic compounds contain positive ions (cations) whose charge is balanced by negative ions (anions). The formulas of compounds or molecular elements give the number of atoms of each element chemically combined in the molecule. In the case of ionic compounds, the formula tells us the ratio of cations and anions present. The formula represents one formula unit rather than one molecule. This information is summarized as follows:

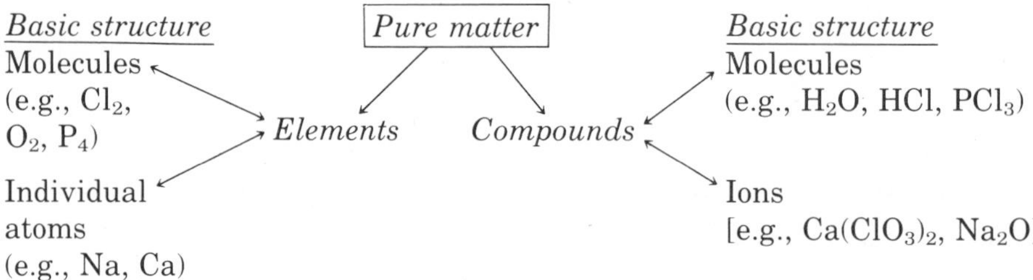


**Nuclear Reactions** The nuclei of all elements heavier than bismuth (atomic number 83) are unstable. That is, they disintegrate or decay to the nuclei of other elements and give off radiation. There are three major types of radiation: (1) alpha particles, which are helium nuclei; (2) beta particles, which are electrons; and (3) gamma rays, which are high-energy forms of light.

Radioactive isotopes decay at widely different rates. A measure of the rate of decay is called the half-life. Half-lives vary from billions of years for isotopes such as $^{238}U$ to fractions of a second for others.

Radiation affects living tissue. When radiation passes through matter, it causes ionization that damages or kills cells. Alpha and beta radiation do not cause damage (unless ingested) because of low penetration. Gamma radiation is extremely penetrating and thus can cause considerable damage from exposure.

Nuclear changes can be brought about by artificial means in which nuclei are combined with the help of particle accelerators. Many new heavy elements as well as medically useful isotopes have been produced in this manner.

Fission is the breaking of certain heavy isotopes into fragments. Such a process can occur through a chain reaction and liberate a large amount of energy. The energy originates from the conversion of mass according to Einstein's law. Fission is controllable and is the energy source in nuclear power reactors. Fusion is the combination of light nuclei to form heavier ones. Again, a mass loss accompanies the reaction, which converts into energy. Fusion is the process that occurs in the hydrogen bomb and the sun. Efforts are underway to control fusion so as to generate power.

The different types of reactions that nuclei undergo that have been discussed are summarized as follows:

| | Reaction | Type |
|---|---|---|
| **1** | $^{240}_{96}Cm \rightarrow ^{236}_{94}Pu + ^{4}_{2}He$ | Alpha radiation |
| **2** | $^{71}_{30}Zn \rightarrow ^{71}_{31}Ga + ^{0}_{-1}e$ | Beta radiation |
| **3** | $^{253}_{99}Es + ^{4}_{2}He \rightarrow ^{256}_{101}Md + ^{1}_{0}n$ | Synthetic element from nuclear reaction |
| **4** | $^{235}_{92}U + ^{1}_{0}n \rightarrow ^{94}_{38}Sr + ^{139}_{54}Xe + 3^{1}_{0}n$ | Fission |
| **5** | $^{2}_{1}H + ^{3}_{1}H \rightarrow ^{4}_{2}He + ^{1}_{0}n$ | Fusion |

## Exercises

### The Composition of the Atom

**3-1** Which of the following were not part of Dalton's atomic theory?
(a) Atoms are the basic building blocks of nature.
(b) Atoms are composed of electrons, neutrons, and protons.
(c) Atoms are reshuffled in chemical reactions.
(d) The atoms of an element are identical.
(e) Different isotopes can exist for the same element.

**3-2** Which of the following describes a neutron?
(a) +1 charge, mass number 1 amu
(b) +1 charge, mass number 0 amu
(c) 0 charge, mass number 1 amu
(d) −1 charge, mass number 0 amu

**3-3** Which of the following describes an electron?
(a) +1 charge, mass number 1 amu
(b) +1 charge, mass number 0 amu
(c) −1 charge, mass number 1 amu
(d) −1 charge, mass number 0 amu

**3-4** The elements O, N, Si, and Ca are among several that are primarily composed of one isotope. Using the table of atomic weights in the inside cover, write the atomic number and mass number of the principal isotope of these elements.

**3-5** Give the mass numbers and atomic numbers of the following isotopes.
(a) $^{27}Al$ (b) $^{133}Cs$ (c) $^{114}Cd$ (d) $^{207}Pb$

**3-6** Give the numbers of protons, neutrons, and electrons in each of the following isotopes.
(a) $^{45}_{21}Sc$ (b) $^{232}_{90}Th$ (c) $^{223}_{87}Fr$ (d) $^{90}_{38}Sr$

**3-7** Three isotopes of uranium are $^{234}_{92}U$, $^{235}_{92}U$, and $^{238}_{92}U$. How many protons, neutrons, and electrons are in each isotope?

**3-8** Using the table of elements inside the front cover, write the symbol of the neutral element with its mass number and atomic number given the following information.
(a) Element symbol is Mo; mass number is 96.
(b) Element is silver, and nucleus contains 61 neutrons.

(c) Atomic number is 14, and nucleus contains 14 neutrons.
(d) Nucleus contains 39 nucleons, of which 20 are neutrons.
(e) Neutral atom contains 76 electrons, and nucleus contains 114 neutrons.

**3-9** Using the table of elements inside the front cover, write the symbol of the neutral element with its mass number and atomic number given the following information.
(a) Element symbol is Ce; nucleus contains 82 neutrons.
(b) Nucleus contains 26 protons and 30 neutrons.
(c) Element is silicon, with 30 nucleons in the nucleus.
(d) Neutral atom has 11 electrons and 12 neutrons in the nucleus.
(e) Element is tungsten; nucleus contains 184 nucleons.

### Atomic Weight

**3-10** A certain element weighs 5.81 times as much as $^{12}C$. What is the atomic weight of the element? What is the element?

**3-11** The atomic weight of a certain element is about $3\frac{1}{3}$ times heavier than $^{12}C$. Give the atomic weight, the name, and the symbol of the element?

**3-12** If the atomic weight of $^{12}C$ is defined as exactly 8 instead of 12, what would be the atomic weight of the following elements to three significant figures? Assume that the elements have the same weights relative to each other as before. That is, a hydrogen still weighs about one-twelfth as much as carbon.
(a) H (b) N (c) Na (d) Ca

**3-13** Assume that the atomic weight of $^{12}C$ is defined as exactly 10 and that the atomic weight of an element is 43.3 amu on this basis. What is the element?

**3-14** Bromine is composed of 50.5% $^{79}Br$ and 49.5% $^{81}Br$. The atomic weight of $^{79}Br$ = 78.92 amu and $^{81}Br$ = 80.92 amu. What is the atomic weight of bromine?

**3-15** Silicon occurs in nature as a mixture of three isotopes: $^{28}Si$ (at. wt. 27.98 amu), $^{29}Si$ (at. wt. 28.98), and $^{30}Si$ (at. wt. 29.97). The mixture is 92.21% $^{28}Si$, 4.70% $^{29}Si$, and 3.09% $^{30}Si$. Calculate the atomic weight of naturally occurring silicon.

**3-16** Naturally occurring Cu is 69.09% $^{63}Cu$ (at. wt. 62.96 amu). The only other isotope present is $^{65}Cu$ (at. wt. 64.96 amu). What is the atomic weight of copper?

* **3-17** Chlorine occurs in nature as a mixture of $^{35}Cl$ and $^{37}Cl$. If the atomic weight of $^{35}Cl$ is approximately 35.0 amu and that of $^{37}Cl$ is 37.0 amu and the atomic weight of the mixture as it occurs in nature is 35.5 amu, what is the proportion of the two isotopes?

* **3-18** The atomic weight of the element gallium is 69.72 amu. If it is composed of two isotopes $^{69}Ga$ (at. wt. 68.926 amu) and $^{71}Ga$ (at. wt. = 70.925 amu), what is the percent of $^{69}Ga$ present?

**3-19** Using the periodic table, determine the atomic number and the atomic weight of each of the following elements.
(a) Re (b) Co (c) Br (d) Si

### Formulas, Molecules, and Ions

**3-20** Which of the following is the formula of a diatomic element and which is that of a diatomic compound?
(a) $NO_2$ (d) $(NH_4)_2S$
(b) CO (e) $N_2$
(c) $K_2O$ (f) $CO_2$

**3-21** Determine the number of atoms of each element in the formula of each of the following compounds.
(a) $C_6H_4Cl_2$
(b) $C_2H_5OH$ (ethyl alcohol)
(c) $CuSO_4 \cdot 9H_2O$ ($H_2O$'s are parts of a single molecule)
(d) $C_9H_8O_4$ (aspirin)
(e) $Al_2(SO_4)_3$
(f) $(NH_4)_2CO_3$

**3-22** What is the total number of atoms in each molecule or formula unit for the compounds listed in Problem 3-21?

**3-23** How many carbon atoms are in each molecule or formula unit of the following compounds?

(a) $C_8H_{18}$ (octane in gasoline)
(b) $NaC_7H_4O_3NS$ (saccharin)
(c) $Fe(C_2O_4)_2$
(d) $Al_2(CO_3)_3$

**3-24** Write the formula of each of the following compounds.
(a) Sulfur dioxide (one sulfur and two oxygen atoms)
(b) Carbon dioxide
(c) Sulfuric acid (two hydrogens, one sulfur, and four oxygens)
(d) Calcium perchlorate (one calcium and two $ClO_4^-$ ions)
(e) Ammonium phosphate (three $NH_4^+$ ions and one $PO_4^{3-}$ ion)
(f) Acetylene (two carbons and two hydrogens)

3-25 Write the formula of each of the following compounds.
(a) Phosphorus pentachloride (one phosphorus and five chlorines)
(b) Dinitrogen tetroxide (two nitrogens and four oxygens)
(c) Butane (four carbons and ten hydrogens)
(d) Sodium sulfate (two $Na^+$ ions and one $SO_4^{2-}$ ion)
(e) Potassium bicarbonate (one potassium, one hydrogen, one carbon, and three oxygens)

**3-26** Neon gas is composed of individual atoms, and oxygen gas is composed of $O_2$ molecules. Describe how these two elements may appear on the atomic or molecular level. In the gaseous state the particles are comparatively far apart.

3-27 The compound HF contains covalent bonds, and the compound KF contains ionic bonds. Describe how the basic particles of these two compounds may appear.

**3-28** How many total protons and electrons are present in each of the following ions?
(a) $K^+$ (b) $Br^-$ (c) $S^{2-}$ (d) $NO_2^-$
(e) $Al^{3+}$ (f) $NH_4^+$

3-29 How many total protons and electrons are present in each of the following ions?
(a) $Ba^{2+}$ (b) $N^{3-}$ (c) $Cr^{3+}$ (d) $NO^+$
(e) $SO_4^{2-}$

**3-30** Write the symbol of the element, mass number, atomic number, and electrical charge given the following information.
(a) An isotope of Sr contains 36 electrons and 52 neutrons.
(b) An isotope contains 24 protons, 28 neutrons, and 21 electrons.
(c) An isotope contains 36 electrons and 45 neutrons and has a $-2$ charge.
(d) An isotope of nitrogen contains 7 neutrons and 10 electrons.
(e) An isotope contains 54 electrons and 139 nucleons and has a $+3$ charge.

3-31 Write the symbol of the element, mass number, atomic number, and electrical charge given the following information.
(a) An isotope of Sn contains 68 neutrons and 48 electrons.
(b) An isotope contains 204 nucleons and 78 electrons and has a $+3$ charge.
(c) An isotope contains 45 neutrons and 36 electrons and has a $-1$ charge.
(d) An isotope of aluminum has 14 neutrons and a $+3$ charge.

## Nuclear Radiation

**3-32** Complete the following nuclear equations involving alpha and beta particles.
(a) ${}^{214}_{83}Bi \rightarrow {}^{214}_{84}Po +$ ______
(b) ${}^{90}_{37}Rb \rightarrow$ ______ $+ {}^{0}_{-1}e$
(c) ${}^{235}_{92}U \rightarrow$ ______ $+ {}^{4}_{2}He$
(d) ______ $\rightarrow {}^{41}_{21}Sc + {}^{0}_{-1}e$
(e) ______ $\rightarrow {}^{210}_{82}Pb + {}^{0}_{-1}e$

3-33 Complete the following nuclear equations involving alpha and beta particles.
(a) ${}^{239}_{93}Np \rightarrow$ ______ $+ {}^{0}_{-1}e$
(b) ${}^{226}_{88}Ra \rightarrow {}^{222}_{86}Rn +$ ______
(c) ______ $\rightarrow {}^{235}_{92}U + {}^{4}_{2}He$
(d) ${}^{32}_{15}P \rightarrow {}^{32}_{16}S +$ ______

**3-34** Write a nuclear equation that includes all isotopes and particles from the following information.
(a) ${}^{230}_{90}Th$ decays to form ${}^{226}_{88}Ra$.
(b) ${}^{214}_{84}Po$ emits an alpha particle.
(c) ${}^{210}_{84}Po$ emits a beta particle.

(d) An isotope emits an alpha particle and forms $^{235}_{92}U$.
(e) $^{14}_{6}C$ decays to form $^{14}_{7}N$.

* **3-35** The decay series of $^{238}_{92}U$ to $^{206}_{92}Pb$ involves alpha and beta emissions in the following sequence: $\alpha$, $\beta$, $\beta$, $\alpha$, $\alpha$, $\alpha$, $\alpha$, $\alpha$, $\beta$, $\alpha$, $\beta$, $\beta$, $\beta$, $\alpha$. Identify all isotopes formed in the series.

## Nuclear Decay and Half-life

**3-36** The half-life of a certain isotope is 10 years. If we start with a 10.0-g sample of the isotope, how much is left after 20 years?

**3-37** Start with 12.0 g of a certain radioactive isotope. After 11 years only 3.0 g is left. What is the half-life of the isotope?

**3-38** The radioactive isotope $^{90}_{38}Sr$ can accumulate in bones, where it replaces calcium. It emits a high-energy beta particle, which eventually can cause cancer.
(a) What is the product of the decay of $^{90}_{38}Sr$?
(b) How long would it take for a 0.10-mg sample of $^{90}_{38}Sr$ to decay to where only $2.5 \times 10^{-2}$ mg was left? The half-life of $^{90}_{38}Sr$ is 25 years.

**3-39** The radioactive isotope $^{131}_{53}I$ accumulates in the thyroid gland. On the one hand, this can be useful in detecting diseases of the thyroid and even in treating cancer at that location. On the other hand, exposure to excessive amounts of this isotope, such as from a nuclear power plant, can *cause* cancer of the thyroid. $^{131}_{53}I$ emits a beta particle with a half-life of 8.0 days. What is the product of the decay of $^{131}_{53}I$? If one started with $8.0 \times 10^{-6}$ g of $^{131}_{53}I$, how much would be left after 32 days?

**3-40** The isotope $^{14}_{6}C$ is used to date fossils of formerly living systems such as prehistoric animal bones. The radioactivity due to $^{14}_{6}C$ of a sample of the fossil is compared with the radioactivity of currently living systems. The longer something has been dead, the lower will be the radioactivity due to $^{14}_{6}C$ of the sample. If the radioactivity of a sample of bone from a mammoth was one-fourth of the radioactivity of the current level, how old was the fossil? ($t_{1/2}$ = 5760 years.)

**3-41** Of the three types of radiation, which is the most ionizing? Which is the most penetrating? Which is the most damaging to living cells?

## Nuclear Reactions

**3-42** When $^{235}_{92}U$ and $^{239}_{94}Pu$ undergo fission, a variety of reactions actually take place. Complete the following.
(a) $^{235}_{92}U + ^{1}_{0}n \rightarrow$ ______ $+ ^{146}_{58}Ce + 3^{1}_{0}n$
(b) $^{235}_{92}U + ^{1}_{0}n \rightarrow ^{90}_{37}Rb + ^{142}_{55}Cs +$ ______
(c) $^{239}_{94}Pu + ^{1}_{0}n \rightarrow ^{141}_{56}Ba +$ ______ $+ 2^{1}_{0}n$

**3-43** Complete the following nuclear reactions.
(a) $^{35}_{17}Cl + ^{1}_{0}n \rightarrow$ ______ $+ ^{1}_{1}H$
(b) $^{27}_{13}Al +$ ______ $\rightarrow ^{25}_{12}Mg + ^{4}_{2}He$
(c) $^{27}_{13}Al + ^{4}_{2}He \rightarrow$ ______ $+ ^{1}_{0}n$
*(d) $^{238}_{92}U + 15^{1}_{0}n \rightarrow ^{253}_{100}Es +$ ______
(e) $^{244}_{96}Cm + ^{12}_{6}C \rightarrow$ ______ $+ 2^{1}_{0}n$
(f) ______ $+ ^{2}_{1}H \rightarrow ^{238}_{93}Np + ^{1}_{0}n$
(g) $^{242}_{94}Pu + ^{22}_{10}Ne \rightarrow ^{260}_{104}Ku +$ ______

**3-44** Complete the following nuclear reactions.
(a) $^{249}_{96}Cm +$ ______ $\rightarrow ^{260}_{103}Lr + 4^{1}_{0}n$
(b) $^{15}_{7}N + ^{1}_{1}H \rightarrow$ ______ $+ ^{4}_{2}He$
(c) $^{27}_{13}Al +$ ______ $\rightarrow ^{25}_{12}Mg + ^{4}_{2}He$
(d) $^{249}_{98}Cf +$ ______ $\rightarrow ^{263}_{106}X + 4^{1}_{0}n$

**3-45** What is the difference between fission and fusion? What is the source of the energy of these two processes?

**3-46** The fissionable isotope $^{239}_{94}Pu$ is made from the abundant isotope of uranium, $^{238}_{92}U$, in nuclear reactors. When $^{238}_{92}U$ absorbs a neutron from the fission process, eventually $^{239}_{94}Pu$ is formed. This is the principle of the "breeder reactor," although $^{239}Pu$ is formed in all reactors. Complete the following:

$^{238}_{92}U + ^{1}_{0}n \rightarrow$ ______ $+ ^{0}_{-1}e \rightarrow$
$^{239}_{94}Pu +$ ______

# Review Test on Chapters 1–3*

*The following multiple-choice questions have one correct answer.*

1. A substance completely fills any size container. The substance is a:
   (a) gas (b) solid (c) liquid
2. A substance has a distinct melting point and density. It cannot be decomposed into chemically simpler substances. The substance is a:
   (a) heterogeneous mixture (c) compound
   (b) solution (d) element
3. A sample of a certain liquid has uniform properties, but the temperature at which the liquid boils slowly changes as it boils away. The liquid sample is a:
   (a) heterogeneous mixture (c) compound
   (b) solution (d) element
4. Which of the following is the symbol for silicon?
   (a) Se (b) S (c) Si (d) Sc
5. Lead dioxide is the name of
   (a) two elements (c) a compound
   (b) a mixture (d) a solution
6. Which of the following is a chemical property of iodine?
   (a) It melts at 114 °C.
   (b) It has a density of 4.94 g/mL.
   (c) When heated, it forms a purple vapor.
   (d) It forms a compound with sodium.
7. A charged car battery has ______ energy which can be converted into ______ energy.
   (a) chemical, electrical (c) light, electrical
   (b) heat, mechanical (d) mechanical, chemical
8. Which of the following relates to the precision of a measurement?
   (a) the number of significant figures
   (b) the location of the decimal point
   (c) the closeness to the actual value
   (d) the number of zeros that are significant
9. Which of the following is not an exact relationship?
   (a) 1 mile = 5280 ft (c) $10^3$ m = 1 km
   (b) 0.62 mile = 1 km (d) 4 qt = 1 gal
10. How many significant figures are in the measurement 0.0880 m?
    (a) 5 (b) 4 (c) 3 (d) 2
11. The number $785 \times 10^{-8}$ expressed in standard scientific notation is:
    (a) $7.85 \times 10^{-10}$ (c) $7.85 \times 10^{-6}$
    (b) $78.5 \times 10^{-7}$ (d) $0.785 \times 10^{-6}$
12. One kiloliter is the same as:
    (a) $10^6$ mL (b) $10^{-3}$ L (c) $10^3$ mL
    (d) 10 L
13. The temperature of a sample of matter relates to:
    (a) its density (c) its physical state
    (b) its specific heat (d) the intensity of heat
14. Which of the following is the lowest possible temperature?
    (a) 0 K (b) 0 °C (c) 0 °F (d) −273 K
15. Which of the following has the highest density?
    (a) a truckload of feathers (c) a cup of water
    (b) an ounce of gold (d) a gallon of gasoline
16. Which of the following has a charge of +1 and a mass of 1 amu?
    (a) a neutron (c) a proton
    (b) an electron (d) a helium nucleus
17. Isotopes have:
    (a) the same mass number and the same atomic number

*Answers to all questions and problems in the review tests are given in Appendix H.

(b) the same mass number but different atomic numbers
(c) different mass numbers and different atomic numbers
(d) different mass numbers but the same atomic number

**18.** Which of the following is false about the isotope $^{32}_{16}S$?
(a) It has 16 protons.
(b) It has 16 neutrons.
(c) It has 16 electrons.
(d) It has 16 nucleons.

**19.** Which of the following isotopes is used as the standard for atomic weight?
(a) $^{12}C$ (b) $^{16}O$ (c) $^{13}C$ (d) $^{1}H$

**20.** Which of the following is a diatomic molecule of a compound?
(a) NO (b) $NH_4Cl$ (c) $H_2$ (d) $CO_2$

**21.** Which of the following is not a basic particle of an element?
(a) an atom (b) a molecule (c) an ion

**22.** Which of the following is a polyatomic cation?
(a) CO (b) $K^+$ (c) $SO_4^{2-}$ (d) $NH_4^+$

**23.** What would be the electrical charge on a sulfur atom containing 18 electrons?
(a) 2− (b) 1− (c) 0 (d) 2+

**24.** One formula unit of the compound $Al_2(CO_3)_3$ contains:
(a) nine oxygen atoms
(b) three oxygen atoms
(c) six aluminum atoms
(d) six carbon atoms

**25.** What would be the formula of an ionic compound made up of $Ba^{2+}$ ions and $Cl^-$ ions?
(a) $Ba_2Cl$ (b) $BaCl_2$ (c) BaCl
(d) $BaCl_3$

### Problems

*Carry out the following calculations.*

**1.** Round off the following numbers to three significant figures.
(a) 173.8 (b) 0.0023158 (c) 18,420

*Carry out the following calculations. Express your answer to the proper decimal place or number of significant figures.*

**2.** 1.09 cm + 0.078 cm + 16.0021 cm
**3.** 4.2 in. × 16.33 in.
**4.** 398 g ÷ 31 mL
**5.** $0.892 \times 10^4$ cm $\times 0.0022 \times 10^6$ cm
**6.** $7784 \times 10^{-6}$ kg ÷ 0.56 kg/mL

*Express the answer to the following in standard scientific notation. Assume that the numbers are measurements so that the answer should be expressed to the proper decimal place or number of significant figures.*

**7.** $(92.8 \times 10^6) \times (1.3 \times 10^4)$
**8.** $(0.4887 \times 10^{-4}) \div (0.03 \times 10^6)$
**9.** $(4.0 \times 10^6)^2$

*Make the following conversions.*

**10.** 189 cm to in.
**11.** 43 qt to L
**12.** 197 miles/hr to km/hr
**13.** 6.74 lb/ft to g/cm

*Solve the following problems.*

**14.** A cube of metal measures 2.2 cm on a side. It weighs 47.68 g. What is its density?
**15.** What is the volume of a 678-g sample of a substance that has a density of 11.3 g/mL?
**16.** What is the temperature in °C if the reading is 112 °F?
**17.** If the temperature on the Celsius scale increases by 27 degrees, what is the equivalent increase on the Fahrenheit scale?
**18.** The temperature of a certain liquid increases by 12.0 °C if 16.5 calories of heat is added to a 20.0-g sample. What is the specific heat of the liquid?
**19.** The atomic weight of an element is 10.15 times that of the defined standard. What is the atomic weight of the element, and what is the name and symbol of the element?
**20.** The atoms of a certain element are distributed between two isotopes; 65% of the atoms have a mass number of 116.0, and the remainder have a mass number of 112.0. Using mass number as an approximation of atomic weight, calculate the atomic weight of the element to tenths of an amu.

Fireworks are colorful because of the presence of certain elements.

# Chapter 4
# The Arrangement of the Electrons in the Atom

## Purpose of Chapter 4

In Chapter 4 we describe the arrangement of electrons in the atom so as to understand the basis of the periodic properties of the elements, their atomic spectra, and eventually chemical bonds between elements.

## Objectives for Chapter 4

After completion of this chapter, you should be able to:

1. Describe how to locate elements with similar chemical properties in the periodic table. (4-1)
2. Distinguish between the properties of a continuous and a discrete spectrum. (4-2)
3. Describe the relationship between the wavelength of light and energy. (4-2)
4. List the important assumptions of Bohr's model for the hydrogen atom. (4-3)
5. Give the principal quantum number designation, the letter designation, and the capacity of the first five shells. (4-4)
6. Give the four sublevels found in known elements and their order of energy. (4-5)
7. Show the distribution of electrons in the sublevels of a specified shell. (4-5)
8. Demonstrate the Aufbau principle by listing the normal order of filling of sublevels. (4-6)
9. Write the electron configuration for the atoms of a specified element. (4-6)
10. Relate electron configuration of the elements to the periodic table. (4-6)
11. Use only the periodic table and noble gas core shorthand to give the expected electron configuration of a specified element. (4-6)
12. Describe what is meant by an orbital. (4-7)
13. Give the number of orbitals in each type of sublevel. (4-7)
14. Describe the shapes of *s*, *p*, and *d* orbitals. (4-8)
15. Demonstrate the Pauli exclusion principle and Hund's rule by writing box diagrams of the outer electrons for a specified element. (4-9)

It would at first seem impossible for a chemist (let alone a beginning student of chemistry) to know a great deal about each of the 108 known elements. However, there is one fortunate fact that helps simplify the study of chemistry: elements belong to groups or families. Like members of a human family, each element in a chemical family has its unique characteristics but, more importantly, each element has many chemical properties in common with other elements in its family. Although this makes the study of chemistry much more convenient, it raises an interesting question: What is common to a group of elements that explains their chemical similarities? As we will see in this chapter, the answer lies in the arrangement of the electrons in the atoms of these elements.

Historically, to understand the nature of electrons in atoms a sophisticated model or theory was needed that emphasized electrons. This model was needed not only to explain the existence of families of elements but other mysteries seemingly related to electrons in atoms. One such mystery was the discovery that under certain conditions atoms emit a pattern of light (called an emission spectrum) that is unique and charactertistic of the element much as fingerprints are unique to each individual.

Rutherford's nuclear model of the atom, discussed in Chapter 3, was a start, but it did not go far enough. In 1913, one of Rutherford's students, Niels Bohr, from Denmark, expanded on the nuclear model and provided an ingenious description of the electrons in the atom. The immediate success of Bohr's model was assured when he was able to explain the known spectrum of hydrogen atoms. In addition to this, Bohr's model laid the foundation for further developments that would eventually be consistent with the existence of chemical families. In this chapter our main goal will be to develop our current theories of the nature of electrons in atoms. This effort will reward us not only with a general knowledge of the chemical relationships of the elements but also a foundation to discuss another important fundamental phenomenon—the chemical bond.

Before Bohr's model of the atom and subsequent modifications are discussed, we will set the stage with some background on two puzzles that we have mentioned: the existence of families of elements and their emission spectra.

## 4-1 The Periodic Table of the Elements

Before Bohr's model of the atom, scientists were well aware that certain elements display close chemical similarities. *This is demonstrated by arranging the elements by increasing atomic number, with elements in the same family or group falling into vertical columns.* This arrangement is known as the **periodic table** and is shown inside the back cover.

An example of this vertical relationship, which you will recognize, is found in group IB: copper, silver, and gold. These familiar metals are used for coins and jewelry because of their low chemical reactivity as well as

**Figure 4-1** METALS IN WATER. Sodium is a metal that reacts violently with water. The copper, silver, and gold in coins, however, are completely unreactive in water. Here sodium (on the left), reacts vigorously with water while the coins are unaffected.

their lustrous beauty and comparative rarity (see Figure 4-1). These elements can be used for this purpose because they do not rust as does iron or become covered with a dull coating as does aluminum. On the other hand, lithium, sodium, and potassium are also metals, but they are very reactive (see Figure 4-1). They easily form compounds with the oxygen in the air and dissolve in water (explosively in the case of Na and K). Notice that these closely related elements are all members of group IA. The chemical relationships among the elements have been known and studied for more than 150 years, and the periodic table (with, of course, fewer elements) has been displayed for a little more than a century.

Elements in the groups or vertical columns appear in a somewhat repeating or periodic relationship. For example, the number of elements that come *between* the elements of group IA are 2, 8, 8, 18, 18, and 32. This is one example of the many periodic properties of nature. The sun and the moon rise and set in a regular, periodic manner. The seasons of the year, tides in the oceans, and final exams all occur in predictable and repeating cycles. Thus elements not only exist in families, but they do so in a periodic manner. An important goal of this chapter is to establish the reason for this *periodicity* of elements.

The information implied in this table is very important to the chemist. It is well worth the time to discuss it in some detail. We will do this in Chapter 5 after we answer the question: Why are elements periodic?

## 4-2 The Emission Spectra of the Elements

When any substance is heated to a high enough temperature, it glows. The tungsten filament in an incandescent light bulb glows as a result of the heat

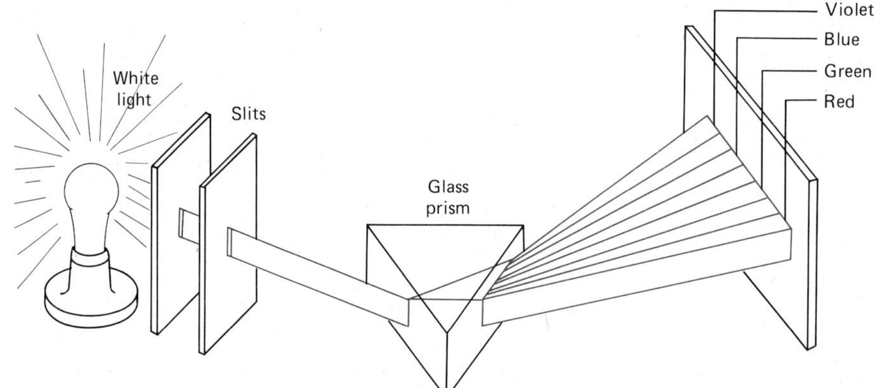


**Figure 4-2** THE SPECTRUM OF INCANDESCENT LIGHT. Incandescent light is composed of all colors of the rainbow.

generated by the flow of electricity through the filament. Let's take a close look at the white light that comes from the bulb. *When a beam of the light is allowed to pass through a glass prism, the light breaks down into a rainbow of colors called its* **spectrum.** *Since one color blends gradually into another, this is known as a* **continuous spectrum** (see Figure 4-2). Raindrops in the air also serve as prisms after a rainstorm, breaking white sunlight into the rainbow.

Light is a form of energy, like electricity and heat. Light, however, dissipates in all directions with a wave nature. This property of light is analogous to the waves formed when a pebble lands in a calm lake. Light waves, however, travel at a speed of $3.0 \times 10^{10}$ cm/sec (186,000 miles/sec). A property of a wave is called its **wavelength.** *It is the distance between two peaks on a wave,* as shown below. Wavelength is designated by the Greek symbol $\lambda$ (lambda).

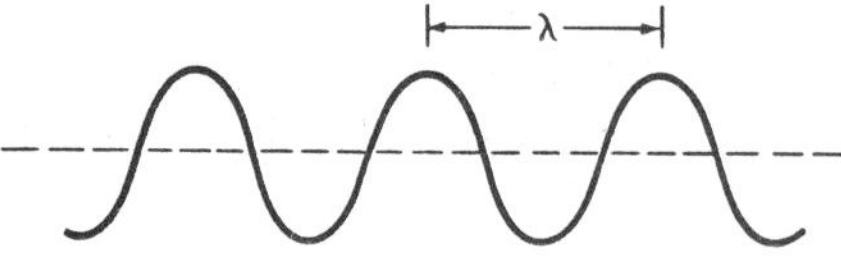


Each specific color of light in the rainbow has a specific $\lambda$. In the rainbow, red has the longest $\lambda$ and violet has the shortest. When light has a somewhat longer $\lambda$ than red, it is invisible to the eye and is called **infrared** (beneath red) light. When light has a somewhat shorter $\lambda$ than violet, it is also invisible and is called **ultraviolet** (beyond violet) light.

*The energy of light is inversely proportional to its wavelength* (see Appendix B for a discussion of proportionalities). That is,

$$E \propto \frac{1}{\lambda} \qquad E = \frac{hc}{\lambda}$$

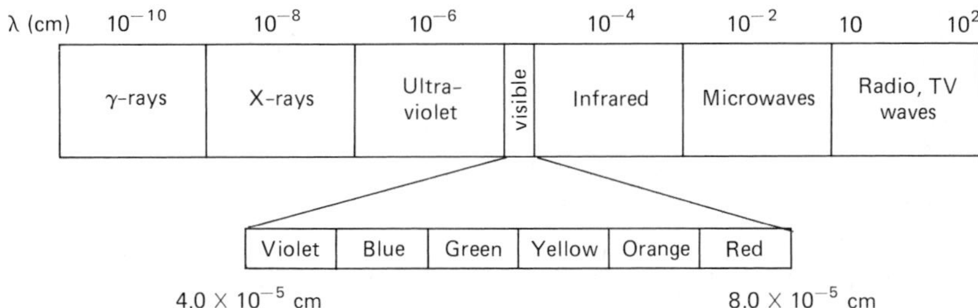


**Figure 4-3** THE SPECTRUM OF LIGHT. Light is classified according to its wavelength.

where $h$ is a constant of proportionality and is known as Planck's constant, and $c$ is also a constant and is equal to the speed of light.

**Review Appendix B-3 on proportionalities.**

Notice that the larger the value of λ, the smaller the energy, and vice versa. Therefore, red light, with the longer λ, has less energy than violet light. Indeed, ultraviolet light has enough energy to damage living organisms. X-rays and gamma rays have extremely small values of λ and are accordingly extremely powerful. (Medical x-rays, however, are given in small doses in order to minimize the damage.) In Figure 4-3 these and other forms of light are shown in terms of relative wavelengths.

As shown in Figure 4-2, white light from a glowing tungsten filament is a combination of all wavelengths and energies of light in the visible range. When an element is heated to such an extent that it breaks into individual gaseous atoms, the spectrum produced is different from the continuous spectrum of the rainbow. In this case, only definite or discrete colors are produced. Recall that if the light produced has a specific color (λ), it has a specific energy ($E$). The spectrum of a gaseous atomic element (known as its *atomic emission spectrum*) is particular and characteristic of that element. You may have observed this phenomenon in the chemicals that create colors when added to the logs in a fireplace or to the powder in fireworks. Lithium compounds create a red color, boron compounds create a green color, and sodium compounds create a yellow color. *Since only certain colors are produced by the atoms of hot, gaseous elements, their atomic spectra are referred to as* **discrete** *or* **line spectra.**

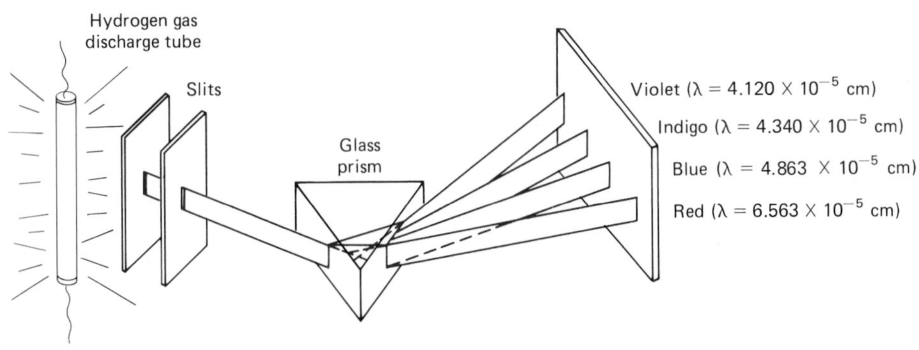


**Figure 4-4** THE ATOMIC SPECTRUM OF HYDROGEN IN THE VISIBLE RANGE. The atomic spectrum of an element is composed of discrete colors.

Hydrogen, the simplest of the elements, has the most orderly atomic spectrum, as shown in Figure 4-4. In a sense, this made the hydrogen spectrum the most puzzling to the scientists.

The fact that light, as a form of energy, is emitted by atomic hydrogen in a discrete and yet orderly manner was also a matter that required an explanation.

## 4-3 A Model for the Electrons in the Atom

By 1913 the nuclear model of the atom proposed by Rutherford was generally accepted by scientists. In that year Neils Bohr set out to explain the origin of the atomic spectrum of hydrogen by expanding on this model. Bohr's contribution would provide not only a reasonable picture of what forces hold the atom together but also the mathematical relations corresponding to the known experimental facts about the hydrogen atom.

First, Bohr proposed that the electrons revolve around the nucleus in stable, circular orbits (see Figure 4-5). The attractive forces between the negative electron and the positive nucleus would be exactly balanced by the centrifugal force* of the orbiting electron. This was a controversial and bold statement by Bohr, because classical physics predicted that this would not be possible—the electron would collapse into the nucleus. Bohr neatly sidestepped this problem by postulating (suggesting without proof as a necessary condition) that classical physics does not apply in the small dimensions of the atom.

*The force on a rotating object that tends to move it radially outward. For example, when a discus thrower lets go of the discus, it goes straight out from the point of release.

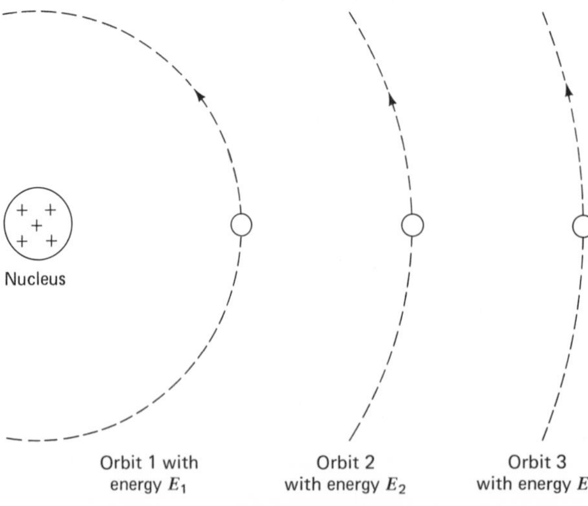


**Figure 4-5** BOHR'S MODEL OF THE HYDROGEN ATOM. In this model the electron can exist only in definite or discrete energy states.

Second, Bohr postulated that the orbits in which electrons reside can be located only at certain, definite distances from the nucleus. An electron cannot exist anywhere between these available orbits unless it is in the process of jumping between one and another. Electrons in such a stationary orbit would have a definite or discrete energy. *Since only discrete energy states are available to the electron in an atom, the electronic energy is said to be* **quantized** *according to Bohr's theory.*

The discrete energy levels in the atom can be compared to the stairway in your home. In Figure 4-6 you can see that between floor A and floor B there are five steps that are available energy levels. Since you cannot stand (with both feet together) between steps, you can see that only these five energy states are possible between floors. *A stairway is thus quantized.* On the other hand, a ramp would be continuous, as all positions or energy levels between floors are possible.

*When an electron resides in the available orbit that is lowest in energy (which is also the closest to the nucleus), the electron is said to be in the* **ground state.** The electron can jump from the ground state to a higher energy level farther from the nucleus if it is supplied exactly the right amount of energy to make the jump. (In the analogy to the stairway, it takes a definite amount of energy for a person to climb one step.) The electron then has potential energy and is said to be in an **excited state.** When an electron is in an excited state, it can fall back down to a lower level; but the electron must then give up the same amount of energy that it previously absorbed. The energy that is emitted by this process comes off in the form of light. Since the orbits have discrete energy, the difference in energy between the levels is fixed. This in turn means that the energy of the light is discrete, with a discrete wavelength and hence a specific color. (See Figure 4-7.)

Bohr did significantly more than just present a model to explain discrete spectra. From the mathematical relationships suggested by his model, he was able to calculate the wavelengths of the light emitted by the hydrogen atom. The value of his theory was proven when the calculated values corresponded with the values measured from actual experiments (one series of lines is shown in Figure 4-4).

Because of the beautiful simplicity of Bohr's model of the atom, modern

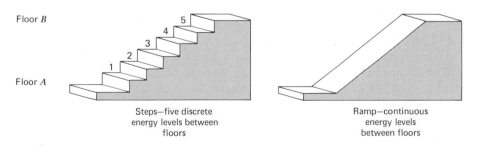

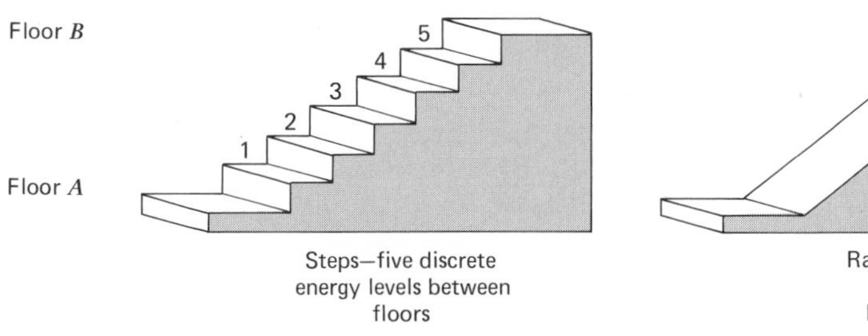


**Figure 4-6** DISCRETE VERSUS CONTINUOUS ENERGY LEVELS. The steps represent discrete energy levels, the ramp continuous energy levels.

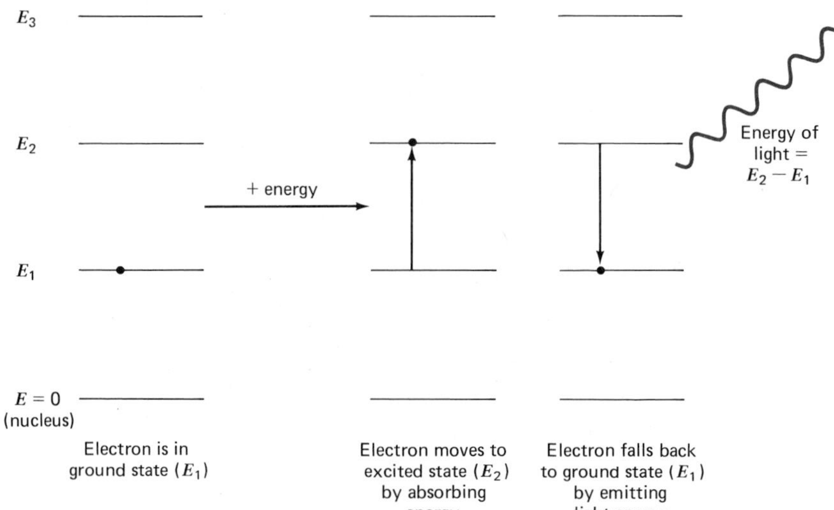


**Figure 4-7** LIGHT FROM THE HYDROGEN ATOM. An electron in an excited state emits energy in the form of light when it drops to a lower energy level.

scientists still use his picture in certain situations. However, our theories have now gone well beyond this model. For one thing, Bohr's model was not correct for atoms with more than one electron. That essentially limits the theory to hydrogen. Also, new experiments and new knowledge of the nature of matter have forced us away from viewing electrons as particles orbiting at definite distances from the nucleus with definite speeds and energies. Now we have a mathematical model that more accurately describes the energies and locations of the electrons but leaves no convenient picture for us to visualize. This mathematical descripition is a branch of physics and chemistry known as **wave mechanics.** Despite the complexity of wave mechanics, the results can be easily grasped and put to use in the effort to explain the chemical nature of the elements. What follows then, are results of calculations of wave mechanics.

## 4-4 Shells

*Electrons reside in regions of space referred to as* **shells** or **principal energy levels.** The shell that lies closest to the nucleus has the lowest energy, and each successive shell from the nucleus is higher in energy. There is a definite amount of energy involved when an electron moves from one shell to another. In this respect, shells in an atom are analogous to dormitories near a campus. A dormitory is a defined region of space where students reside. In a series of dorms on a hill, obviously the one lowest on the hill requires the least amount of effort to reach and is thus considered the *lowest* in energy (see Figure 4-8). Also analogous to electrons in shells, a definite amount of energy is needed for a student to move uphill from dorm 1 to dorm 2.

The shells in an atom are labeled in two ways (see Table 4-1). Histor-

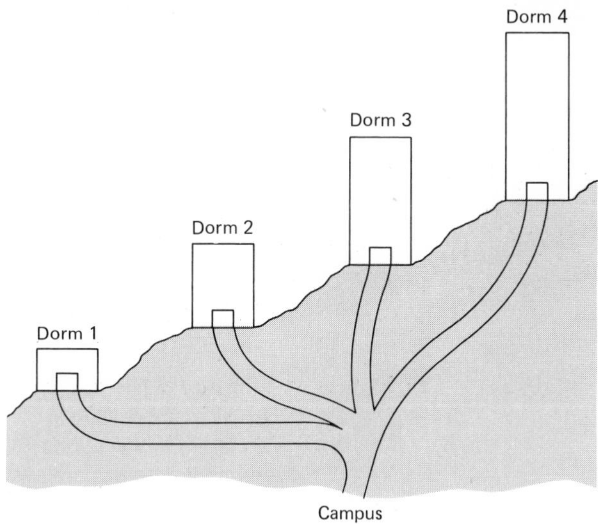


**Figure 4-8** ENERGY LEVELS OF DORMS. The dorms on the hill represent different energy levels.

ically, the letters *K, L, M, N,* and so on, were used, with the *K* shell being closest to the nucleus (and the lowest in energy), the *L* shell next, and so on. The modern way is to label the shells by means of a number ($n$). *These numbers are called the* **principal quantum numbers** *and describe the energy of the shell.* The $n = 1$ shell corresponds to the *K* shell, $n = 2$ to the *L* shell, and so on. A result of Bohr's theory that is still valid is that each shell holds $2n^2$ electrons. You can see that the shells get larger the farther they are from the nucleus. This result is reasonable, since the farther a shell is from the nucleus the greater the volume of space that is available to the electrons.

**See Problems 4-1 through 4-8.**

## 4-5 Sublevels

Let's look again at the analogy of the dorms. If the careful student wants to live in the dorm of lowest energy, he or she would naturally look to the one lowest on the hill with space available. But another point should be raised as to the energy required to get to a certain room. That question is, "On what floor is the available room?" Since you can count on the elevators being full, on the top floor, or out of order, it is not far-fetched to assume that the student will have to use the stairs. In our example, the first dorm

**Table 4-1** Shells

| Letter designation | *K* | *L* | *M* | *N* | *O* |
|---|---|---|---|---|---|
| Quantum number designation | 1 | 2 | 3 | 4 | 5 |
| Capacity ($2n^2$) | 2 | 8 | 18 | 32 | 50 |

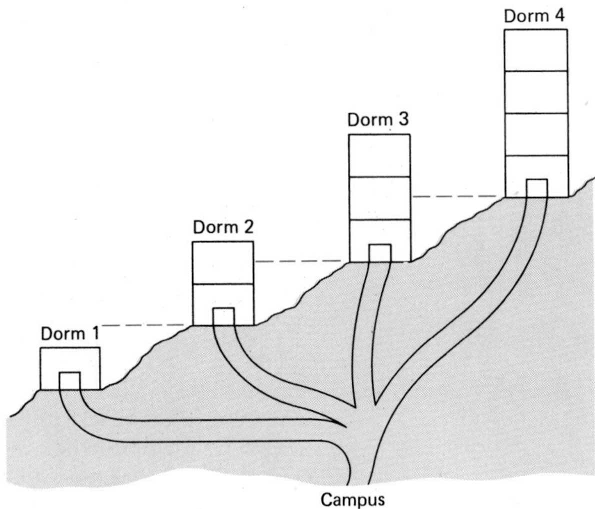


**Figure 4-9** ENERGY LEVELS WITHIN A DORM. The floors in the dorms represent different energy levels within each dorm.

has only one floor, the second has two, the third has three, and so on. (This unusual situation is for the sake of the analogy.) Thus, not only do the students need to know how far up the hill they must walk but how many flights of stairs must be climbed once they are in the dorm. Each floor within a dorm is at a different energy level (see Figure 4-9).

Just as each dorm has floors, *a shell or principal energy level has* **sublevels.** The sublevels found in the currently known elements are labeled in order of increasing energy:

$$s < p < d < f$$

Each shell has the number of sublevels equal to the value of its $n$ quantum number. Thus, analogous to the dorms, the first shell ($n = 1$) has one sublevel (an $s$), the $n = 2$ shell has two (an $s$ and a $p$), the $n = 3$ shell has three (an $s$, a $p$, and a $d$), and the $n = 4$ shell has four (an $s$, a $p$, a $d$, and an $f$). Each sublevel has a different electron capacity. The $s$ sublevel holds 2 electrons, the $p$ holds 6 electrons, the $d$ holds 10 electrons, and the $f$ holds 14 electrons. This information has been summarized for the first four shells in Table 4-2. Notice how the total capacity of a shell is distributed among the sub-

**Table 4-2** Shells and Sublevels

| Shell | 1 | 2 | | 3 | | | 4 | | | |
|---|---|---|---|---|---|---|---|---|---|---|
| Sublevel | $s$ | $s$ | $p$ | $s$ | $p$ | $d$ | $s$ | $p$ | $d$ | $f$ |
| Sublevel capacity | 2 | 2 | 6 | 2 | 6 | 10 | 2 | 6 | 10 | 14 |
| Shell capacity | 2 | 8 | | 18 | | | 32 | | | |

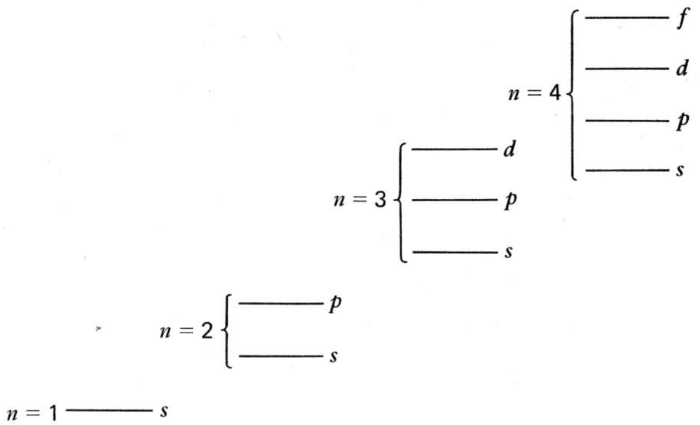

$n = 4$: $f$, $d$, $p$, $s$
$n = 3$: $d$, $p$, $s$
$n = 2$: $p$, $s$
$n = 1$: $s$

Figure 4-10 ENERGY LEVELS IN THE ATOM. Each successive shell contains one additional sublevel.

levels present within the shell. The information in Table 4-2 should be committed to memory before proceeding.

In Figure 4-10, the first four shells and their sublevels are shown in order of increasing energy, analogous to the dorms shown in Figure 4-9.

## 4-6 Electron Configuration of the Elements

We are now ready to examine the electron arrangement or configuration of all the electrons in an atom. The logical place to start is with the one-electron atom, hydrogen, and work on from there. Two principles will guide us:

1 A shell will be designated by the prinicipal quantum number *n*, a sublevel by the appropriate letter designation, and a superscript number indicating the number of electrons in that particular sublevel.

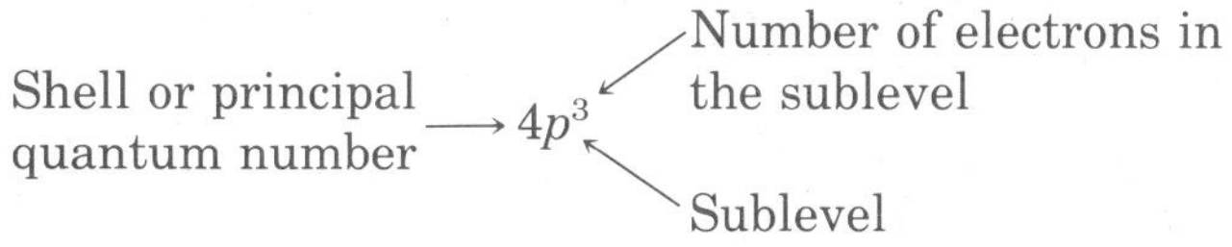

Shell or principal quantum number $\longrightarrow 4p^3$ — Number of electrons in the sublevel; Sublevel

2 *Electrons fill the lowest available energy levels first (shell and sublevel).* This is known as the **Aufbau principle.** It is analogous to the student who wants the available room of lowest energy. Obviously, the most desirable room is on the lowest available floor in the lowest available dorm on the hill.

The first element, H, has one electron, which goes into the first shell, which has only the *s* sublevel. The electron configuration for the one electron is

$$1s^1$$

The next element, He, has two electrons. Since the first shell has room for two electrons, the electron configuration is

$$1s^2$$

Next comes Li with three electrons. Two fill the first shell, but the third must move on to the second shell. The second shell has two sublevels, the *s* and *p*, but as you know the *s* is the lower in energy. The electron configuration of Li is

$$1s^2 2s^1$$

The next seven elements complete the filling of the second shell:

| | | | | | |
|---|---|---|---|---|---|
| Be | $1s^2 2s^2$ | C | $1s^2 2s^2 2p^2$ | O | $1s^2 2s^2 2p^4$ |
| B | $1s^2 2s^2 2p^1$ | N | $1s^2 2s^2 2p^3$ | F | $1s^2 2s^2 2p^5$ |
| | | | | Ne | $1s^2 2s^2 2p^6$ |

If you will remember, one of our goals was to explain the reason for the periodic table. An early observation may lead to this end. He and Ne are in a vertical column, indicating that they are members of the same family or group of elements. This family is known as the noble gases. As a whole, the noble gases have little or no tendency to enter into chemical reactions to form chemical bonds. (They are sometimes referred to as the inert gases.) Notice that the electron configuration of He and Ne indicate that they do have something in common. That is, they both have "filled shells." Perhaps that is the reason for the particular behavior of the noble gases. This idea can be tested later as we look at the electron configuration of heavier elements. As we proceed, we can adopt a shorthand way of writing that saves us from repeating all of the filled inner sublevels. A noble gas in brackets will be equivalent to all of the electrons of that noble gas (e.g., [Ne] = $1s^2 2s^2 2p^6$).

The electron configuration of the next eight elements are:

| | | | |
|---|---|---|---|
| Na | $[Ne]3s^1$ | P | $[Ne]3s^2 3p^3$ |
| Mg | $[Ne]3s^2$ | S | $[Ne]3s^2 3p^4$ |
| Al | $[Ne]3s^2 3p^1$ | Cl | $[Ne]3s^2 3p^5$ |
| Si | $[Ne]3s^2 3p^2$ | Ar | $[Ne]3s^2 3p^6$ |

Let's make some more observations. Note that although Ar is in the noble gas group, the element does not have a filled shell. The first premise must be modified or discarded. However, we might suggest that perhaps filled outer *s* and *p* sublevels is what is characteristic of a noble gas (except for He, which doesn't have a 1*p* sublevel). Notice also that Li and Na have something in common, and that is one electron beyond a noble gas. Since

Li and Na are also in the same group, it appears at this point that similar electron configurations relate to the periodic relationship of the elements.

The next element after Ar presents a problem. Normally we might expect K, element 19, to have one electron in a $3d$ sublevel. In fact it has its nineteenth electron in the $4s$ sublevel. How could this be possible? For the answer, let's go back to the analogy shown in Figure 4-9. Notice that although dorm 4 is higher up the hill than dorm 3, all of the floors in dorm 4 are not necessarily higher than those in dorm 3. In fact, the top floor of dorm 3 is higher than the ground floor of dorm 4. The sharp student would thus choose the ground floor of dorm 4 in preference to the top floor of dorm 3. In the case of elements a similar phenomenon is true. For the neutral atom the $4s$ sublevel is lower in energy than the $3d$ sublevel and fills first. After the $4s$ sublevel, the $3d$ fills, followed by the $4p$. In Figure 4-10, notice that the $4s$ sublevel is lower in energy than the $3d$ sublevel.

K $[Ar]4s^1$

Ca $[Ar]4s^2$

Sc $[Ar]4s^23d^1$

·
·
·

Zn $[Ar]4s^23d^{10}$

Ga $[Ar]4s^23d^{10}4p^1$

·
·
·

Kr $[Ar]4s^23d^{10}4p^6$

Notice that our observations on periodicity seem valid. Li, Na, and K, which are all in the same group, have an $ns^1$ electron configuration ($n$ is a principal quantum number). Kr, a noble gas, has filled outer $4s$ and $4p$ sublevels, which was suggested earlier to be characteristic of this family of elements. *Apparently, vertical columns of elements have the same sublevel electron configuration but different shells.*

At this point, a scheme may be helpful to remember the order of filling of the sublevels. As you go to shells of higher energy, the shells get larger and much more mixing of sublevels occurs. A way to remember is illustrated in Figure 4-11. Write in horizontal columns all of the sublevels that exist starting with the ls. Under ls put $2s$, followed by $2p$, and so on. Draw a stair-step down on the right. Now draw a diagonal arrow through each corner of the stairstep. The top arrow points to the first sublevel filled. The second arrow points to the second sublevel, the third arrow points to the $2p$ followed by the $3s$, and so on.

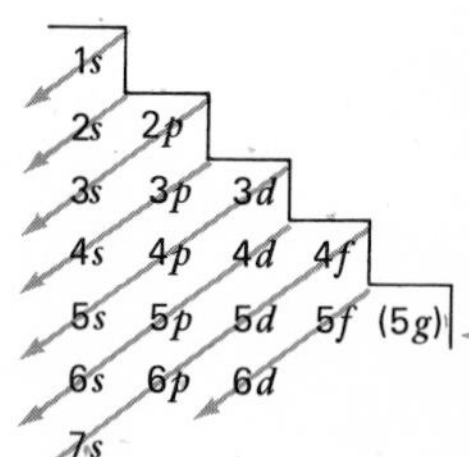


**Figure 4-11** THE ORDER OF FILLING OF SUBLEVELS. The arrows indicate the order of filling.

**Example 4-1**
Write the electron configuration for iron.

**Solution**
From inside the front cover, we find that iron has an atomic number of 26, which means the neutral atom has 26 electrons. Write the sublevels in the order of filling plus a running summation of the total number of electrons involved.

| Sublevel | Number of Electrons in the Sublevel | Total Number of Electrons |
|---|---|---|
| $1s$ | 2 | 2 |
| $2s$ | 2 | 4 |
| $2p$ | 6 | 10 |
| $3s$ | 2 | 12 |
| $3p$ | 6 | 18 |
| $4s$ | 2 | 20 |
| $3d$ | 10 | 30(over 26) |

Iron therefore has six electrons past the filled $4s$ sublevel, but it does not completely fill the $3d$ sublevel. The electron configuration for iron is written out as

$$1s^2 2s^2 2p^6 3s^2 3p^6 4s^2 3d^6$$

or, if we use noble gas shorthand,

$$[\mathrm{Ar}]4s^2 3d^6$$

After some practice, the easiest way to predict the electron configuration of the elements is by use of the periodic table. In this endeavor the fact that vertical columns usually have the same electron sublevel configuration can be used to our advantage. In Figure 4-12 the sublevel configuration is listed above the group and the shell ($n$) is listed in the box for the element. With practice, one can quickly give the electron configuration of any element with only a periodic table for reference. It should be pointed out that certain elements do deviate from the filling predicted by Figure 4-11, especially those elements where $d$ and $f$ sublevels are partially filled. For example, Cr (at. no. 24) is actually $[\mathrm{Ar}]4s^1 3d^5$ rather than the expected $[\mathrm{Ar}]4s^2 3d^4$. Another example of an exception to the normal filling is the element La (at. no. 57). Figure 4-12 indicates that La has the outer sublevel configuration $6s^2 5d^1$ rather than $6s^2 4f^1$. Figure 4-12 is correct. Although these exceptions are important in the discussion of the chemistry of these elements, it is not important at this point to know all of the exceptions to the normal order of filling.

| $s^1$ | $s^2$ | $d^1$ | $d^2$ | $d^3$ | $d^{5*}$ | $d^5$ | $d^6$ | $d^7$ | $d^8$ | $d^{10*}$ | $d^{10}$ | $s^2p^1$ | $s^2p^2$ | $s^2p^3$ | $s^2p^4$ | $s^2p^5$ | $s^2p^6$ ($s^2$ for 1) |
|---|---|---|---|---|---|---|---|---|---|---|---|---|---|---|---|---|---|
| 1 | | | | | | | | | | | | | | | | | 1 |
| 2 | 2 | | | | | | | | | | | 2 | 2 | 2 | 2 | 2 | 2 |
| 3 | 3 | | | | | | | | | | | 3 | 3 | 3 | 3 | 3 | 3 |
| 4 | 4 | 3 | 3 | 3 | 3 | 3 | 3 | 3 | 3 | 3 | 3 | 4 | 4 | 4 | 4 | 4 | 4 |
| 5 | 5 | 4 | 4 | 4 | 4 | 4 | 4 | 4 | 4 | 4 | 4 | 5 | 5 | 5 | 5 | 5 | 5 |
| 6 | 6 | 5 | 5 | 5 | 5 | 5 | 5 | 5 | 5 | 5 | 5 | 6 | 6 | 6 | 6 | 6 | 6 |
| 7 | 7 | 6 | 6 | 6 | | | | | | | | 7 | 7 | 7 | 7 | 7 | 7 |
| 8 | 8 | | | | | | | | | | | | | | | | |

| $f^1$ | $f^2$ | $f^3$ | $f^4$ | $f^5$ | $f^6$ | $f^7$ | $f^8$ | $f^9$ | $f^{10}$ | $f^{11}$ | $f^{12}$ | $f^{13}$ | $f^{14}$ |
|---|---|---|---|---|---|---|---|---|---|---|---|---|---|
| 4 | 4 | 4 | 4 | 4 | 4 | 4 | 4 | 4 | 4 | 4 | 4 | 4 | 4 |
| 5 | 5 | 5 | 5 | 5 | 5 | 5 | 5 | 5 | 5 | 5 | 5 | 5 | 5 |

*This has an $ns^1$ configuration rather than $ns^2$.

**Figure 4-12** ELECTRON CONFIGURATION AND THE PERIODIC TABLE. The electron configuration of an element can be determined from its position in the periodic table. (The value of "$n$" is shown in the boxes.)

**Example 4-2**
Write the electron configuration for (a) vanadium and (b) lead.

**Solution**
(a) In the front cover of the book, note that vanadium (V) has atomic number 23. Locate V in the periodic table shown in Figure 4-12. V has the electron configuration

$$1s^22s^22p^63s^23p^64s^23d^3$$

Or, starting with the previous noble gas

$$[\mathrm{Ar}]4s^23d^3$$

(b) Lead (Pb) has atomic number 82. Locate Pb in the periodic table shown in Figure 4-12. Using the noble gas shorthand, notice that the following sublevels come after Xe in lead:

$$[\mathrm{Xe}]6s^25d^{10}4f^{14}6p^2$$

**Example 4-3**
What element has the following electron configuration? $[\mathrm{Kr}]5s^24d^{10}5p^5$

**Solution**
In Figure 4-12, locate $s^2p^5$ in the upper right. Atomic number 53 has the given electron configuration. Atomic number 53 is the element iodine.

See Problems 4-9 through 4-31.

## 4-7 Orbitals

If one were to attempt to locate a student in a dorm, more information would be needed than just the particular dorm and the floor. One would have to know the student's room number. Just as floors are made up of individual rooms, sublevels are composed of individual orbitals. *An* **orbital** *is the region of space in which there is the highest probability of finding the electron.* Like a two-person room, each orbital can hold a maximum of two electrons. An *s* sublevel is like a one-room floor, so it is all one orbital holding up to two electrons. The *p* sublevel is made up of three separate orbitals holding a total of six electrons. A *d* sublevel is composed of five orbitals, holding two each for a maximum capacity of 10 electrons, and finally the *f* sublevel is composed of seven orbitals holding a total of 14 electrons. (See Figure 4-13.)

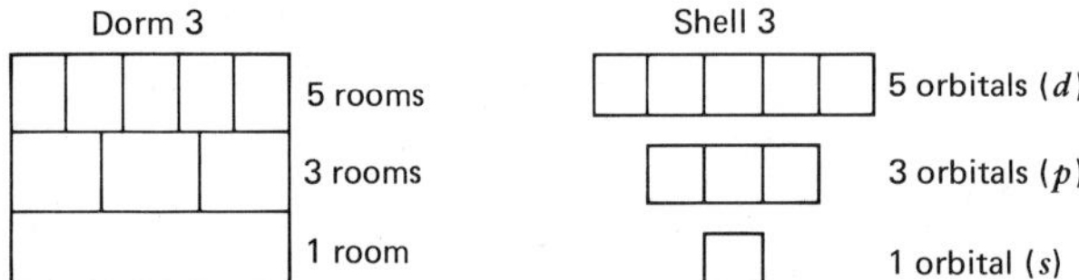


**Figure 4-13** DORM ROOMS AND ORBITALS. The rooms on each floor of a dorm represent different regions of space at the same energy level, which is analogous to the orbitals of a sublevel.

The energy of an electron is determined only by the shell and sublevel. Under normal conditions all orbitals within a certain sublevel (e.g., the 4*p*) have the same energy. In the dorm analogy, once you have reached the floor, it doesn't matter whether you go to the room on the right, left, or straight ahead. The same amount of energy is required.

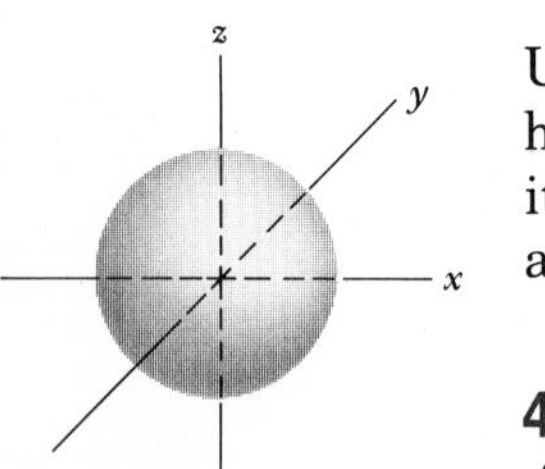


**Figure 4-14** AN *s* ORBITAL. An *s* orbital has a spherical shape.

## 4-8 The Shapes of Orbitals

An important topic of chemistry concerns the shape or geometry of molecules. The geometry of molecules is related to the shape of the orbitals of the electrons involved in bonding.

An *s* orbital has a perfectly spherical shape. Electrons in *s* orbitals are free to occupy any position within the sphere (see Figure 4-14).

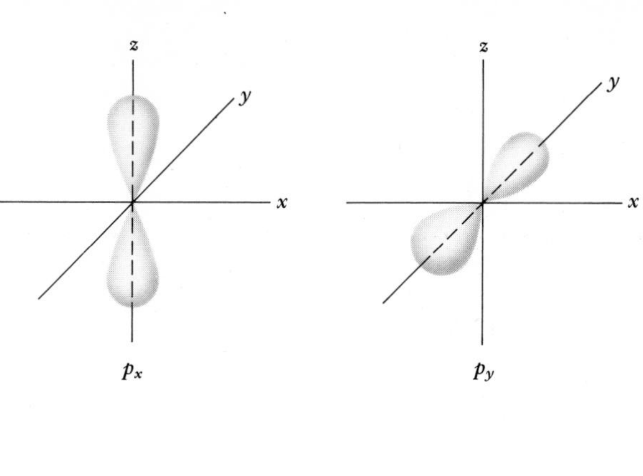


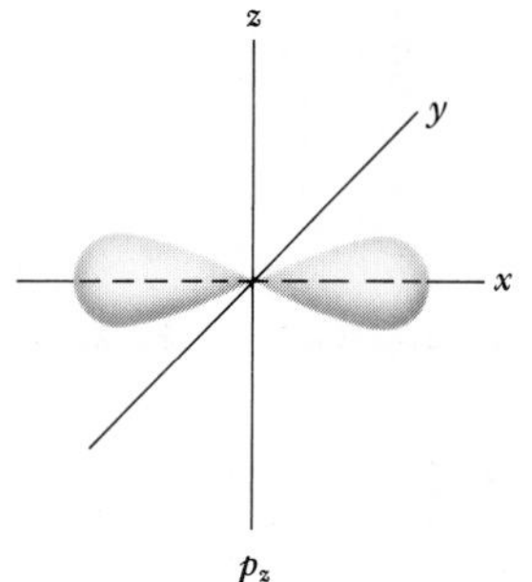


**Figure 4-15** THE *p* ORBITALS. Each *p* orbital has two lobes.

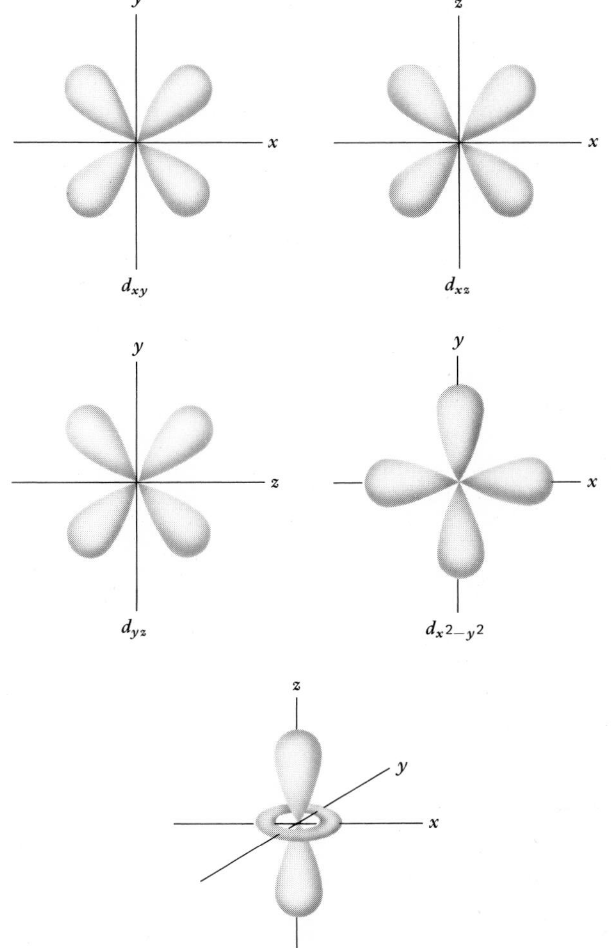


**Figure 4-16** THE *d* ORBITALS. Except for the $d_{z^2}$ the *d* orbitals have four lobes.

The three different $p$ orbitals, on the other hand, are roughly shaped like a weird baseball bat with two fat ends called "lobes." If the axis through one of the $p$ orbitals is defined as the $x$ axis, the orbital is designated as the $p_x$ orbital. The other two $p$ orbitals are at 90° angles to the $p_x$ orbital and to each other. They are referred to as the $p_y$ and $p_z$ orbitals (see Figure 4-15).

The $d$ orbitals are more complex. Four of the orbitals consist of four lobes each. The fifth consists of two lobes along the $z$ axis and a torus (a doughnut shape) in the $xy$ plane (see Figure 4-16). No matter how many lobes an orbital has, however, it still holds no more than two electrons.

The $f$ orbitals are even more complex. However, since they are rarely involved in bonding, their shapes are not usually important to the chemist.

The shapes of the orbitals that have been shown represent the regions of space where the probability of finding the electron in a particular orbital is highest. In this respect, it is like locating students in their rooms. The highest probability is at the desk or in bed, but there is a probability that the students won't be there at all. Likewise, there is a very small probability that the electron in an orbital is not in the region indicated by the shape of the orbital. In summary, these pictures of orbitals represent a "cloud" of maximum probability. It is *not* proper in the modern sense to think of an electron orbiting the nucleus in some path within the orbital.

## 4-9 Box Diagrams of Electrons

In a previous section we described the electron configuration of the elements by showing the distribution of electrons in shells and sublevels. We now wish to show how electrons in sublevels are distributed in the orbitals. To represent an orbital in an atom we will employ a box ( □ ) for each separate orbital. Each electron is represented as an arrow pointing either up or down (↑ or ↓). An electron in an atom behaves as a spinning particle just as the earth spins on its axis as it rotates around the sun. When two electrons occupy the same orbital, the two electrons have opposite or "paired" spins. This phenomenon is represented by the two arrows pointing in opposite directions. This is called the **Pauli exclusion principle,** *which states that no two electrons can reside in the same region of space (orbital) with the same spin* (represented by the direction of the arrow). Thus two electrons in one orbital are represented as

[↑↓]

The box diagrams of the first four elements are

| | 1s | | | 1s | 2s |
|---|---|---|---|---|---|
| H | [↑] | | Li | [↑↓] | [↑] |
| He | [↑↓] | | Be | [↑↓] | [↑↓] |

The box diagrams of the next six elements are

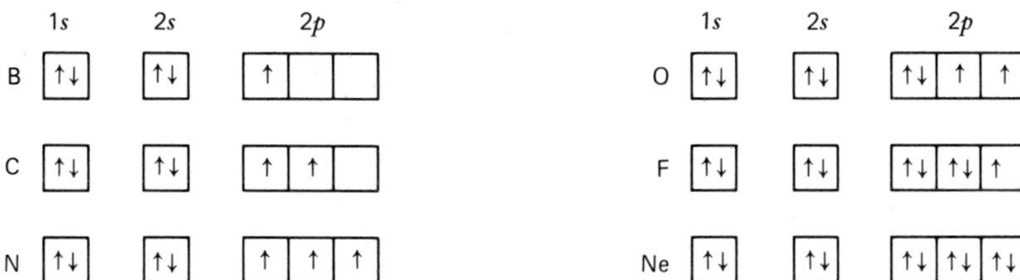


Notice the element carbon has two unpaired electrons in two different $2p$ orbitals. This follows **Hund's rule,** *which states that electrons occupy separate orbitals of the same energy (same sublevel) with parallel spins.* This result is understandable, since electrons, having the same charge, would prefer to get as far away from each other as possible within the same sublevel. In the dorm analogy, a similiar situation develops between two students of the same charge (sex). Given two unoccupied rooms in the same building on the same floor, the two students will probably prefer private rooms. Pairing occurs when no more empty rooms are available.

The same phenomenon occurs with elements where the $d$ sublevel is partially occupied. For example Fe (at. no. 26) has the electron configuration [Ar] $4s^2 3d^5$. All five electrons in the $3d$ sublevel occupy separate orbitals. Thus there are five unpaired electrons in Fe as illustrated by the box diagram

| | | $4s$ | $3d$ |
|---|---|---|---|
| Fe | [Ar] | ↑↓ | ↑ ↑ ↑ ↑ ↑ |

### Example 4-4

Show the box diagrams for the electrons beyond the noble gas core for the elements S, Ni, Cr, and Pr.

### Solution

First determine the electron configuration for the element and then show the box diagram for the outer orbitals beyond the noble gas core.

| | | | |
|---|---|---|---|
| | | $3s$ | $3p$ |
| S [Ne] $3s^2 3p^4$ | [Ne] | ↑↓ | ↑↓ ↑ ↑ |
| | | $4s$ | $3d$ |
| Ni [Ar] $4s^2 3d^8$ | [Ar] | ↑↓ | ↑↓ ↑↓ ↑↓ ↑ ↑ |
| | | $4s$ | $3d$ |
| Cr [Ar] $4s^1 3d^5$ | [Ar] | ↑ | ↑ ↑ ↑ ↑ ↑ |

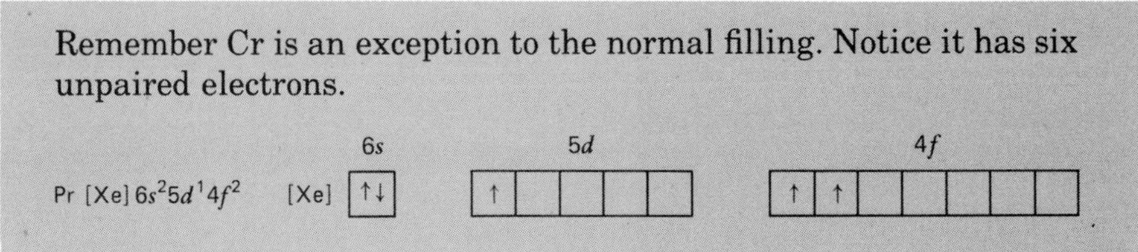


See Problems 4-32 through 4-42.

## 4-10 Chapter Summary

Two interesting phenomena about the elements were discussed in this chapter. The first is that elements can be arranged into groups or families of elements with similar properties in what is called the periodic table. The second concerns the discrete colors or spectra of their hot gaseous atoms. Niels Bohr in 1913 provided a model that explained the second puzzle (at least for hydrogen atoms) and gave some rationale for the first point. Bohr's model suggested that electrons are in orbits or shells of discrete energy. Electron transitions from high-energy orbits or excited states to lower-energy orbits or to the ground state resulted in emission of light of discrete energy. Since energy of light is related to its wavelength, the light would be of a definite wavelength. If the wavelength is in the visible part of the spectrum, the emitted light has a definite color.

Bohr's model is still useful, although in modern theories the electron is viewed as having a certain probability of a location in space rather than a definite location as suggested by Bohr. The modern theories of the electron suggest that an electron in an atom is described by four criteria: (1) shell or principal energy level, (2) sublevel, (3) orbital, and (4) electron spin.

The shell and sublevel describe the energy of an electron. Shells are designated in order of increasing energy by the quantum number $n$.

$$n = 1 < n = 2 < n = 3 < n = 4, \text{etc.}$$

The four sublevels found in known elements are designated by letters in order of increasing energy

$$s < p < d < f$$

The number of electrons each shell can hold is $2n^2$, and the number of sublevels in each shell is $n$. For example, the third shell ($n = 3$) holds 18 electrons and contains three sublevels ($s$, $p$, and $d$). Electrons in atoms of elements fill these sublevels according to the Aufbau principle, which means that the lowest in energy fills first. The electron configuration of an element tells us the electron population of the various sublevels for the atoms of an element. The order of filling of sublevels becomes somewhat garbled as we go to higher energy shells. The best way to remember the order of filling is by reference to the periodic table. Vertical columns have the same sublevel configuration but have consecutively larger values of $n$. This then is the reason for these chemical relationships and periodicity. *Families of elements lie in vertical columns and have the same population of their sublevels beyond the previous noble gas.*

At this point the original two questions are answered, but there are two other aspects of the electron in an atom that can be described. Each sublevel that exists has one or more orbitals. Orbitals are regions of space where there is the greatest probability of finding the electron. Each separate orbital can hold up to two electrons. An *s* sublevel consists of just one spherically shaped orbital. The *p* sublevel consists of three propeller-shaped orbitals oriented at angles of 90° from each other. The *d* sublevel consists of five orbitals, and the *f* sublevel consists of seven orbitals. The information on shells, sublevels, and orbitals is summarized for the first four shells as follows. Each orbital is represented as a box.

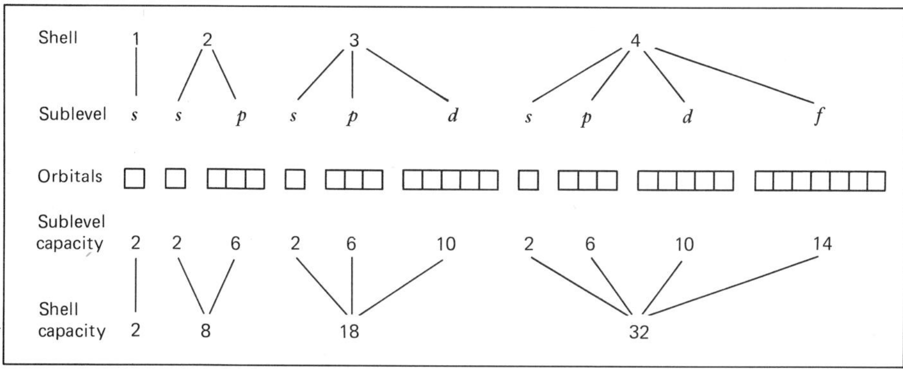


The final property concerning the electron in an atom is its spin. The Pauli exclusion principle states that electrons in the same orbital must have opposite or paired spins. Hund's rule states that electrons occupy separate orbitals of the same energy with their spins parallel rather than pair up in the same orbital. All of this can be summarized by the box diagram for the outer electrons of the element As (at. no. 33).

| | $4s$ | $3d$ | $4p$ |
|---|---|---|---|
| As [Ar] $4s^2 3d^{10} 4p^3$ | ↑↓ | ↑↓ ↑↓ ↑↓ ↑↓ ↑↓ | ↑ ↑ ↑ |

## Exercises

**4-1** Referring to the periodic table, determine which of the following elements is expected to be chemically similar to calcium?
(a) K (b) Sr (c) Cl (d) Mg (e) Al

**4-2** The elements fluorine, chlorine, and bromine all exist as diatomic molecules.

Explain this behavior in terms of chemical families.

**4-3** Ultraviolet light causes sunburns, but visible light does not. Explain.

**4-4** The $n = 8$ and the $n = 9$ shells are very close in energy. Using the Bohr model, describes how the wavelengths of light compare as an electron falls from these two shells to the $n = 1$ shell.

**4-5** The $n = 3$ shell is of considerably greater energy than the $n = 2$ shell. Using the Bohr model, describe how the wavelengths of light compare as an electron falls from these two shells to the $n = 1$ shell.

**4-6** What is the total capacity of the fourth shell from the nucleus?

4-7 What is the total capacity of the sixth shell from the nucleus?

**4-8** An electron in the lithium atom is in the third energy level. Is the electron in the ground or excited state? Can the atom emit light, and if so, how?

## Sublevels and Electron Configurations

**4-9** Which of the following sublevels do not exist? 6*s*, 3*p*, 2*d*, 5*f*, 1*p*, 4*d*, 3*f*

**4-10** What is the electron capacity of each sublevel in the $n = 4$ shell?

4-11 What is the electron capacity of each sublevel in the $n = 3$ and the $n = 2$ shells?

4-12 Explain how a sublevel in one shell can be lower in energy than a sublevel in a lower shell.

**4-13** Theoretically, there should be a fifth sublevel in the $n = 5$ shell. This would be called the *g* sublevel. What is the electron capacity of the *g* sublevel?

**4-14** Which of the following sublevels fills first?

(a) 6*s* or 6*p* (c) 4*p* or 4*f*
(b) 5*d* or 5*p* (d) 4*f* or 4*d*

4-15 Which of the following sublevels is lower in energy?

(a) 6*s* or 5*p* (d) 4*d* or 5*p*
(b) 6*s* or 4*f* (e) 4*f* or 6*d*
(c) 5*s* or 4*d* (f) 3*d* or 4*p*

**4-16** Write the following sublevels in order of increasing energy: 4*s*, 5*p*, 4*p*, 5*s*, 4*f*, 4*d*, 6*s*.

**4-17** Write the total electron configuration for each of the following elements.

(a) Mg (b) Ge (c) Pd (d) Si

4-18 Write the electron configuration that is implied by [Ar].

**4-19** Using the noble gas shorthand, write the electron configuration for each of the following elements.

(a) S (b) Zn (c) Pu (d) I

4-20 Using the noble gas shorthand, write the electron configuration for each of the following elements.

(a) Se (b) Tl (c) Pr (d) Ag

**4-21** What elements have the following electron configurations? Use only the periodic table.

(a) $1s^2 2s^2 2p^5$
(b) $[Ar]4s^2 3d^{10} 4p^1$
(c) $[Xe]6s^2$
(d) $[Xe]6s^2 5d^1 4f^7$
(e) $[Ar]4s^1 3d^{10}$ (exception to rules)

4-22 What elements have the following electron configurations? Use only the periodic table.

(a) $[He]2s^2 2p^6$ (c) $[Ar]4s^1$
(b) $[Xe]6s^2 4f^{14} 5d^3$ (d) $[Kr]5s^2 4d^5$

**4-23** If any of the elements in Problem 4-21 belong to a numerical group (e.g., IA) in the periodic table, indicate the group.

**4-24** What group in the periodic table has the following general electron configuration?

(a) $ns^2np^2$ (*n* is the principal quantum number)
(b) $ns^2np^6$
(c) $ns^1(n-1)d^{10}$ (e.g., $4s^1 3d^{10}$)
(d) $ns^2(n-1)d^1(n-2)f^2$ (two elements)

**4-25** Write the symbol of the element that corresponds to the following.

(a) the first element with a *p* electron
(b) the first element with a filled 3*p* sublevel
(c) three elements with one electron in the 4*s* sublevel
(d) the first element with one *p* electron that also has a filled *d* sublevel
(e) the element after Xe that has two electrons in a *d* sublevel

**4-26** Write the symbol of the element that corresponds to the following.
(a) the first element with a half-filled *p* sublevel
(b) the element with three electrons in a 4*d* sublevel
(c) the first two elements with a filled 3*d* sublevel
(d) the element with three electrons in a 5*p* sublevel

**4-27** Write the general electron configuration for the following groups.
(a) IIA (b) IIB (c) VIA (d) IVB

**4-28** How does He differ from the other elements in group 0?

4-29 Which of the following elements fits the electron configuration $ns^2(n-1)d^{10}np^4$?
(a) Cr (b) Te (c) S (d) O (e) Si

**4-30** What would be the atomic number of the element with one electron in the 5*g* sublevel?

* 4-31 Element number 114 has not yet been made, but scientists believe that it will form a comparatively stable isotope. What would be its electron configuration and in what group would it be?

## Orbitals and Box Diagrams

**4-32** How many orbitals are in the following sublevels?
(a) 2*p* (b) 4*d* (c) 2*d* (d) 3*s* (e) 5*f*

**4-33** How many orbitals are in the $n = 3$ shell?

**4-34** How many orbitals would be in the *g* sublevel? (See Problem 4-13.)

4-35 What is meant by the "shape" of an orbital?

**4-36** Describe the shape of the 4*p* orbitals.

4-37 Describe the shape of the 3*s* orbital.

**4-38** Given the following possible box diagrams for the ground state of nitrogen. One is excluded by the Aufbau principle, one by the Pauli exclusion principle, one by Hund's rule, and one is correct. Indicate the incorrect ones and the principle or rule that is violated.

| | 1*s* | 2*s* | 2*p* | | |
|---|---|---|---|---|---|
| (1) | ↑↓ | ↑↓ | ↑↓ | ↑ | |
| (2) | ↑↓ | ↑↓ | ↑ | ↑ | ↑ |
| (3) | ↑↓ | ↑ | ↑↓ | ↑ | ↑ |
| (4) | ↑↓ | ↑↑ | ↑ | ↑ | ↑ |

**4-39** Write the box diagrams for electrons beyond the noble gas core for the following elements.
(a) S (b) V (c) Br (d) Pm

4-40 Write the box diagrams for electrons beyond the noble gas core for the following elements.
(a) As (b) Ar (c) Tc (d) Tl

**4-41** How many unpaired electrons are in the atoms of elements in groups IIB, VB, VIA, VIIA; an atom with electron configuration given by $[ns^1(n-1)d^5]$; and Pm (at. no. 61)?

4-42 There are three groups in the periodic table that have all of their electrons paired. What are the groups?

A periodic table is almost always in view wherever chemistry is taught or practiced.

# Chapter 5
# The Periodic Nature of the Elements

## Purpose of Chapter 5

In Chapter 5 we describe the origin and meaning of the periodic table, and some trends and properties of the common elements.

## Objectives for Chapter 5

After completion of this chapter, you should be able to:

1. Give a brief discussion of the origin of the periodic table and describe how it was first constructed. (Intro.)
2. Locate on the periodic table those elements existing as gases, solids, and liquids at 25 °C. (5-1)
3. Locate on the periodic table those elements that are metals, nonmetals, and metalloids. (5-1)
4. Locate on the periodic table the elements in the first seven periods. (5-2)
5. Give the characteristics of the electron configuration of the four general categories of elements. (5-2)
6. Locate on the periodic table the seven representative element groups and describe some properties characteristic of each group. (5-2)
7. Give the general trends of the atomic radii of the representative elements. (5-3)
8. Give the general trends in the first ionization energies for the representative elements. (5-4)
9. Predict whether the formation of a specified monatomic cation is feasible. (5-4)
10. Give the general trends in electron affinity for the representative elements. (5-5)
11. Predict whether the formation of a specified monatomic anion is feasible. (5-5)

In a quick trip through any chemistry building or laboratory it is hard not to notice a periodic table or two hanging on a wall. The importance of this table lies not with its use as adornment but with the implicit information it contains. A glance at the table tells us much about the properties of an element just from noting its location in the table. The table is as handy to a chemist as a brush is to a painter.

The existence of related elements has been known for around 200 years, but it wasn't until 1869 that the forerunner of the modern periodic table was first introduced. In that year, two scientists, Dimitri Mendeleev of Russia and Lothar Meyer of Germany, independently presented a table of the elements arranged in order of increasing atomic weights (the concept of atomic number was still unknown). When the elements were ordered in this way, it was found that, for the most part, elements in families appeared in a regular or periodic manner. Mendeleev, however, is usually credited with solving two problems that caused difficulty with this ordering. First, if elements are ordered by increasing atomic weight and elements with similar chemical properties are placed in vertical columns, Mendeleev found that it was necesesary to leave some spaces blank! For example, a space was left under silicon and above tin for a yet undiscovered element, which Mendeleev called "eka-silicon." Later, an element, germanium, was discovered with properties as predicted by the location of the blank space. The second problem involved some misfits if the elements were ordered according to atomic weights. For example, tellurium is heavier than iodine, although Mendeleev realized that iodine clearly belonged under bromine and tellurium under selenium, and not vice versa. Mendeleev simply reversed the order, suggesting that perhaps the atomic weights were in error. We now know, of course, that the elements in the periodic table are listed in order of increasing atomic number instead of atomic weight. We also know that the basis for the periodic nature of the elements lies in their outer electron configurations, as discussed in Chapter 4.

As background for this chapter, you should be familiar with the electron configurations of the elements, especially for any electrons beyond the previous noble gas. This has all been discussed in Chapter 4.

## 5-1 The Physical Properties of the Elements

The 108 known elements come in all physical states and a variety of colors. For example, at room temperature chlorine is a light green gas, bromine is a red-brown liquid, and iodine is a purple crystalline solid.* Carbon is either a black solid (graphite) or a colorless crystal (diamond). In Figure 5-1 the right side of the periodic table is shown in a manner that indicates the

*Most naturally occurring solids exist as crystals that have a defined and regular three-dimensional shape such as a cube (table salt and table sugar) or a tetrahedron (naturally occurring diamond).

Solid

Gas

Liquid

| | | | | | | |
|---|---|---|---|---|---|---|
| | | | | | 1 $H_2$* | 2 He |
| | 5 B | 6 C | 7 $N_2$ | 8 $O_2$ | 9 $F_2$ | 10 Ne |
| | 13 Al | 14 Si | 15 $P_4$ | 16 $S_8$ | 17 $Cl_2$ | 18 Ar |
| 30 Zn | 31 Ga | 32 Ge | 33 As | 34 Se | 35 $Br_2$ | 36 Kr |
| 48 Cd | 49 In | 50 Sn | 51 Sb | 52 Te | 53 $I_2$ | 54 Xe |
| 80 Hg | 81 Tl | 82 Pb | 83 Bi | 84 Po | 85 $At_2$ | 86 Rn |

*$H_2$ indicates diatomic molecules at 25 °C.

**Figure 5-1** THE PHYSICAL STATES OF THE ELEMENTS.

physical state and formula of the elements as they are found at 25 °C. The temperature must be specified, since at very high temperatures all elements are gases and at very low temperatures all elements are liquids or solids. Notice that, under these conditions, eleven of the elements are gases (lightly shaded), two are liquids (darkly shaded), and all the rest are solids, including the elements at the left end of the chart, which is not shown. At 29 °C, however, two solids (cesium and gallium) melt to become liquids.

Perhaps the most basic division of the elements, at least to the chemist, is between metals and nonmetals. You may be familiar with certain metallic properties such as the ability to conduct electricity and heat, but other properties such as hardness and luster are not solely metallic properties. Of more importance to the chemist are the differences in chemical properties between metals and nonmetals. The basis of this fundamental difference will be evident before this chapter is completed.

In the periodic table shown in Figure 5-2, the heavy stairstep line separates the metals (light) from the nonmetals (heavy shade). Elements on the borderline have properties of both classes. For example, silicon is a nonmetal but has some metallic properties. Germanium, which is directly under silicon, is a metal with some nonmetallic properties. Most of these borderline elements are sometimes referred to as metalloids (lightly shaded elements). As you can see, most known elements (79%) are metals. All elements not shown are metals.

## 5-2 Periods and Groups

**Periods** *are horizontal rows of elements in the periodic table. Each period is composed of all of the elements between two noble gases.* The periods and the sublevels that compose each are listed in Table 5-1.

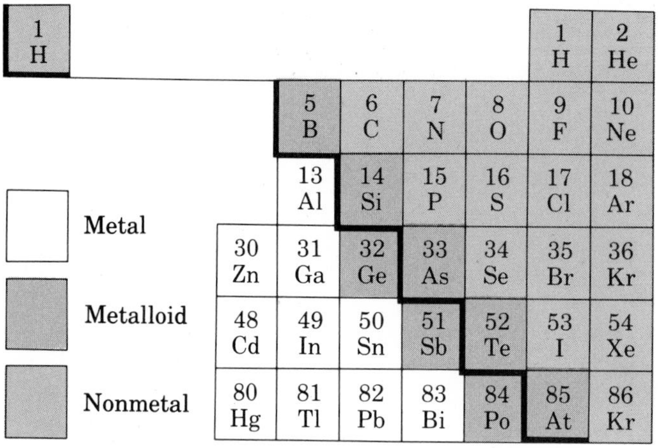

| 1 H | | | | | | | |
|---|---|---|---|---|---|---|---|
| | | | | | | 1 H | 2 He |
| | | 5 B | 6 C | 7 N | 8 O | 9 F | 10 Ne |
| | | 13 Al | 14 Si | 15 P | 16 S | 17 Cl | 18 Ar |
| | 30 Zn | 31 Ga | 32 Ge | 33 As | 34 Se | 35 Br | 36 Kr |
| | 48 Cd | 49 In | 50 Sn | 51 Sb | 52 Te | 53 I | 54 Xe |
| | 80 Hg | 81 Tl | 82 Pb | 83 Bi | 84 Po | 85 At | 86 Kr |

Metal

Metalloid

Nonmetal

**Figure 5-2** METALS, NONMETALS AND METALLOIDS.

**See Problems 5-1 through 5-4.**

**Groups** *of elements* (or **families**) *are located in vertical columns in the periodic table.* Elements in groups have the same number and arrangement of electrons in their outermost sublevel, as was shown in Figure 4-12. The groups of elements can be classified into four general categories.

1 The **representative elements** (also known as **main group elements**) are metals and nonmetals in which *s* and *p* sublevels are being filled. They are shown as the A group elements.

2 The **noble** or **rare gases** (group 0 or VIII) are nonmetals at the end of each period. Except for He all have filled outer *s* and *p* sublevels.

3 The **transition elements** are metals in which the *d* sublevel is being filled. They are shown on the periodic table as the B group elements.

4 The **inner transition elements** are metals in which the *f* sublevel is being filled. The elements in which the 4*f* sublevel is being filled are known as the **lanthanides** or **rare earths** (at. no. 58 to 71) and the elements in which the 5*f* sublevel is being filled are known as the **ac-**

**Table 5-1** Periods

| Period | Number of Electrons | Sublevels | Noble Gas at End of Period |
|---|---|---|---|
| 1 | 2 | 1*s* | He |
| 2 | 8 | 2*s*, 2*p* | Ne |
| 3 | 8 | 3*s*, 3*p* | Ar |
| 4 | 18 | 4*s*, 3*d*, 4*p* | Kr |
| 5 | 18 | 5*s*, 4*d*, 5*p* | Xe |
| 6 | 32 | 6*s*, 5*d*, 4*f*, 6*p* | Rn |
| 7 | 20 | 7*s*, 6*d*, 5*f* | ? |

**See Problems 5-5 and 5-6.**

**tinides** (at. no. 90 to 103). There is very little chemical similarity *between* the lanthanides and actinides, but there is a strong similarity *among* the lanthanides and *among* the actinides.

**See Problems 5-7 and 5-8.**

Since much of the chemistry that will be discussed in this book involves the representative elements, they will be singled out for a formal introduction. Most of these elements will seem like old friends as we get to know them better. We will also have a little more to say about the noble gas elements and the transition metals.

**Group IA** These metals follow the noble gases and are known as the **alkali metals.** They are all soft, shiny metals but are very reactive with both air and water. All of these elements react with group VIIA elements to form ionic compounds with the general formula MF, MCl, MBr, and MI. The properties of this group vary smoothly from top to bottom. (See Figure 5-3.)

Hydrogen, like the other elements in this group, has one electron in an *s* sublevel. However, *hydrogen is not an alkali metal.* It is actually a nonmetal with little in common with the other members of the group other than the formula of some of its compounds (e.g., HCl and LiCl). Hydrogen is a unique element with little periodic relationship to any other element. It is often given a position all to itself in the periodic table.

**Group IIA** The group IIA elements are known as the **alkaline earth metals.** They are not as reactive as the alkali metals, since most of the metals can exist in air (see Figure 5-4). The elements form compounds with oxygen with the general formula MO. With group VIIA elements they have the general formula $MCl_2$, $MBr_2$, and so on. Beryllium chemistry is somewhat unique compared with the chemistry of the rest of the group. Magnesium forms a strong, lightweight **alloy** (*a homogeneous mixture of metals*). It is also used in flashbulbs because of the bright light it gives off when ignited.

**Group IIIA** In group IIIA there is a significant difference between the nonmetal at the top of the group (boron) and the next element, aluminum. Because of its unique chemistry, boron chemistry is usually described separately from that of the other members of the group. Aluminum is an important metal in this group because of its combination of low density and high strength. Aluminum is also the most common metal in the earth's crust (the outer layers including the atmosphere). Compounds of aluminum are components of most clays (see Figure 5-5).

**Group IVA** **Organic chemistry** is a major branch of chemistry that focuses mainly on the millions of compounds formed by carbon, the lightest member of group IVA. The complex organic compounds involved in life processes are studied in another but closely allied branch of chemistry called **biochem-**

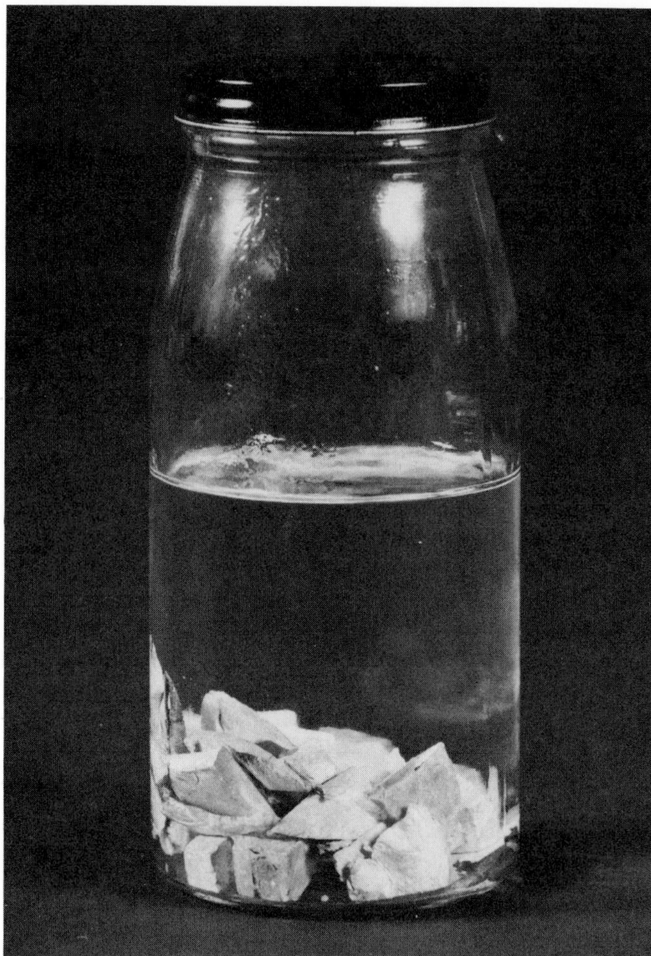

**Figure 5-3** GROUP IA, THE ALKALI METALS. Sodium is a typical alkali metal which must be stored under mineral oil because of its chemical reactivity.

**istry.** Pure carbon occurs in two crystalline forms as an element: graphite and diamond. *Different forms of a pure element are called* **allotropes** (see Figure 5-6).

Silicon and germanium, known as semiconductors, are metalloids and find use in transistors because of their ability to conduct a limited amount of electricity. Silicon, a major component of sand, is also a plentiful element in the earth's crust. Tin and lead have properties typical of other metals.

**Group VA** Group VA shows the most startling change in properties from top to bottom of any group. At the top nitrogen exists as a diatomic gas, $N_2$. It is important as the major component (about 80%) of our atmosphere as well as a component of living organisms. Phosphorus is a nonmetal solid that exists as $P_4$ molecules in a common allotrope, called white phosphorus.

**Figure 5-4** GROUP IIA, THE ALKALINE EARTH METALS. Magnesium is a typical alkaline earth metal. It is not as reactive as the alkali metals and can be stored in the open air. Shown here is a piece of magnesium ribbon.

It is very reactive in air, burning violently with the evolution of large amounts of heat. Arsenic and antimony are metalloids, but bismuth is definitely metallic. (See Figure 5-7.)

**Group VIA** Group VIA (see Figure 5-8) is sometimes (actually very rarely) referred to as the **chalcogens.** As in group VA, there is a large variation in properties from top to bottom. At the top, oxygen has a unique chemistry, and for this reason is usually discussed separately from the other elements in the group (like other elements in the second period). Oxygen is the most

**Figure 5-5** GROUP IIIA. In this group boron (left) is a nonmetal but aluminum (right) and other members of the group are metals.

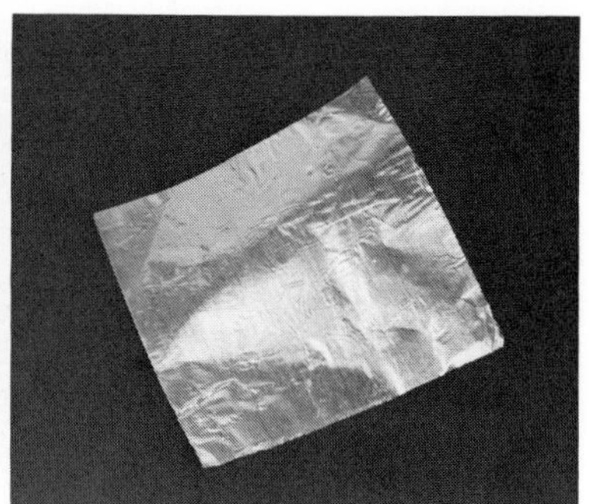

**Figure 5-6** GROUP IVA. Carbon is found in two crystalline allotropes, graphite and diamond (left and center). Tin is a typical metal (right).

abundant element in the earth's crust and makes up about 20% of the atmosphere. It exists mainly as $O_2$ in the air, but there is another important allotrope of oxygen, $O_3$, which is known as **ozone.** Ozone in the upper atmosphere is essential to life on earth, since it absorbs damaging ultraviolet light from the sun. In the lower atmosphere, however, it is a toxic, pungent-smelling pollutant. Oxygen forms compounds either directly or indirectly with all other elements except helium, neon, and argon. Sulfur and selenium are also nonmetals and form many important compounds. Tellurium and polonium are both metalloids, although the chemistry of pollonium has not been studied extensively. It is a highly radioactive and unstable element that occurs in nature only in trace amounts, making it difficult to study.

**Group VIIA** Like the alkali metals, group VIIA elements show a strong chemical similarity from top to bottom. This group, known as the **halogens,**

**Figure 5-7** Phosphorus (left) is a reactive nonmetal and must be stored under water. Bismuth (right) is a typical metal.

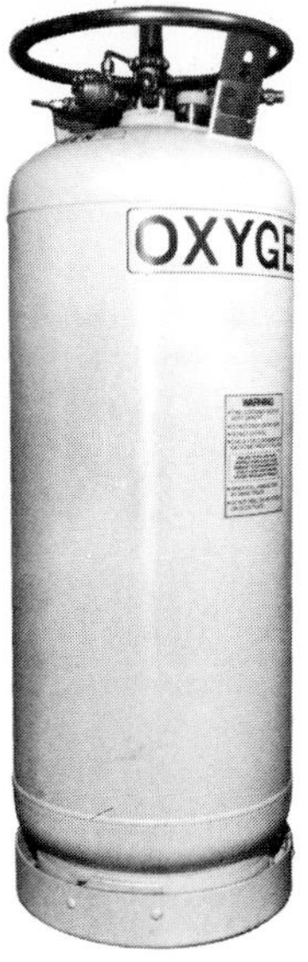

**Figure 5-8** GROUP VIA. The two famous members of this group are oxygen, which is a gas, and sulfur, which is a nonmetal solid.

are all nonmetals and exist as diatomic molecules in the elemental state. Astatine is classified as a metalloid, but because of its radioactivity it is also difficult to study. All of the halogens are very reactive elements. Fluorine is especially reactive and, like oxygen, forms compounds with all other elements except helium, neon, and argon. At room temperature, fluorine and chlorine are gases, bromine is a liquid, and iodine is a solid (see Figure 5-9).

Sometimes hydrogen is also shown in group VIIA as well as IA. However,

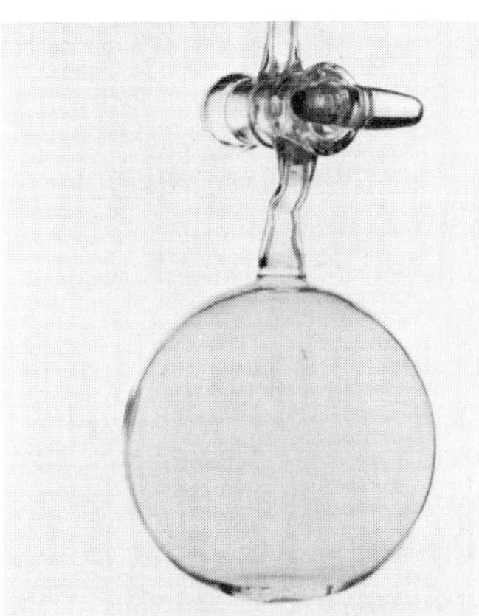

**Figure 5-9** GROUP VIIA, THE HALOGENS. At room temperature chlorine (left) is a gas, bromine (center) is a liquid, and iodine (right) is a solid. The elements of this group are very reactive chemically. (Left, Time/Life Books, Inc.)

hydrogen is not a halogen although it does have some chemical resemblance to the group VIIA elements. Hydrogen can form a negative ion and exists as a diatomic gas, $H_2$. Hydrogen is not nearly as reactive as the halogens, however. It is sometimes placed in group VIIA because, like the halogens, it is the element before a noble gas.

**Group 0** This group is known as the **noble** or **rare gases.** At one time this group was also referred to as the inert gases because it was thought that these elements could not form chemical bonds and existed only as individual atoms. This is still true for He, Ne, and Ar as these elements find use where noncombustible gases are needed. In 1962, a scientist proved that Xe forms compounds with the two most reactive elements, fluorine and oxygen. Later it was discovered that Rn and Kr show similar behavior by forming bonds with oxygen and fluorine. Despite the existence of these compounds the noble gases as a group are still considered the most unreactive elements.

**Transition Elements** Many of our most familiar metals are transition elements (IB–VIIIB) such as chromium, iron, nickel, silver, and gold. There are some chemical similarities between elements in an A group and those in the corresponding B group, such as elements in IIA and IIB. For the most part, however, transition elements are much less reactive than representative metals. The chemistry of transition elements is somewhat more complex than the representative elements, so it is harder to generalize. For example, manganese forms the following compounds with oxygen: MnO, $Mn_2O_3$, $MnO_2$, and $Mn_2O_7$. A discussion of the chemistry of transition elements is beyond the scope of this text.

**See Problems 5-9 through 5-20.**

## 5-3 Periodic Trends: Atomic Radius

The radius of an atom is the distance from the nucleus to the outermost electrons. This distance is in fact neither clearly defined nor easily measured. As mentioned in Chapter 4, modern theory of the atom discourages a model of the atom in which an electron orbits the nucleus at a fixed distance. Instead, the electrons have a certain probability of being at a range of distances from the nucleus. Therefore, the radius does not actually have a fixed value. Besides the theoretical problem, there is a more practical problem. It is possible to determine the distance between the centers of two atoms in a solid, but it is difficult to decide where one atom starts and the other ends. Nevertheless, consistent values for the radii of neutral atoms have been compiled and are listed for the first three rows of representative elements in Figure 5-10. Two units are usually used for such small distances, the nanometer (nm, $10^{-9}$ meter) and the picometer (pm, $10^{-12}$ meter). We will use picometers since the numbers are easier to express (e.g., 37 pm rather than 0.037 nm).

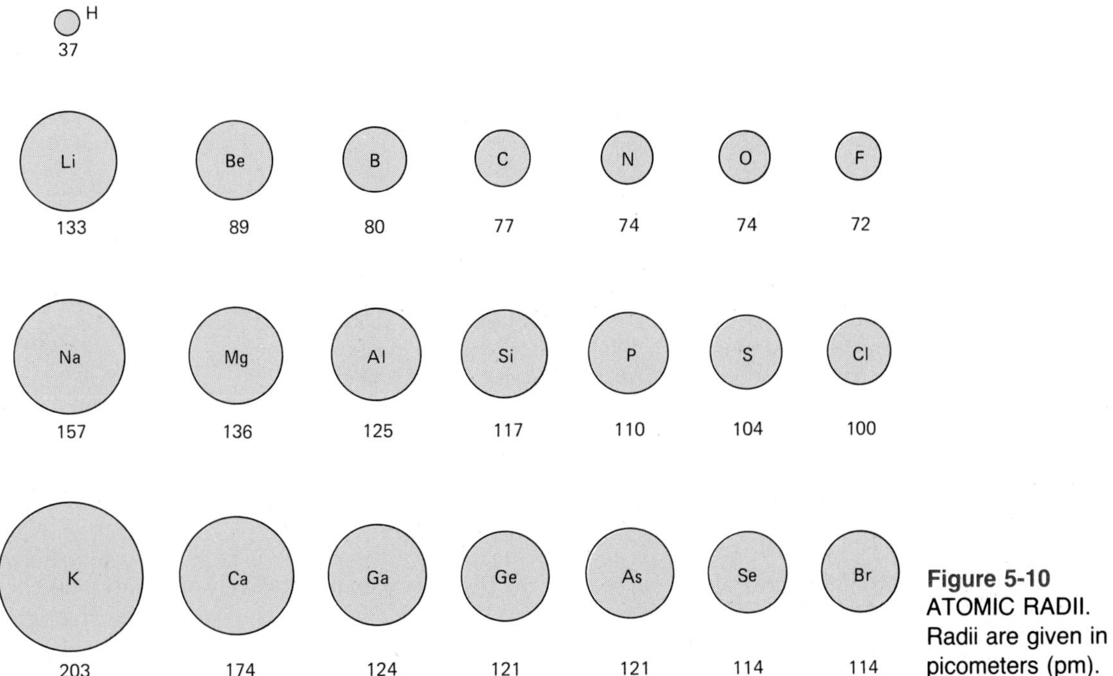


**Figure 5-10** ATOMIC RADII. Radii are given in picometers (pm).

There are two trends that are apparent from the figure.

1 *Horizontally from left to right across a period, the atoms generally get smaller.* Notice that this trend has exceptions (e.g., N and O both have radii of 74 pm). (See Figure 5-10.) How can atoms get smaller despite their having more protons, neutrons, and electrons as we move across a period? Remember that the electrons and the protons in the nucleus have opposite electrical charges and are thus attracted to each other. The stronger the attraction, the more they are pulled together, leading to a smaller atom. Obviously, the outer electrons in a beryllium atom have a greater attraction for the nucleus than the outer electron in the preceding element, lithium.

First let's consider the two inner electrons ($1s^2$) in beryllium and lithium. In both cases these electrons lie inside the outer $2s$ electrons so the two electrons shield or cancel the charge of two protons in the nuclei. The outer $2s$ electron in lithium would thus feel about a $+1$ net charge since the remaining $+2$ charge is canceled by the inner electrons. In the case of beryllium, however, there are two outer electrons that lie in the same sublevel. Since the extra electron in beryllium does not completely shield or cancel the charge of the extra proton, both outer electrons feel about a $+2$ charge on the nucleus. The model used in the previous discussion is somewhat simple compared to the actual situation. For example, more exact calculations of shielding give a value of $+1.3$ for the charge felt by the outer electron in Li and $+1.8$ for each outer electron in Be. In any case, it is

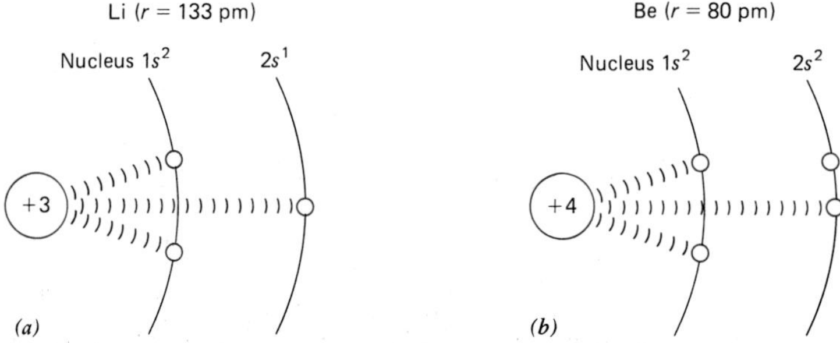


**Figure 5-11** THE RADII OF LITHIUM AND BERYULLIUM. (*a*) One outer 2*s* electron is shielded from the +3 nuclear charge by two inner electrons. The outer electron "feels" about a +3 − 2 = +1 nuclear charge. (*b*) Two outer 2*s* electrons are shielded from the +4 nuclear charge by two inner electrons. The outer electrons "feel" about +4 − 2 = +2 nuclear charge. Thus beryllium is smaller than lithium.

obvious that the outer electrons in Be are more strongly attracted to the nucleus than the outer electron in Li. The electrons that are more strongly attracted are drawn more tightly around the nucleus, making a smaller atom. (See Figure 5-11.)

**2** *Vertically, down a group, the atoms generally get larger.* Each successive element down a group has many more protons (and electrons) than the preceding element. In this case, however, all of the additional electrons lie in an inner shell (energy level). This means that each electron effectively shields or cancels the charge of a proton. Thus each succeeding element feels about the same nuclear charge. The lower element is larger, however, because the outer electron is in a shell (energy level) situated farther from

**See Problems 5-21 through 5-25.**

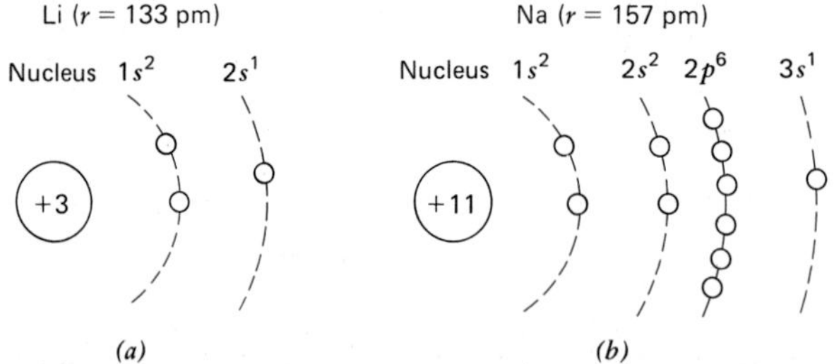


**Figure 5-12** THE RADII OF SODIUM AND LITHIUM. (*a*) One outer 2*s* electron is shielded from +3 charge by two inner electrons. The outer electron "feels" about +3 − 2 = +1 nuclear charge. (*b*) One outer 3*s* electron is shielded from +11 charge by 10 inner electrons. The outer electron "feels" about +11 − 10 = +1 nuclear charge, but it is in a shell further from the nucleus.

the nucleus. This is illustrated in Figure 5-12 in a comparison of the size of lithium and sodium.

In summary, the trends in atomic size are explained by the following facts:

1 Electrons in the same energy level do not entirely cancel or shield an equivalent +1 charge from the other outer electrons. Thus each successive atom feels a greater nuclear charge and is smaller.

2 The outer electrons of elements in the same group feel about the same nuclear charge, but each successive element has electrons in shells farther from the nucleus. Thus each successive element is larger.

## 5-4 Periodic Trends: Ionization Energy

**Ionization energy** *(I.E.) is defined as the energy required to remove an electron from a gaseous atom to form a gaseous ion.* Since the outermost electron is generally the least firmly attached, it is the first to go.

$$M(g) \rightarrow M^{+}(g) + e^{-}$$

This is always an endothermic process (requires energy), since an electron is held in the atom by its attraction to the nucleus. This attractive force must be overcome to separate the electron from the atom, leaving the rest of the atom as a positive ion. *The amount of energy that this process requires depends on how strongly the outermost electron is attracted to the nucleus.* Fortunately, the same reasoning that explains the trends in atomic radii also explains the trends in ionization energy. The ionization energies for the second and third period are shown in Table 5-2. The energy unit abbreviated kJ/mol stands for kilojoules per mole, which is energy per a certain defined quantity of atoms.

Across a period the I.E. generally gets larger, which means increased difficulty in removing an electron. This corresponds to what was discussed in Section 5-3. Since the outer electrons are more strongly attracted to the

**Table 5-2** Ionization Energy of Some Elements

| Element | I.E. (kJ/mol) | Element | I.E. (kJ/mol) |
|---|---|---|---|
| Li | 520 | Na | 496 |
| Be | 900 | Mg | 738 |
| B | 801 | Al | 578 |
| C | 1086 | Si | 786 |
| N | 1402 | P | 1102 |
| O | 1314 | S | 1000 |
| F | 1681 | Cl | 1251 |
| Ne | 2081 | Ar | 1520 |

nucleus, the atoms of each successive element are smaller and the outermost electron harder to remove.

Down a group the I.E. gets smaller for removal of the first electron. As the atoms become larger, the outer electron is less firmly attracted to the nucleus because of the greater distance. Thus the outer electron is easier to remove for each successive element.

*The ease with which an atom forms a positive ion by losing an electron is a fundamental difference between a metal and a nonmetal.* Notice that metallic properties increase to the left and down in the periodic table. This also corresponds to the trend toward lower I.E. Therefore, metals form cations more easily than nonmetals. Nonmetal behavior, to the right and up in the periodic table, is indicated by a comparative difficulty in forming cations. This fundamental trend affects the difference in bonding behavior in the metals and nonmetals, which will be discussed in the next chapter.

The second I.E. for an ion involves the removal of an electron from a +1 ion to form a +2 ion:

$$M^+(g) \rightarrow M^{2+}(g) + e^-$$

In a similar manner, the third I.E. forms a +3 ion and so forth. In all cases it becomes increasingly difficult to remove each succeeding electron. When an electron is removed from an atom, the resulting ion is smaller than the parent atom. (See Figure 5-13.) This is because the remaining electrons do feel a larger share of the nuclear charge. Since the remaining electrons are held more tightly, the ion is smaller and the next electron is harder to remove. The trends in consecutive ionization energies for Na through Al are shown in Table 5-3.

Notice that the second I.E. for Na, the third I.E. for Mg, and the fourth I.E. for Al are all very large compared with the preceding number. For example, it takes 2188 kJ (1450 + 738) to remove the first two electrons from Mg but about three times as much energy to remove the third electron (7732 kJ). In the case of Al, the first three electrons can be removed (to form $Al^{3+}$) at a cost of 5137 kJ, but the fourth would require an input of energy of 11,580 kJ (to form $Al^{4+}$).

There is a good reason for these sudden jumps in energy required to remove a particular electron. It depends on whether the electron is removed from an outer sublevel or a filled inner sublevel. Notice that for Na, the first electron lost comes from the outer 3*s* sublevel but the second must be

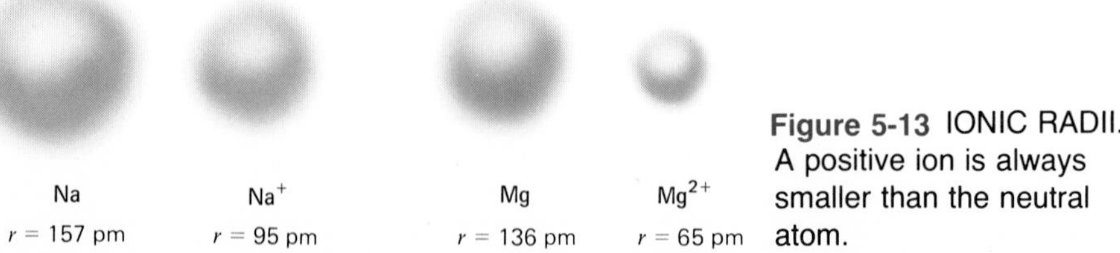


**Figure 5-13** IONIC RADII. A positive ion is always smaller than the neutral atom.

Table 5-3

| Element | First I.E. | Second I.E. | Third I.E. | Fourth I.E. |
|---|---|---|---|---|
| Na | 496 | 4565 | 6912 | 9,540 |
| Mg | 738 | 1450 | 7732 | 10,550 |
| Al | 577 | 1816 | 2744 | 11,580 |

removed from the inner filled 2*p* sublevel. Removal of the first two electrons from Mg to form a $Mg^{2+}$ ion is easy compared with formation of a $Mg^{3+}$ ion, which also would require removal of one electron from the inner 2*p* sublevel. The same argument holds for the difficulty in formation of an $Al^{4+}$ ion. As we will discuss in more detail in the next chapter, *metal cations do exist in ionic compounds.* However, only cations that form with a modest expense in energy are actually capable of existence in compounds. Therefore, a monatomic (one atom) cation is feasible if *(1) it is a metal, and (2) in the case of +2 and +3 ions, an electron was not removed from an inner noble gas core.*

**See Problems 5-26 through 5-33.**

**Example 5-1**
Tell whether the following cations are likely to form and state the reason for your answer. (a) $B^+$ (b) $K^+$ (c) $Sr^{3+}$ (d) $Tl^{3+}$

**Answer**
(a) $B^+$. No. Nonmetals do not form monatomic cations.

(b) $K^+$. Yes. K is a metal, and the one electron removed is the outer 4*s* electron.

(c) $Sr^{3+}$. No. Sr is a metal, but the third electron would have to come from an inner 2*p* sublevel. Elements in Group IIA have only two electrons beyond the noble gas core.

(d) $Tl^{3+}$. Yes. Tl is a metal, and the three electrons removed all come from the outer 6*s* and 6*p* sublevels.

## 5-5 Periodic Trends: Electron Affinity

**Electron affinity** *(E.A.) is defined as the energy released when a gaseous atom adds an electron to form a gaseous ion.*

$$X(g) + e^- \rightarrow X^-(g)$$

The electron affinities of some nonmetals are given in Table 5-4. The trends are somewhat uneven but generally the formation of a negative ion

**Table 5-4** Electron Affinity[a]

| Element | E.A. (kJ/mol) | Element | E.A. (kJ/mol) | Element | E.A. (kJ/mol) |
|---|---|---|---|---|---|
| B | −30 | | | | |
| C | −121 | Si | −140 | | |
| N | +9 | P | −75 | | |
| O | −140 | S | −200 | Se | −160 |
| F | −334 | Cl | −349 | Br | −325 |

[a]A negative value means that the process is exothermic.

becomes more favorable to the right in the periodic table. The *general* trend is to lower electron affinities down a group although we do find that the third period (Si to Cl) is actually higher than the second. Thus a high electron affinity (ease in formation of an anion) is generally a nonmetal property just as a low ionization energy (ease in formation of a cation) is a metallic property. Although it is energetically unlikely to form a nonmetal cation in chemical reactions formation of a nonmetal anion is favorable.

It is also possible that monatomic nonmetal anions can form with a −2 or even a −3 charge (N and P only). Although formation of these ions requires energy, the amount is comparatively small. In order for a nonmetal to add one to three electrons, however, there must be room for the extra electrons in the same outer sublevel as the other electrons. For example, oxygen ([Ne]$2s^2 2p^4$) is a nonmetal with two vacancies in its outer $2p$ sublevel. It can therefore add two electrons to form a −2 anion. If oxygen added a third electron, it would have to advance to the outer $3s$ sublevel. *Nonmetal anions also exist in ionic compounds* as well as metal cations. A monatomic anion is feasible if (1) *it is a representative element nonmetal* and (2) *in the case of −2 and −3 ions the extra electrons can be added to the outer p sublevel.*

**Example 5-2**

Tell whether the following anions are likely and state the reason for your answer. (a) $Ba^-$ (b) $Br^-$ (c) $S^{3-}$ (d) $N^{3-}$

**Answer**

(a) $Ba^-$. No. Metals do not form monatomic anions.

(b) $Br^-$. Yes. Br is a nonmetal and VIIA elements ($ns^2np^5$) all have one vacancy in the outer sublevel.

(c) $S^{3-}$. No. Sulfur is a nonmetal but VIIA elements ($ns^2np^4$) have only two vacancies in the $p$ sublevel.

(d) $N^{3-}$. Yes. Nitrogen is a nonmetal and is a VA element ($ns^2np^3$), which have three vacancies in the $p$ sublevel.

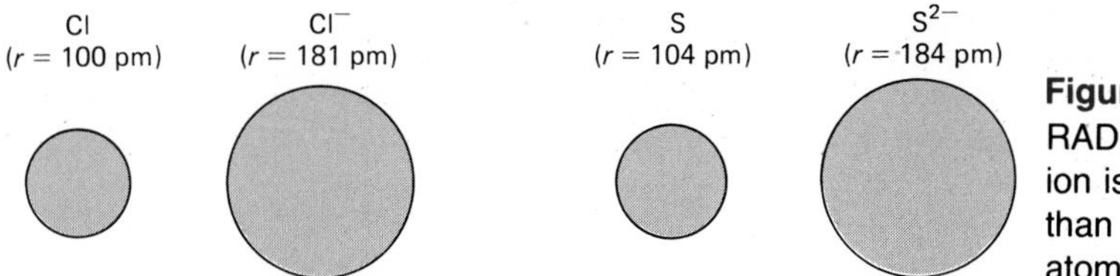


**Figure 5-14** IONIC RADII. A negative ion is always larger than the neutral atom.

We have already seen that cations are smaller than their parent atoms. (See Figure 5-13.) Conversely, anions are considerably *larger* than their parent atoms because the nuclear charge felt by the outer electrons is reduced somewhat because it is spread out over more electrons. (See Figure 5-14.)

**See Problems 5-34 through 5-43.**

## 5-6 Chapter Summary

This chapter presents for your consideration, the pride of chemistry, the periodic table. As your knowledge of chemistry grows, you will appreciate this table more and more as a tool to provide quick information.

Two terms used in the periodic table are periods and groups. A period is a series of elements starting after a noble gas and ending with the next noble gas. Groups are vertical columns of elements. As discussed in the previous chapter, vertical columns are composed of elements (a family) that have the same sublevel electron configuration but different shells ($n$). Some of the information concerning each group of the representative elements (A groups) and the other three categories are summarized as follows.

| Group Number | Outer Sublevel Configuration | Name | Comments |
|---|---|---|---|
| IA | $ns^1$ | Alkali metals | Many properties in common for the group from top to bottom. Very reactive |
| IIA | $ns^2$ | Alkaline earths | Less reactive metals and larger variation in properties from top to bottom than IA |
| IIIA | $ns^2np^1$ | — | Nonmetal at the top, others are metals. Large variation in properties |
| IVA | $ns^2np^2$ | — | Large variation in properties. One nonmetal, two metalloids, and two metals |
| VA | $ns^2np^3$ | — | Wide range in properties among the group. Two nonmetals, two metalloids, and one metal |

| Group Number | Outer Sublevel Configuration | Name | Comments |
|---|---|---|---|
| VIA | $ns^2np^4$ | Chalcogens | Some similarities among nonmetals at the top |
| VIIA | $ns^2np^5$ | Halogens | All nonmetals with many similar properties from top to bottom |
| 0 | $ns^2np^6$ | Noble gases | Nonmetal gases at the end of each period |
| IB–VIIIB | $ns^2(n - 1)d^x$ | Transition elements | Many familiar metals, but complex chemistry |
| — | $6s^25d^14f^x$ | Lanthanides | Horizontal rather than vertical similarities |
| — | $7s^26d^15f^x$ | Actinides | |

Finally three important trends associated with the periodic table were discussed. These are atomic radii, ionization energy, and electron affinity. The consequences of these trends are important, as they indicate the distinctly different behavior of metals and nonmetals towards ion formation. This will be developed further in the next chapter. The information on trends and consequences is summarized as follows.

| | Physical State | Periodic Trends | Consequences |
|---|---|---|---|
| Metals | All solids except for one liquid | 1. Larger radii<br>2. Smaller first ionization energy<br>3. Smaller electron affinity | Can form cations ($M^+$) relatively easy. Can also form $M^{2+}$ and $M^{3+}$ if electrons available outside noble gas core |
| Nonmetals | One liquid, 11 gases, 10 solids | 1. Smaller radii<br>2. Larger first ionization energy<br>3. Larger electron affinity | Can form anions ($X^-$). Can also form $X^{2-}$ and $X^{3-}$ if there is room in the outer *p* sublevel |

## Exercises

### The Periodic Table

**5-1** Referring to Figure 5-1, tell which of the following are gases at room temperature.
(a) Ne (b) S (c) B (d) Cl (e) Br (f) N (g) Na

**5-2** Referring to Figure 5-1, tell which of the following elements exist as diatomic molecules under normal conditions.
(a) N (b) C (c) Ar (d) F (e) H (f) B (g) Xe (h) Hg

**5-3** Referring to Figure 5-2, tell which of the following elements are metals.
(a) Ru (b) Sn (c) Hf (d) Te (e) Ar (f) B (g) Se (h) W

**5-4** Which, if any, of the elements in Problem 5-3 can also be classified as a metalloid?

**5-5** How many elements would be in the seventh period if it were complete?

**5-6** How many elements would be in a complete eighth period? (*Hint:* Consider the 5*g* sublevel.)

**5-7** Classify the following elements into one of the four main categories of elements.
(a) Fe (b) Te (c) Pm (d) La (e) Xe (f) H (g) In

**5-8** Classify the following elements into one of the four main categories of elements.
(a) Se (b) U (c) Ni (d) Sr (e) Zn (f) I (g) Er

**5-9** Why is hydrogen sometimes classified with the IA elements? How is H different from the IA elements? Why is hydrogen sometimes classified with the VIIA elements?

**5-10** Identify the following elements using the periodic table and the information in Figures 5-1 and 5-2.
(a) a nonmetal, monatomic gas in the third period
(b) a transition metal that is a liquid
(c) a diatomic gas in group VA
(d) a metalloid in the fourth period that has three electrons in its outer sublevel
(e) the second metal in the second period
(f) the only member of a group that is a metal

**5-11** Identify the following elements using the periodic table and the information in Figures 5-1 and 5-2.
(a) a nonmetal, diatomic liquid
(b) the last element in the third period
(c) a metalloid that is more metal than nonmetal and has only two electrons in an outer *p* sublevel
(d) a nonmetal, diatomic solid
(e) a metal with one *p* electron and no *d* electrons

* **5-12** Assume that elements heavier than element number 106 will eventually be made. If so, what would be the atomic number and group number of the next nonmetal?

**5-13** Which of the following elements are halogens?
(a) $O_2$ (b) $P_4$ (c) $I_2$ (d) $N_2$ (e) Li (f) $H_2$ (g) $Br_2$

**5-14** Which of the following elements are alkaline earth metals?
(a) Sr (b) C (c) B (d) Be (e) Na (f) K

**5-15** Which of the following elements are transition elements?
(a) In (b) Ti (c) Ca (d) Xe (e) Pd (f) Eu (g) Ag

**5-16** Which of the following elements is the most chemically reactive?
(a) $O_2$ (b) $Cl_2$ (c) Na (d) $F_2$ (e) Xe

**5-17** Which of the following elements does not form chemical bonds?
(a) Au (b) $F_2$ (c) Ne (d) $O_2$ (e) Xe (f) Li

**5-18** Which of the following is an allotrope of oxygen?
(a) O (b) $O^{2-}$ (c) $H_2O$ (d) $O_3$ (e) $O_4$

**5-19** Which of the following is an allotrope of carbon?

(a) diamond (b) ozone (c) bronze
(d) carbon monoxide

* **5-20** It has been argued that the elements Lu and Lr should be placed in group IIIB rather than La and Ac because of a better fit of their chemical properties to the other IIIB elements, Sc and Y. Do Lu and Lr have the proper electron configurations for group IIIB? What is the one difference in the electron configuration of La and Lu?

### Periodic Trends

**5-21** In each of the following pairs of elements, which has the largest radius?
(a) As or Se (b) Ru or Rh
(c) Sr or Ba (d) F or I

**5-22** In each of the following pairs of elements, which has the largest radius?
(a) Tl or Pb (b) Sc or Y (c) Pr or Ce
(d) P or Br

**5-23** Four elements have the following radii: 117 pm, 122 pm, 129 pm, and 134 pm. The elements are V, Cr, Nb, and Mo. Assign the proper radius to the element.

**5-24** Four elements have the following radii: 180 pm, 154 pm, 144 pm, and 141 pm. The elements are In, Sn, Tl, and Pb. Assign the proper radius to the element.

* **5-25** Zirconium and hafnium are in the same group and have almost the same radius despite the general trends. As a result the two elements have almost identical chemical and physical properties. The fact that these elements have almost the same radius is due to the *lanthanide contraction.* With the knowledge that atoms get progressively smaller as a sublevel is being filled, can you explain this phenomenon? (*Hint:* Follow all of the expected trends between the two elements.)

**5-26** Which of the following elements has the higher ionization energy?
(a) Ti or V (b) P or Cl (c) Mg or Sr
(d) Fe or Os (e) B or Br

**5-27** Which of the following elements has the higher ionization energy?
(a) B or Al (b) Ba or Cs (c) Ga or S
(d) As or Sb

**5-28** Four elements have the following first ionization energies in kJ/mol: 762, 709, 579, and 558. The elements are Ga, Ge, In, and Sn. Assign the proper ionization energy to the element.

**5-29** Four elements have the following first ionization energies in kJ/mol: 869, 941, 1010, and 1140. The elements are Se, Br, Te, and I. Assign the proper ionization energy to the element.

**5-30** The first four ionization energies for Ga are 578.8 kJ (per mol), 1979 kJ, 2963 kJ, and 6200 kJ. How much energy is required to form each of the following ions: $Ga^{+}$, $Ga^{2+}$, $Ga^{3+}$, and $Ga^{4+}$? Why does the formation of $Ga^{4+}$ require a comparatively large amount of energy?

**5-31** The first five ionization energies for carbon are 1086 kJ (per mol), 2353 kJ, 4620 kJ, 6223 kJ, and 37,830 kJ. How much energy is required to form the following ions: $C^{+}$, $C^{2+}$, $C^{3+}$, $C^{4+}$, and $C^{5+}$. In actual fact even $C^{+}$ does not form in compounds. Compare the energy required to form this ion with that needed to form some metal ions and explain.

**5-32** Tell which of the following monatomic cations are feasible and may exist in compounds. If an ion is not likely, give the reason.
(a) $Cs^{+}$ (b) $Rb^{2+}$ (c) $Ne^{+}$
(d) $Tl^{4+}$ (e) $Sc^{3+}$ (f) $Te^{2+}$

**5-33** Tell which of the following monatomic cations are feasible and may exist in compounds. If an ion is not likely, give the reason.
(a) $Sc^{4+}$ (b) $Se^{+}$ (c) $Ca^{2+}$
(d) $Rb^{+}$

**5-34** Tell which of the following monatomic anions are feasible and may exist in compounds. If an ion is not likely, give the reason.
(a) $F^{2-}$ (b) $Se^{2-}$ (c) $Be^{-}$ (d) $N^{3-}$

**5-35** Tell which of the following monatomic anions are feasible and may exist in compounds. If an ion is not likely, give the reason.
(a) $I^{-}$ (b) $Bi^{3-}$ (c) $K^{-}$ (d) $Te^{2-}$

**5-36** Which of the following atoms would most easily form a cation?
(a) B (b) Al (c) Si (d) C

**5-37** Which of the following atoms would not be able to form a 2+ cation?
(a) Sr (b) Li (c) B (d) Ba

**5-38** Which of the following would most likely form an anion?
(a) Be (b) Al (c) Ga (d) I

**5-39** Which of the following atoms would not be likely to form a 2− anion?
(a) Se (b) Po (c) Cl (d) H

**5-40** Noble gases form neither anions or cations. Why?

**5-41** Arrange the following ions and atoms in order of increasing radii.
(a) Mg (b) $S^{2-}$ (c) S (d) $Mg^{2+}$
(e) $K^+$ (f) $Se^{2-}$

**5-42** Arrange the following ions and atoms in order of increasing radii.
(a) Br (b) $K^+$ (c) K (d) $I^-$
(e) $Br^-$ (f) $Ca^{2+}$

* **5-43** On the planet Zerk, the periodic table of elements is slightly different from ours. On Zerk, there are only two *p* orbitals, so a *p* sublevel holds only four electrons. There are only four *d* orbitals, so a *d* sublevel holds only eight electrons. Everything else is the same as on earth, such as the order of filling (1*s*, 2*s*, etc.) and what is characteristic of noble gases, metals, and nonmetals. Construct a Zerkian periodic table using numbers for elements up to element number 50. Then answer these questions.

(a) How many elements are in the second period? In the fourth period?
(b) What are the atomic numbers of the noble gases at the ends of the third and fourth periods?
(c) What is the atomic number of the first inner transition element?
(d) Which element is most likely to be a metal: element number 5 or element number 11; element number 17 or element number 27?
(e) Which element has the largest radius: element number 12 or element number 13; element number 6 or element number 12?
(f) Which element has a higher ionization energy: (1) element number 7 or element number 13; (2) element number 7 or element number 5; (3) element number 7 or element number 9?
(g) Which ions are reasonable?
(1) $16^+$ (2) $9^{2+}$ (3) $7^+$ (4) $13^-$
(5) $17^{4+}$ (6) $15^+$ (7) $1^-$

Both of these compounds, water and salt, are essential to life yet they are fundamentally different types of compounds.

# Chapter 6
# The Nature of the Chemical Bond

## Purpose of Chapter 6

In Chapter 6 we describe how the electron configuration of an element influences the type of bond it forms and the formulas of its compounds.

## Objectives for Chapter 6

After completion of this chapter, you should be able to:

1. Write Lewis dot structures for the atoms of any representative element. (6-2)
2. Demonstrate the octet rule by determining the charge acquired by a specified atom of a representative element. (6-1, 6-3)
3. Write the formula of binary ionic compounds formed by representative elements. (6-4)
4. Demonstrate the octet rule by determining the formulas of some simple binary covalent compounds. (6-1, 6-5)
5. Distinguish between single, double, and triple covalent bonds. (6-7)
6. Determine the number of valence electrons present in a polyatomic ion. (6-8)
7. Write Lewis structures for a variety of molecules and ions. (6-9)
8. Write equivalent resonance structures for specified molecules or ions. (6-10)
9. Give the general periodic trends in electronegativity. (6-11)
10. Determine the direction of a bond dipole from the electronegativities of the elements. (6-11)
11. Predict ionic bonding from differences in electronegativities. (6-11)
12. Predict polarity of molecules from a knowledge of the molecular geometry. (6-12)

Water and table salt are two of the most common (and essential) compounds in our lives. But why are these particular compounds what they are? Why does the formula for water happen to be $H_2O$ but not $H_3O$ or $HO_2$? Why do crystals of salt consist of ions but crystals of ice consist of molecules? Why do the elements in both of these substances form compounds but some other elements have little or no tendency to form chemical bonds at all? These are fundamental questions concerned with the bonding of elements to form compounds. The answers to these questions are very logical if we look to the basis for bond formation, which lies in the configuration of the electrons in the atom. This chapter describes how electron configuration helps explain and even predict the types of bonds formed between elements.

The background necessary for this chapter includes:

1 The nature of ionic and molecular compounds (Sections 3-2 and 3-3)

2 Electron configuration of the elements (Section 4-6)

3 The relationship of electron configuration to the periodic table (Section 4-6)

4 The formation of cations (Section 5-5) and the formation of anions (Section 5-6)

## 6-1 Bond Formation and the Representative Elements

In a way, people are much like the atoms of which they are made. Both people and atoms are generally social creatures, forming bonds with others because of a mutual need for a more stable arrangement. Unlike most people, however, there is a family or group of elements that have little or no tendency to join with others to form bonds. As mentioned previously, these are the noble gas elements, which all exist as individual gaseous atoms under normal conditions.

Elements that form chemical bonds do so through some alteration of their atoms' electron configuration. Since noble gases form few chemical bonds, it seems reasonable to assume that there is a special stability connected with their electron configuration as it is. As we will find in this chapter, *representative elements (and a few transition elements) form many bonds in such a manner that the electron configuration of their atoms is altered to become like that of a noble gas element.*

In Chapter 4 we showed that noble gas elements are characterized by filled outer *s* and *p* sublevels (i.e., $ns^2np^6$). Since this is a total of eight electrons (except for He), it is referred to as an **octet** of electrons. The types of bonds and formulas of many compounds of the representative elements can be understood and even predicted by reference to the octet rule. The **octet rule** *states that atoms of the representative elements form bonds in* *such a manner so as to have access to eight outer electrons.* The outer *s* and *p* electrons in the atoms of an element are referred to as the **valence electrons.** Representative elements that border helium in the periodic table are influenced by a "duet" rule. These elements (H, Li, and Be) can alter their electron configuration to be like helium, which was only two outer electrons.

An octet of valence electrons can be obtained in three ways.

1 A metal may lose one to three electrons to form a cation with the electron configuration of the previous noble gas.

2 A nonmetal may gain one to three electrons to form an anion with the electron configuration of the next noble gas.

3 Atoms (usually two nonmetals) may share electrons with other atoms to obtain access to the number of electrons in the next noble gas.

We have already discussed the relative ease of formation of cations and anions in Chapter 5. The formation of cations and the formation of anions thus complement each other to form ionic compounds. Case 3 produces molecular compounds in which atoms are held together by covalent bonds.

## 6-2 Lewis Dot Structures

The bonding of elements of the representative elements focuses mainly on the valence electrons, which are those electrons in the outer *s* and *p* sublevels. **Lewis dot structures*** *of these elements represent the valence electrons as dots around the symbol of the element.* Since bonding of these elements involves access to eight electrons (four pairs), the electrons are represented as one or two dots on the four sides of the element's symbol. Although the valence electrons come from two different sublevels (*s* and *p*), only the total number of these electrons is important for discussions of bonding. Thus it is convenient to place one electron on each side of the symbol first (groups IA through IVA) and then represent pairs of electrons (group VA through 0).† In this manner, the Lewis dot structures of the first four periods are shown in Figure 6-1. Notice that the group number also represents the number of dots or outer electrons for a neutral atom of the representative elements.

*Named after the American chemist G. N. Lewis (1875–1946), who developed this theory of bonding.

†This procedure of representing the *s* electrons as unpaired for bonding purposes has a basis in fact. For example, although we learned in Chapter 4 that carbon has two paired electrons ($2s^2$) and two unpaired electrons (in $2p$ orbitals), it forms bonds as if all four valence electrons were in equivalent or what are referred to as "hybridized" orbitals.

**See Problems 6-1 through 6-4.**

| IA | IIA | IIIA | IVA | VA | VIA | VIIA | O |
|---|---|---|---|---|---|---|---|
| H | | | | | | | He |
| Li | Be· | B· | ·C· | ·N· | ·O: | :F: | :Ne: |
| Na | Mg· | Al· | ·Si· | ·P· | ·S: | :Cl: | :Ar: |
| K | Ca· | Ga· | ·Ge· | ·As· | ·Se: | :Br: | :Kr: |

Figure 6-1
DOT STRUCTURES.

## 6-3 The Formation of Ions

In Section 5-5 the formation of positive ions was discussed. When one electron is removed from a sodium atom, it was found that the process does not require a large amount of energy compared with other ionizations such as the removal of an electron from a nonmetal or even a second electron from sodium. As shown below, using a dot structure for Na, the electron that is lost is the outer $3s^1$ electron, leaving a cation ($Na^+$) that has an octet of electrons, which is the same electron configuration of the previous noble gas, neon.

$$\dot{\mathrm{Na}} \rightarrow \mathrm{Na}^+ + e^-$$

$$[\mathrm{Ne}]3s^1 \qquad [\mathrm{Ne}]$$

All of the atoms of the group IA *metals* can lose one electron to form a +1 ion with the electron configuration of the preceding noble gas.

The element magnesium can lose two electrons, to leave an ion with the electron configuration of Ne:

$$\dot{\mathrm{Mg}}\cdot \rightarrow \mathrm{Mg}^{2+} + 2e^-$$

$$[\mathrm{Ne}]3s^2 \qquad [\mathrm{Ne}]$$

All other metals in group IIA can lose two electrons to form a +2 ion with the configuration of the preceding noble gas.

Group IIIA *metals* can lose three electrons to form a +3 ion:*

$$\dot{\underset{\cdot}{\mathrm{Al}}}\cdot \rightarrow \mathrm{Al}^{3+} + 3e^-$$

$$[\mathrm{Ne}]3s^23p^1 \qquad [\mathrm{Ne}]$$

*Ions such as $Tl^{3+}$, $Ga^{3+}$, and $Zn^{2+}$ have a filled *d* sublevel in addition to a noble gas configuration. This is sometimes referred to as a pseudo-noble gas configuration. The filled *d* sublevel does not seem to affect the stability of these ions. In this text we will not distinguish between a noble gas and pseudo-noble gas electron configuration. Transition metals also form positive ions, but for the most part these ions do not relate to a noble gas configuration. Some of these ions will be discussed in Chapter 7.

In this group, boron is *not* a metal and does not form a $+3$ ion. Its bonds are covalent in nature and will be discussed later.

Four electrons would have to be removed from a group IVA atom to attain a noble gas configuration. As mentioned in the previous chapter, each successive electron is harder to remove. Therefore, metal ions with a $+4$ charge are not generally considered to exist in ionic compounds because of the large energy requirement in their formation. The elements Ge, Sn, and Pb bond either by electron sharing or by forming a $+2$ ion that does not follow the octet rule.

In summary, *metals of the representative elements may lose up to three electrons to form ions with a noble gas electron configuration.*

Positive ions do not form independently, as the lost electrons must go somewhere. The electrons are added by atoms that form negative ions. In Chapter 5 we learned that this process is more favorable for nonmetals.

Atoms of group VIIA elements (including H) are all one electron short of a noble gas configuration. If they add one electron they will then have an octet of electrons or the same electron configuration of the next noble gas. The resulting ion will then have a $-1$ charge. (Remember that hydrogen, an exception to the octet rule, needs only a total of two electrons to achieve the electron configuration of He.)

$$e^- + \mathrm{H}\cdot \rightarrow \mathrm{H}:^-$$
$$1s^1 \qquad 1s^2 = [\mathrm{He}]$$

$$e^- + \ :\ddot{\underset{\cdot}{\mathrm{Cl}}}: \rightarrow \ :\ddot{\underset{\cdot\cdot}{\mathrm{Cl}}}:^-$$
$$[\mathrm{Ne}]3s^23p^5 \quad [\mathrm{Ne}]3s^23p^6 = [\mathrm{Ar}]$$

Group VIA nonmetals can attain a noble gas configuration by adding two electrons to form a $-2$ ion:

$$2e^- + \ \cdot\ddot{\underset{\cdot}{\mathrm{O}}}: \rightarrow \ :\ddot{\underset{\cdot\cdot}{\mathrm{O}}}:^{2-}$$
$$[\mathrm{He}]2s^2p^4 \qquad [\mathrm{He}]2s^22p^6 = [\mathrm{Ne}]$$

In group VA, nitrogen and phosphorus can add three electrons to form $-3$ ions:

$$3e^- + \ \cdot\ddot{\underset{\cdot}{\mathrm{N}}}\cdot \rightarrow \ :\ddot{\underset{\cdot\cdot}{\mathrm{N}}}:^{3-}$$
$$[\mathrm{He}]2s^22p^3 \qquad [\mathrm{He}]2s^22p^6 = [\mathrm{Ne}]$$

For the most part, group IVA nonmetals bond by electron sharing instead of ion formation. Although there is some evidence for the $C^{4-}$ ion, formation of such a highly charged ion would be an energetically unfavorable process.

In summary, *nonmetals may add up to three electrons to form ions with a noble gas electron configuration.*

**See Problems 6-5 through 6-16.**

## 6-4 Formulas of Binary Ionic Compounds

Metals part with electrons with an input of energy (ionization energy), and nonmetals add an electron with the evolution of energy (electron affinity). Although the energy involved in the overall process involves more than just the ionization energy of the metal and the electron affinity of the nonmetal, you can appreciate the basis for a compatible marriage between metals and nonmetals. For example, the combination of sodium and chlorine produces a **binary** (*composed of two elements*) ionic compound:

$$Na + \cdot\ddot{\underset{..}{Cl}}: \rightarrow Na^+ :\ddot{\underset{..}{Cl}}:^-$$

Formula = NaCl

The transfer of the one electron results in two ions with equal but opposite charges. Both ions have noble gas electron configurations. When lithium combines with oxygen, however, two lithium atoms are required to satisfy the need of oxygen for two electrons. The anion formed has a $-2$ charge from the added electrons. Notice that the resulting compound is electrically neutral, as all compounds must be. The two $+1$ ions balance the $-2$ charge on the oxygen [i.e., $2(+1) - 2 = 0$]. The chemical formula for the compound formed is $Li_2O$.

$$2\,Li\cdot + \cdot\ddot{\underset{..}{O}}: \rightarrow 2(Li^+)\ :\ddot{\underset{..}{O}}:^{2-}$$

Formula = $Li_2O$

When calcium combines with bromine, the opposite situation exists. Two Br atoms are needed to take up the two electrons from one Ca.

$$\cdot Ca\cdot + 2\,\cdot\ddot{\underset{..}{Br}}: \rightarrow Ca^{2+}2\left(:\ddot{\underset{..}{Br}}:^-\right)$$

Formula = $CaBr_2$

When aluminum combines with oxygen, it is somewhat more complex to follow the transfer of the three electrons of aluminum onto the oxygens, which need two each. In this case, two Al's and three O's are needed to achieve a charge balance of zero and to satisfy the octet rule.

$$2\,Al + 3\,\ddot{\underset{..}{O}}: \rightarrow 2(Al^{3+})3\left(:\ddot{\underset{..}{O}}:^{2-}\right)$$

Formula = $Al_2O_3$

To predict the formulas of binary ionic compounds, only the group numbers of the two elements are needed to give the charges on the respective ions. First, write the appropriate ions, including the charges adjacent to each other (IA, 1+; IIA, 2+; IIIA, 3+; VA, 3−; VIA, 2−; VIIA, 1−). The numerical value of the charge becomes the subscript of the other element as shown by the arrows. Notice that this method predicts the formula of the compound containing calcium and oxygen to be $Ca_2O_2$. Formulas of ionic compounds, however, should be expressed with the simplest whole numbers for the subscripts. The formula is expressed as CaO.

$$Na^{1+}\ N^{3-} = Na_3N$$

$$Ga^{3+}\ S^{2-} = Ga_2S_3$$

$$Ca^{2+}\ O^{2-} = Ca_2O_2 = CaO \quad \text{(write the simplest formula)}$$

**Example 6-1**

What is the formula of the ionic compound formed between

(a) aluminum and fluorine and (b) barium and sulfur?

**Solution**

(a) Aluminum is in group IIIA and fluorine is in group VIIA, and they have the dot structures

$$\cdot\dot{\text{Al}}\cdot \qquad \cdot\ddot{\underset{\cdot\cdot}{\text{F}}}\text{:}$$

To have a noble gas configuration (an octet) the Al, a metal, must lose all three outer electrons to form a +3 ion. Three fluorine atoms are needed to add one electron each to form three −1 ions. Notice that each fluorine can add only one electron, which gives the $F^-$ ion an octet. The compound formed is

$$Al^{3+}3(:\ddot{\underset{\cdot\cdot}{F}}:^-) = \underline{\underline{AlF_3}}$$

We could also determine the formula from the charges of the respective ions formed. Al becomes $Al^{3+}$ (group IIIA) and F becomes $F^{1-}$ (group VIIA).

$$Al^{+3}\ F^{-1} = \underline{\underline{AlF_3}}$$

(b) Barium is in group IIA and sulfur is in group VIA, and they have the dot structures

$$\dot{\text{Ba}}\cdot \qquad \ddot{\underset{\cdot\cdot}{\text{S}}}\text{:}$$

One Ba atom gives up two electrons and one S atom takes up two electrons, forming the compound

$$Ba^{2+}:\ddot{\underset{\cdot\cdot}{S}}:^{2-} = \underline{\underline{BaS}}$$

We could also determine the formula from the charges of the respective ions formed. Ba becomes $Ba^{2+}$ and S becomes $S^{2-}$

$$Ba^{2+}\ S^{2-} = Ba_2S_2 = \underline{\underline{BaS}}$$

Besides single atoms, two or more atoms can also bond together by covalent bonds with the whole unit behaving as an ion (e.g., $CO_3^{2-}$). These are called **polyatomic ions** and are discussed later in this chapter.

As mentioned in Chapter 3, ionic compounds do not consist of discrete molecular units. That is, in a crystal of a compound such as NaCl one $Na^+$ ion is not attached to any one $Cl^-$ ion. As shown in Figure 6-2, each $Na^+$ ion is surrounded by six $Cl^-$ ions and each $Cl^-$ ion is surrounded by six $Na^+$ ions in a three-dimensional arrangement called a **lattice**. To understand the relative sizes of the ions in Figure 6-2 recall from Chapter 5 that cations are smaller than their parent atoms and that anions are larger than their parent atoms. Thus anions are considerably larger than cations. The electrostatic forces of attraction between adjacent oppositely charged ions are known as ionic bonds and hold the crystal together. Like NaCl, *ionic compounds are solids at room temperature.*

## 6-5 The Covalent Bond

As we mentioned, ionic bonds form between two elements that are complementary. One element has a strong attraction for electrons (high electron affinity), and the other has a weak grasp on its outer electrons (low ionization energy). The strong dominates the weak, and the electrons are exchanged. But what if neither of the atoms gives up electrons easily? Bonds still form, but another suitable arrangement must be made to attain a noble gas configuration for the elements involved. If one element is not strong enough to take the electrons away from the other, the alternative arrangement is to share electrons.

When fluorine combines with a metal, it can reach an octet by complete removal of an electron from the metal to form $F^-$. If it combines with another F atom with the same electron affinity and ionization energy, a sharing arrangement is necessary:

$$:\ddot{\underset{\cdot\cdot}{F}}\cdot \; \cdot\ddot{\underset{\cdot\cdot}{F}}: \rightarrow \quad :\ddot{\underset{\cdot\cdot}{F}}:\ddot{\underset{\cdot\cdot}{F}}:$$

Shared pair of electrons
(one from each F)

See Problems 6-17 through 6-21.

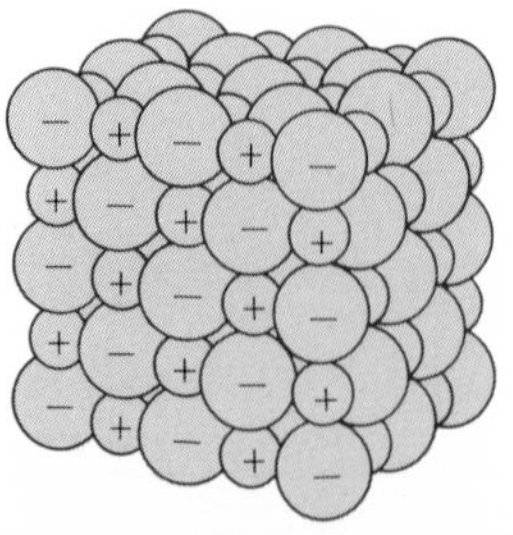


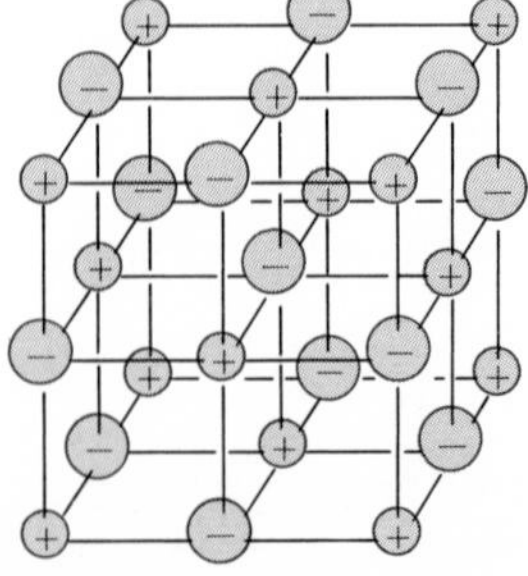


**Figure 6-2** AN IONIC SOLID. Each cation (small spheres) is surrounded by six anions. Each anion (large spheres) is surrounded by six cations.

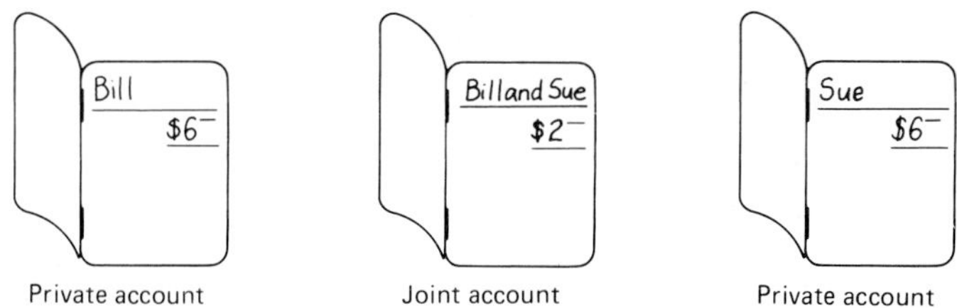


**Figure 6-3** INCREASE YOUR MONEY BY SHARING. The money in a joint account is analogous to electrons in a covalent bond.

The two fluorine atoms attain an octet of electrons by means of a **covalent bond,** *which is a shared pair of electrons between two atoms.* (In this case, one electron in the shared pair originates from each F.*)

There is a simple analogy to the covalent bond. A husband and wife each have $7, but both feel the need to tell the world that they have access to $8, no more, no less. They will have to cooperate to accomplish the goal. Each must hold $6 in a private bank account and each must contribute $1 toward a $2 joint bank account, which is shared. Both can then truthfully say that they *have attained access* to exactly $8 (see Figure 6-3).

Just as Bill and Sue each have access to $8, each fluorine atom has access to eight electrons—six of its own and two shared in the bond.

The concept of Lewis dot structures can now be extended to include molecules. There are several variations of how Lewis structures are represented for molecules. Pairs of electrons are sometimes represented as a pair of dots (:) or as a dash (—). In this text, we will use a pair of dots to represent unshared pairs (called lone pairs) of electrons on an atom and a dash to represent a pair of electrons shared between atoms. In this way the two different environments of electrons (shared and unshared) can be distinguished.

By writing Lewis structures in accordance with the octet rule, we will be able either to (1) justify the formula of a given compound or (2) predict the formula of a compound between two elements. (Although we will not pursue this topic to a large extent in this text, the shape or geometry of most simple molecules can also be predicted from their Lewis structures.)

The Lewis structure of $F_2$ is illustrated as follows.

Total of 14 outer electrons (7 from each F)

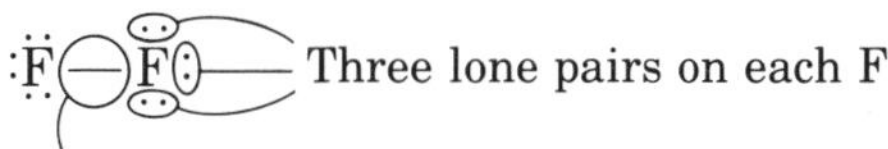


*An F atom has on unpaired electron in a $2p$ orbital. Formation of a covalent bond pairs the electrons in the two F atoms so that the $F_2$ molecule has all electrons paired. Although most atoms of the representative elements have unpaired electrons, most molecules or ions formed from these elements do not have unpaired electrons.

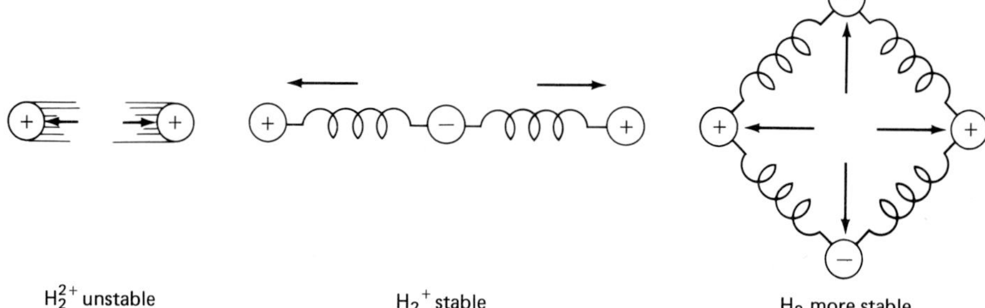


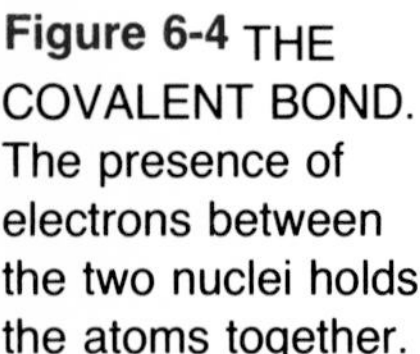
**Figure 6-4** THE COVALENT BOND. The presence of electrons between the two nuclei holds the atoms together.

All of the halogens exist as diatomic molecules for the same reason as does $F_2$. Hydrogen, which forms the simplest of all molecules, also exists as a diatomic gas with one covalent bond between atoms:

$$H—H$$

Notice again that hydrogen does not follow the octet rule. Its outer sublevel holds only two electrons, which is the electron configuration of He.

How does the covalent bond hold the two hydrogen atoms together? Taken alone, the two nuclei (represented as $H_2^{2+}$ in Figure 6-4) would repel each other because of their like charges (→ represents a repulsive force). The presence of one electron (with a negative charge) between the two hydrogen nuclei (represented as $H_2^+$) changes the situation. The nuclei are now held together (at a certain distance) by the two nucleus-electron attractions. These two attractions are stronger than the nucleus-nucleus repulsion. [(ꝏꝏ) represents an attractive force in Figure 6-4]. If there are two electrons between the two hydrogen atoms as in the $H_2$ molecule, the nuclei are held together even tighter by the *four* nucleus–electron attractions even though the second electron adds an electron–electron repulsion.

## 6-6 Simple Binary Molecules

The formulas of some simple binary molecules can be predicted by writing out their Lewis structures and applying the octet rule. A compound between hydrogen and fluorine is correctly predicted to have the formula HF, since each atom is only one electron short of a noble gas structure.

$$H\cdot \; :\ddot{\underset{\cdot\cdot}{F}}\cdot \rightarrow H—\ddot{\underset{\cdot\cdot}{F}}:$$

The simplest compound between hydrogen and oxygen has the formula $H_2O$:

$$H\cdot \quad H\cdot \quad \cdot\ddot{\underset{\cdot}{O}}: \rightarrow \begin{array}{c} H—\ddot{O}: \\ | \\ H \end{array}$$

In a similar manner, the Lewis structures of the simplest N, H and C, H compounds are shown below. (See also Figure 6-5.)

$$3\,H\cdot + \cdot\ddot{N}\cdot \rightarrow H{-}\ddot{N}(H){-}H \qquad 4\,H\cdot + \cdot\dot{C}\cdot \rightarrow H{-}C(H)(H){-}H$$

ammonia ($NH_3$) methane ($CH_4$)

Writing the Lewis structures of simple molecules is easy if you become familiar with the following facts. We will expand on these facts or "rules" later for more complex molecules.

1 Representative elements in the second period (Li to F) never attain more than eight outer electrons. There are some cases where they have fewer than eight. Nonmetals in the third, fourth, and fifth periods can have access to more than eight in certain compounds that will not be discussed here.

2 Hydrogen never has access to more than two electrons. Therefore, H bonds to only one atom at a time.

3 Generally, the atoms in a molecule have a very symmetrical and compact skeletal arrangement. Thus, $SO_3$ has an S surrounded by three O's,

```
  O
  S
 O O
```

rather than structures such as

```
S O O O    O S O O    S O
                      O O
```

In most cases, the first atom in a formula is the central atom, with the other atoms bound to it.

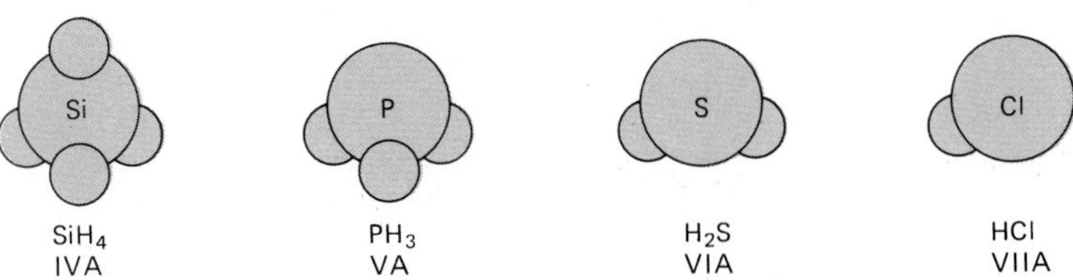


**Figure 6-5** FORMULAS OF HYDROGEN COMPOUNDS. The formulas of some simple hydrogen compounds can be predicted from the octet rule.

**Example 6-2**
Write the Lewis structure for $PF_3$.

**Procedure**
Arrange the atoms in a symmetrical manner. In this case the P will be surrounded by three F's:

```
F P F
  F
```

Now add the outer electrons. The P has five (represented by X's) and each F has seven (represented by dots).

```
:F:P:F:
  :F:
```

Finally, we can represent the bonded electrons as dashes.

```
:F—P—F:
   |
  :F:
```

**Example 6-3**
Write the Lewis structure for $N_2H_4$ (hydrazine).

**Procedure**
Arrange the atoms in a symmetrical manner. Notice that all of the H's must be on the outside, since they bond to only one atom at a time.

```
H N N H
  H H
```

Now add the electrons. Each H contributes one and each N contributes 5. The electrons from the H's are represented by X's and the electrons from N's are represented by dots for clarity. As far as the molecule is concerned, however, all electrons are identical.

```
H:N:N:H  →  H—N—N—H
  H H         |  |
              H  H
```

**See Problems 6-22 through 6-24.**

More examples of Lewis structures will be worked in Section 6-9 after we have developed the concept of the multiple bond.

## 6-7 The Multiple Covalent Bond

In Figure 6-2 an analogy to the covalent bond illustrated how two people could each claim access to $8 with only a total of $14 between them. What if they only had $12 between them? In this case they could still claim $8 each if they shared $4 in the joint account and kept $4 each in private accounts.

Two atoms may also share four electrons (two pairs) to achieve an octet of valence electrons. *The sharing of two pairs of electrons (between the same two atoms) is known as a* **double bond.** An example of a molecule containing a double bond is illustrated by the Lewis structure of $SO_2$:

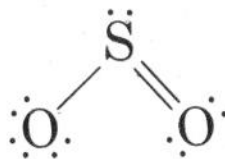

Notice that all atoms in the molecule have access to eight electrons.*

The $N_2$ molecule has a total of 10 outer electrons available to achieve octets. In this case, each N holds only one lone pair to itself and shares six electrons (three from each N). *When three pairs of electrons are shared between atoms, it is known as a* **triple bond.** The Lewis structure for $N_2$ is

$$:N{\equiv}N:$$

Only C, N, and O among the representative elements are ever involved in triple bonds. Triple and quadruple bonds (eight shared electrons) also exist in bonds between certain transition metals.

## 6-8 Polyatomic Ions

As we mentioned earlier, *two or more atoms can bond together by covalent bonds to form what is known as a* **polyatomic ion.** The charge on the ion is not located on any particular atom in the ion but belongs to the entire structure. A simple example of a polyatomic ion is the hypochlorite ion, $ClO^-$. The existence of the $-1$ charge tells us that the ion contains one electron *in addition* to those from neutral Cl and O atoms. The total number of outer electrons involved in the $ClO^-$ ion is calculated as follows:

*It would seem that $O_2$, with 12 valence electrons, would make an excellent example of the simplest molecule with a double bond as follows:

$$:\ddot{O}{=}\ddot{O}:$$

This Lewis structure implies that all of the electrons in $O_2$ are paired. However, experiments show that $O_2$ has two unpaired electrons. Although writing Lewis structures works very well in explaining the bonding in most simple molecules, it should be kept in mind that it is simply the representation of a theory. In this case, the theory just doesn't work. Thus $O_2$ is usually not represented by a Lewis structure. Other theories on bonding work well to explain the bonding of $O_2$, but these will not be discussed here.

| | |
|---|---|
| From a neutral Cl | 7 |
| From a neutral O | 6 |
| Additional electrons indicated by charge | 1 |
| Total number of electrons | 14 |

Two atoms bonded together with 14 electrons has a Lewis structure like $F_2$ with a single covalent bond:

$$\left[:\ddot{\underset{..}{Cl}}—\ddot{\underset{..}{O}}:\right]$$

Notice that the brackets indicate that the total ion has a −1 charge. The extra electron has not been specifically identified, since electrons are all identical and the charge belongs to the ion as a whole.

See Problems 6-25 and 6-26

## 6-9 Writing Lewis Structures

Writing correct Lewis structures is an important skill in basic chemistry, since it is necessary in later courses such as organic chemistry and biochemistry. Fortunately, the whole process becomes straightforward (with some practice, of course) if one follows a few basic rules. These rules that follow have been expanded from those mentioned on page 147.

1 Check to see whether any ions are involved in the compounds. Write out any ions present.
   (a) Metal–nonmetal binary compounds are mostly ionic.
   (b) If IA or IIA metals (except Be) are part of the formula, ions are present. For example KClO is $K^+ClO^-$ because K is a *IA element and forms only a +1 ion.* If K is +1, the ClO must be −1 to have a neutral compound. Likewise $Ba(NO_3)_2$ contains ions because Ba is a *IIA element and forms only a +2 ion.* In order to maintain neutrality each $NO_3$ ion must have a −1 charge. The ions are represented as $Ba^{2+}$ $2(NO_3{}^-)$.

2 For a molecule, add up all of the outer (or valence) electrons of the neutral atoms. For an ion, add (if negative) or subtract (if positive) the number of electrons indicated by the charge.

3 Arrange the atoms of the molecule or ion in a skeletal arrangement. As mentioned, this is usually the most symmetrical and compact arrangement. Remember that H can bond to only one atom at a time.

4 Put a dash representing a shared pair of electrons between adjacent atoms that have covalent bonds (not between ions). Subtract the electrons used for this (two for each bond) from the total calculated in Step 2.

5 Distribute the remaining electrons among the atoms so that no atom has more than eight electrons.

6 Check all atoms for an octet (except H). If an atom has access to fewer than eight electrons, put an electron pair from an adjacent atom into a double bond. Each double bond increases by two the number of electrons available to the atom needing electrons. Remember that you cannot satisfy an octet for an atom by adding any electrons at this point.

**Example 6-4**
Write the Lewis structure for $NCl_3$.

1 This is a binary compound between two nonmetals. Therefore, it is not ionic.

2 The total number of electrons available for bonding is

$$\begin{array}{rr} \text{N} \quad 1 \times 5 = & 5 \\ \text{Cl} \quad 3 \times 7 = & \underline{21} \\ \text{Total outer electrons} = & 26 \end{array}$$

3 The skeletal arrangement is

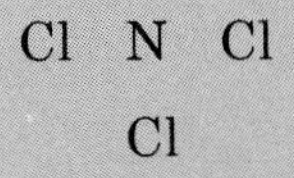

```
Cl  N  Cl

    Cl
```

4 Use six electrons to form bonds:

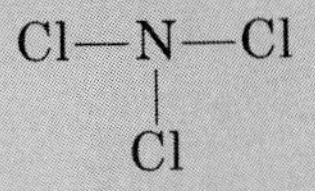

```
Cl—N—Cl
   |
   Cl
```

5 Distribute the remaining 20 electrons (26 − 6 = 20):

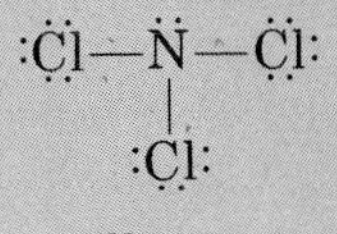

```
:C̤̈l—N̈—C̤̈l:
     |
    :C̤l:
```

6 Check to make sure that all atoms satisfy the octet rule.

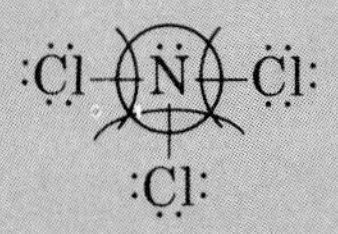

```
:C̤̈l—(N̈)—C̤̈l:
      |
     :C̤l:
```

**Example 6-5**
Write the Lewis structure for $K_2S$.

1 This is a binary compound between a group IA metal and a group VIA nonmetal. It is therefore ionic. Since K is a +1 ion, the S

is a −2 ion. The Lewis structure is simply that of an ionic compound (no dashes between ions).

$$2(K^+)\left[:\ddot{\underset{..}{S}}:\right]^{2-}$$

**Example 6-6**
Write the Lewis structure for $N_2O_4$.

1 This is not ionic.

2 The total number of outer electrons available is

| | | |
|---|---|---|
| N | 2 × 5 = | 10 |
| O | 4 × 6 = | 24 |
| Total | | 34 |

3 The symmetrical skeletal structure is

```
O   O
N   N
O   O
```

4 Add 10 electrons for the five bonds:

```
O       O
 \     /
  N—N
 /     \
O       O
```

5 Add the remaining 24 electrons (34 − 10 = 24):

```
:Ö:       :Ö:
   \     /
    N—N
   /     \
:Ö:       :Ö:
```

6 Check for octets. The oxygens are all right, but both N's need access to one more pair of electrons each. Therefore, make one double bond for each N using a lone pair of electrons from an adjacent O.

```
:Ö:       :Ö:          :Ö:        Ö.
   \     /                \     //
    N—N        →           N—N
   /     \                //    \
:Ö:       :Ö:          .Ö       :Ö:
```

**Example 6-7**
Write the Lewis structure for $CaCO_3$.

1 This is an ionic compound composed of $Ca^{2+}$ and $CO_3^{2-}$ ions (since you know that Ca is in group IIA, it must have a +2 charge, therefore the polyatomic anion must be −2). A Lewis structure can be written for $CO_3^{2-}$.

2 For the $CO_3^{2-}$ ion the total number of outer electrons available is:

$$\begin{aligned} \text{C} \quad 1 \times 4 &= 4 \\ \text{O} \quad 3 \times 6 &= 18 \\ \text{From charge} &= \underline{2} \\ \text{Total electrons} &= 24 \end{aligned}$$

3, 4 The skeletal structure with bonds is

```
O   O
 \ /
  C
  |
  O
```

5 Add the remaining 18 electrons (24 − 6 = 18):

```
[ :Ö·   ·Ö: ]2-
[    \ /    ]
[     C     ]
[     |     ]
[    :Ọ:    ]
```

6 The C needs two more electrons, so one double bond is added using one lone pair from one oxygen:

```
   :Ö:                 [    :Ö:    ]2-
    |                  [     |     ]
    C        → Ca2+    [     C     ]
   / \                 [    / \\   ]
 :Ọ·   ·Ọ:             [ :Ọ·    Ö· ]
```

**Example 6-8**
Write the Lewis structure for $H_2SO_4$.

1 All three atoms are nonmetals, which means that all bonds are covalent.

2 The total number of outer electrons available is

$$\begin{array}{lrr} H & 2 \times 1 = & 2 \\ S & 1 \times 6 = & 6 \\ O & 4 \times 6 = & \underline{24} \\ \text{Total} & & 32 \end{array}$$

**3** In most molecules containing H and O the H is bound to an O and the O to some other atom, which in this case is S. The skeletal structure is

```
      O
H  O  S  O  H
      O
```

**4** Add 12 electrons for the six bonds:

```
        O
        |
H—O—S—O—H
        |
        O
```

**5** Add the remaining 20 electrons (32 − 12 = 20):

```
         ..
        :O:
         |
    ..   |   ..
H—O—S—O—H
    ..   |   ..
         |
        :O:
         ..
```

**6** All octets are satisfied.

---

**Example 6-9**
Write the Lewis structure for NO.

**1** It is not ionic.

**2** There are 5 + 6 = 11 outer electrons.

**3, 4, 5** With a bond and the other electrons the structure is

```
      ..
·N—O:
 ..   ..
```

**6** As written, the N has access to five electrons. By making one double bond, the nitrogen can be brought up to seven electrons. However, there is no way that an odd number of electrons can be arranged so that all atoms have access to the even number of eight electrons. In this case one of the atoms does not have an octet, as shown below:

```
 .    ..
·N=O:
```

**See Problems 6-27 through 6-33.**

## 6-10 Resonance Hybrids

The Lewis structure for $SO_2$ is

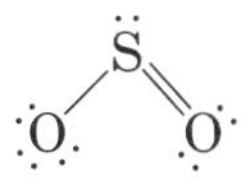

Experiments indicate that the length and strength of a double bond are different from those of a single bond between the same two elements. However, the above structure by itself implies something that isn't true according to available evidence. It implies that one S—O bond is different from the other. In fact, both S—O bonds are known to be identical and have length and strength halfway between a single and a double bond. Since there is no way to illustrate this fact with one Lewis structure, it must be done with two. The two structures below (connected by a double-headed arrow) are known as **resonance structures.** The actual structure is a **hybrid** of the resonance structures.

(1) (2) (1) (2)

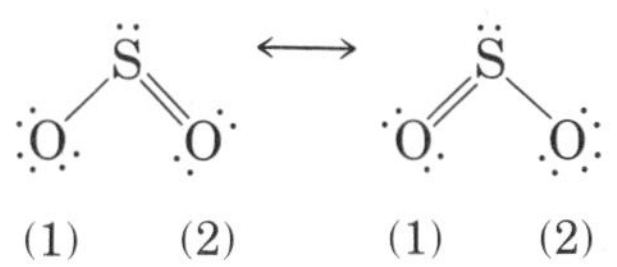

*Resonance structures exist for compounds when equally correct Lewis structures can be written without changing the basic skeletal geometry or changing the position of any atoms.*

**Example 6-10**

Write equivalent resonance structures for the $CO_3^{2-}$ ion.

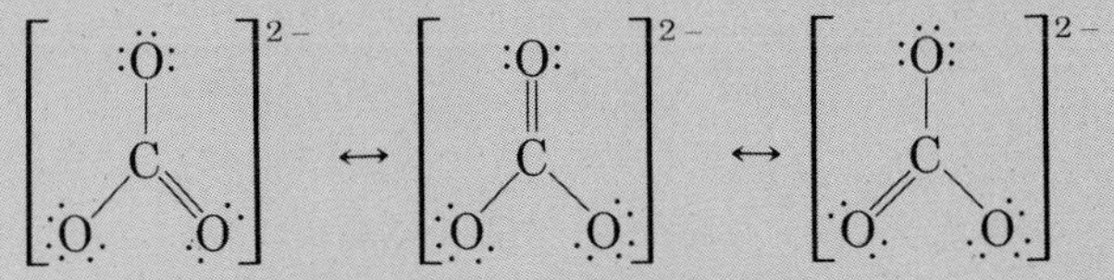

The three resonance structures indicate that the C—O bond has one-third double bond properties and two-thirds single bond properties.

The word "resonance" is an unfortunate choice. It implies that the S—O bond is changing (resonating) between a double and a single bond and that what is seen is actually just an average of the two extremes. This is not the case; the S—O bond actually *exists* full-time as half double and half single. In this sense a resonance hybrid is like the hybrid tomato that you may grow in the garden or buy in the grocery store. The tomato has a sweet taste and large size but not because it is changing rapidly back and forth

**See Problems 6-34 through 6-38.**

between a large tomato and a small, sweet tomato. It exists full-time with properties of its two parents.

## 6-11 Electronegativity and Polarity

When electrons are shared between the atoms of different elements, they are not shared equally. To some extent one atom pulls the electrons closer to itself and away from the other atom. *The ability of an atom of an element to attract electrons to itself in a covalent bond is known as the element's* **electronegativity**. Electronegativity is a periodic property that increases up and to the right in the periodic table. Thus a comparatively large value for electronegativity is a property characteristic of a nonmetal. The values of electronegativity listed in Table 6-1 were calculated by Linus Pauling (winner of two Nobel Prizes and of current vitamin C fame). Although more refined values are now available, the actual values are not as important as how the electronegativity of one element compares with another. In a covalent bond, the more electronegative atom pulls the electrons closer to itself and thus acquires a partial negative charge (symbolized by $\delta^-$). The other atom is left with an equal but opposite positive charge (symbolized by $\delta^+$).

*A covalent bond that has a partial separation of charge because of unequal sharing of electrons is known as* a **polar covalent bond.** A polar bond has

**Table 6-1** Electronegativity

Increases →

Decreases ↓

| 1<br>H<br>2.1 | | | | | | | | | | | | | | | | |
|---|---|---|---|---|---|---|---|---|---|---|---|---|---|---|---|---|
| 3<br>Li<br>1.0 | 4<br>Be<br>1.5 | | | | | | | | | | | 5<br>B<br>2.0 | 6<br>C<br>2.5 | 7<br>N<br>3.0 | 8<br>O<br>3.5 | 9<br>F<br>4.0 |
| 11<br>Na<br>0.9 | 12<br>Mg<br>1.2 | | | | | | | | | | | 13<br>Al<br>1.5 | 14<br>Si<br>1.8 | 15<br>P<br>2.1 | 16<br>S<br>2.5 | 17<br>Cl<br>3.0 |
| 19<br>K<br>0.8 | 20<br>Ca<br>1.0 | 21<br>Sc<br>1.3 | 22<br>Ti<br>1.5 | 23<br>V<br>1.6 | 24<br>Cr<br>1.6 | 25<br>Mn<br>1.5 | 26<br>Fe<br>1.8 | 27<br>Co<br>1.8 | 28<br>Ni<br>1.8 | 29<br>Cu<br>1.9 | 30<br>Zn<br>1.6 | 31<br>Ga<br>1.6 | 32<br>Ge<br>1.8 | 33<br>As<br>2.0 | 34<br>Se<br>2.4 | 35<br>Br<br>2.8 |
| 37<br>Rb<br>0.8 | 38<br>Sr<br>1.0 | 39<br>Y<br>1.2 | 40<br>Zr<br>1.4 | 41<br>Nb<br>1.6 | 42<br>Mo<br>1.8 | 43<br>Tc<br>1.5 | 44<br>Ru<br>2.2 | 45<br>Rh<br>2.2 | 46<br>Pd<br>2.2 | 47<br>Ag<br>2.4 | 48<br>Cd<br>1.7 | 49<br>In<br>1.7 | 50<br>Sn<br>1.8 | 51<br>Sb<br>1.9 | 52<br>Te<br>2.1 | 53<br>I<br>2.5 |
| 55<br>Cs<br>0.7 | 56<br>Ba<br>0.9 | 57<br>La<br>1.1 | 72<br>Hf<br>1.3 | 73<br>Ta<br>1.5 | 74<br>W<br>1.7 | 75<br>Re<br>1.9 | 76<br>Os<br>2.2 | 77<br>Ir<br>2.2 | 78<br>Pt<br>2.2 | 79<br>Au<br>2.4 | 80<br>Hg<br>1.9 | 81<br>Ti<br>1.8 | 82<br>Pb<br>1.8 | 83<br>Bi<br>1.9 | 84<br>Po<br>2.0 | 85<br>At<br>2.2 |
| 87<br>Fr<br>0.7 | 88<br>Ra<br>0.9 | 89<br>Ac<br>1.1 | | | | | | | | | | | | | | |

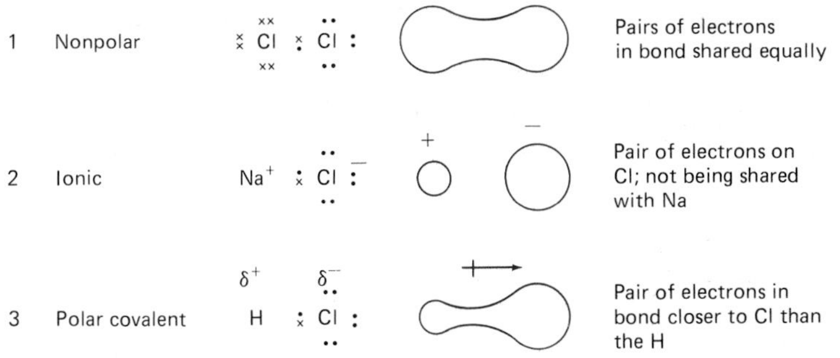

1 Nonpolar  $:\ddot{Cl}:\ddot{Cl}:$  Pairs of electrons in bond shared equally

2 Ionic  $Na^+ :\ddot{Cl}:^-$  Pair of electrons on Cl; not being shared with Na

3 Polar covalent  $\overset{\delta^+}{H} :\overset{\delta^-}{\ddot{Cl}}:$  Pair of electrons in bond closer to Cl than the H

**Figure 6-6** NONPOLAR, IONIC, AND POLAR COVALENT BONDS.

a negative end and a postive end and is said to contain a **dipole** (two poles).* The dipole of a bond is represented by an arrow pointing from the positive to the negative end (+⟶).

$$\overset{\delta^+}{X}—\overset{\delta^-}{Y}$$

$\delta^+ \; \delta^-$ —Representation of polarity

+⟶ —Representation of bond dipole

Notice from Table 6-1 that fluorine is the most electronegative element. This means that fluorine has the strongest attraction for electrons in a bond and will always be the negative end of a dipole. Oxygen is the second most electronegative element and will be the negative end of a dipole in all bonds except those with fluorine.

The greater the difference in electronegativity between two atoms, the greater the polarity (the bond has a larger dipole). As a matter of fact, if the difference between two elements is around 1.9 to 2.0, the bond is generally ionic, meaning that one atom has gained complete control of the electron pair in the bond to acquire a full negative charge.*

In summary, when two atoms compete for a pair of electrons in a bond, there are three things that can happen:

1 Both atoms share the electrons equally.

2 One atom takes the pair of electrons completely to itself, forming an ionic bond.

3 The two atoms share the electrons but not equally.

The molecule $Cl_2$ illustrates case 1, as shown in Figure 6-6. Since both atoms obviously have the same electronegativity, the electron pair is shared

*The earth is polar (contains a dipole), with a north and south magnetic pole.

*This is not a flawless way for predicting whether a bond is ionic or covalent. However, it does point out that in many cases the borderline between ionic and covalent character is not necessarily sharp.

equally between the two atoms. Each Cl has access to its three lone pairs of electrons plus exactly one-half of the electron pair in the bond.

**Case 1:** *Each Cl*

| | |
|---|---|
| Unshared electrons | = 6 |
| Portion of shared electrons ($\frac{1}{2} \times 2$) | = 1 |
| Total electrons | = 7 |

The seven electrons leave each Cl (group VIIA) exactly neutral. Thus a bond between atoms of the same element are said to be **nonpolar.** If elements have the same or nearly the same electronegativity (e.g., C and S), the bonds between these atoms are essentially nonpolar.

The ionic compound NaCl illustrates case 2. As shown in Figure 6-5, the pair of electrons is not shared but is with the Cl. The Cl has three lone pairs of electrons plus the two electrons in question.

**Case 2:** *Each Cl*

| | |
|---|---|
| Unshared electrons | = 6 |
| Portion of shared electrons ($1 \times 2$) | = 2 |
| Total electrons | = 8 |

Since seven valence electrons leave a Cl atom neutral, eight electrons gives Cl a charge of $-1$. The Na has an equal but opposite charge of $+1$. Notice that there is a large difference in electronegativity ($3.0 - 0.9 = 2.1$) between the Cl and the Na, leading to a prediction of ionic character for the bond on this basis.

In between these extremes lie the many bonds formed between two atoms that share electrons but not equally. In HCl (case 3), the Cl has a greater electronegativity than H but not enough to form an ionic bond. The Cl has three lone pairs of electrons plus more than (>) one-half of the electrons in the bond. This gives the Cl atom a partial negative charge and leaves the H atom with a partial positive charge. (See Figure 6-6.)

**Case 3:** *Each Cl*

| | |
|---|---|
| Unshared electrons | = 6 |
| Portion of shared electrons ($>\frac{1}{2} \times 2$) | = >1 |
| Total electrons | = >7 |

**Example 6-11**

Referring to Table 6-1, rank the following bonds in order of increased polarity. Write the atom with the positive end of the dipole first. Indicate if any of the bonds are predicted to be ionic on the basis of electronegativity difference.

Ba—Br, C—N, Be—F, B—H, Be—Cl

**Solution**

Calculate the difference in electronegativity between the elements.

| | | |
|---|---|---|
| Ba—Br | $2.8 - 0.9 = 1.9$ | Ba positive (less electronegative) |
| C—N | $3.0 - 2.5 = 0.5$ | C positive |

Be—F 4.0 − 1.5 = 2.5 Be positive
Be—H 2.1 − 2.0 = 0.1 B positive
Be—Cl 3.0 − 1.5 = 1.5 Be positive

B—H < C—N < Be—Cl < Ba—Br < Be—F

The difference in electronegativity suggests that Ba—Br and Be—F are ionic bonds.

**See Problems 6-39 through 6-44.**

## 6-12 Molecular Polarity

The polarity of molecules is determined by:

1 The polarity of the bonds in the molecule and

2 The geometry of the molecule*

In a molecule such as $CO_2$, both C—O bonds have a dipole as shown below. (Since O is more electronegative than C, the C is the positive end and the O negative.)

$$\overset{\longleftarrow\!+}{}\;\overset{+\!\longrightarrow}{}$$
$$:\ddot{O}=C=\ddot{O}:$$

The size of each dipole is the same, but the direction† is exactly opposite since $CO_2$ is a linear molecule. *Thus the bond dipoles cancel, leaving the molecule nonpolar.* On the other hand, COS is also linear, but the dipoles in opposite directions are unequal (in fact, Table 6-1 predicts that the C—S bond is nonpolar since C and S have the same electronegativity). Since the bond dipoles do not cancel, there is a resultant molecular dipole. *The* **molecular dipole** *is the net or resultant effect of all of the individual bond dipoles.*

Bond dipole

$$S=C\overset{+\!\longrightarrow}{=}O$$
$$\underset{+\!\longrightarrow}{}$$

Molecular dipole

The relation of individual bond dipoles to the total molecular dipole is analogous to the effect of two people pulling on a block of concrete, as illustrated in Figure 6-7. The force exerted by each person is analogous to a bond dipole, and the net effect (the resultant) of their efforts is analogous to the molecular dipole. Two people of equal strength pulling in opposite

*The geometry of simple molecules can be predicted from either the Lewis structure or a knowledge of the shapes of what are called "hybridized" orbitals. This topic is not covered in this text, so the geometry of the sample molecules should be accepted as fact.

†The size and direction together make up what is known as the dipole moment of the bond.

**Figure 6-7** FORCES AT 180°. If forces are equal but opposite (180°), they cancel. If the forces are not equal, there is a resultant force.

directions (an angle of 180°) will cancel each other's force and there will be no net movement.

Other examples of molecules that are nonpolar although the bonds are polar include:

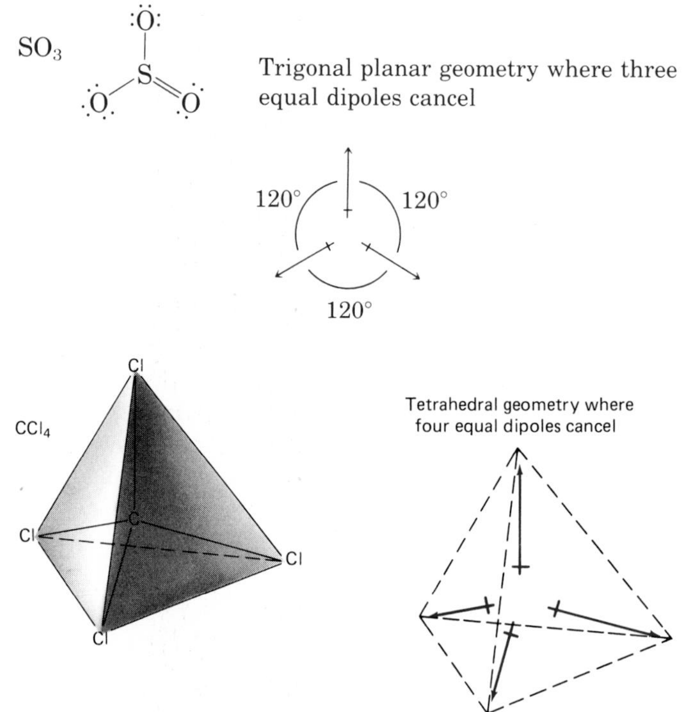


Many molecules exists in which the equal bond dipoles do not cancel. Ordinary water is an important example of this. Although an $H_2O$ molecule has two equal bond dipoles, they are not at an angle of 180° and thus do not cancel. If $H_2O$ were a linear molecule, it would be nonpolar and have vastly different properties than those discussed in Chapter 11.

Bond dipoles at an angle less than 180° is a situation analogous to two people pulling on a block of concrete at an angle as illustrated in Figure 6-8. Both the direction and resultant force depend on the strength of the

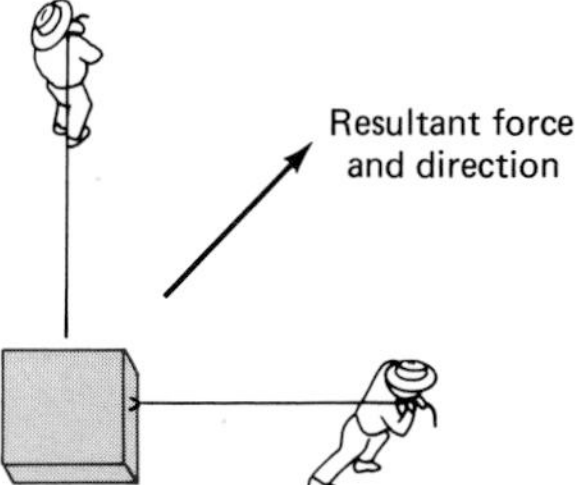


**Figure 6-8** FORCES AT LESS THAN 180°. Both the direction and resultant force depends on the angle between the two people.

two people and the angle between them. (The resultant force and direction of movement also depend on their comparative strengths, if different.) In a similar manner, the resultant dipole for the $H_2O$ molecule is determined by the size of each H—O dipole and the angle between the two dipoles:

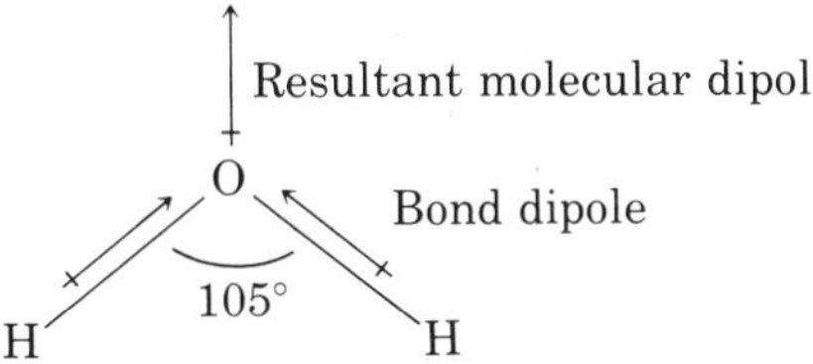


In later chapters the importance of the polarity of molecules will become evident. The physical state and other physical properties as well as chemical properties of compounds are determined to a large extent by the polarity of the molecules. We will refer back to this discussion at that time.

**See Problems 6-45 through 6-51.**

## 6-13 Chapter Summary

In this chapter we examine one of the most important periodic properties of the elements—the nature of its bonds. This chapter shows us why certain elements combine to form ionic compounds and others combine to form molecular compounds. The bonding between representative elements is the easiest to understand because we can make generalizations based on the following three premises:

1 Noble gases do not readily form chemical bonds.

2 Noble gases do not form bonds (are stable) because they have an octet of outer or valence electrons. This corresponds to filled outer *s* and *p* sublevels.

3 Other atoms of representative elements achieve the same stability of noble gases by gaining, losing, or sharing electrons with other atoms to attain an octet of electrons.

To follow the changes in electron configuration of elements as bonds are formed, Lewis dot structures were introduced. Dot structures emphasize only the valence electrons.

Our first look at bonding concerned binary compounds involving a metal and a nonmetal. In this case metals can attain an octet by losing electrons and nonmetals by gaining electrons. Two nonmetals form binary compounds by electron sharing, which is known as covalent bonding. This is summarized as follows:

| Elements | Type of Bond | Comments |
|---|---|---|
| Metal–nonmetal | Ionic | Nonmetal can form $-1$, $-2$, or $-3$ ion. Metal can form $+1$, $+2$, or $+3$ ion. |
| Nonmetal–nonmetal | Covalent | Single, double, or triple bonds used to form octet. |

The construction of Lewis structures of compounds was discussed and the procedural rules given. The rules are summarized for three compounds as follows:

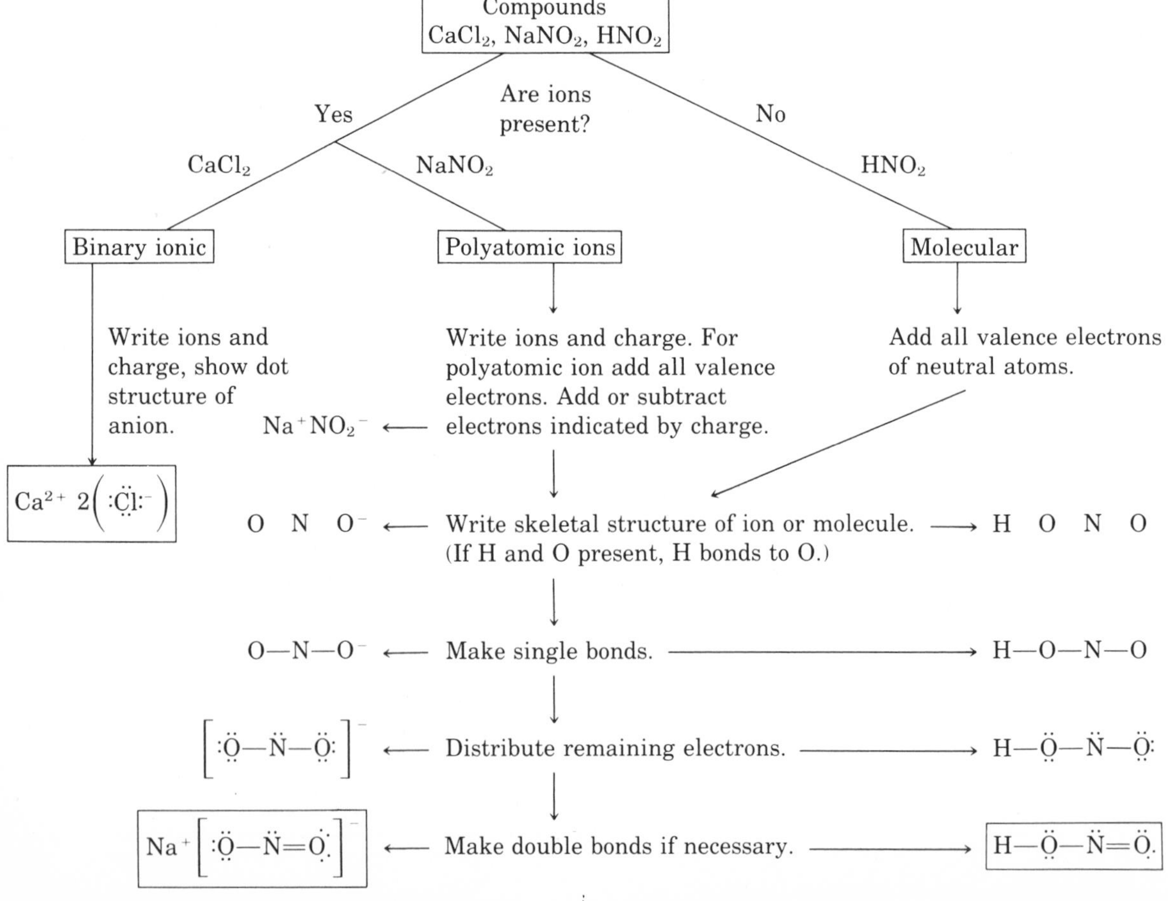

Resonance structures were introduced as a method of illustrating what one Lewis structure does not do. That is, they indicate that certain bonds in one molecule are identical and exist as a hybrid of the equivalent resonance structures. Electronegativity is a periodic property that gives us information about the polarity of covalent bonds. Large differences in electronegativity between two atoms in a bond may indicate ionic nature. Any difference indicates that the bond is polar covalent. Polar bonds contain a positive center and a negative center known as a dipole.

Many important physical properties of a compound are determined by the overall or net polarity of the molecule. If the geometry and magnitude of the dipoles is such that the bond dipole forces cancel, the molecule is nonpolar. If there is a resultant molecular dipole, the molecule is polar.

## Exercises

### Dot Structures of Elements

**6-1** Write Lewis dot structures for:
(a) Ca (b) Sb (c) Sn (d) I (e) Ne (f) Bi
(g) All group VIA elements

**6-2** Identify the group from the given dot structure.
(a) ·Ṁ· (b) ·Ẍ· (c) Ȧ·

**6-3** Why are only outer electrons represented in dot structures?

**6-4** Which of the following dot structures are incorrect?

(a) Ṗb· (d) Ċs
(b) ·B̈i· (e) ·T̈e·
(c) :Ḧe: (f) ·Ṫl·

### Binary Ionic Compounds

**6-5** From the periodic table predict which pairs of elements can combine to form ionic bonds.

(a) H and Cl (d) Al and F
(b) S and Sr (e) B and Cl
(c) H and K (f) Xe and F

**6-6** From the periodic table predict which pairs of elements can combine to form ionic bonds.

(a) Ba and I (d) Sn and Sb
(b) Cs and Se (e) P and S
(c) C and O (f) Cs and P

**6-7** Which of the ions do not have a noble gas electron configuration?

(a) $Sr^{2+}$ (e) $In^{+}$
(b) $S^{-}$ (f) $Pb^{2+}$
(c) $Cr^{2+}$ (g) $Ba^{2+}$
(d) $Te^{2-}$ (h) $Tl^{3+}$

**6-8** The ions $Li^{+}$, $Be^{2+}$, and $H^{-}$ do not follow the octet rule. Why?

**6-9** Write Lewis dot structures for the following ions.
(a) $K^{+}$ (b) $O^{-}$ (c) $I^{-}$ (d) $P^{3-}$ (e) $Ba^{+}$
(f) $Xe^{+}$ (g) $Sc^{3+}$

**6-10** What is the origin of the octet rule? How does the octet rule relate to *s* and *p* sublevels and to noble gases?

**6-11** Which of the ions listed in Problem 6-9 do not follow the octet rule?

**6-12** Write the charge on the following atoms that would give the element a noble gas configuration.
(a) Mg (b) Ga (c) Br (d) S (e) P

**6-13** Write the charge on the following atoms that would give the element a noble gas configuration.
(a) Rb (b) Ba (c) Te (d) N

**6-14** Write six ions that have the same electron configuration as Ne.

**6-15** Write five ions that have the same electron configuration as Kr.

**6-16** The $Tl^{3+}$ ion does not have the same electron configuration as Xe although it lost its three outermost electrons. Explain.

**6-17** Complete the following table with formulas of the ionic compounds that form between the anion and cation shown.

| Cation/Anion | $Br^-$ | $S^{2-}$ | $N^{3-}$ |
|---|---|---|---|
| $Cs^+$ | CsBr | | |
| $Ba^{2+}$ | | | |
| $In^{3+}$ | | | |

**6-18** Write the formula of the compound formed between the following nonmetals with the metal calcium.
(a) I (b) O (c) N (d) Te (e) F

**6-19** Write the formula of the compound formed between the following metals with the nonmetal sulfur.
(a) Be (b) Cs (c) Ga (d) Sr

**6-20** Most transition metal ions cannot be predicted by reference to the octet rule. Determine the charge on the metal cation from the charge on the anion for the following compounds.
(a) $Cr_2O_3$ (b) $FeF_3$ (c) MnS (d) CoO (e) $NiBr_2$ (f) VN

**6-21** Why isn't a formula unit of $BaCl_2$ referred to as a molecule?

### Lewis Structures of Compounds

**6-22** From their electron configuration, predict the simplest formula formed from the combination of the following two elements.

(a) H and Se (d) Cl and O
(b) H and Ge (e) N and Cl
(c) Cl and F (f) C and Br

**6-23** From a consideration of the octet rule, which of the following compounds are impossible?
(a) $PH_3$ (b) $Cl_3$ (c) $SCl_2$ (d) $NBr_4$ (e) $H_3O$

**6-24** Write Lewis structures for the following.
(a) $C_2H_6$ (b) $H_2O_2$ (c) $NF_3$ (d) $SCl_2$ (e) $C_2H_6O$ (There are two correct answers; both have the lone pairs on the oxygen.)

**6-25** Determine the charge on each of the following polyatomic anions from the charge on the cation.
(a) $K_2SO_4$ (b) $Ca(ClO_3)_2$ (c) $Al_2(SO_3)_3$ (d) $Ca_3(PO_4)_2$ (e) $CaC_2$

**6-26** Determine the charge on each of the following polyatomic anions from the charge on the cation.

(a) $NaClO_2$ (c) $AlAsO_4$
(b) $SrSeO_3$ (d) $Ca(H_2PO_4)_2$

**6-27** Write Lewis structures for:
(a) CO (b) $SO_3$ (c) KCN (d) $H_2SO_3$ (both H's on different O's)

**6-28** Write Lewis structures for:
(a) $N_2O$ (b) $Ca(NO_2)_2$ (c) $AsCl_3$ (d) $H_2S$ (e) $CH_2Cl_2$ (f) $NH_4^+$

**6-29** Write Lewis structures for:
(a) $Cl_2O$ (b) $SO_3^{2-}$ (c) $C_2H_4$ (d) $H_2CO$ (e) $BF_3$ (f) $NO^+$

**6-30** Write Lewis structures for:
(a) $CO_2$ (b) $BaCl_2$ (c) $NO_3^-$ (d) $SiCl_4$ (e) $C_2H_2$ (f) $O_3$

**6-31** Write Lewis structures for:
(a) $Cs_2Se$ (b) $CH_3CO_2^-$ (all H's on one C, both O's on the other C) (c) $LiClO_3$ (d) $N_2O_3$ (e) $PBr_3$

* **6-32** Refer to Problem 5-43. Use the periodic table from the planet Zerk.
(1) What are the simplest formulas of compounds formed between the following elements? (Example: Between 7 and 7 is $7_2$.)
(a) 1 and 7 (b) 1 and 3 (c) 1 and 5 (d) 7 and 9 (e) 7 and 13 (f) 10 and 13 (g) 6 and 7 (h) 3 and 6
(2) Write Lewis structures for all of the above. Indicate which are ionic. (Remember that on Zerk there will be something different than an octet rule.)

**6-33** Which of the following compounds contains both ionic and covalent bonds?
(a) $H_2SO_3$ (b) $K_2SO_4$ (c) $K_2S$ (d) $H_2S$ (e) $C_2H_6$ (f) $BaCl_2$

### Resonance Structures

**6-34** Write all equivalent resonance structures (if any) for each of the following.
(a) $SO_3$ (b) $NO_2^-$ (c) $SO_3^{2-}$

**6-35** Write all equivalent resonance structures (if any) for the following.
(a) $NO_3^-$ (b) $N_2O_4$

**6-36** Write the equivalent resonance structures for the $H_3BCO_2^{2-}$ anion. The skeletal structure for the ion is

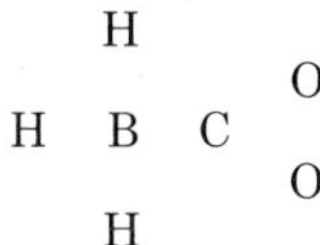


**6-37** What is meant by a resonance hybrid? What is implied about the nature of the C—O bond from the resonance structures in Problem 6-36?

* **6-38** A possible Lewis structure for $CO_2$ involves a triple bond between the C and O. Write the two resonance structures involving the triple bond. What is implied about the nature of the CO bond by these two structures? How does this relate to the common Lewis structure for $CO_2$ involving two double bonds?

### Electronegativity and Polarity

**6-39** Rank the following elements in order of increasing electronegativity. B, Ba, Be, C, Cl, Cs, F, O

**6-40** For bonds between the following elements, indicate the positive end of the dipole by a $\delta^+$ and the negative end by a $\delta^-$. Also indicate with a dipole arrow the direction of the dipole.

| | |
|---|---|
| (a) N—H | (f) S—Se |
| (b) B—H | (g) C—B |
| (c) Li—H | (h) Cs—N |
| (d) F—O | (i) C—S |
| (e) O—Cl | |

**6-41** Rank the bonds in Problem 6-40 in order of increasing polarity.

**6-42** Predict whether the following two elements will form an ionic or covalent bond on the basis of difference in electronegativity.
(a) Sc—Br (b) Cu—B (c) B—Br
(d) Al—F

**6-43** Predict whether the following two elements will form an ionic or covalent bond on the basis of difference in electronegativity.
(a) Al—Cl (b) Ca—I (c) K—O
(d) Mn—Te

**6-44** Which of the following bonds would be nonpolar?
(a) I—F (b) I—I (c) C—H (d) N—Cl
(e) B—N

**6-45** How can a molecule be nonpolar if it contains polar bonds?

**6-46** From the geometry of the following, determine which molecules are polar.
(a) $CS_2$ S—C—S linear
(b) COS S—C—O linear
(c) $SF_2$ S (F F) bent
(d) $CCl_4$ C (Cl Cl Cl Cl) tetrahedron
(e) $CHCl_3$ C (H Cl Cl Cl) tetrahedron

**6-47** From the geometry of the following, determine which molecules are polar.
(a) $BF_3$ B (F F F) trigonal planar
(b) $BF_2Cl$ B (Cl F F) trigonal planar
(c) $NH_3$ N (H H H) trigonal pyramid
(d) $NH_2Cl$ N (H Cl H) trigonal pyramid

**6-48** Compare the expected molecular polarity of $H_2O$ and $H_2S$. Assume both molecules are bent with the same angle.

**6-49** Compare the expected molecular polarities of $CH_4$ and $CH_2F_2$. Assume that both molecules are tetrahedral.

**6-50** Compare the expected molecular polarities of $CHCl_3$ and $CHF_3$. Assume both molecules are tetrahedral.

**6-51** $CO_2$ is a nonpolar molecule, but CO is polar. Explain.

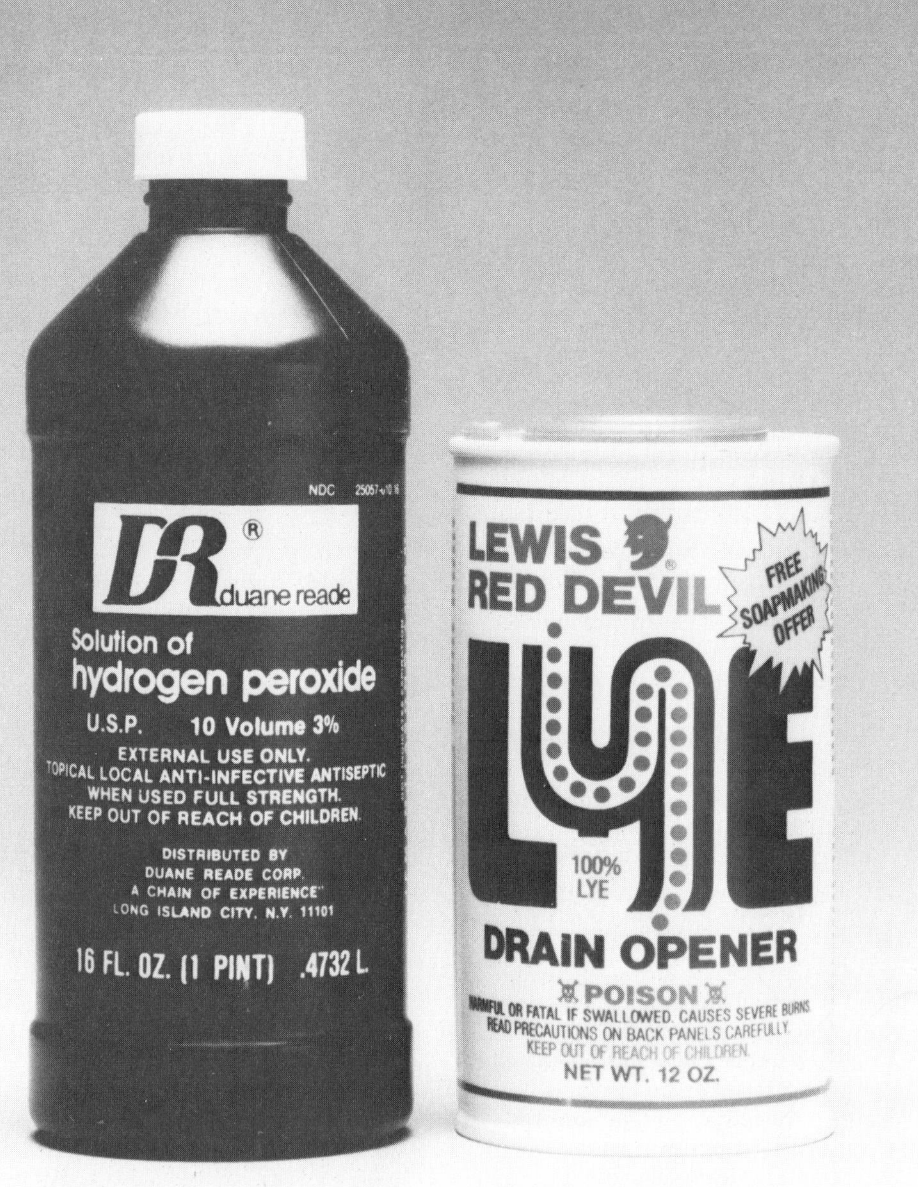


The names of many compounds are familiar because they are the ingredients of common products.

# Chapter 7
# The Naming of Compounds

## Purpose of Chapter 7

In Chapter 7 we introduce the language of chemistry so that you can properly name compounds from formulas and write formulas from names.

## Objectives for Chapter 7

After completion of this chapter, you should be able to:

1. Determine the oxidation state of an element in a compound. (7-1)
2. Identify metals having only one oxidation state. (7-2)
3. Name metal-nonmetal binary compounds and write formulas given a name. (7-2)
4. Apply the Stock method for naming compounds with metals that have variable oxidation states. (7-2)
5. Write the names and formulas for the ions listed in Table 7-3. (7-3)
6. Use the Greek prefixes in Table 7-4 to name nonmetal-nonmetal binary compounds and write formulas given a name. (7-4)
7. Name binary and oxyacids and write formulas given an acid name. (7-5)

Lately, it seems that when we hear the names of chemicals in the news there is an implication of mystery and even evil. We hear of nitrates and nitrites in our meats, dioxins in the soil, sulfur dioxide and carbon monoxide in the air, and sulfuric acid and phosphates in the lakes. These, of course, are very real problems, yet we ourselves are made of chemicals. Many of these have names even stranger than those that we have come to fear. Water, sodium chloride, and amino acids are names of chemicals, and these are essential to life. Perhaps if we are able to understand how chemical compounds are named (nomenclature) the whole subject will be put into balance.

A hundred years ago there were not all that many known compounds, so the citizen as well as the chemist did not need much order in nomenclature. Thus many compounds, such as lye, baking soda, ammonia, and water, have *common* names dating from long ago. These names tell little, if anything, about the elements of which they are composed. (The common names of some well-known compounds are given in Problem 7-27) This, of course, would never do in the naming of the millions of compounds that have now been identified. Obviously, a more orderly approach was needed. Since 1921, a group of chemists who make up an organization called IUPAC (International Union of Pure and Applied Chemistry) has met regularly to discuss the naming of compounds. From this group comes *systematic* methods of chemical nomenclature.

This chapter involves the naming of inorganic compounds, which includes all those compounds except organic compounds. (Organic compounds are composed of carbon, hydrogen, and sometimes other elements.) This is essentially the language of chemistry. Like any language, a good deal of memorization is needed. With some basic ground rules, however, what might otherwise be an impossible task becomes quite reasonable and orderly.

As background to this chapter you should be familiar with the differences between metals and nonmetals and between ionic and covalent compounds as discussed in the preceding chapter.

## 7-1 Oxidation States

Since two elements can form more than one compound (e.g., $FeO$ and $Fe_2O_3$), a method of keeping track of the electrons used in bond formation is needed. This is done by reference to the **oxidation states** of the elements. *The oxidation state* of an element is the charge the atoms of that element would have if the electrons in each bond are assigned to the more electronegative element.* A familiarity with oxidation states will not only help in the naming of compounds in this chapter but will also provide a basis for our discussion of chemical reactions involving changes in oxidation states in Chapter 14.

The calculation of oxidation states of elements in compounds can be simplified by a knowledge of the following rules.

*Also referred to as the **oxidation number**.

1 The oxidation state of an element is zero (e.g., $O_2$, $Br_2$, $P_4$).

2 The oxidation state of a monatomic ion is the same as the charge on that ion (e.g., $Na^+$ = +1 oxidation state, $O^{2-}$ = −2, $Al^{3+}$ = +3).
(a) Alkali metals are always +1.
(b) Alkaline earth metals are always +2.

3 The halogens have a −1 oxidation state in binary (two-element) compounds whether ionic or covalent *when bound to a less electronegative element.*

4 Oxygen in a compound is usually −2. Certain compounds (which are rare), called peroxides or superoxides, contain O in a lower negative oxidation state. Oxygen is positive when bound to F.

5 Hydrogen in a compound is usually +1. When combined with a less electronegative element, usually a metal, H has a −1 oxidation state (e.g., LiH).

6 The oxidation states of all of the elements in a neutral compound add to zero. In the case of polyatomic ions, the sum of the oxidation states equals the charge on the ion.

The following examples illustrate the assignment of oxidation states.

---

**Example 7-1**
What are the oxidation states of the elements in FeO?
**Rule 6:** The oxidation states of the two elements add to zero.

$$(\text{ox. state Fe}) + (\text{ox. state O}) = 0$$

**Rule 4:** Since O is −2,

$$\text{Fe} + (-2) = 0$$

$$\text{Fe} = \underline{\underline{+2}} \quad \text{(This is an ionic compound.)}$$

---

**Example 7-2**
What is the oxidation state of the N in $N_2O_5$?
**Rule 6:** The oxidation states add to zero, as shown by the equation

$$2(\text{ox. state N}) + 5(\text{ox. state O}) = 0$$

**Rule 4:** Since O is −2,

$$2\text{N} + 5(-2) = 0$$

$$2\text{N} = +10$$

$$\text{N} = \underline{\underline{+5}} \quad \text{(This is a covalent compound.)}$$

---

**Example 7-3**

What is the oxidation state of the S in $H_2SO_3$?

**Rule 6:** The oxidation states add to zero.

$$2(\text{ox. state H}) + (\text{ox. state S}) + 3(\text{ox. state O}) = 0$$

**Rules 4 and 5:** H is usually +1 and O is usually −2.

$$2(+1) + S + 3(-2) = 0$$

$$S = \underline{\underline{+4}}$$

**Example 7-4**

What is the oxidation state of the As in $AsO_4^{3-}$?

**Rule 6:** The oxidation states add to the charge on the ion.

$$(\text{ox. state As}) + 4(\text{ox. state O}) = -3$$

$$As + 4(-2) = -3$$

$$As = \underline{\underline{+5}}$$

**See Problems 7-1 through 7-8.**

## 7-2 Naming Metal–Nonmetal Binary Compounds

Most metal–nonmetal binary compounds are classified as ionic compounds. There are two kinds of metals, however. In the first kind, such as the metals in groups IA and IIA and Al in group IIIA, the metal forms only one oxidation state. The second kind of metals form more than one oxidation state and will be discussed shortly.

In both writing and naming a compound, the metal comes first and nonmetal second. *Name the metal using its full English name and the anion with its English root plus* ide. Table 7-1 lists all of the monatomic anions and their names.

**Table 7-1** Monatomic Anions

| Element | Ion | Name of Anion | Element | Ion | Name of Anion |
|---|---|---|---|---|---|
| hydrogen | $H^-$ | hydride | sulfur | $S^{2-}$ | sulfide |
| fluorine | $F^-$ | fluoride | tellurium | $Te^{2-}$ | telluride |
| chlorine | $Cl^-$ | chloride | selenium | $Se^{2-}$ | selenide |
| bromine | $Br^-$ | bromide | nitrogen | $N^{3-}$ | nitride |
| iodine | $I^-$ | iodide | phosphorus | $P^{3-}$ | phosphide |
| oxygen | $O^{2-}$ | oxide | carbon | $C^{4-}$ | carbide |

**Example 7-5**
Name the following binary ionic compounds: KCl, $Li_2S$, and $Mg_3N_2$.

**Answer**

| | |
|---|---|
| KCl | potassium chloride |
| $Li_2S$ | lithium sulfide |
| $Mg_3N_2$ | magnesium nitride |

**Example 7-6**
Write the formulas of the following compounds: aluminum fluoride, calcium selenide, and potassium phosphide.

**Answer**
Aluminum is in group IIIA, so it has a +3 charge in compounds. Fluorine is in group VIIA, so it has a −1 charge in compounds.

$$Al^{(3+)} \; F^{(-1)} \quad = AlF_3$$

Calcium is in group IIA, so it has a +2 charge in compounds. Selenium is in group VIA, so it has a −2 charge in compounds.

$$Ca^{(2+)} \; Se^{(2-)} \quad = Ca_2Se_2 = CaSe$$

Potassium is in group IA, so it has a +1 charge in compounds. Phosphorus is in group VA so it has a −3 charge in compounds.

$$K^{(+1)} \; N^{(3-)} \quad = K_3N$$

Iron (in group VIII) forms $FeCl_2$ and $FeCl_3$. As a result, the name iron chloride does not distinguish between the two compounds. Currently, the most common method used to differentiate between the two compounds is called the **Stock method.** *In this method the oxidation state of the metal is listed with Roman numerals in parentheses after the name of the element. The* **older method** *of naming these ions uses the root of the English name of the metal plus* ous *for the lower oxidation state and* ic *for the higher oxidation state.* If its symbol is derived from a Latin name, the Latin root is used. Several metals with variable oxidation states are shown using both methods of nomenclature in Table 7-2.

The Stock method is the more convenient of the two, since it is not necessary to memorize the oxidation state corresponding to each name. This method will be emphasized in this text.

**Table 7-2** Metals with Variable Oxidation States

| Metal | Stock Method | Old Method | Metal | Stock Method | Old Method |
|---|---|---|---|---|---|
| thallium | thallium(I) | thallous | lead | lead(II) | plumbous |
| | thallium(III) | thallic | | lead(IV) | plumbic |
| iron | iron(II) | ferrous | tin | tin(II) | stannous |
| | iron(III) | ferric | | tin(IV) | stannic |
| chromium | chromium(II) | chromous | copper | copper(I) | cuprous |
| | chromium(III) | chromic | | copper(II) | cupric |
| cobalt | cobalt(II) | cobaltous | gold | gold(I) | aurous |
| | cobalt(III) | cobaltic | | gold(III) | auric |

As discussed in Chapter 6, most metal-nonmetal binary compounds are ionic. Where the metal is in a +4 oxidation state or higher (e.g., $SnBr_4$ and $Mn_2O_7$), the compounds generally have properties more typical of molecular compounds that are neutral compounds containing only covalent bonds. It should be noted then that an oxidation state for a metal does not necessarily indicate the presence of an ion.

**Example 7-7**
Name the following compounds: $SnBr_4$, $CoF_3$, $Au_2O$.

**Answer**
Refer to Section 7-1 to determine the oxidation state of the metal ion.

$SnBr_4$ tin(IV) bromide* or stannic bromide

$CoF_3$ cobalt(III) fluoride or cobaltic fluoride

$Au_2O$ gold(I) oxide or aurous oxide

*Pronounced tin-four bromide.

**Example 7-8**
Write formulas for lead(IV) oxide, iron(II) chloride, chromium(III) sulfide, and manganese(VII) oxide.

**Answer**
lead(IV) oxide — If lead is in a +4 oxidation state, two $O^{2-}$ ions are needed to form a neutral compound.

$$Pb^{4+} O^{2-} = Pb_2O_4 = PbO_2$$

iron(II) chloride — Two chorides are needed to balance the +2 iron

$$Fe^{2+}\ Cl^{1-} = \underline{\underline{FeCl_2}}$$

chromium(III) sulfide

$$Cr^{3+}\ S^{2-} = \underline{\underline{Cr_2S_3}}$$

manganese(VII) oxide

$$Mn^{7+}\ O^{2-} = \underline{\underline{Mn_2O_7}}$$

**See Problems 7-9 through 7-18.**

## 7-3 Naming Compounds with Polyatomic Ions

Polyatomic ions contain two or more atoms covalently bonded together. The group of atoms has an electrical charge and thus behaves in much the same manner as the monatomic ions discussed previously. Table 7-3 contains the more common polyatomic anions (and one cation) encountered in chemistry.

In naming metal-polyatomic anion compounds, we follow essentially the same rules as before. The metal is named and written first (followed by the oxidation state if appropriate), and then the name of the anion is given or written.

There is some systemization possible that will help in learning Table 7-3. In many cases *the anions are composed of oxygen and one other element. These anions are thus called* **oxyanions.** When there are two oxyanions of

**Table 7-3** Polyatomic Ions

| Ion | Name | Ion | Name |
|---|---|---|---|
| $C_2H_3O_2^-$ | acetate | $HSO_3^-$ | hydrogen sulfite or bisulfite |
| $NH_4^+$ | ammonium | $OH^-$ | hydroxide |
| $CO_3^{2-}$ | carbonate | $ClO^-$ | hypochlorite |
| $ClO_3^-$ | chlorate | $NO_3^-$ | nitrate |
| $ClO_2^-$ | chlorite | $NO_2^-$ | nitrite |
| $CrO_4^{2-}$ | chromate | $C_2O_4^{2-}$ | oxalate |
| $CN^-$ | cyanide | $ClO_4^-$ | perchlorate |
| $Cr_2O_7^{2-}$ | dichromate | $MnO_4^-$ | permanganate |
| $HCO_3^-$ | hydrogen carbonate or bicarbonate | $PO_4^{3-}$ | phosphate |
| $HSO_4^-$ | hydrogen sulfate or bisulfate | $SO_4^{2-}$ | sulfate |
|  |  | $SO_3^{2-}$ | sulfite |

the same element (e.g., $SO_3^{2-}$ and $SO_4^{2-}$), they, of course, have different names. The anion with the element other than oxygen in the *higher* oxidation state uses the root of the element plus *ate*. The anion with the element in the *lower* oxidation state uses the root of the element plus *ite*. For example:

| | | |
|---|---|---|
| $SO_4^{2-}$ | sulfate | (S is in +6 ox. state) |
| $SO_3^{2-}$ | sulfite | (S is in +4 ox. state) |

The four oxyanions for Cl are

| | | |
|---|---|---|
| $ClO_4^-$ | perchlorate | (Cl is in +7 ox. state) |
| $ClO_3^-$ | chlorate | (Cl is in +5 ox. state) |
| $ClO_2^-$ | chlorite | (Cl is in +3 ox. state) |
| $ClO^-$ | hypochlorite | (Cl is in +1 ox. state) |

Notice that the two oxyanions of intermediate oxidation states are assigned the *ate* (+5) and the *ite* (+3) endings. The highest oxidation state (+7) also received a prefix *per* in addition to the *ate* ending. The lowest oxidation state (+1) receives the prefix *hypo* in addition to the *ite* ending. Oxyanions of Br and I follow the same pattern.

**Example 7-9**
Name the following compounds: $K_2CO_3$ and $Fe_2(SO_4)_3$.

**Answer**

| | |
|---|---|
| $K_2CO_3$ | potassium carbonate |
| $Fe_2(SO_4)_3$ | iron(III) sulfate |

**Example 7-10**
Give formulas for barium acetate, ammonium sulfate, thallium(III) nitrate, and manganese(III) phosphate.

**Answer**

barium acetate

Barium is in group IIA, so it has a +2 charge. Acetate is the $C_2H_3O_2^-$ ion.

$Ba^{(2+)}\ C_2H_3O_2^{(1-)} = Ba(C_2H_3O_2)_2$

(If more than one polyatomic ion is in the formula, enclose the ion in parentheses.)

| | |
|---|---|
| ammonium sulfate | From Table 7-3, ammonium = $NH_4^+$, sulfate = $SO_4^{2-}$. |
| | $NH_4^{1+}$ $SO_4^{2-}$ = $\underline{\underline{(NH_4)_2SO_4}}$ |
| thallium(III) nitrate | $Tl^{3+}$ $NO_3^{1-}$ = $\underline{\underline{Tl(NO_3)_3}}$ |
| manganese(III) phosphate | $Mn^{3+}$ $PO_4^{3-}$ = $Mn_3(PO_4)_3$ = $\underline{\underline{MnPO_4}}$ |

**Example 7-11**
Arsenic forms two oxyanions with the formulas $AsO_3^{3-}$ and $AsO_4^{3-}$. Name the two anions using the root *arsen*.

**Answer**

$AsO_3^{3-}$ is arsen<u>ite</u> (As is in the +3 oxidation state).

$AsO_4^{3-}$ is arsen<u>ate</u> (As is in the +5 oxidation state).

**Example 7-12**
Referring to the nomenclature for Cl, name the following: $NaBrO_4$ and $Ca(IO)_2$.

**Answer**
$NaBrO_4$
The Br is in the +7 oxidation state. Thus it is named in an analogous manner to Cl in the +7 oxidation state. Since $ClO_4^-$ is the *perchlorate* ion, the $BrO_4^-$ ion is the *perbromate* ion.

<u>sodium perbromate</u>

$Ca(IO)_2$
The I is in the +1 oxidation state. Thus the $IO^-$ ion is analogous to the $ClO^-$ ion.

<u>calcium hypoiodite</u>

**See Problems 7-19 through 7-29.**

## 7-4 Naming Nonmetal–Nonmetal Binary Compounds

In the case of all binary compounds, the procedure is to write the element that is the metal or is the closest to being a metal (the least electronegative) first and the other element second. For example, we write CO rather than OC and $H_2O$ rather than $OH_2$. Exception to this rule are ammonia, which is written $NH_3$, and methane (and other hydrogen–carbon compounds), which

**Table 7-4** Greek Prefixes

| Number | Prefix | Number | Prefix | Number | Prefix |
|---|---|---|---|---|---|
| 1 | mono | 5 | penta | 8 | octa |
| 2 | di | 6 | hexa | 9 | nona |
| 3 | tri | 7 | hepta | 10 | deca |
| 4 | tetra | | | | |

is written $CH_4$. This is simply because it's always been done that way for these well-known compounds.

The compounds are also named with the least electronegative element (most metallic) first when its English name is used. The more nonmetallic element is second, with its root plus *ide* as discussed before. Since nonmetals (except F) have variable oxidation states, some means must be provided to specify particular compounds. The Stock system is *not* preferred in this case, because it can still be ambiguous. For example, both $NO_2$ and $N_2O_4$ would have the Stock names of nitrogen(IV) oxide. In these compounds use is made of the Greek prefixes shown in Table 7-4 to indicate numbers of atoms in a compound.

The use of prefixes is illustrated by the names of the oxides of nitrogen listed in Table 7-5.

## 7-5 Naming Acids

Acids are a class of compounds so named because of a particular type of chemical reaction they undergo, particularly when present in water solutions. These reactions are discussed in Chapter 13. Most neutral compounds that behave as acids are hydrogen compounds of most of the anions listed in Tables 7-1 and 7-3. Unlike metal compounds of these anions, however, *acids are covalent in the pure state.*

**Table 7-5** The Oxides of Nitrogen

| Formula | Name | Oxidation State of N |
|---|---|---|
| $N_2O$ | dinitrogen monoxide (sometimes referred to as nitrous oxide) | +1 |
| NO | nitrogen monoxide (sometimes referred to as nitric oxide) | +2 |
| $N_2O_3$ | dinitrogen trioxide | +3 |
| $NO_2$ | nitrogen dioxide | +4 |
| $N_2O_4$ | dinitrogen tetroxide[a] | +4 |
| $N_2O_5$ | dinitrogen pentoxide[a] | +5 |

[a]The "a" is omitted from *tetra* and *penta* for ease in pronunciation.

**See Problems 7-30 through 7-35.**

**Table 7-6** Binary Acids

| Anion | Formula of Acid | Compound Name | Acid Name |
|---|---|---|---|
| $Cl^-$ | $HCl$ | hydrogen chloride | hydrochloric acid |
| $F^-$ | $HF$ | hydrogen fluoride | hydrofluoric acid |
| $I^-$ | $HI$ | hydrogen iodide | hydroiodic acid |
| $S^{2-}$ | $H_2S$ | hydrogen sulfide | hydrosulfuric acid |

**Binary acids** *are hydrogen compounds of most of the monatomic anions listed in Table 7-1.* The binary acids are privileged to have two names depending on whether they are referred to in the pure state or as an acid in aqueous solution. The compound name is like any other binary compound previously discussed. The acid name is obtained by adding the prefix *hydro* to the anion root and changing the *ide* ending to *ic* followed by the word *acid.* Both names are illustrated in Table 7-6.

The following are *not* generally considered to be binary acids: $H_2O$, $NH_3$, $CH_4$, and $PH_3$.

In addition to binary acids, **oxyacids** *are hydrogen compounds of most of the oxyanions listed in Table 7-3.* To name an oxyacid, we use the root of the anion to form the name of the acid. If the name of the oxyanion ends in *ate,* it is changed to *ic* followed by the word *acid.* If the name of the anion ends in *ite,* it is changed to *ous* plus the word *acid.* Hydrogen compounds of oxyanions are usually named only as an acid (e.g., $HNO_3$ is referred to only as *nitric acid,* not as hydrogen nitrate). Naming of oxyacids is illustrated in Table 7-7.

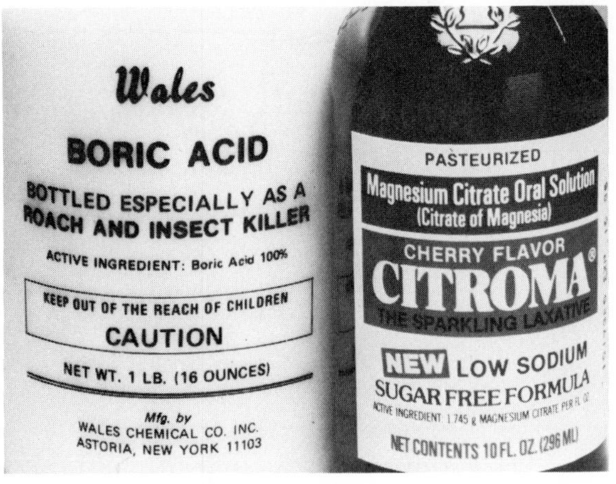

Figure 7-1 COMPOUNDS AND PRODUCTS. Ionic compounds, molecular compounds, and acids are all used in various products.

**Table 7-7** Oxyacids

| Anion | Name of Anion | Formula of Acid | Name of Acid |
|---|---|---|---|
| $C_2H_3O_2^-$ | acetate | $HC_2H_3O_2$ | acetic acid |
| $CO_3^{2-}$ | carbonate | $H_2CO_3$ | carbonic acid |
| $NO_3^-$ | nitrate | $HNO_3$ | nitric acid |
| $PO_4^{3-}$ | phosphate | $H_3PO_4$ | phosphoric acid |
| $ClO_2^-$ | clorite | $HClO_2$ | chlorous acid |
| $ClO_4^-$ | perchlorate | $HClO_4$ | perchloric acid |
| $SO_3^{2-}$ | sulfite | $H_2SO_3$ | sulfurous acid |
| $SO_4^{2-}$ | sulfate | $H_2SO_4$ | sulfuric acid |

**Example 7-13**

Name the following acids: $H_2Se$, $H_2C_2O_4$, and HClO.

**Answer**

| | |
|---|---|
| $H_2Se$ | hydroselenic acid |
| $H_2C_2O_4$ | oxalic acid |
| HClO | hypochlorous acid |

**Example 7-14**

Give formulas for the following: permanganic acid, chromic acid, and acetic acid.

**Answer**

| | |
|---|---|
| permanganic acid | $HMnO_4$ |
| chromic acid | $H_2CrO_4$ |
| acetic acid | $HC_2H_3O_2$ |

## 7-6 Chapter Summary

This entire chapter can be summarized by the following flow chart for the naming of compounds.

**Naming Compounds Flow Chart**

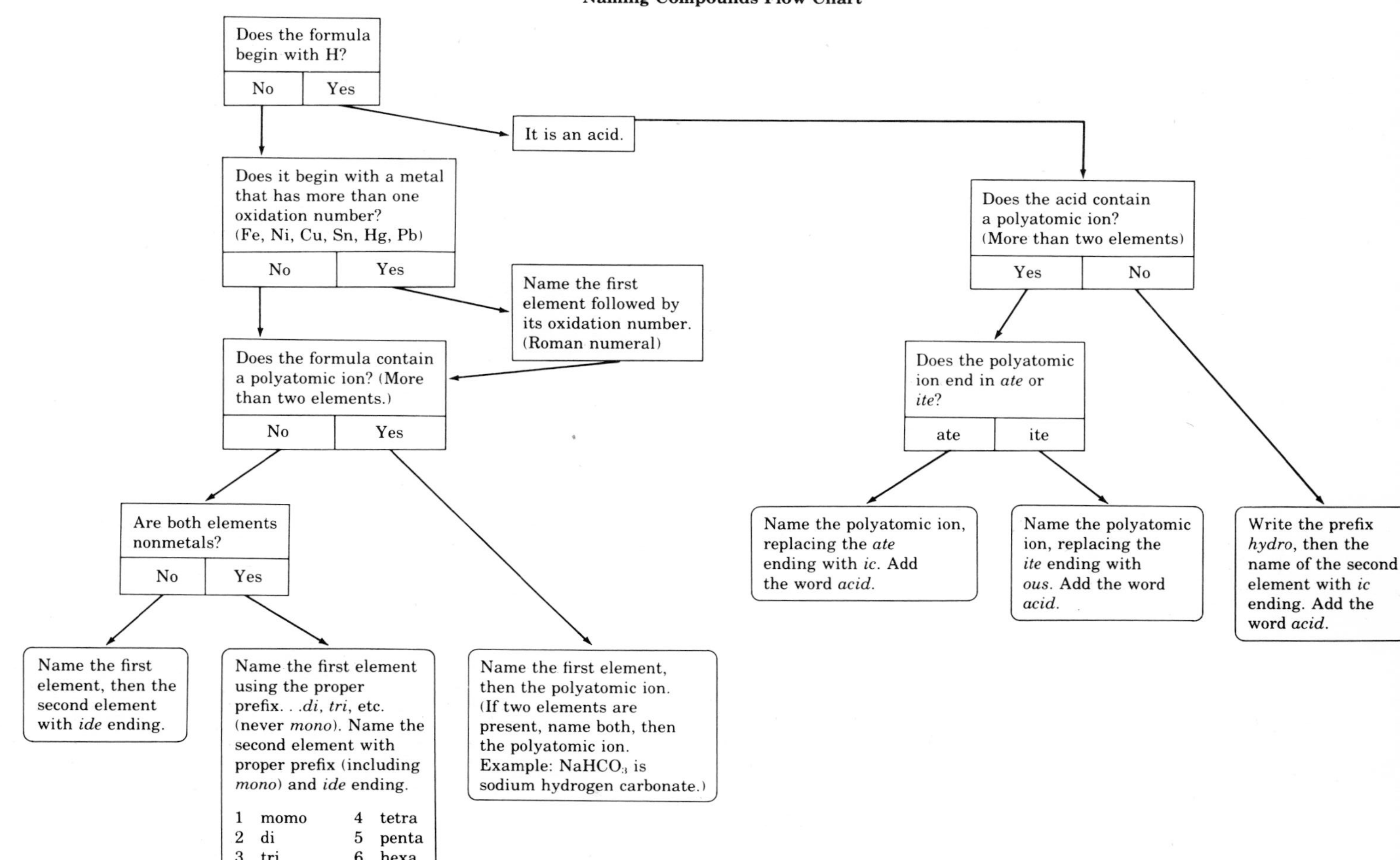


Source: Reprinted from D. Robson, *J. Chem. Ed.*, 1983, **60**, 131.

## Exercises

### Oxidation States

**7-1** Give the oxidation states of the elements in each of the following compounds.
(a) $PbO_2$ (b) $P_4O_{10}$ (c) $C_2H_2$ (d) $N_2H_4$ (e) LiH (f) $BCl_3$ (g) $Rb_2Se$ (h) $Bi_2S_3$

**7-2** Give the oxidation states of the elements in each of the following compounds.
(a) $ClO_2$ (b) $XeF_2$ (c) CO (d) $O_2F_2$ (e)$Mn_2O_3$ (f) $Bi_2O_5$

**7-3** Which of the following elements form *only* the +1 oxidation state in compounds?
(a) Li (b) H (c) Ca (d) Cl (e) K (f) Al (g) Rb

**7-4** Which of the following elements form *only* the +2 oxidation state in compounds?
(a) O (b) B (c) Be (d) Sr (e) Sc (f) Hg (g) Ca

**7-5** What is the only oxidation state of Al in its compounds?

**7-6** What is the oxidation state of each of the following?
(a) the P in $H_3PO_4$
(b) the C in $H_2C_2O_4$
(c) the Cl in $ClO_4^-$
(d) the Cr in $CaCr_2O_7$
(e) the S in $SF_6$
(f) the N in $CsNO_3$
(g) the Mn in $KMnO_4$

**7-7** What is the oxidation state of each of the following?
(a) the Se in $SeO_3^{2-}$
(b) the I in $H_5IO_6$
(c) the S in $Al_2(SO_3)_3$
(d) the Cl in $HClO_2$
(e) the N in $(NH_4)_2S$

**7-8** Nitrogen exists in nine oxidation states. Order the following compounds in order of increasing oxidation state of N.
$K_3N$, $N_2O_4$, $N_2$, $NH_2OH$, $N_2O$, $Ca(NO_3)_2$, $N_2H_4$, $N_2O_3$, NO

### Metal-Nonmetal Binary Compounds

**7-9** Name the following compounds.
(a) LiF (b) BaTe (c) $Sr_3N_2$ (d) $BaH_2$ (e) $AlCl_3$

**7-10** Name the following compounds
(a) $CaI_2$ (b) FrF (c) BeSe (d) $Mg_3P_2$ (e) RaS

**7-11** Give formulas for the following compounds.
(a) rubidium selenide
(b) strontium hydride
(c) radium oxide
(d) aluminum carbide
(e) beryllium fluoride

**7-12** Give formulas for the following compounds.
(a) potassium hydride
(b) cesium sulfide
(c) potassium phosphide
(d) barium telluride

**7-13** Name the following compounds using the Stock method.
(a) $Bi_2O_5$ (b) SnS (c) $SnS_2$ (d) $Cu_2Te$ (e) TiC

**7-14** Name the following compounds using the Stock method.
(a) $CrI_3$ (b) $TiCl_4$ (c) $IrO_4$ (d) $MnH_2$ (e) $NiCl_2$

**7-15** Give formulas for the following compounds.
(a) copper(I) sulfide
(b) vanadium(III) oxide
(c) silver(I) bromide
(d) nickel(II) carbide
(e) chromium(VI) oxide

**7-16** Give formulas for the following compounds.
(a) yttrium(III) hydride
(b) lead(IV) chloride
(c) bismuth(V) fluoride
(d) palladium(II) selenide

**7-17** From the magnitude of the oxidation states of the metals predict which of the compounds in Problems 7-13 and 7-15 are primarily covalent.

**7-18** From the magnitude of the oxidation states of the metal predict which of the compounds in Problems 7-14 and 7-16 are primarily covalent.

## Compounds with Polyatomic Ions

**7-19** Which of the following is the chlorate ion?
(a) $ClO_2^-$ (b) $ClO_4^-$ (c) $ClO_3^-$
(d) $Cl_3O^-$ (e) $ClO_3^+$

**7-20** Which of the following ions have a −2 charge?
(a) sulfate (d) carbonate
(b) nitrite (e) sulfite
(c) chlorite (f) phosphate

**7-21** What is the name and formula of the most common polyatomic cation?

**7-22** Which of the following oxyanions contain four oxygen atoms?
(a) nitrate (e) phosphate
(b) permanganate (f) oxalate
(c) perchlorate (g) carbonate
(d) sulfite

**7-23** Name the following compounds. Use the Stock method where appropriate.
(a) $CrSO_4$ (e) $(NH_4)_2CO_3$
(b) $Al_2(SO_3)_3$ (f) $NH_4NO_3$
(c) $Fe(CN)_2$ (g) $Bi(OH)_3$
(d) $RbHCO_3$

**7-24** Name the following compounds. Use the Stock method where appropriate.
(a) $Na_2C_2O_4$ (b)$CaCrO_4$ (c) $Fe_2(CO_3)_3$
(d) $Cu(OH)_2$

**7-25** Give formulas for the following compounds.
(a) magnesium permanganate (e) indium(III) bisulfate
(b) cobalt(II) cyanide (f) iron(III) oxalate
(c) strontium hydroxide (g) ammonium dichromate
(d) thallium(I) sulfite (h) mercury(I) acetate [The mercury(I) ion exists as $Hg_2^{2+}$.]

**7-26** Give formulas for the following compounds.
(a) zirconium(IV) phosphate (e) radium hydrogen sulfate
(b) sodium cyanide (f) beryllium phosphate
(c) thallium(I) nitrite (g) chromium(III) hypochlorite
(d) nickel(II) hydroxide

**7-27** Give the systematic name for each of the following.

| Common Name | Formula | Systematic Name |
|---|---|---|
| (a) Table salt | $NaCl$ | |
| (b) Baking soda | $NaHCO_3$ | |
| (c) Marble or limestone | $CaCO_3$ | |
| (d) Lye | $NaOH$ | |
| (e) Chile saltpeter | $NaNO_3$ | |
| (f) Sal ammoniac | $NH_4Cl$ | |
| (g) Alumina | $Al_2O_3$ | |
| (h) Slaked lime | $Ca(OH)_2$ | |
| (i) Caustic potash | $KOH$ | |

**7-28** The perxenate ion has the formula $XeO_6^{4-}$. Write formulas of compounds of perxenate with the following.
(a) calcium (b) potassium (c) aluminum

* **7-29** Name the following compounds. In these compounds an ion is involved that is not in Table 7-3. However, the name can be determined by reference to other ions of the central element or from ions in Table 7-3 where the central atom is in the same group.
(a) $PH_4F$ (b) $KBrO$ (c) $Co(IO_4)_3$
(d) $CaSiO_3$ (e) $AlPO_3$ (f) $CrMoO_4$

## Nonmetal-Nonmetal Binary Compounds

**7-30** The following two elements combine to make binary compounds. Which element should be written and named first?
(a) Si and S (b) F and I (c) H and B
(d) Kr and F (e) H and F

**7-31** The following two elements combine to

make binary compounds. Which element should be written and named first?
(a) S and P (b) O and Cl (c) O and F
(d) As and Cl

**7-32** Name the following.
(a) $CS_2$ (b) $BF_3$ (c) $P_4O_{10}$ (d) $Br_2O_3$ (e) $SO_3$ (f) $Cl_2O$ (g) $PCl_5$ (h) $SF_6$

**7-33** Name the following.
(a) $PF_3$ (b) $I_2O_3$ (c) $ClO_2$ (d) $AsF_5$
(e) $SeCl_4$ (f) $SiH_4$

**7-34** Write the formulas for each of the following.
(a) tetraphosphorus hexoxide (b) carbon tetrachloride (c) iodine trifluoride (d) dichlorine heptoxide (e) sulfur hexafluoride (f) xenon dioxide

**7-35** Write formulas for each of the following.
(a) xenon trioxide (b) sulfur dichloride (c) dibromine monoxide (d) carbon disulfide (e) diboron hexahydride (also known as diborane)

## Acids

**7-36** Name the following acids.
(a) HCl (b) $HNO_3$ (c) HClO (d) $HMnO_4$
(e) $HIO_4$ (f) HBr

**7-37** Write formulas for each of the following acids.
(a) cyanic acid (b) hydroselenic acid (c) chlorous acid (d) carbonic acid (e) hydroiodic acid (f) acetic acid

**7-38** Write formulas for each of the following acids.
(a) oxalic acid (b) nitrous acid (c) dichromic acid (d) phosphoric acid

* **7-39** Refer to the ions in Problems 7-28 and 7-29. Write the acid names for the following.
(a) HBrO (b) $HIO_4$ (c) $H_3PO_3$
(d) $HMoO_4$ (e) $H_4XeO_6$

**7-40** Write the formulas and the names of the acids formed from the arsenite ($AsO_3^{3-}$) ion and the arsenate ($AsO_4^{3-}$) ion.

# Review Test on Chapters 4–7*

*The following multiple choice questions have one correct answer.*

1. Which of the following are the sublevels present in the third (n = 3) shell?
   (a) *s, p* (b) *p, d, f* (c) *s, p, d* (d) *s, p, d, f*
2. What is the electron capacity of the 4*f* sublevel?
   (a) 6 (b) 7 (c) 14 (d) 8 (e) 10
3. Which of the following elements has the electron configuration [Kr] $5s^2 4d^2$?
   (a) Sn (b) Zr (c) Ti (d) Pr (e) Sr
4. What is the electron configuration of In (at. no. 49)?
   (a) [Kr] $4s^2 4p^1$ (d) [Kr] $4s^2 3d^{10} 4p^1$
   (b) [Kr] $5s^2 4d^1$ (e) [Kr] $5s^2 4d^6 5p^1$
   (c) [Kr] $5s^2 4d^{10} 5p^1$
5. Silicon has two unpaired electrons. This is a result of:
   (a) Hund's rule (c) the Aufbau principle
   (b) the Pauli exclusion principle (d) Bohr's model
6. How many orbitals are in a *d* sublevel?
   (a) 5 (b) 3 (c) 10 (d) 14 (e) 2
7. Which of the following is the box diagram for electrons in the outer orbitals of a sulfur atom?

(a) 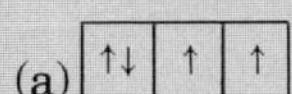
(b) ↑↓ | ↓ | ↑
(c) 
(d) ↑↓
(e) 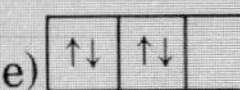

8. The following is the shape of what type of orbital?

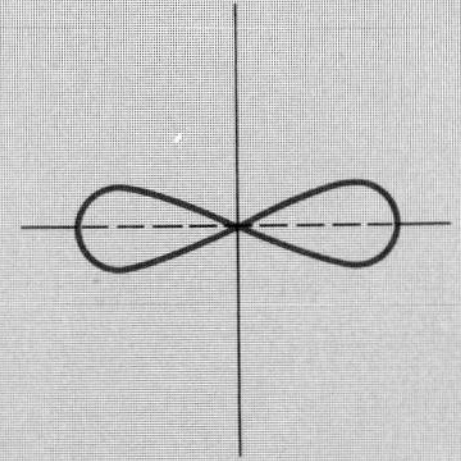

   (a) *s* (b) *p* (c) *d* (d) *f*
9. Which of the following elements is an alkali metal?
   (a) Sc (b) Al (c) Cl (d) K (e) H
10. The element U (at. no. 92) is in which classification of elements?
    (a) representative (b) transition (c) noble gas (d) inner transition
11. Which of the following is an allotrope of carbon?
    (a) ozone (b) diamond (c) carbon dioxide (d) sugar (e) quartz
12. Which of the following groups of elements have the general electron configuration $ns^2np^5$?
    (a) halogens (b) alkaline earths (c) noble gases (d) Group VA (e) transition elements
13. Which of the following atoms has the smallest radius?
    (a) As (b) Se (c) Sb (d) Te
14. Which of the following atoms has the highest first ionization energy?
    (a) N (b) O (c) P (d) S
15. Which of the following ions requires the most energy to form?
    (a) $K^+$ (b) $Ca^{3+}$ (c) $Al^{3+}$ (d) $F^-$ (e) $O^{2-}$
16. Which of the following ions violates the octet rule?
    (a) $I^{2-}$ (b) $S^{2-}$ (c) $Br^-$ (d) $N^{3-}$ (e) $Te^{2-}$
17. Which of the following two elements would be expected to form an ionic bond?
    (a) Mg, Ca (b) B, F (c) Mg, Br (d) S, O (e) C, O
18. Which of the following is the correct formula for the compound formed between beryllium and bromine?
    (a) BeBr (b) $Br_2Be$ (c) $BBr_2$ (d) $BeBr_2$ (e) $Be_2Br$
19. Which of the following two elements would be most likely to form a covalent bond?
    (a) Al, F (b) K, Sn (c) As, Br (d) K, H
20. Which of the following is the correct

*Answers to all questions and problems in the review tests are given in Appendix H.

formula for the simplest compound formed between As and Cl?
(a) $Cl_2As$ (b) $AsCl_2$ (c) $AsCl_3$ (d) AsCl (e) $AsCl_4$

21. Which of the following compounds contains both ionic and covalent bonds?
(a) $BaCO_3$ (b) $H_2CO_3$ (c) $HNO_3$ (d) $K_3N$ (e) $CCl_4$

22. Which of the following is the most electronegative element?
(a) B (b) Na (c) O (d) Cl (e) N

23. Which of the following bonds is the most polar?
(a) C—O (b) B—Cl (c) Be—Cl (d) Be—F (e) N—N

24. Which of the following molecules is polar?
(a) $N_2O_4$ (b) NO (c) $CO_2$ (d) $BF_3$

25. Which of the following metals has more than one oxidation state?
(a) Na (b) Be (c) Bi (d) Al (e) Ra

26. What is the oxidation state of the N in $NO_2$?
(a) +2 (b) +3 (c) −4 (d) +4 (e) +6

27. What is the oxidation state of the S in $Na_2S_2O_3$?
(a) +4 (b) +2 (c) −4 (d) +1 (e) −2

28. Which of the following is the formula for the permanganate ion?
(a) $MnO_4$ (b) $MnO_4^{2-}$ (c) $MnO_3^{2-}$ (d) $MgO_4^-$ (e) $MnO_4^-$

29. Which of the following is the commonly accepted name for $SeO_3$?
(a) selenium(VI) oxide (b) selenium oxide (c) selenium dioxide (d) selenium trioxide

30. Which of the following is the name for $HClO_3$?
(a) hydrogen chlorate (b) chlorous acid (c) perchloric acid (d) chloric acid (e) hydrochloric acid

## Problems

1. Answer the following questions about the element aluminum (Al).
   (a) What is its group number and general classification?
   (b) What is its electron configuration?
   (c) Is it a solid, liquid, or gas?
   (d) Is it a metal or nonmetal?
   (e) What is its electrical charge when present as an ion?
   (f) What noble gas has the same electron configuration as its ion?
   (g) What is the simplest formula when it forms a binary compound with (1) sulfur, (2) bromine, and (3) nitrogen?
   (h) What are the names of the three compounds in question (g)?
2. Answer the following questions about the element nitrogen (N).
   (a) What is its group number and general classification?
   (b) What is its electron configuration?
   (c) How many unpaired electrons are in an atom of nitrogen?
   (d) Is it a metal or nonmetal?
   (e) Does it exist as a solid, liquid, or gas under normal conditions?
   (f) What is the formula of the element?
   (g) What is the Lewis structure of the element?
   (h) What is its electrical charge when present as a monatomic ion?
   (i) What noble gas has the same electron configuration as its monatomic ion?
   (j) What is its polarity (negative or positive) when covalently bound to (1) oxygen and (2) boron?
   (k) What is the simplest formula when it forms a binary compound with (1) magnesium, (2) lithium, and (3) fluorine?
   (l) What are the names of the three compounds in question (k)?
   (m) What is the Lewis structure of the compound with fluorine in part (k)?
3. Answer the following with the symbol of the appropriate element.
   (a) a diatomic gas in nature but whose atoms have two unpaired electrons
   (b) a monatomic gas in nature but the atoms do not have an octet of electrons
   (c) the first element with a filled *d* sublevel
   (d) a nonmetal that forms a −2 ion that has a filled 4*d* sublevel
   (e) a metal that forms a +3 ion with a noble gas configuration by loss of a 4*d* electron

(f) a metal with a half-filled *p* sublevel but no *f* electrons
(g) the most chemically reactive element
(h) a transition element with 24 electrons beyond the previous noble gas
(i) a solid nonmetal whose atoms can form a −3 ion
(j) a metal that forms a +2 ion with the electron configuration of Xe

4. For each of the following compounds write (1) the formula in ionic form if ions are present, (2) the name of the compound, and (3) the Lewis structure.

| Formula | Ionic Form | Name | Lewis Structure |
|---|---|---|---|
| NaClO | $Na^+ClO^-$ | sodium hypochlorite | $Na^+ \left[ :\ddot{\underset{..}{Cl}}-\ddot{\underset{..}{O}}: \right]^-$ |
| (a) $MgSO_4$ | | | |
| (b) $HNO_2$ | | | |
| (c) $LiNO_3$ | | | |
| (d) $Co_2(CO_3)_3$ | | | |
| (e) $Cl_2O$ | | | |

5. For each of the following compounds write (1) the formula of the compound, (2) the formula in ionic form if ions are present, and (3) the Lewis structure.

| Name | Formula | Ionic Form | Lewis Structure |
|---|---|---|---|
| (a) dinitrogen trioxide | | | |
| (b) chromium(III) sulfite | | | |
| (c) iron(II) hydroxide | | | |
| (d) strontium oxalate | | | |
| (e) hydroiodic acid | | | |

There are more atoms in a single glass of water than grains of sand in the picture.

# Chapter 8
# Quantitative Relationships: The Mole

## Purpose of Chapter 8

In Chapter 8 we introduce the unit of amount in chemistry, the mole, which relates a certain number of atoms, molecules, or formula units to the mass of an element or compound.

## Objectives for Chapter 8

After completion of this chapter, you should be able to:

1. Describe the unit known as the mole and tell why it is needed in chemistry. (Intro., 8-1)
2. Write the molar mass of any element from the periodic table. (8-2)
3. Calculate the mass of the same number of atoms of one element given the mass of a different element. (8-2)
4. Convert between moles, mass, and number of atoms of any element. (8-2)
5. Calculate the formula weight of a specified compound. (8-3)
6. Convert between moles, mass, and number of molecules or formula units of a compound. (8-3)
7. Calculate the percent composition of a compound from its formula. (8-4)
8. Distinguish between an empirical and a molecular formula. (8-5)
9. Calculate the empirical formula of a compound from its percent composition or weight composition. (8-5)
10. Use the data from chemical analysis to establish the molecular formula of a compound. (8-5, 8-6)

Thus far, our efforts have been directed toward a description of the most basic forms of matter. Now we turn our attention to the quantitative relations of chemistry. For this effort, the material in Chapter 2 and the mathematical review available in the Appendix will help to make this work surprisingly straightforward. In this chapter, you will find that it is especially helpful to work problems assigned in the margin as you proceed. Throughout the chapter, one concept builds directly on another.

In Chapter 3, the small size of the atom was discussed. Even with the most powerful microscope, atoms cannot be seen except for fuzzy images of the larger atoms. Indeed, in order to have enough carbon atoms to be able to see a small, black smudge the size of the period at end of this sentence, it would take about $10^{15}$ atoms (one quadrillion atoms). You would not be able to count that far in a lifetime.

The atomic weights of the elements were also discussed in Chapter 3. The mass or weight of *one* atom of $^{12}_{6}C$ was defined as exactly 12 amu (atomic mass units). From this the mass of other isotopes could be determined by comparison to $^{12}_{6}C$, and from an element's natural abundance of isotopes, the average atomic weight of the element was obtained. The weight unit *amu,* which was introduced at that time, is of no use in laboratory situations, however, because it represents such a tiny weight. For example, the weight of *one* hydrogen atom, which is about 1 amu, in grams is

$$1.01 \text{ amu} = 1.68 \times 10^{-24} \text{ g}$$

Obviously, we will not be able to deal with known numbers of atoms in a laboratory situation by an actual count. However, we can use weights that represent equivalent or known numbers of atoms or molecules. To do this, however, we need a counting unit that represents a huge number of particles or objects.

## 8-1 The Mole

It would be convenient if one of the familiar units such as dozen or gross could be used with atoms. Unfortunately, they are of little help. Instead of about $10^{14}$ atoms, there would be about $10^{13}$ dozen atoms, or about $10^{12}$ gross of atoms. Obviously, this hardly simplifies the numbers involved. Thus it is necessary to introduce a new unit to deal with very large numbers. This unit is called the **mole** (the SI symbol is "mol"). The value of 1 mol is

$$1.00 \text{ mol} = 6.02 \times 10^{23} \text{ objects or particles}$$

One mole of iron atoms in the form of nails weighs about 2 oz, which is plenty to deal with in a laboratory situation.

*The* **molar number,** *$6.02 \times 10^{23}$, is also known as* **Avogadro's number.** It is not an exact and defined number such as 12 in one dozen or 144 in one gross. It is actually known to more significant figures than the three that are shown and used here.

**Figure 8-1** MOLES OF ELEMENTS. One mole of Cu, Fe, and Hg contain the same number of atoms.

## 8-2 The Molar Mass of the Elements

The mole is defined by a mass as well as a number. *The mass of 1 mol of an element is the atomic weight expressed in grams. This is known as the* **molar mass** of an element. From the periodic table you can see that the mass of 1 mol of oxygen atoms is 16.0 g (to three significant figures), the mass of 1 mol of helium atoms is 4.00 g, and the mass of 1 mol of uranium atoms is 238 g.

One mole of atoms implies two things: the molar *mass,* which is different for each element, and the molar *number,* which is the same for each element. (See Figure 8-1.)

In many of the calculations to come it is not necessary to know the actual numbers of atoms involved. It is very helpful, however, to realize when we are dealing with the same number of atoms of different elements. As an example, from the periodic table, notice that one helium atom (4.00 amu) weighs one-third as much as one carbon atom (12.0 amu). As shown

**Table 8-1** Weight Relation of C and He

| C | He | Number of Atoms of Each Element Present |
|---|---|---|
| 12.0 amu | 4.00 amu | 1 |
| 24.0 amu | 8.00 amu | 2 |
| 360 amu | 120 amu | 30 |
| 12.0 g | 4.00 g | $6.02 \times 10^{23}$ |
| 12.0 lb | 4.00 lb | $2.73 \times 10^{26}$ |
| 24.0 ton | 8.00 ton | $1.09 \times 10^{30}$ |

in Table 8-1, whenever helium and carbon are present in a 4:12 (1:3) weight ratio, regardless of the units, the same number of atoms are present in each case.

**Example 8-1**

The formula of the compound magnesium sulfide (MgS) indicates that there is one atom of Mg for every atom of S. What weight of sulfur is combined with 46.0 lb of magnesium?

**Procedure**

From the atomic weights in the periodic table, notice that there are the same number of magnesium atoms in 24.3 g of magnesium as there are sulfur atoms in 32.1 g of sulfur. There is also the same number of atoms in 24.3 lb of magnesium as there are in 32.1 lb of sulfur. This statement can be converted into two conversion factors, which we can use to convert a weight of one element to an equivalent weight of the other.

$$(1)\ \frac{24.3 \text{ lb of Mg}}{32.1 \text{ lb of S}} \qquad (2)\ \frac{32.1 \text{ lb of S}}{24.3 \text{ lb of Mg}}$$

Use factor 2 to convert weight of Mg to an equivalent weight of S.

$$46.0 \cancel{\text{ lb of Mg}} \times \frac{32.1 \text{ lb of S}}{24.3 \cancel{\text{ lb of Mg}}} = \underline{\underline{60.8 \text{ lb of S}}}$$

See Problems 8-1 through 8-8.

There are the same number of atoms in 60.8 lb of sulfur as in 46.0 lb of magnesium.

Review the unit-factor method in Appendix D and Chapter 2.

As you can see, there are a number of relationships that are implied with the mole. Before illustrating the large number of problems possible from these relationships, let's preview just how familiar the problems will be with a simple, everyday, dozen oranges. We will then work problems involving the less familiar mole of atoms. In these problems the value of the unit-factor method of working problems will become apparent.

In these problems, assume that there is such a thing as a standard orange (all oranges are exactly the same). The "dozen mass" (the weight of 12) of those particular oranges is 3.60 lb. The relationships given or implied by this statement are as follows:

$$1 \text{ doz oranges} = 12 \text{ oranges} \qquad 1 \text{ doz oranges} = 3.60 \text{ lb}$$

From these relationships, four conversion factors can be written:

$$(1)\ \frac{1\text{ doz oranges}}{12\text{ oranges}} \qquad (2)\ \frac{12\text{ oranges}}{\text{doz oranges}}$$

$$(3)\ \frac{3.60\text{ lb}}{\text{doz oranges}} \qquad (4)\ \frac{1\text{ doz oranges}}{3.60\text{ lb}}$$

**Example 8-2a**
What is the weight of exactly 3 dozen oranges?

**Procedure**
Convert dozen to weight. Use a conversion factor that has what you want (lb) in the numerator and what you wish to cancel (doz) in the denominator. This is factor 3. In shorthand this can be written:

$$\text{dozen} \xrightarrow{\text{factor 3}} \text{weight}$$

**Solution**

$$3.00\ \cancel{\text{doz oranges}} \times \frac{3.60\text{ lb}}{\cancel{\text{doz oranges}}} = \underline{\underline{10.8\text{ lb}}}$$

**Example 8-3a**
How many dozens of oranges are there is 18.0 lb?

**Procedure**
This is the reverse of the previous problem, so use the inverse of the previous factor, which is factor 4.

$$\text{weight} \xrightarrow{\text{factor 4}} \text{dozen}$$

**Solution**

$$18.0\ \cancel{\text{lb}} \times \frac{1\text{ doz}}{3.60\ \cancel{\text{lb}}} = \underline{\underline{5.00\text{ doz oranges}}}$$

**Examples 8-4a**
What is the weight of 142 oranges?

**Procedure**
Since a factor that relates number to weight is not available, use two factors in a two-step process. Note that you can go from number of dozens (factor 1) and from dozens to weight (factor 3).

$$\text{number} \xrightarrow{\text{factor 1}} \text{dozen} \xrightarrow{\text{factor 3}} \text{weight}$$

**Solution**

$$142\ \cancel{\text{oranges}} \times \frac{1\ \cancel{\text{doz oranges}}}{12\ \cancel{\text{oranges}}} \times \frac{3.60\text{ lb}}{\cancel{\text{doz oranges}}} = \underline{\underline{42.6\text{ lb}}}$$

**Example 8-5a**
How many individual oranges are there in 73.5 lb?

**Procedure**
This is the reverse of the previous problem, so convert what's given to dozens (factor 4) and then dozens to individual oranges (factor 2).

$$\text{weight} \xrightarrow{\text{factor 4}} \text{dozen} \xrightarrow{\text{factor 2}} \text{number}$$

**Solution**

$$73.5\ \cancel{\text{lb}} \times \frac{1\ \cancel{\text{doz oranges}}}{3.60\ \cancel{\text{lb}}} \times \frac{12\ \text{oranges}}{\cancel{\text{doz oranges}}} = \underline{\underline{245\ \text{oranges}}}$$

**Review scientific notation in Appendix C.**

Now, instead of working with dozens and "dozen mass," let's work some very similar problems with moles and molar mass. Review scientific notation in Appendix C if necessary. Notice that Example 8-2a is the analog of Example 8-2b, and so forth.

In these problems you will work with sodium atoms instead of oranges. The molar mass (the weight of $6.02 \times 10^{23}$ atoms) of Na is 23.0 g. The relationships given or implied by this statement are as follows:

$$1\ \text{mol of Na} = 6.02 \times 10^{23}\ \text{atoms} \qquad 1\ \text{mol of Na} = 23.0\ \text{g}$$

From these relationships, four conversion factors can be written:

$$(1)\ \frac{1\ \text{mol Na}}{6.02 \times 10^{23}\ \text{atoms Na}} \qquad (2)\ \frac{6.02 \times 10^{23}\ \text{atoms Na}}{\text{mol Na}}$$

$$(3)\ \frac{23.0\ \text{g}}{\text{mol Na}} \qquad (4)\ \frac{1\ \text{mol Na}}{23.0\ \text{g}}$$

**Example 8-2b**
What is the weight of exactly 3 mol of Na?

**Procedure**
Convert moles to weight. Use a conversion factor that has what you want (g) in the numerator and what you wish to cancel (mol) in the denominator. This is factor 3. In shorthand, this can be written:

$$\text{moles} \xrightarrow{\text{factor 3}} \text{weight}$$

**Solution**

$$3.00\ \cancel{\text{mol Na}} \times \frac{23.0\ \text{g}}{\cancel{\text{mol Na}}} = \underline{\underline{69.0\ \text{g}}}$$

**Example 8-3b**
How many moles are there in 34.5 g of Na?

**Procedure**
This is the reverse of the previous problem, so use the inverse of the previous factor, which is factor 4.

$$\text{weight} \xrightarrow{\text{factor 4}} \text{moles}$$

**Solution**

$$34.5\ \cancel{g} \times \frac{1 \text{ mol Na}}{23.0\ \cancel{g}} = \underline{\underline{1.50 \text{ mol Na}}}$$

**Example 8-4b**
What is the weight of $1.20 \times 10^{24}$ atoms of Na?

**Procedure**
Since a factor that relates number to weight is not available, use two factors in a two-step process. Note that you can go from number to moles (factor l) and from moles to weight (factor 3).

$$\text{number} \xrightarrow{\text{factor 1}} \text{moles} \xrightarrow{\text{factor 3}} \text{weight}$$

**Solution**

$$1.20 \times 10^{24}\ \cancel{\text{atoms Na}} \times \frac{1\ \cancel{\text{mol Na}}}{6.02 \times 10^{23}\ \cancel{\text{atoms Na}}} \times \frac{23.0 \text{ g}}{\cancel{\text{mol Na}}} = \underline{\underline{46.0\text{g}}}$$

**Example 8-5b**
How many individual atoms are there in 11.5 g of Na?

**Procedure**
This is the reverse of the previous problem, so convert what's given to moles (factor 4) and then moles to individual atoms (factor 2).

$$\text{weight} \xrightarrow{\text{factor 4}} \text{moles} \xrightarrow{\text{factor 2}} \text{number}$$

**Solution**

$$11.5\ \cancel{g} \times \frac{1\ \cancel{\text{mol Na}}}{23.0\ \cancel{g}} \times \frac{6.02 \times 10^{23} \text{ atoms Na}}{\cancel{\text{mol Na}}}$$

$$= \underline{\underline{3.01 \times 10^{23} \text{ atoms Na}}}$$

At this point you should pay special attention to the problems at the end of this chapter indicated in the margin. It is essential that you be able to work these types of problems backward and forward. If you master these problems, the rest of this chapter and the next should flow smoothly. For additional practice on problems of this type, refer to Appendix D, especially Examples D-12, D-13, D-18, D-19, D-21, and D-22.

**See Problems 8-9 through 8-17.**

## 8-3 The Molar Mass of Compounds

*The* **formula weight** *of a compound is determined from the number of atoms and the atomic weight of each element indicated by the formula.* As mentioned in previous chapters, the formula of a compound may represent discrete molecular units* or ratios of ions in ionic compounds (referred to as a formula unit). In this text we will use formula weight to cover either possibility. The following examples illustrate the computation of formula weights.

**Example 8-6**

What is the formula weight of $CO_2$?

| Atom | Number of Atoms in Molecule | | Atomic Weight | | Total Weight of Atom in Molecule |
|---|---|---|---|---|---|
| C | 1 | × | 12.0 amu | = | 12.0 amu |
| O | 2 | × | 16.0 amu | = | 32.0 amu |
| | | | | | 44.0 amu |

The formula weight of $CO_2$ is

44.0 amu

**Example 8-7**

What is the formula weight of $Fe_2(SO_4)_3$?

| Atom | Number of Atoms in Formula Unit | | Atomic Weight | | Total Weight of Atom in Formula Unit |
|---|---|---|---|---|---|
| Fe | 2 | × | 55.8 amu | = | 112 amu |
| S | −3 | × | 32.1 amu | = | 96.3 amu |
| O | 12 | × | 16.0 amu | = | 192 amu |
| | | | | | 400 amu |

The formula weight of $Fe_2(SO_4)_3$ is

400 amu

**See Problems 8-18 and 8-19.**

*Where molecules are present, the formula weight is often referred to as the molecular weight.

**Figure 8-2** MOLES OF COMPOUNDS. One mole of $NaHCO_3$, $H_2O$, and NaCl contain the same number of molecules or formula units.

*The weight of one mole of molecules or ionic formula units is referred to as the* **molar mass** *of the compound. (The units of molar mass are thus g/mol.) The molar mass is the formula weight expressed in grams and represents the weight of $6.02 \times 10^{23}$ formula units. (See Figure 8-2.)*

To illustrate the rather significant amount of information that has been covered so far, let's look at a sample compound. A molecule of sulfuric acid ($H_2SO_4$) contains two atoms of hydrogen, one atom of sulfur, and four atoms of oxygen. One mole of $H_2SO_4$ has the composition shown in Table 8-2.

**Table 8-2** The Composition of One Mole of $H_2SO_4$

| Total | | Components | |
|---|---|---|---|
| 1.00 mol of molecules | | 2 mol of H atoms | 2.02 g<br>$1.20 \times 10^{24}$ H atoms |
| $6.02 \times 10^{23}$ molecules | $H_2SO_4$ | 1 mol of S atoms | 32.1 g<br>$6.02 \times 10^{23}$ S atoms |
| 98.1 g | | 4 mol of O atoms | 64.0 g<br>$2.41 \times 10^{24}$ O atoms |

**Example 8-8**
How many moles of each atom are present in 0.345 mol of $Al_2(CO_3)_3$? What is the total number of moles of atoms present?

**Procedure**
In this problem, notice that there are 2 Al atoms, 3 C atoms, and 9 O atoms in each formula unit of compound. Therefore, in 1 mol of compound, there are 2 mol of Al, 3 mol of C, and 9 mol of O atoms. This can be expressed with conversion factors as

$$\frac{2 \text{ mol Al}}{\text{mol } Al_2(CO_3)_3} \qquad \frac{3 \text{ mol C}}{\text{mol } Al_2(CO_3)_3} \qquad \frac{9 \text{ mol O}}{\text{mol } Al_2(CO_3)_3}$$

**Solution**

$$\text{Al} \quad 0.345 \cancel{\text{mol } Al_2(CO_3)_3} \times \frac{2 \text{ mol Al}}{\cancel{\text{mol } Al_2(CO_3)_3}} = \underline{\underline{0.690 \text{ mol Al}}}$$

$$\text{C} \quad 0.345 \cancel{\text{mol } Al_2(CO_3)_3} \times \frac{3 \text{ mol C}}{\cancel{\text{mol } Al_2(CO_3)_3}} = \underline{\underline{1.04 \text{ mol C}}}$$

$$\text{O} \quad 0.345 \cancel{\text{mol } Al_2(CO_3)_3} \times \frac{9 \text{ mol O}}{\cancel{\text{mol } Al_2(CO_3)_3}} = \underline{\underline{3.10 \text{ mol O}}}$$

Total moles of atoms present:

$$0.690 \text{ mol Al} + 1.04 \text{ mol C} + 3.10 \text{ mol O} = 4.83 \text{ mol atoms}$$

**Example 8-9**
What is the weight of 0.345 mol of $Al_2(CO_3)_3$, and how many individual ionic formula units does this amount represent?

**Procedure**
This problem is worked much like the examples in which we were dealing with moles of atoms rather than compounds. First, find the formula weight of the compound as illustrated in Examples 8-6 and 8-7. The formula weight of the compound is 234 amu, so the molar mass is 234 g/mol.

**Solution**

$$0.345 \cancel{\text{mol } Al_2(CO_3)_3} \times \frac{234 \text{ g}}{\cancel{\text{mol } Al_2(CO_3)_3}} = \underline{\underline{80.7 \text{ g}}}$$

$$0.345 \cancel{\text{mol } Al_2(CO_3)_3} \times \frac{6.02 \times 10^{23} \text{ formula units}}{\cancel{\text{mol } Al_2(CO_3)_3}}$$

$$= \underline{\underline{2.08 \times 10^{23} \text{ formula units}}}$$

It is important to be specific when discussing moles. For example, with the element chlorine be careful to distinguish whether you are discussing a mole of atoms (Cl) or a mole of molecules ($Cl_2$). *Notice that there are two Cl atoms per $Cl_2$ molecule and 2 mol of Cl atoms per mole of $Cl_2$ molecules.*

$$\text{1 mol of Cl atoms} \begin{cases} 35.5 \text{ g} \\ 6.02 \times 10^{23} \text{ atoms} \end{cases}$$

$$\text{1 mol } Cl_2 \text{ molecules} \begin{cases} 71.0 \text{ g } (35.5 \text{ g} + 35.5 \text{ g}) \\ 6.02 \times 10^{23} \text{ molecules } (1.20 \times 10^{24} \text{ atoms}) \end{cases}$$

This situation is analogous to the difference between a dozen sneakers and a dozen *pairs* of sneakers. Both the weight and the actual number of the dozen pair of sneakers is double the dozen sneakers.

**See Problems 8-20 through 8-28.**

## 8-4 Percent Composition

In the use of nitrogen fertilizers an important consideration is the weight of nitrogen actually added to the soil. There are several products available, but they differ as to the percent by weight of nitrogen in the compounds used. For example, pure ammonia gas is advertized as the most effective since it is 82% nitrogen. Solid ammonium nitrate is more convenient to handle but is only 35% by weight nitrogen. The percent by weight (also known as the percent composition) of a compound represents the weight of each element per 100 g (or any other weight unit) of compound. It can be calculated from the formula of a compound. For example, the molar mass of $CO_2$ is 44.0 g/mol, which is composed of 12.0 g of carbon (one mole) and 32.0 g of oxygen (two moles). The percent composition is calculated by dividing the total weight of each component element by the total weight (molar mass) of the compound times 100%.

$$\frac{\text{Total wt. of component atom}}{\text{Total wt. (molar mass)}} \times 100\% = \text{percent composition}$$

In $CO_2$, the percent composition of C is

$$\frac{12.0 \text{ g C}}{44.0 \text{ g } CO_2} \times 100\% = \underline{\underline{27.3\% \text{ C}}}$$

and the percent composition of O is

$$\frac{32.0 \text{ g O}}{44.0 \text{ g } CO_2} \times 100\% = \underline{\underline{72.7\% \text{ O}}}$$

(Or, since there are only two components, if one is 27.3% the other must be $100 - 27.3 = 72.7\%$. Right?)

**Example 8-10**
What is the percent composition of all of the elements in limestone, $CaCO_3$?

**Procedure**
Find the molar mass and convert the *total* weight of each element to percent of the molar mass.

**Solution**
For $CaCO_3$:

$$\text{Ca} \quad 1\ \cancel{\text{mol Ca}} \times \frac{40.1\ \text{g}}{\cancel{\text{mol Ca}}} = 40.1\ \text{g}$$

$$\text{C} \quad 1\ \cancel{\text{mol C}} \times \frac{12.0\ \text{g}}{\cancel{\text{mol C}}} = 12.0\ \text{g}$$

$$\text{O} \quad 3\ \cancel{\text{mol O}} \times \frac{16.0\ \text{g}}{\cancel{\text{mol O}}} = \underline{48.0\ \text{g}}$$

$$\text{Molar mass} = 100.1\ \text{g/mol}\ CaCO_3$$

$$\%\ \text{Ca} \quad \frac{40.1\ \text{g Ca}}{100.1\ \text{g}\ CaCO_3} \times 100\% = \underline{\underline{40.0\%}}$$

$$\%\ \text{C} \quad \frac{12.0\ \text{g C}}{100.1\ \text{g}\ CaCO_3} \times 100\% = \underline{\underline{12.0\%}}$$

$$\%\ \text{O} \quad \frac{48.0\ \text{g O}}{100.1\ \text{g}\ CaCO_3} \times 100\% = \underline{\underline{48.0\%}}$$

**Example 8-11**
What is the percent composition of all the elements in borazine, $B_3N_3H_6$?

**Procedure**
Find the molar mass and convert the *total* weight of each element to percent of the molar mass.

**Solution**
For $B_3N_3H_6$:

$$\text{B} \quad 3\ \cancel{\text{mol B}} \times \frac{10.8\ \text{g}}{\cancel{\text{mol B}}} = 32.4\ \text{g}$$

$$\text{N} \quad 3\ \cancel{\text{mol N}} \times \frac{14.0\ \text{g}}{\cancel{\text{mol N}}} = 42.0\ \text{g}$$

$$\text{H} \quad 6 \cancel{\text{mol H}} \times \frac{1.01 \text{ g}}{\cancel{\text{mol H}}} = \underline{6.1 \text{ g}}$$

$$\text{Molar mass} = 80.5 \text{ g/mol } B_3N_3H_6$$

$$\%\text{ B} \quad \frac{32.4 \text{ g B}}{80.5 \text{ g } B_3N_3H_6} \times 100\% = \underline{\underline{40.2\%}}$$

$$\%\text{ N} \quad \frac{42.0 \text{ g N}}{80.5 \text{ g } B_3N_3H_6} \times 100\% = \underline{\underline{52.2\%}}$$

$$\%\text{ H} \quad \frac{6.1 \text{ g H}}{80.5 \text{ g } B_3N_3H_6} \times 100\% = \underline{\underline{7.6\%}}$$

The percent composition can now be used to calculate the weight of a certain element in a quantity of compound.

**Example 8-12**
What is the percent composition of nitrogen in ammonium nitrate and ammonia? What is the weight of nitrogen in 150 lb of each of these compounds?

**Procedure**
Express the percent composition in appropriate fraction form (i.e., in this case, lb of N per 100 lb of compound). The weight is calculated as follows:

$$\text{weight of N} = \begin{matrix}\text{total weight} \\ \text{of compound}\end{matrix} \times \text{fraction of N}$$

**Solution**
For ammonium nitrate ($NH_4NO_3$):

$$\begin{aligned}\text{Molar mass} &= (2 \times \text{N}) + (4 \times \text{H}) + (3 \times 0) \\ &= 80.0 \text{ g/mol}\end{aligned}$$

$$\text{Percent nitrogen} = \frac{28.0 \text{ g N}}{80.0 \text{ g } NH_4NO_3} \times 100\% = \underline{\underline{35.0\%}}$$

$$\text{Fraction of nitrogen} = \frac{35.0 \text{ lb N}}{100 \text{ lb } NH_4NO_3}$$

$$\begin{aligned}\text{Weight nitrogen} &= 150 \text{ lb } \cancel{NH_4NO_3} \times \frac{35.0 \text{ lb N}}{100 \text{ lb } \cancel{NH_4NO_3}} \\ &= \underline{\underline{52.5 \text{ lb N}}}\end{aligned}$$

For ammonia ($NH_3$):

$$\text{Molar mass} = (\text{N}) + (3 \times \text{H}) = 17.0 \text{ g/mol}$$

$$\text{Percent nitrogen} = \frac{14.0 \text{ g N}}{17.0 \text{ g } NH_3} \times 100\% = \underline{\underline{82.4\%}}$$

$$\text{Fraction of nitrogen} = \frac{82.4 \text{ lb N}}{100 \text{ lb } NH_3}$$

$$\text{Weight nitrogen} = 150 \text{ } \cancel{\text{lb } NH_3} \times \frac{82.4 \text{ lb N}}{100 \text{ } \cancel{\text{lb } NH_3}}$$

$$= \underline{\underline{124 \text{ lb N}}}$$

## 8-5 Empirical and Molecular Formulas

We have shown that we can calculate the percent composition given the formula. Is the reverse possible? Not quite. Notice that in Example 8-11 the formula of borazine is $B_3N_3H_6$. This is called the **molecular formula** *since it represents the actual number of atoms in each molecular unit.* It is not the simplest whole-number ratio of atoms, however. *The simplest whole-number ratio of atoms in a compound is called the* **empirical formula.** In the case of borazine, the numerical subscripts can be divided through by 3 to leave the empirical formula, $BNH_2$. Since the percent composition is the *same* for an empirical formula as for a molecular formula, the best we can do is calculate the empirical formula from the percent composition or the weight composition.

To calculate the empirical formula from the percent composition, three steps are necessary:

1 Convert percent composition to an actual weight.

2 Convert weight to moles.

3 Find the whole-number ratio of the moles of different atoms. This is best illustrated by the following examples.

**Example 8-13**
What is the empirical formula of laughing gas, which is 63.6% nitrogen and 36.4% oxygen?

**Procedure**
To convert percent to a specific weight remember that percent means parts per 100. Therefore, if we simply assume that we have a 100-g

quantity of compound, the percent converts to a specific weight as follows:

$$63.6\%\ \text{N} = \frac{63.6\ \text{g N}}{100\ \text{g compound}} \qquad 36.4\%\ \text{O} = \frac{36.4\ \text{g O}}{100\ \text{g compound}}$$

We now convert the weights to moles of each element.

$$63.6\ \text{g}\,\cancel{\text{N}} \times \frac{1\ \text{mol N}}{14.0\ \text{g}\,\cancel{\text{N}}} = 4.54\ \text{mol N in 100 g}$$

$$36.4\ \text{g}\,\cancel{\text{O}} \times \frac{1\ \text{mol O}}{16.0\ \text{g}\,\cancel{\text{O}}} = 2.28\ \text{mol O in 100 g}$$

*The ratio of N to O atoms will be the same as the ratio of the moles of N to O (or the ratio of the number of dozens or any other unit).* The formula cannot remain fractional, since only whole numbers of atoms are present in a compound. To find the whole-number ratio of moles, divide through by the smallest number of moles, which in this case is 2.28 mol of O.

$$\text{N}\ \frac{4.54}{2.28} = 2.0 \qquad \text{O}\ \frac{2.28}{2.28} = 1.0$$

The empirical formula of the compound is

$$\underline{\underline{N_2O}}$$

---

**Example 8-14**

A sample of ordinary rust is composed of 2.78 g of iron and 1.19 g of oxygen. What is the empirical formula of rust?

**Procedure**

Convert the weight of each element to moles

$$2.78\ \text{g}\,\cancel{\text{Fe}} \times \frac{1\ \text{mol Fe}}{55.8\ \text{g}\,\cancel{\text{Fe}}} = 0.0498\ \text{mol Fe}$$

$$1.19\ \text{g}\,\cancel{\text{O}} \times \frac{1\ \text{mol O}}{16.0\ \text{g}\,\cancel{\text{O}}} = 0.0744\ \text{mol O}$$

Divide through by the smallest number of moles.

$$\text{Fe}\ \frac{0.0498}{0.0498} = 1.0 \qquad \text{O}\ \frac{0.0744}{0.0498} = 1.5$$

This time we're not quite finished, since $FeO_{1.5}$ still has a fractional number that must be cleared. (You should keep at least two significant figures in these numbers, so as not to round off a number like 1.5 to 2.) This fractional number can be cleared by multiplying both

**See Problems 8-40 through 8-51.**

subscripts by a number that produces whole numbers only. In this case, multiply both subscripts by 2 and you have the empirical formula.

$$Fe_{(1\times 2)}O_{(1.5\times 2)} = \underline{\underline{Fe_2O_3}}$$

In order to determine the molecular formula of a compound, the molar mass (g/mol) and the mass of one empirical unit (g/emp. unit) are needed. The ratio of these two quantities must be a whole number. This whole number ($a$), is the number of empirical units in 1 mol.

$$a = \frac{\text{molar mass}}{\text{emp. mass}} = \frac{\text{X g/mol}}{\text{Y g/emp. unit}} = 1, 2, 3, \text{etc., emp. unit/mol}$$

Multiply the subscripts of the atoms in the empirical formula by "$a$" and you have the molecular formula. For example, the empirical formula of benzene is CH and its molar mass is 78 g/mol. The empirical mass (formula mass of CH) is 13.0 g/emp. unit.

$$a = \frac{\text{78 g/mol}}{\text{13 g/emp. unit}} = \text{6 emp. unit/mol}$$

The molecular formula is

$$C_{(1\times 6)}H_{(1\times 6)} = C_6H_6$$

**Example 8-15**

A pure phosphorus-oxygen compound is 43.7% phosphorus and the remainder oxygen. The molar mass is 284 g/mol. What are the empirical and molecular formulas of this compound?

**Procedure**

First, find the empirical formula. Convert percent to a weight.

$$43.7\%\ P = \frac{\text{43.7 g P}}{\text{100 g compound}} \qquad \%\ O = 100\% - 43.7\% = 56.3\%$$

$$56.3\%\ O = \frac{\text{56.3 g O}}{\text{100 g compound}}$$

Convert the weight to moles.

$$43.7\ \text{g P} \times \frac{\text{1 mol P}}{\text{31.0 g P}} = 1.41\ \text{mol P}$$

$$56.3\ \text{g O} \times \frac{\text{1 mol O}}{\text{16.0 g O}} = 3.52\ \text{mol O}$$

Divide through by the smallest number of moles.

$$\text{P} \quad \frac{1.41}{1.41} = 1.0 \qquad \text{O} \quad \frac{3.52}{1.41} = 2.5$$

Multiply both numbers by "2" to remove the fraction.

$$PO_{2.5} = P_{(1\times 2)}O_{(2.5\times 2)} = \underline{\underline{P_2O_5}}$$

To find the molecular formula, first compute the empirical mass, which is

$$\begin{array}{lrr} \text{P} & 2 \times 31.0\text{ g} = & 62.0\text{ g} \\ \text{O} & 5 \times 16.0\text{ g} = & \underline{80.0\text{ g}} \\ & & 142\text{ g/emp. unit} \end{array}$$

$$a = \frac{284\text{ g/mol}}{142\text{ g/emp. unit}} \qquad a = 2\text{ emp. unit/mol}$$

The molecular formula is

$$P_{(2\times 2)}O_{(2\times 5)} = \underline{\underline{P_4O_{10}}}$$

## 8-6 The Use of Empirical and Molecular Formulas

In March 1983 the American Chemical Society recorded its six millionth unique chemical compound. All of these compounds have been characterized and reported in the chemical literature. (See Figure 8-3.) Obviously, since the molecular formula is of primary importance in the identification of a new compound, the types of calculations performed in this chapter are frequently carried out. Before a research chemist reports the discovery of a new compound, several items must be firmly established: (1) purity, (2) empirical formula, (3) molecular formula, and (4) molecular structure (order and geometry of the atoms in the molecule). A listing of some physical and chemical properties of the new substance may also be reported.

Before we move on, let's discuss a little more about how (1) through (4) are established.

**1** *Is it a pure substance?* This can be determined by examination of physical properties. A mixture of salt and sugar looks pure (as any joker who has put salt in the sugar jar knows), but heating the mixture soon reveals the truth. The sugar crystals soon melt and decompose, but the salt crystals do not. Pure substances have definite and sharp melting and boiling points, mixtures do not (see Chapter 1).

If the substance is pure and the chemists have established the melting and/or boiling point, they should then take a look in a handbook listing physical properties of known compounds to make sure they haven't just

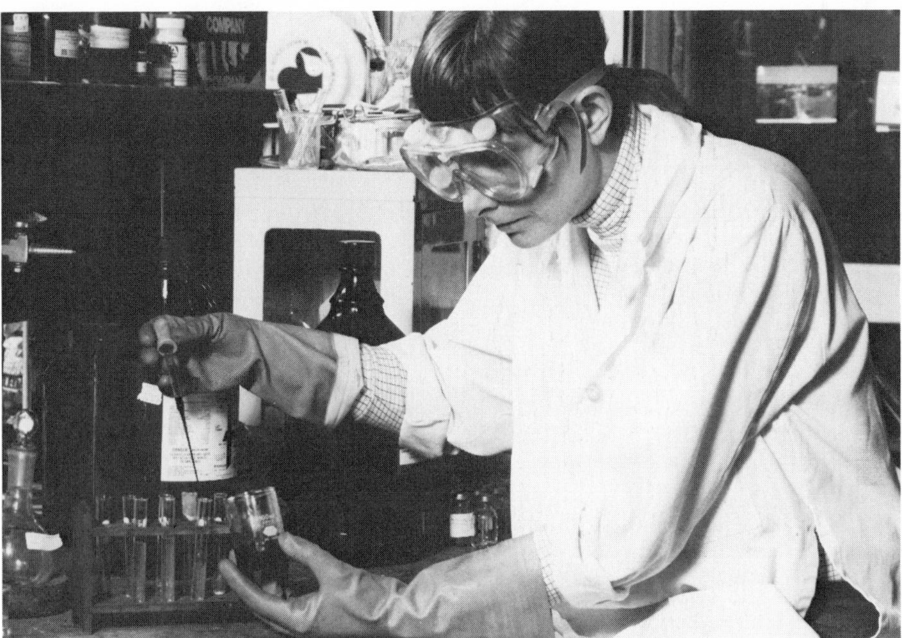

**Figure 8-3**
A RESEARCH LABORATORY. The preparation and identification of new compounds occurs frequently in chemical laboratories.

rediscovered something old. If, as often happens, they have, then back to the lab. If the compound still appears to be original, then on to the next step.

**2** *What is the empirical formula of the new compound?* Most chemists take the easy but smart way out and send a sample of their new pure substance to a commercial analytical laboratory, which reports back on percent composition of the elements requested. To actually obtain percent composition, many methods are used depending on the particular element; but the most commonly requested elements, carbon and hydrogen, are determined by analysis of combustion products. Such determination is illustrated in Problems 8-51 and 8-57.

**3** *What is the molecular formula of the new compound?* With the empirical formula now available, one next needs the molar mass to obtain the molecular formula. The molar mass can be obtained commercially like percent composition, or it may be obtained by several straightforward laboratory experiments. The determination of the molar mass of an unknown substance is usually performed in the general chemistry laboratory.

**4** *What is the structure of the molecule?* Molecules can have the same molecular formula but different structures. For example, $C_2H_6O$ is the formula for both dimethyl ether and ethyl alcohol (see Problem 6-9). Determination of structure can take a few minutes or a few years depending on the nature of the molecule, its complexity, and the instruments necessary. Nobel Prizes have been awarded for determination of structure of molecules (e.g., Watson and Crick for the structure of DNA).

With all of the above information in hand, chemists should feel confident enough to face their peers with their discovery.

See Problems 8-52 through 8-57.

## 8-7 Chapter Summary

The atom is so small that it takes literally an uncountable number of atoms to work with in a laboratory. One person could not count the atoms in a single tiny crystal of sugar even if several billion years were available. Regardless of the problem, we still need to be able to measure equivalent numbers of atoms of elements. Since an actual count is out of the question, we use a weight that contains a known quantity of atoms. The standard quantity used is the SI unit known as the mole (mol) and it represents $6.02 \times 10^{23}$ atoms, molecules, or formula units of the element or compound. By use of atomic weights from the periodic table the mass of one mole can be determined. The information on the mole is summarized as follows.

| Unit | Number | Mass |
|---|---|---|
| 1.00 mol | $6.02 \times 10^{23}$ atoms | Atomic weight of element in grams |
| | $6.02 \times 10^{23}$ molecules | Formula weight of molecular compound in grams |
| | $6.02 \times 10^{23}$ formula units | Formula weight of ionic compound in grams |

A major emphasis of this chapter is to become comfortable with conversions between moles and number and, more importantly, between moles and weight. The conversions are summarized as follows. Notice that moles can be converted to either a number (using Avogadro's number as a conversion factor) or a weight (using the molar mass as a conversion factor).

$$\text{Number} \xrightarrow{\text{Avogadro's number}} \text{Moles} \xleftrightarrow{\text{Molar mass}} \text{Weight}$$

Percent composition can be obtained from the formula of the compound and atomic weights of the elements. If the percent composition or weight composition is given, the empirical formula can be calculated. The empirical formula is the simplest whole number of atoms in a compound. The empirical formula is determined according to the following steps.

Percent composition → weight in grams of elements → moles of atoms → mole ratio of atoms → whole number mole ratio of atoms → empirical formula

The molecular formula may be the same or a multiple of the empirical formula. It is obtained by comparing the molar mass of the compound to the empirical mass.

## Exercises

### Relative Numbers of Atoms

**8-1** A piece of pure gold weighs 145 g. What would the same number of silver atoms weigh?

**8-2** A chunk of pure iron weighs 212 lb. What would the same number of carbon atoms weigh?

**8-3** Some copper pennies weigh 16.0 g; the same number of atoms of a precious metal weigh 49.1 g. What is the metal?

**8-4** In the compound CuO, what weight of copper is present for each 18.0 g of oxygen?

**8-5** In the compound NaBr, what weight of sodium is present for each 295 g of bromine?

* **8-6** In a compound containing one atom of carbon and one atom of another element, it is found that 25.0 g of carbon is combined with 33.3 g of the other element. What is the element and the formula of the compound?

**8-7** In the compound $MgBr_2$, what weight of bromine is present for each 46.0 g of magnesium? (Remember, there are two bromines per magnesium.)

* **8-8** In the compound $SO_3$, what weight of sulfur is present for each 80.0 lb of oxygen?

### Moles of Atoms

**8-9** Fill in the blanks. Use the unit-factor method to determine the answers.

| Element | Weight in Grams | Number of Moles | Number of Atoms |
|---|---|---|---|
| S | 8.00 | 0.250 | $1.50 \times 10^{23}$ |
| (a) P | 14.5 | | |
| (b) Rb | | 1.75 | |
| (c) Al | | | $6.02 \times 10^{23}$ |
| (d) | 363 | | $3.01 \times 10^{24}$ |
| (e) Ti | | | 1 |

**8-10** Fill in the blanks.

| Element | Weight in Grams | Number of Moles | Number of Atoms |
|---|---|---|---|
| (a) Na | 16.0 | _____ | _____ |
| (b) Mg | _____ | $3.01 \times 10^{4}$ | _____ |
| (c) _____ | 43.2 | _____ | $2.41 \times 10^{24}$ |
| (d) K | $5.75 \times 10^{-12}$ | _____ | _____ |
| (e) Ne | _____ | _____ | $7.66 \times 10^{22}$ |

**8-11** What is the weight in grams of each of the following?
(a) 1.00 mol of Cu
(b) 0.50 mol of S
(c) $6.02 \times 10^{23}$ atoms of Ca

**8-12** What is the weight in grams of each of the following?
(a) $6.02 \times 10^{22}$ atoms of Ca
(b) 15.0 mol of Li
(c) $3.40 \times 10^{-6}$ mol of B

**8-13** How many individual atoms are in each of the following?
(a) 32.1 mol of sulfur
(b) 32.1 g of sulfur
(c) 32.0 g of oxygen

**8-14** How many moles are in each of the following?
(a) 281 g of silicon
(b) $7.34 \times 10^{25}$ atoms of phosphorus
(c) 19.0 atoms of fluorine

**8-15** Which has more atoms, 50.0 g of Al or 50.0 g of Fe?

**8-16** Which contains more Ni, 20.0 g, $2.85 \times 10^{23}$ atoms, or 0.450 mol?

**8-17** Which contains more Cr, 0.025 mol, $6.0 \times 10^{21}$ atoms or 1.5 g of Cr?

### Formula Weight

**8-18** What is the formula weight of each of the following (to three significant figures)?
(a) $KClO_2$
(b) $SO_3$
(e) $Na_2CO_3$
(f) $CH_3COOH$

(c) $N_2O_5$ (g) $Fe_2(CrO_4)_3$
(d) $H_2SO_4$

8-19 What is the formula weight of each of the following?
(a) $CuSO_4 \cdot 6H_2O$ (include the $H_2O$s) (c) $Cr_2(C_2O_4)_3$
(b) $B_4H_{10}$ (d) $Cl_2O_7$

### Moles of Molecules or Formula Units

8-20 Fill in the blanks in the table. Use the unit-factor method to determine the answers.

| Molecules | Weight in Grams | Number of Moles | Number of Molecules or Formula Units |
|---|---|---|---|
| $N_2O$ | 23.8 | 0.542 | $3.26 \times 10^{23}$ |
| (a) $H_2O$ | | 10.5 | |
| (b) $BF_3$ | | | $3.01 \times 10^{21}$ |
| (c) $SO_2$ | 14.0 | | |
| (d) $K_2SO_4$ | | $1.20 \times 10^{-4}$ | |
| (e) $SO_3$ | | | $4.50 \times 10^{24}$ |
| (f) $N(CH_3)_3$ | 0.450 | | |

8-21 Fill in the blanks.

| Molecules | Weight in Grams | Number of Moles | Number of Molecules or Formula Units |
|---|---|---|---|
| (a) $O_3$ | | 0.0651 | |
| (b) $NO_2$ | 23.8 | | |
| (c) $Cl_2O_3$ | | | $9.50 \times 10^{20}$ |
| (d) $UF_6$ | | | 12 |

8-22 How many moles of each type of atom are present in 2.55 mol of grain alcohol, $C_2H_6O$? What is the total number of moles of atoms present? What is the weight of each element present? What is the total weight?

8-23 How many moles are there in 28.0 g of $Ca(ClO_3)_2$? How many moles of each element are present? How many total moles of atoms are present?

8-24 How many moles are there in 49.0 g of $K_2Cr_2O_7$? How many moles of each element are present? How many total moles of atoms are present?

8-25 What weight of each element is present in 1.50 mol of $H_2SO_3$?

8-26 What weight of each element is present in 8.95 mol of boric acid ($H_3BO_3$)?

8-27 How many moles of $O_2$ are there in $1.20 \times 10^{22}$ $O_2$ molecules? How many moles of oxygen atoms? What is the weight of oxygen molecules? What is the weight of oxygen atoms?

8-28 How many moles of $Cl_2$ molecules are in 0.710 g of $Cl_2$? How many moles of Cl atoms are present? What is the weight of Cl atoms present?

### Percent Composition

8-29 What is the percent composition of a compound composed of 1.375 g of N and 3.935 g of O?

8-30 A sample of a compound weighs 4.86 g and is composed of silicon and oxygen. What is the percent composition if 2.27 g of the weight is silicon?

8-31 The weight of a sample of a compound is 7.44 g. Of that weight 2.88 g is potassium, 1.03 g is nitrogen, and the remainder is oxygen. What is the percent composition of the elements?

8-32 What is the percent composition of all of the elements in the following compounds? (a) $C_2H_6O$ (b) $C_3H_6$ (c) $C_9H_{18}$ (d) $Na_2SO_4$ (e) $(NH_4)_2CO_3$

8-33 What is the percent composition of all of the elements in the following compounds? (a) $HNO_3$ (b) $Al_2(CO_3)_3$ (c) $Cl_2O_3$ (d) $(NH_4)_2HPO_4$

**8-34** What is the percent composition of all of the elements in $Na_2B_4O_7 \cdot 10H_2O$ (borax)?

8-35 What is the percent composition of all of the elements in saccharin ($C_7H_5SNO_3$)?

**8-36** What weight of carbon is in a 125-g quantity of $Na_2C_2O_4$?

**8-37** What weight of phosphorus is in 25.0 lb of $Na_3PO_4$?

8-38 What weight of copper is in 275 kg of $CuCO_3$?

**8-39** Iron is recovered from $Fe_2O_3$. How many pounds of iron can be recovered from each ton of the iron ore ($Fe_2O_3$)? (1 ton = 2000 lb)

## Empirical Formulas

**8-40** Which of the following are not empirical formulas?
(a) $N_2O_4$ (b) $Cr_2O_3$ (c) $H_2S_2O_3$
(d) $H_2C_2O_4$ (e) $Mn_2O_7$

**8-41** Convert the following mole ratios of elements to empirical formulas.
(a) 0.25 mol of Fe and 0.25 mol of S
(b) 1.88 mol of Sr and 3.76 mol of I
(c) 0.32 mol of K, 0.32 mol of Cl, and 0.96 mol of O
(d) 1.0 mol of I and 2.5 mol of O
(e) 2.0 mol of Fe and 2.66 mol of O
(f) 4.22 mol of C, 7.03 mol of H, and 4.22 mol of Cl

8-42 Convert the following mole ratios of elements to empirical formulas.
(a) 1.20 mol of Si and 2.40 mol of O
(b) 0.045 mol of Cs and 0.022 mol of S
(c) 1.0 mol of X and 1.20 mol of Y
(d) 3.11 mol of Fe, 4.66 mol of C, and 14.0 mol of O

**8-43** What is the empirical formula of a compound that has the percent composition 63.1% oxygen and 36.8% nitrogen?

8-44 What is the empirical formula of a compound that has the following percent composition: 41.0% K, 33.7% S, and 25.3% O?

**8-45** In an experiment it was found that 8.25 g of potassium combines with 6.75 g of $O_2$. What is the empirical formula of the compound?

8-46 Orlon is composed of very long molecules with a percent composition of 26.4% N, 5.66% H, and 67.9% C. What is the empirical formula for orlon?

**8-47** A compound is 21.6% Mg, 21.4% C, and 57.0% O. What is the empirical formula of the compound?

**8-48** A compound is composed of 9.90 g of carbon, 1.65 g of hydrogen, and 29.3 g of chlorine. What is the empirical formula of the compound?

8-49 A compound is composed of 0.46 g of Na, 0.52 g of Cr, and 0.64 g of O. What is the empirical formula of the compound?

**8-50** A compound is composed of 24.1% nitrogen, 6.90% hydrogen, 27.6% sulfur, and the remainder oxygen. What is the empirical formula of the compound?

***8-51** A hydrocarbon (a compound that contains only carbon and hydrogen) was burned and the products of the combustion were collected and weighed. All of the carbon present in the original compound is now present in 1.20 g of $CO_2$. All of the hydrogen is present in 0.489 g of $H_2O$. What is the empirical formula of the compound? (*Hint:* Remember that all of the moles of C atoms in $CO_2$ and H atoms in $H_2O$ came from the original compound.)

## Molecular Formulas

**8-52** A compound has the following percent composition: 20.0% C, 2.2% H, and 77.8% Cl. The molar mass of the compound is 545 g/mol. What is the molecular formula of the compound?

8-53 A compound is composed of 1.65 g of nitrogen and 3.78 g of sulfur. If its molar mass is 184 g/mol, what is its molecular formula?

**8-54** A compound was reported in 1967 to have a percent composition of 18.7% B, 20.7% C, 5.15% H, and 55.4% O. Its molar mass was found to be about 115 g/mol. What is the molecular formula of the compound?

8-55 A new compound reported in 1970 has a percent composition of 34.9% K, 21.4% C, 12.5% N, 2.68% H, and 28.6% O. It has a molar mass of about 224 g/mol. What is its molecular formula?

**8-56** A new compound reported in 1982 has a molar mass of 834 g/mol. A 20.0-g sample of the compound contains 18.3 g of iodine and the remainder carbon. What is the molecular formula of the compound?

*8-57 A 0.500-g sample of a compound containing C, H, and O was burned and the products collected. The combustion produced 0.733 g of $CO_2$ and 0.302 g of $H_2O$. The molar mass of the compound is 60.0 g/mol. What is the molecular formula of this compound? (*Hint:* Find the weight of C and H in the original compound; the remainder of the 0.500 g will be O.)

These pellets of ore are to be processed to produce iron. The amount of iron that can be produced depends upon the amount of ore available and the proportion of iron in the ore.

# Chapter 9
# Chemical Reactions: Equations and Quantities

## Purpose of Chapter 9

In Chapter 9 we use balanced chemical equations to illustrate types of reactions and quantitative calculations.

## Objectives for Chapter 9

After completion of this chapter, you should be able to:

**1** Describe the information represented by a chemical equation. (9-1)
**2** Balance simple equations. (9-1)
**3** Classify chemical reactions among the five types listed. (9-2)
**4** Use the balanced equation to obtain mole relationships among reactants and products (reaction factors). (9-3)
**5** Make the following stoichiometric conversions:
  **(a)** Mole to mole (9-3)
  **(b)** Mole to weight (9-3)
  **(c)** Weight to weight (9-3)
  **(d)** Weight to number (9-3)
**6** Calculate the percent yield from the actual yield and the theoretical yield. (9-4)
**7** Calculate the percent purity of a sample from the yield of a product. (9-4)
**8** Identify the limiting reactant, the reactant in excess, and the yield from given amounts of reactants. (9-5)

If you are like most students, you have probably given some thought as to how much time you must put in to get an "A" out of this course. This question is not really too much different from questions such as how many gallons of gasoline can be extracted from a barrel of crude oil or how much natural gas ($CH_4$) can be produced from a ton of coal or even how many eggs are needed to make one cake. Your grade, the gasoline, the gas, and the cake are all products of an endeavor and as such are limited by the input of ingredients.

In a chemical reaction, the maximum amount of products relates to the ingredients (reactants) by what is called "the chemical equation." The main purpose of this chapter is to explore and develop the extensive quantitative relationships implied by these equations. Before we begin the study of the quantitative aspects, however, we will carefully examine the construction and the balancing of an equation and then briefly discuss some of the types of chemical reactions that can be illustrated by an equation.

The background for this chapter mainly involves a thorough knowledge of Chapter 8 concerning mole relationships.

## 9-1 Chemical Equations

Gaseous hydrogen molecules combine with gaseous oxygen molecules in a 2:1 ratio to produce liquid water. A much simpler way to represent all of this information is with a chemical equation. *A* **chemical equation** *is the symbolic representation of a chemical reaction.* Our example appears as follows:

$$2H_2(g) + O_2(g) \rightarrow 2H_2O(l)$$

Let's go back a few steps and evolve this chemical equation from the start in order to appreciate fully all that is shown. At first, to represent the names of the elements and the product compound we write

$$H + O \rightarrow H_2O$$

(Yield is represented as $\rightarrow$.) It is necesary, however, to represent the hydrogen and oxygen in the chemical state as they are found in nature at normal temperature and pressure or as found at the actual reaction conditions. As discussed previously, both of these elements exist as diatomic molecules. Including this information, we have

$$H_2 + O_2 \rightarrow H_2O$$

At this point, recall that matter cannot be created or destroyed in a chemical reaction. If you look closely at the preceding equation, we apparently did just that, because there is one less oxygen atom on the right as on the left. *Equations must be* **balanced,** *which means that all atoms on the left of the yield sign* (the **reactants**) *must be present on the right* (the **products**). The equation cannot be balanced by changing the subscripts of a compound, because that changes its identity. For example, the equation would be balanced if the $H_2O$ were changed to $H_2O_2$. However, $H_2O_2$ is hydrogen peroxide, which is not the same thing as water. *An equation is balanced by introducing* **coefficients.** In this case, a 2 in front of the $H_2O$ solves the original oxygen problem but unbalances the hydrogens:

$$H_2 + O_2 \rightarrow 2H_2O$$

This problem can be solved easily. Simply return to the left and put a coefficient of 2 in front of the $H_2$ and the equation is balanced:

$$2H_2 + O_2 \rightarrow 2H_2O$$

Finally, the physical states of the reactants and products under the reaction conditions are sometimes added in parentheses after the molecule. To indicate the states or reaction conditions, use the following.

| | |
|---|---|
| $(g)$ = gas | $(l)$ = liquid |
| $(s)$ = solid | $(aq)$ = aqueous or water solution |

So now we have

$$2H_2(g) + O_2(g) \rightarrow 2H_2O(l)$$

Other symbols sometimes used in chemical equations are ↑ (an arrow up) for a gas that is evolved in the products and ↓ (an arrow down) for a solid formed in the products. The symbol Δ (delta) over the arrow indicates that the reactants must be heated before a reaction occurs. For example,

$$Ba(s) + H_2SO_4(aq) \rightarrow BaSO_4 \downarrow + H_2 \uparrow$$

As you can see, the simple chemical equation contains a wealth of information in shorthand. When you practice balancing equations at the end of the chapter, remember the following.

1 The subscripts of a compound are fixed; they cannot be changed to balance an equation.

2 The coefficients used should be the smallest whole numbers possible.

3 The coefficient multiplies every number in the formula. For example, $2KClO_2$ indicates the presence of two atoms of K, two atoms of Cl, and four atoms of O.

Although there are no set rules for balancing simple equations, there are some guidelines that generally apply:

1 First, balance the element (other than H or O) having the greatest number of atoms in a molecule as either a reactant or product.

2 Next, balance any other atoms other than H or O.

3 Balance either H or O next. Try balancing the one that appears in the fewest number of compounds first and then the other.

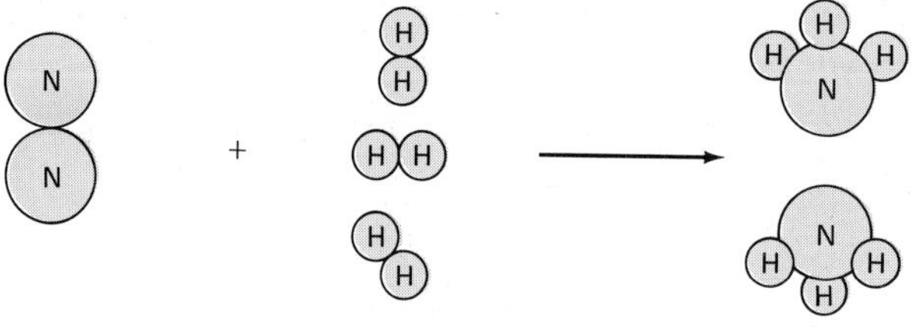


**Figure 9-1** NITROGEN PLUS HYDROGEN YIELDS AMMONIA. In the chemical reaction, the atoms are simply rearranged into different molecules.

**Example 9-1**

Write a balanced chemical equation from the following word equation: Nitrogen gas reacts with hydrogen gas to produce ammonia gas.

**Procedure**

The unbalanced chemical equation is

$$N_2(g) + H_2(g) \rightarrow NH_3\ (g)$$

Consider the N in $N_2$. There are two N's on the left, so two are required on the right. Add a 2 before the $NH_3$.

$$N_2(g) + H_2(g) \rightarrow 2NH_3(g)$$

There are now six hydrogens on the right but only two on the left. Put a 3 before the $H_2$ on the left and the equation is balanced (see Figure 9-1).

**Solution**

$$\underline{\underline{N_2(g) + 3H_2(g) \rightarrow 2NH_3(g)}}$$

**Example 9-2**

Balance the following equation:

$$B_2H_6(g) + H_2O(l) \rightarrow H_3BO_3(aq) + H_2(g)$$

**Procedure**

Consider the B in $B_2H_6$. There are two B's on the left, so two are required on the right. Add a 2 before the $H_3BO_3$.

$$B_2H_6(g) + H_2O(l) \rightarrow 2H_3BO_3(aq) + H_2(g)$$

Since oxygen appears in fewer compounds than hydrogen, balance oxygen next. Add a 6 before the $H_2O$ on the left to balance the 6 O's on the right.

$$B_2H_6(g) + 6H_2O(l) \rightarrow 2H_3BO_3(aq) + H_2(g)$$

Only hydrogen remains. On the left, 18 hydrogens have been locked in. On the right, 6 hydrogens have been locked in with the $2H_3BO_3$. We will need 12 additional hydrogens on the right, which can be provided by placing a 6 before the $H_2$.

**Solution**

$$B_2H_6(g) + 6H_2O(l) \rightarrow 2H_3BO_3(aq) + 6H_2(g)$$

**Example 9-3**
Balance the following equation.

$$C_3H_8O(l) + O_2(g) \rightarrow CO_2(g) + H_2O(l)$$

Other than H and O, there is only carbon. Add a 3 before the $CO_2$ on the right to balance the three carbons in the $C_3H_8O$ molecule.

$$C_3H_8O(l) + O_2(g) \rightarrow 3CO_2(g) + H_2O(l)$$

Hydrogen appears in fewer compounds than oxygen, so it should be balanced next. Add a 4 before the $H_2O$ on the right.

$$C_3H_8O(l) + O_2(g) \rightarrow 3CO_2(g) + 4H_2O(l)$$

Now notice that 10 O's are needed on the left. One O is with the $C_3H_8O$, so 9 O's or $\frac{9}{2}$ $O_2$'s are needed.

$$C_3H_8O(l) + \tfrac{9}{2}\,O_2 \rightarrow 3CO_2(g) + 4H_2O(l)$$

Since we are expressing only whole numbers in a balanced equation, all coefficients should be multiplied by 2 so as to clear of fractions. The final balanced equation is:

$$2C_3H_8O(l) + 9O_2(g) \rightarrow 6CO_2(g) + 8H_2O(l)$$

**See Problems 9-1 through 9-5.**

Equations can be somewhat more complex than those illustrated by these examples and in the problems. In fact, balancing such equations by inspection is tedious, if not impossible. However, in Chapter 14 we will find that there are systematic methods for balancing the more complicated equations.

## 9-2 Types of Chemical Reactions

There is a variety of ways to classify chemical reactions according to some common characteristic. In this chapter we will attempt to group chemical reactions into five simple types. Although these classifications cover most reactions discussed in this text, there are some that do not fit easily into any of these types. Also, in later chapters we will introduce other classifications that suit our purposes at that time. The five types of reactions that we will employ at this time are: (1) combination reactions, (2) decomposition reactions, (3) combustion reactions, (4) single replacement or substitution reactions, and (5) double displacement reactions. Examples of each type follow.

**Combination Reactions** Combination reactions concern the synthesis of one compound from the elements or from a union of other compounds. In both cases, however, only one product is formed. (See Figure 9-2.)

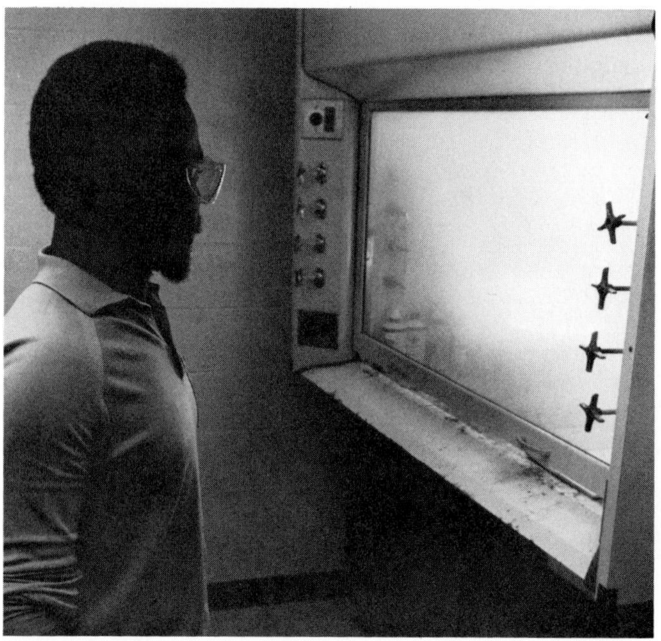

**Figure 9-2** A COMBINATION AND A COMBUSTION REACTION. When magnesium burns in air, the reaction can be classified as either a combination or a combustion reaction.

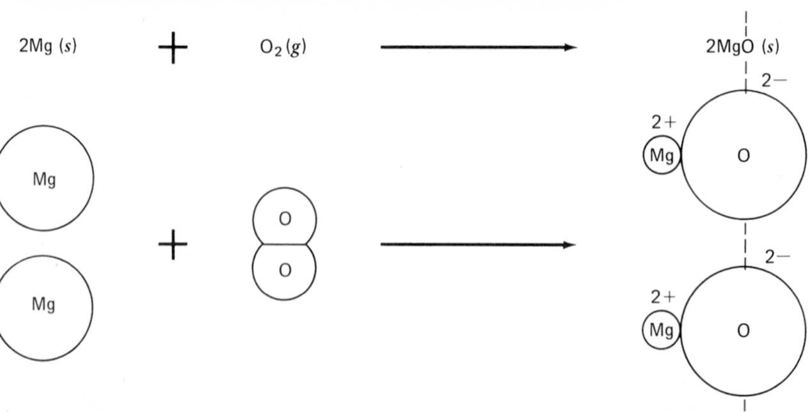

$$2Na(s) + Cl_2(g) \rightarrow 2NaCl(s)$$

$$C(s) + O_2(g) \rightarrow CO_2(g)$$

$$CaO(s) + CO_2(g) \rightarrow CaCO_3(s)$$

$$SO_2(g) + H_2O(l) \rightarrow H_2SO_3(aq)$$

**Decomposition Reactions** Decomposition is simply the opposite of combination. In decomposition reactions one compound decomposes or breaks down into two or more elements or into simpler compounds. In most cases decomposition reactions take place only at high temperature (indicated by Δ). (See Figure 9-3.)

$$2HgO(s) \xrightarrow{\Delta} 2Hg(l) + O_2(g)$$

$$CaCO_3(s) \xrightarrow{\Delta} CaO(s) + CO_2(g)$$

$$2KClO_3(s) \xrightarrow{\Delta} 2KCl(s) + 3O_2(g)$$

**Combustion Reactions** Combustion, better known as burning, is the reaction of a compound or element with oxygen. Notice, however, that the combustion of elements can also be classified as combination reactions. (See Figure 9-2.)

$$C(s) + O_2(g) \rightarrow CO_2(g)$$

$$2Mg(s) + O_2(g) \rightarrow 2MgO(s)$$

When compounds containing carbon and hydrogen burn in sufficient oxygen, carbon dioxide and water are formed.

$$CH_4(g) + 2O_2(g) \rightarrow CO_2(g) + 2H_2O(l)$$

$$2C_3H_8O(l) + 9O_2(g) \rightarrow 6CO_2(g) + 8H_2O(l)$$

When insufficient oxygen is present (as in the combustion of gasoline in an automobile engine), some carbon monoxide (CO) also forms.

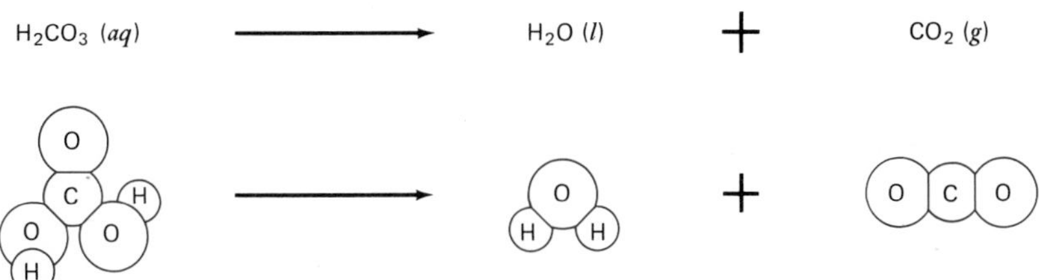

**Figure 9-3** A DECOMPOSITION REACTION. The fizz of carbonated water is a result of a decomposition reaction.

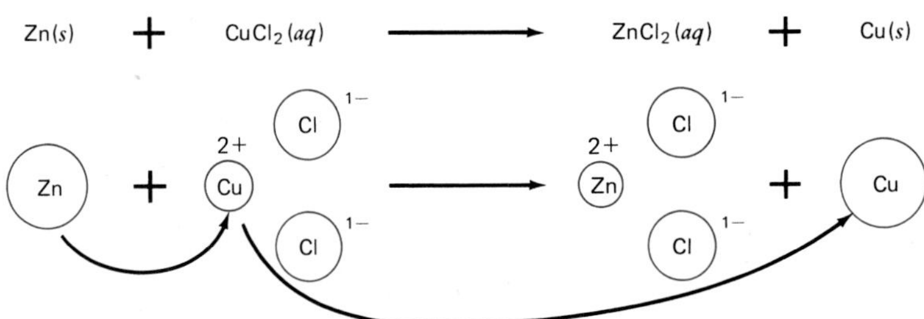


**Figure 9-4** A SINGLE REPLACEMENT REACTION. The formation of a layer of copper on a piece of zinc is a single replacement reaction.

**Single Replacement or Substitution Reactions** Single replacement or substitution reactions involve the substitution of one element in a compound for another. These reactions usually take place in aqueous solution. They will be discussed in more detail in Chapter 14. (See Figure 9-4.)

$$Zn(s) + CuCl_2(aq) \rightarrow ZnCl_2(aq) + Cu(s)$$
(Zn in, Cu out)

$$Mg(s) + 2HCl(aq) \rightarrow MgCl_2(aq) + H_2(g)$$
(Mg in, $H_2$ out)

$$2Na(s) + 2H_2O(l) \rightarrow 2NaOH(aq) + H_2(g)$$
(Na in, $H_2$ out)

**Double Displacement Reactions** Most common examples of this type of reaction also occur in aqueous solution. In this case two compounds react to form two others. In a single replacement reaction we saw a situation analogous to a football player on the bench replacing a player on the field. A double replacement reaction is analogous to two players on the field changing positions. One type of double displacement reaction occurs because of the formation of a solid product from reactants that are in solution. This is also known as a **metathesis** reaction and will be discussed in Chapter 12. The following is an example of such a reaction. (See also Figure 9-5.)

$$(NH_4)_2S(aq) + Pb(NO_3)_2(aq) \rightarrow PbS(s) + 2NH_4NO_3(aq)$$

A second type of double displacement reaction involves formation of a molecular compound from ionic reactants in solution. Most of these reactions are also known as **acid-base reactions** and will be discussed in more detail in Chapter 13. An example of such a reaction is as follows:

**See Problems 9-6 through 9-9.**

$$2HNO_3(aq) + Ca(OH)_2(aq) \rightarrow Ca(NO_3)_2(aq) + 2H_2O(l)$$

## 9-3 Stoichiometry

Let us now look at the following balanced equation and list some of the quantitative relationships that are directly implied.

Figure 9-5 A DOUBLE DISPLACEMENT REACTION. The formation of solid silver chloride when solutions of silver nitrate and sodium chloride are mixed is a double displacement reaction.

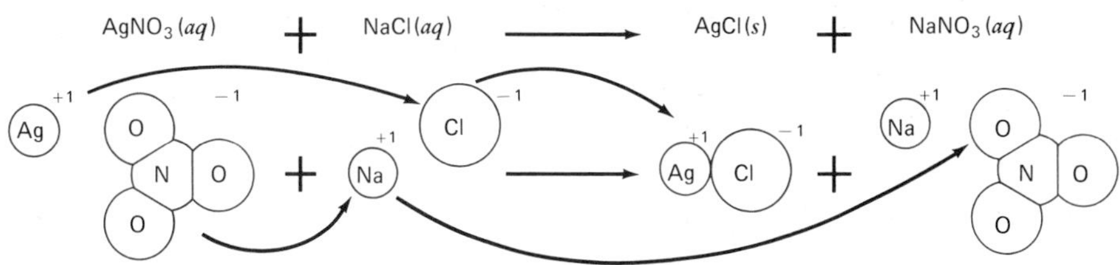


| | $N_2(g)$ | $+\ 3H_2(g)$ | $\rightarrow 2NH_3(g)$ |
|---|---|---|---|
| 1 | 1 molecule | + 3 molecules | → 2 molecules |
| 2 | 12 molecules | + 36 molecules | → 24 molecules |
| 3 | 1 dozen molecules | + 3 dozen molecules | → 2 dozen molecules |
| 4 | $6.02 \times 10^{23}$ molecules | $+\ 18.1 \times 10^{23}$ molecules | $\rightarrow 12.0 \times 10^{23}$ molecules |
| 5 | 1 mol | + 3 mol | → 2 mol |
| 6 | 28 g | + 6 g | → 34 g |

We will be concerned primarily with the last three relationships, as they concern laboratory situations.

Balanced equations give the necessary relationships to convert moles, grams, or number of molecules of one reactant or product into the equivalent number of moles, grams, or number of molecules of another reactant or product. *The quantitative relationship among reactants and products is known as* **stoichiometry.**

First we will examine the relationship of moles in the above equation. The equation tells us:

1 mol of $N_2$ produces 2 mol of $NH_3$

3 mol of $H_2$ produces 2 mol of $NH_3$

1 mol of $N_2$ reacts with 3 mol of $H_2$

From these three relationships six conversion factors can be written. *These conversion factors that originate from a balanced chemical equation will be called* **reaction factors.** They are exact numbers.

$$(1)\ \frac{1 \text{ mol } N_2}{2 \text{ mol } NH_3} \quad (3)\ \frac{1 \text{ mol } N_2}{3 \text{ mol } H_2} \quad (5)\ \frac{3 \text{ mol } H_2}{2 \text{ mol } NH_3}$$

**See Problems 9-10 and 9-11.**

$$(2)\ \frac{2 \text{ mol } NH_3}{1 \text{ mol } N_2} \quad (4)\ \frac{3 \text{ mol } H_2}{1 \text{ mol } N_2} \quad (6)\ \frac{2 \text{ mol } NH_3}{3 \text{ mol } H_2}$$

We can now use these factors to make the following stoichiometry conversions:

**1** Mole ↔ mole

**2** Mole ↔ weight

**3** Weight ↔ weight

**4** Weight ↔ number

This is illustrated by the following examples.

---

### Mole–Mole

**Example 9-4**
How many moles of $NH_3$ can be produced from 5.00 mol of $H_2$?

**Procedure**
Convert moles of what's given ($H_2$) to moles of what's requested ($NH_3$). Use reaction factor 6, which has what's requested in the numerator and what's given in the denominator.

**Scheme**

| *Given* | | *Requested* |
|---|---|---|
| moles ($H_2$) | reaction factor 6 ⟶ | moles ($NH_3$) |

**Solution**

$$5.00 \text{ mol H}_2 \times \frac{2 \text{ mol NH}_3}{3 \text{ mol H}_2} = \underline{\underline{3.33 \text{ mol NH}_3}}$$

### Mole–Weight

**Example 9-5**
How many moles of $NH_3$ can be produced from 33.6 g of $N_2$?

**Procedure**
First convert weight of what's given to moles ($N_2$) and then, as in Example 9-4, convert to moles of what's requested ($NH_3$) using reaction factor 2.

**Scheme**

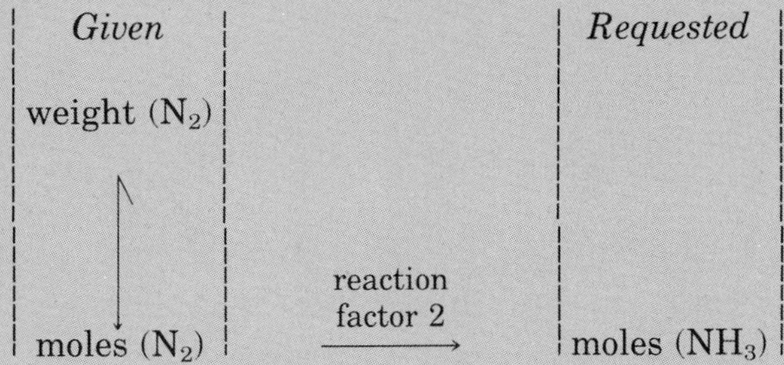


**Solution**

$$33.6 \text{ g N}_2 \times \frac{1 \text{ mol N}_2}{28.0 \text{ g N}_2} \times \frac{2 \text{ mol NH}_3}{1 \text{ mol N}_2} = \underline{\underline{2.40 \text{ mol NH}_3}}$$

### Weight–Weight

**Example 9-6**
What weight of $H_2$ is needed to produce 119 g of $NH_3$?

**Procedure**
As in Example 9-5, (1) convert weight to moles of what's given, (2) convert moles of what's given to moles of what's requested using reaction factor 5, and (3) convert moles of what's requested to weight.

**Scheme**

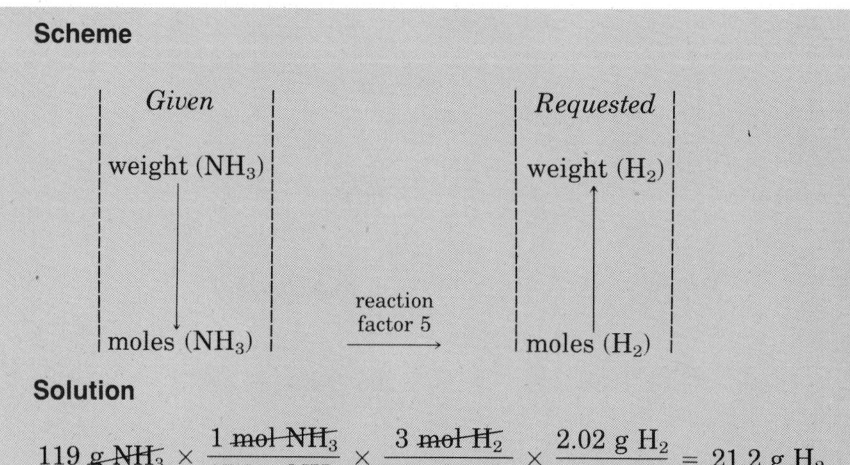


**Solution**

$$119\ \text{g}\ NH_3 \times \frac{1\ \text{mol}\ NH_3}{17.0\ \text{g}\ NH_3} \times \frac{3\ \text{mol}\ H_2}{2\ \text{mol}\ NH_3} \times \frac{2.02\ \text{g}\ H_2}{\text{mol}\ H_2} = \underline{\underline{21.2\ \text{g}\ H_2}}$$

**Weight–Number**

**Example 9-7**
How many individual molecules of $N_2$ are needed to react with 17.0 g of $H_2$?

**Procedure**
This problem is similar to Example 9-6 except that in the final step we convert moles to number of molecules instead of a weight.

**Scheme**

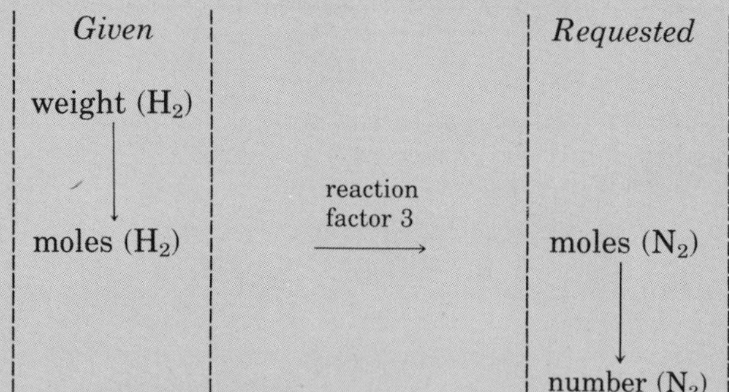


**Solution**

$$17.0\ \text{g}\ H_2 \times \frac{1\ \text{mol}\ H_2}{2.02\ \text{g}\ H_2} \times \frac{1\ \text{mol}\ N_2}{3\ \text{mol}\ H_2}$$
$$\times \frac{6.02 \times 10^{23}\ \text{molecules}\ N_2}{\text{mol}\ N_2} = \underline{\underline{1.69 \times 10^{24}\ \text{molecules}\ N_2}}$$

**Example 9-8**
What weight of $NH_3$ is produced by $4.65 \times 10^{22}$ molecules of $H_2$?

**Procedure**
This example is similar to Example 9-7 except that in the first step we convert number of molecules to moles of what's given.

**Scheme**

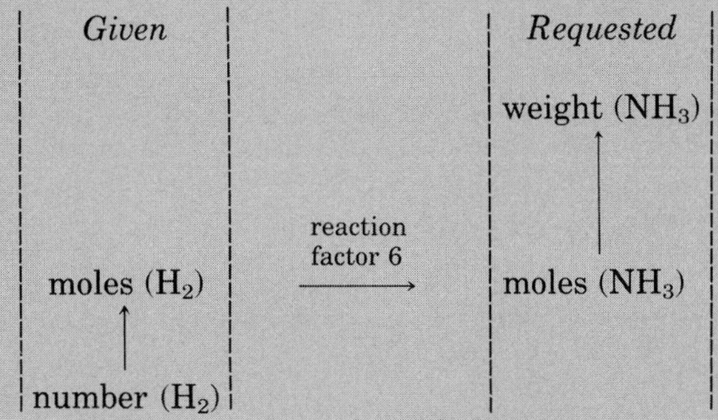


**Solution**

$$4.65 \times 10^{22} \text{ molecules } H_2 \times \frac{1 \text{ mol } H_2}{6.02 \times 10^{23} \text{ molecules } H_2} \times \frac{2 \text{ mol } NH_3}{3 \text{ mol } H_2} \times \frac{17.0 \text{ g } NH_3}{\text{mol } NH_3} = \underline{\underline{0.875 \text{ g } NH_3}}$$

The general procedure for working stoichiometry problems covering all of the possibilities discussed so far is shown in Figure 9-6. In words, the general procedure is as follows:

1 Write down (a) what is given and (b) what is requested.

2 If weight or a number of molecules is given, convert to moles in the first step.

3 Using the correct reaction factor from the *balanced* equation, convert moles of what is given to moles of what is requested in the second step.

4 If a weight or a number of molecules is requested, convert from moles of what is requested in the third step.

As you work the problems, first map out a procedure as shown in the examples. Set up the problems, making sure that all of the factors are correct and the appropriate units cancel, and finally *carefully* do the math. If you have worked stoichiometry problems previously using proportions, you may

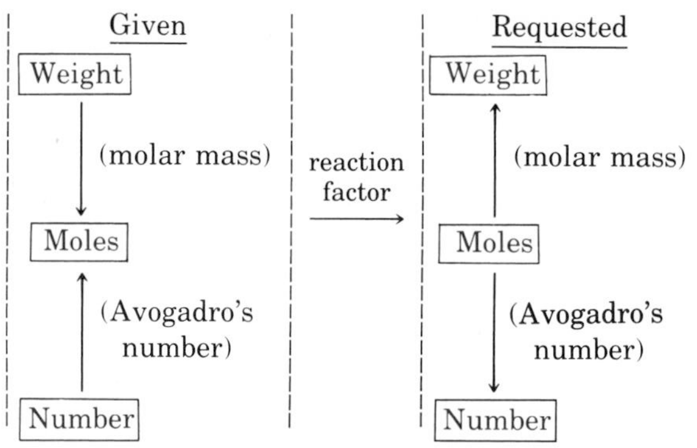


**Figure 9-6**
GENERAL PROCEDURE FOR STOICHIOMETRY PROBLEMS.

find the procedures outlined here slightly more time consuming; but you will also find these procedures more dependable and organized in obtaining the correct answer. In chemistry that's important.

Before proceeding, we will follow the general procedure with two more examples of stoichiometry problems.

**Example 9-9**

Some sulfur is present in coal in the form of pyrite, $FeS_2$ (also known as "fool's gold"). When it burns, it pollutes the air with the combustion product $SO_2$, as shown by the following chemical equation:

$$4FeS_2(s) + 11O_2(g) \rightarrow 2Fe_2O_3(s) + 8SO_2(g)$$

How many moles of $Fe_2O_3$ are produced from 145 g of $O_2$?

**Procedure**

Given: 145 g of $O_2$. Requested: ___?___ mol of $Fe_2O_3$.
Convert grams of $O_2$ to moles of $Fe_2O_3$.

**Scheme**

$$\boxed{\text{weight of } O_2} \xrightarrow{32.0 \text{ g/mol}} \text{moles of } O_2 \longrightarrow \boxed{\text{moles of } Fe_2O_3}$$

From the equation, a reaction factor relating $O_2$ to $Fe_2O_3$ is needed. The factor will have what's requested in the numerator.

$$\frac{2 \text{ mol } Fe_2O_3}{11 \text{ mol } O_2}$$

**Solution**

$$145 \text{ g } O_2 \times \frac{1 \text{ mol } O_2}{32.0 \text{ g } O_2} \times \frac{2 \text{ mol } Fe_2O_3}{11 \text{ mol } O_2} = \underline{\underline{0.824 \text{ mol } Fe_2O_3}}$$

**Example 9-10**
From the equation used in the preceding example, determine the weight of $SO_2$ that is produced from the combustion of 38.8 g of $FeS_2$.

**Procedure**
Given: 38.8 g of $FeS_2$. Requested: __?__ g of $SO_2$.
Convert grams of $FeS_2$ to grams of $SO_2$.

**Scheme**

| weight of $FeS_2$ | | weight of $SO_2$ |
|---|---|---|
| ↓ 120 g/mol | | ↑ 64.1 g/mol |
| moles of $FeS_2$ | ⟶ | moles of $SO_2$ |

The reaction factor from the equation that relates moles of given to moles of requested (in the numerator) is

$$\frac{8 \text{ mol } SO_2}{4 \text{ mol } FeS_2}$$

**Solution**

$$38.8 \text{ g } FeS_2 \times \frac{1 \text{ mol } FeS_2}{120 \text{ g } FeS_2} \times \frac{8 \text{ mol } SO_2}{4 \text{ mol } FeS_2}$$

$$\times \frac{64.1 \text{ g } SO_2}{\text{mol } SO_2} = \underline{\underline{41.5 \text{ g } SO_2}}$$

**See Problems 9-12 through 9-22.**

## 9-4 Percent Yield

In the reaction between $N_2$ and $H_2$ to produce $NH_3$ we have treated the reaction as if all the $N_2$ and $H_2$ present reacts to form $NH_3$. In fact, this isn't the case. If a mixture of $N_2$ and $H_2$ is allowed to react until no additional $NH_3$ is formed, there will still be some unreacted $H_2$ and $N_2$ present. Thus, in many reactions only a certain portion of the reactants convert to products. *The amount of product that is obtained in a reaction is known as the* **actual yield.** *The amount of product that would form if at least one of the reactants is completely consumed is known as the* **theoretical yield.** *The* **percent yield** *is the actual yield in grams divided by the theoretical yield in grams times 100%.*

$$\frac{\text{Actual yield}}{\text{Theoretical yield}} \times 100\% = \text{Percent yield}$$

Certain reactions are incomplete because a reaction can occur where products combine to re-form the original reactants. This is simply the reverse of the original reaction. The two reactions offset each other at some point, leaving appreciable quantities of all species present. This is referred to as a point of equilibrium. A reaction that reaches equilibrium is indicated by double arrows as follows:

$$N_2(g) + 3H_2(g) \rightleftharpoons 2NH_3(g)$$

We will have more to say about equilibrium reactions in later chapters.

**Example 9-11**
In a certain experiment a 4.70-g quantity of $H_2$ is allowed to react with $N_2$; a 12.5-g quantity of $NH_3$ is formed. What is the percent yield based on the $H_2$?

**Procedure**
Find the weight of $NH_3$ that would form if all the 4.70 g of $H_2$ is converted to $NH_3$ (the theoretical yield). Using the actual yield (12.5 g), find the percent yield.

**Scheme**

$$\boxed{\text{weight of } H_2} \qquad \boxed{\text{weight of } NH_3}$$

$$\downarrow 2.02 \text{ g/mol} \qquad \uparrow 17.0 \text{ g/mol}$$

$$\text{moles of } H_2 \longrightarrow \text{moles of } NH_3$$

The reaction factor needed is

$$\frac{2 \text{ mol } NH_3}{3 \text{ mol } H_2}$$

**Solution**

$$4.70 \cancel{\text{g } H_2} \times \frac{1 \cancel{\text{mol } H_2}}{2.02 \cancel{\text{g } H_2}} \times \frac{2 \cancel{\text{mol } NH_3}}{3 \cancel{\text{mol } H_2}} \times \frac{17.0 \text{ g } NH_3}{\cancel{\text{mol } NH_3}}$$

$$= \underline{\underline{26.4 \text{ g } NH_3}} \quad \text{(theoretical yield)}$$

$$\frac{12.5 \text{ g}}{26.4 \text{ g}} \times 100\% = \underline{\underline{47.3\% \text{ yield}}}$$

In addition to situations where one reaction reaches equilibrium and does not go to completion there are cases where other reactions can occur between the same reactants. This, of course, reduces the yield of a certain product. A well-known example occurs in the automobile engine when incomplete combustion of octane ($C_8H_{18}$) produces carbon monoxide rather than carbon dioxide.

Complete combustion: $2C_8H_{18}(g) + 25O_2(g) \rightarrow 16CO_2(g) + 18H_2O(l)$

Incomplete combustion: $2C_8H_{18}(g) + 17O_2(g) \rightarrow 16CO(g) + 18H_2O(l)$

In addition to problems where a pure reactant only partially converts to products, there is a related situation where a reactant converts completely to products but the reactant is part of an impure sample. Analysis of products, however, can be used to determine the percent purity of the original sample. This is illustrated by the following example.

---

**Example 9-12**

A 125-g sample of impure calcium carbonate is heated to drive off all of the $CO_2$ according to the equation

$$CaCO_3(s) \rightarrow CaO(s) + CO_2(g)$$

If 50.6 g of $CO_2$ is collected, what is the purity of the original sample?

**Procedure**

Assume that all $CaCO_3$ in the impure sample decomposes to form CaO and $CO_2$.

1 Find the weight of $CaCO_3$ needed to produce 50.6 g of $CO_2$ using the following scheme.

| weight of $CO_2$ | | weight of $CaCO_3$ |
|---|---|---|
| ↓ 44.0 g/mol | | ↑ 100 g/mol |
| moles of $CO_2$ | ⟶ | moles of $CaCO_3$ |

The reaction factor needed is

$$\frac{1 \text{ mol } CaCO_3}{1 \text{ mol } CO_2}$$

$$50.6 \text{ g } CO_2 \times \frac{1 \text{ mol } CO_2}{44.0 \text{ g } CO_2} \times \frac{1 \text{ mol } CaCO_3}{1 \text{ mol } CO_2} \times \frac{100 \text{ g } CaCO_3}{\text{mol } CaCO_3} = \underline{\underline{115 \text{ g } CaCO_3}}$$

2 Find the percent of the orignal sample that the weight of $CaCO_3$ found in step 1 represents.

$$\frac{115 \text{ g } CaCO_3}{125 \text{ g sample}} \times 100\% = \underline{\underline{92.0\% \text{ pure}}}$$

**See Problems 9-23 through 9-33.**

## 9-5 Limiting Reactant

In the simple reaction

$$2H_2 + O_2 \rightarrow 2H_2O$$

a 4.0-g quantity of $H_2$ reacts completely with 32.0 g of $O_2$. In fact, whenever $H_2$ and $O_2$ are mixed in a 4.0:32 (1:8) weight ratio and a reaction is initiated, all reactants are consumed and only products appear. (This reaction goes to completion, which is a 100% yield.) *When reactants are mixed in exactly the weight ratio determined from the balanced equation, the mixture is said to be* **stoichiometric.**

$$4.0 \text{ g of } H_2 + 32.0 \text{ g of } O_2 \rightarrow 36.0 \text{ g of } H_2O \qquad \text{(stoichiometric)}$$

If a 6.0-g quantity of $H_2$ and 32.0-g quantity of $O_2$ are mixed, there is still only 4.0 g of $H_2$ used so that 2.0 g of $H_2$ remains after the reaction is complete. Thus $H_2$ is in excess, and the amount of $H_2O$ formed is limited by the amount of $O_2$ originally present. The $O_2$ is the limiting reactant. *The* **limiting reactant** *is the reactant that would produce the least amount of products if completely consumed.*

$$6.0 \text{ g of } H_2 + 32.0 \text{ g of } O_2 \rightarrow 36.0 \text{ g of } H_2O + \boxed{2.0 \text{ g of } H_2 \text{ unreacted}}$$
$$(H_2 \text{ in excess, } O_2 \text{ limiting reactant})$$

If a 4.0-g quantity of $H_2$ and a 38.0-g quantity of $O_2$ are mixed, 36.0 g of $H_2O$ will be produced. Now $O_2$ is in excess and $H_2$ is the limiting reactant:

$$4.0 \text{ g of } H_2 + 38.0 \text{ g of } O_2 \rightarrow 36.0 \text{ g of } H_2O + \boxed{6.0 \text{ g of } O_2 \text{ unreacted}}$$
$$(O_2 \text{ in excess, } H_2 \text{ limiting reactant})$$

When quantities of two or more reactants are given, it is necessary to determine which is the limiting reactant (unless they are mixed in exactly stoichiometric amounts). This can be simplified by the following procedures.

1 Determine the number of moles of a product produced by each reactant using the general procedure discussed earlier.

2 The reactant producing the *smallest* yield is the limiting reactant.

**Example 9-13**
Silver tarnishes (turns black) in homes because of the presence of small amounts of $H_2S$ (a rotten-smelling gas that originates from the decay of food). The reaction is as follows:

$$4Ag(s) + 2H_2S(g) + O_2(g) \rightarrow \underset{\text{(black)}}{2Ag_2S(s)} + 2H_2O(l)$$

If 0.145 mol of Ag is present with 0.0872 mol of $H_2S$ and excess $O_2$:

(a) What is the limiting reactant?
(b) What weight of $Ag_2S$ is produced?
(c) What weight of the other reactant in excess remains?

**Procedure**
(a) Convert moles of Ag and $H_2S$ to moles of $Ag_2S$ produced by each.

$$\text{Moles of reactant} \rightarrow \text{moles of } Ag_2S$$

$$\text{Ag} \qquad 0.145\ \cancel{\text{mol Ag}} \times \frac{2\text{ mol }Ag_2S}{4\ \cancel{\text{mol Ag}}} = 0.0725\text{ mol }Ag_2S$$

$$H_2S \qquad 0.0872\ \cancel{\text{mol }H_2S} \times \frac{2\text{ mol }Ag_2S}{2\ \cancel{\text{mol }H_2S}} = 0.0872\text{ mol }Ag_2S$$

Since Ag produces the smaller yield of $Ag_2S$, <u>Ag is the limiting reactant</u>

(b) To find the weight of $Ag_2S$ produced convert mol of $Ag_2S$ to weight.

$$0.0725\ \cancel{\text{mol }Ag_2S} \times \frac{248\text{ g }Ag_2S}{\cancel{\text{mol }Ag_2S}} = \underline{18.0\text{ g }Ag_2S}$$

(c) To find the weight of $H_2S$ in excess, first convert the moles of Ag to moles of $H_2S$. This will be the moles of $H_2S$ reacted.

$$0.145\ \cancel{\text{mol Ag}} \times \frac{2\text{ mol }H_2S}{4\ \cancel{\text{mol Ag}}} = 0.0725\text{ mol }H_2S\text{ reacted}$$

Find the moles of $H_2S$ remaining or unreacted by subtracting moles reacted from the original moles present.

$$0.0872\text{ mol} - 0.0725\text{ mol} = 0.0147\text{ mol }H_2S\text{ unreacted}$$

Finally, convert moles of unreacted $H_2S$ to weight.

$$0.0147\ \cancel{\text{mol }H_2S} \times \frac{34.1\text{ g }H_2S}{\cancel{\text{mol }H_2S}} = \underline{0.501\text{ g }H_2S\text{ in excess}}$$

**Example 9-14**
Methanol ($CH_3OH$) is used as a fuel for racing cars. It burns in the engine according to the following equation.

$$2CH_3OH(l) + 3O_2(g) \rightarrow 2CO_2(g) + 4H_2O(g)$$

If 40.0 g of methanol is mixed with 46.0 g of $O_2$:

(a) What is the limiting reactant?
(b) What is the theoretical yield of $CO_2$?
(c) If 38.0 g of $CO_2$ is actually produced, what is the percent yield?

**Procedure**
(a) Determine the moles of $CO_2$ formed from each of the reactants.

$$\boxed{\text{weight of reactant}} \xrightarrow{\text{molar mass}} \text{moles of reactant} \rightarrow \boxed{\text{moles of } CO_2}$$

$CH_3OH$ $$40.0 \cancel{\text{g } CH_3OH} \times \frac{1 \cancel{\text{mol } CH_3OH}}{32.0 \cancel{\text{g } CH_3OH}} \times \frac{2 \text{ mol } CO_2}{2 \cancel{\text{mol } CH_3OH}} = 1.25 \text{ mol } CO_2$$

$O_2$ $$46.0 \cancel{\text{g } O_2} \times \frac{1 \cancel{\text{mol } O_2}}{32.0 \cancel{\text{g } O_2}} \times \frac{2 \text{ mol } CO_2}{3 \cancel{\text{mol } O_2}} = 0.958 \text{ mol } CO_2$$

Therefore, $\underline{\underline{O_2 \text{ is the limiting reactant}}}$.

(b) The theoretical yield is determined from the amount of product formed *from the limiting reactant*. Thus we simply convert the 0.958 mol of $CO_2$ produced by the $O_2$ to grams.

$$0.958 \cancel{\text{mol } CO_2} \times \frac{44.0 \text{ g } CO_2}{\cancel{\text{mol } CO_2}} = \underline{\underline{42.2 \text{ g } CO_2}}$$

(c) From the theoretical yield and the actual yield determine percent yield.

$$\frac{38.0 \text{ g}}{42.2 \text{ g}} \times 100\% = \underline{\underline{90.0\% \text{ yield of } CO_2}}$$

**See Problems 9-34 through 9-41.**

## 9-6 Chapter Summary

If the symbols of elements are like the letters of chemistry and the formulas of compounds like the words, then the chemical equation is like a complete sentence.

A balanced chemical equation gives the following information.

1 The formulas of reactants and products, including elements (e.g., $H_2$, $P_4$)

2 The physical state of reactants and products [e.g., (*s*), (*g*), (*l*), (*aq*)]

3 The mole ratio of reactants and products

Chemical equations represent many types of reactions. The five types listed in this text are illustrated as follows:

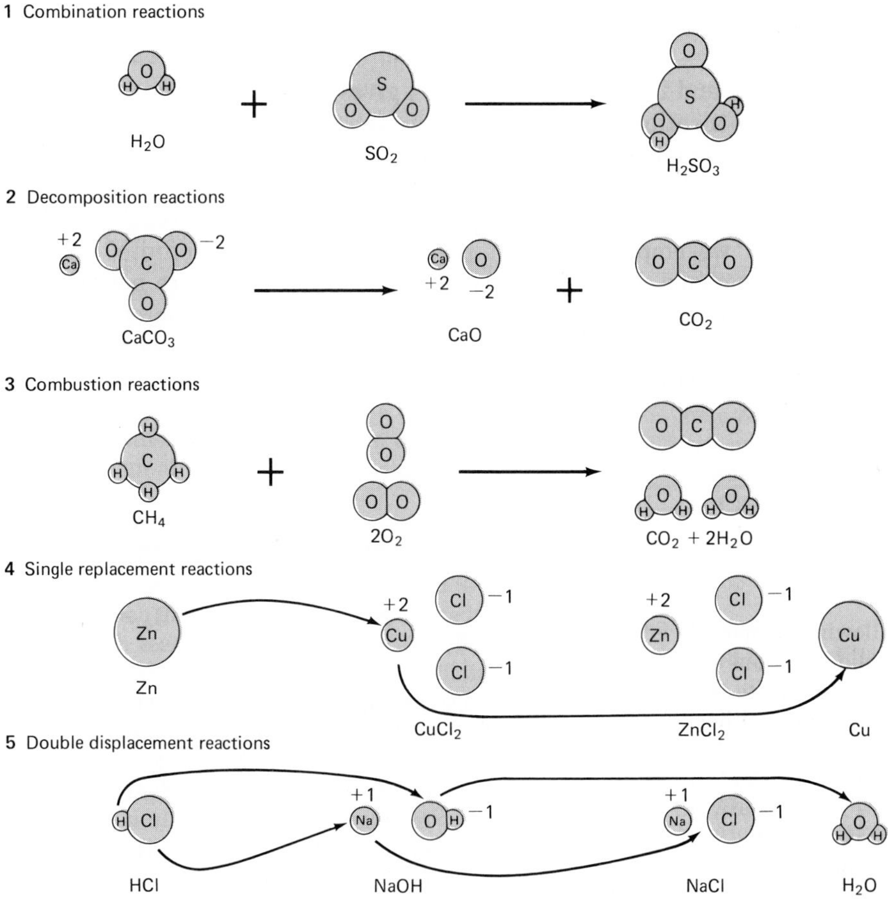


Balanced equations also provide the mole ratios between reactants and products, which are called reaction factors. Reaction factors can then be used to convert quantities of reactants or products into equivalent or stoichiometric quantities of other reactants or products. The conversions performed are summarized as follows.

(a) How many moles of $CO_2$ are produced from the combustion of 0.450 mol of $C_3H_8$? How many moles of $H_2O$? How many moles of $O_2$ are needed?
(b) What weight of $H_2O$ is produced if 0.200 mol of $CO_2$ is also produced?
(c) What weight of $C_3H_8$ is required to produce 1.80 g of $H_2O$?
(d) What weight of $C_3H_8$ is required to react with 160 g of $O_2$?
(e) What weight of $CO_2$ is produced by the reaction of $1.20 \times 10^{23}$ molecules of $O_2$?
(f) How many moles of $H_2O$ are produced if $4.50 \times 10^{22}$ molecules of $CO_2$ are produced?

**9-14** The alcohol component of "gasohol" burns according to the following equation:

$$C_2H_6O(l) + 3O_2(g) \rightarrow 2CO_2(g) + 3H_2O(g)$$

(a) What weight of $O_2$ is needed to produce 5.45 mol of $CO_2$?
(b) What weight of $H_2O$ is produced from $4.58 \times 10^{-4}$ mol of $C_2H_6O$?
(c) What weight of $H_2O$ is produced from 125 g of $C_2H_6O$?
(d) What weight of $CO_2$ is produced if 50.0 g of $H_2O$ is produced?
(e) What weight of $C_2H_6O$ reacts with $8.54 \times 10^{25}$ molecules of $O_2$?

**9-15** In the atmosphere $N_2$ and $O_2$ do not react. In the high temperatures of an automobile engine, however, the following reaction does occur:

$$N_2(g) + O_2(g) \rightarrow 2NO(g)$$

When the NO reaches the atmosphere through the engine exhaust, a second reaction takes place:

$$2NO(g) + O_2(g) \rightarrow 2NO_2(g)$$

The $NO_2$ is a somewhat brownish gas that contributes to the haze of smog and is irritating to the nasal passages and lungs. What weight of $N_2$ is required to produce 155 g of $NO_2$?

**9-16** Manganese metal can be prepared by a reaction called a thermite reaction because it gives off a great amount of heat. In fact, the manganese metal is formed in the molten or liquid state:

$$4Al(s) + 3MnO_2(s) \rightarrow 3Mn(l) + 2Al_2O_3(s)$$

From the reaction, determine what weight of Al is needed to produce 750 g of Mn. How many molecules of $MnO_2$ are used in the process?

* **9-17** Liquid iron is made from iron ore ($Fe_2O_3$) in a three-step process in a blast furnace:

1 $3Fe_2O_3(s) + CO(g) \rightarrow 2Fe_3O_4(s) + CO_2(g)$

2 $Fe_3O_4(s) + CO(g) \rightarrow 3FeO(s) + CO_2(g)$

3 $FeO(s) + CO(g) \rightarrow Fe(l) + CO_2(g)$

What weight of iron would eventually be produced from 125 g of $Fe_2O_3$?

**9-18** Antacids ($CaCO_3$) react with "stomach acid" according to the following equation:

$$CaCO_3(s) + 2HCl(aq) \rightarrow CaCl_2(aq) + CO_2(g)$$

What weight of stomach acid reacts with 1.00 g of $CaCO_3$?

**9-19** Fool's gold (pyrite) is so named because it looks much like gold. When it is placed in aqueous HCl, however, it dissolves and gives off a rotten-smelling gas ($H_2S$). Gold itself does not react with aqueous HCl. From the following equation determine how many individual molecules of $H_2S$ are formed from 0.520 mol of $FeS_2$.

$$FeS_2(s) + 2HCl(aq) \rightarrow FeCl_2(aq) + H_2S(g) + S(s)$$

**9-20** Nitrogen dioxide may form "acid rain" by reaction with water in the air according to the following equation:

$$3NO_2(g) + H_2O(l) \rightarrow 2HNO_3(aq) + NO(g)$$

What weight of nitric acid is produced from 18.5 kg of $NO_2$?

**9-21** The fermentation of sugar to produce ethyl alcohol is represented by the following equation:

$$C_6H_{12}O_6 \rightarrow 2C_2H_5OH + 2CO_2$$

What weight of alcohol is produced from 25.0 mol of sugar?

**9-22** Methane gas can be made from carbon monoxide gas according to the following equation:

$$2CO + 2H_2 \rightarrow CH_4 + CO_2$$

What weight of CO is required to produce $8.75 \times 10^{25}$ molecules of $CH_4$?

## Theoretical and Percent Yield

**9-23** Sulfur trioxide ($SO_3$) is prepared from $SO_2$ according to the following equation:

$$2SO_2(g) + O_2(g) \rightleftharpoons 2SO_3(g)$$

In this reaction not all $SO_2$ is converted to $SO_3$ even with excess $O_2$ present. In a certain experiment, 21.2 g of $SO_3$ was produced from 24.0 g of $SO_2$. What is the theoretical yield of $SO_3$ and the percent yield?

**9-24** The following is a reversible decomposition reaction:

$$2N_2O_5 \rightleftharpoons 4NO_2 + O_2$$

When 15.0 g of $N_2O_5$ was allowed to decompose, it was found that 10.0 g of $NO_2$ formed. What is the percent yield?

**9-25** The following equation represents a reversible combination reaction:

$$P_4O_{10} + 6PCl_5 \rightleftharpoons 10POCl_3$$

If 25.0 g of $PCl_5$ is allowed to react, it is found that there is a 45.0% yield of $POCl_3$. What is the actual yield in grams?

* **9-26** Given the following reaction:

$$2NO_2 + 4H_2 \rightleftharpoons N_2 + 4H_2O$$

What weight of hydrogen is required to produce 250 g of $N_2$ if the yield is 70.0%?

**9-27** Octane in gasoline burns in the automobile engine according to the following equation:

$$2C_8H_{18}(l) + 25O_2(g) \rightarrow 16CO_2(g) + 18H_2O(g)$$

If a 57.0-g sample of octane is burned, 152 g of $CO_2$ is formed. What is the percent yield of $CO_2$?

* **9-28** In Problem 9-27, the $C_8H_{18}$ that is *not* converted to $CO_2$ forms CO. What is the weight of CO formed? (CO is a poisonous pollutant that is converted to $CO_2$ in the car's catalytic converter.)

**9-29** When benzene reacts with bromine, the principal reaction is illustrated as follows:

$$C_6H_6 + Br_2 \rightarrow C_6H_5Br + HBr$$

If the yield of bromobenzene ($C_6H_5Br$) is 65.2%, what weight of bromobenzene is produced from 12.5 g of $C_6H_6$?

* **9-30** A second reaction between $C_6H_6$ and $Br_2$ (see Problem 9-29) produces dibromobenzene ($C_6H_4Br_2$).

(a) Write the balanced equation illustrating this reaction.

(b) If the remainder of the benzene from Problem 9-29 reacted to form dibromobenzene, what is the weight of $C_6H_4Br_2$ produced?

**9-31** A 50.0-g sample of impure $KClO_3$ is decomposed to KCl and $O_2$. If a 12.0-g quantity of $O_2$ is produced, what percent of the sample is $KClO_3$? (Assume that all of the $KClO_3$ present decomposes.)

**9-32** In Example 9-9, $SO_2$ was formed from the burning of pyrite ($FeS_2$) in coal. If a 312-g quantity of $SO_2$ was collected from the burning of 6.50 kg of coal, what percent of the original sample was pyrite?

* **9-33** Copper metal can be recovered from an ore ($CuCO_3$) by a decomposition reaction as illustrated by the following equation:

$$2CuCO_3 \rightarrow 2Cu + 2CO_2 + O_2$$

What is the weight of a sample of ore if it is 47.5% $CuCO_3$ and produces 350 g of Cu? (Assume a complete decomposition of the $CuCO_3$.)

### Limiting Reactant

**9-34** Given the following equation:

$$2Al + 3H_2SO_4 \longrightarrow Al_2(SO_4)_3 + 3H_2$$

If 0.800 mol of Al is mixed with 1.00 mol of $H_2SO_4$, how many moles of $H_2$ are produced? How many moles of one of the reactants remain?

**9-35** Given the following equation:

$$2C_5H_6 + 13O_2 \rightarrow 10CO_2 + 6H_2O$$

If 3.44 mol of $C_5H_6$ is mixed with 20.6 mol of $O_2$, what weight of $CO_2$ is formed?

**9-36** Given the following equation:

$$4NH_3(g) + 3O_2(g) \rightarrow 2N_2(g) + 6H_2O(l)$$

If a 40.0-g sample of $O_2$ is mixed with 1.50 mol of $NH_3$, which is the limiting reactant? What is the theoretical yield (in moles) of $N_2$?

**9-37** Given the following equation:

$$2AgNO_3(aq) + CaCl_2(aq) \rightarrow 2AgCl(s) + Ca(NO_3)_2(aq)$$

If a solution containing 20.0 g of $AgNO_3$ is mixed with a solution containing 10.0 g of $CaCl_2$, which compound is the limiting reactant? What is the theoretical yield (in grams) of AgCl? What weight of one of the reactants remains?

**9-38** Limestone ($CaCO_3$) dissolves in hydrochloric acid as shown by the equation:

$$CaCO_3(s) + 2HCl(aq) \rightarrow CaCl_2(aq) + CO_2(g) + H_2O(l)$$

If a 20.0-g sample of $CaCO_3$ and 10.0 g of HCl are mixed, what weight of $CO_2$ is produced? What weight of one of the reactants remains?

* **9-39** Given the balanced equation:

$$2HNO_3(aq) + 3H_2S(aq) \rightarrow 2NO(g) + 4H_2O(l) + 3S(s)$$

If a 10.0-g quantity of $HNO_3$ is mixed with 5.00 g of $H_2S$, what is the weight of each product and the reactant present in excess after reaction occurs? Assume a complete reaction.

**9-40** Given the following equation:

$$4NH_3(g) + 5O_2(g) \rightarrow 4NO(g) + 6H_2O(l)$$

When an 80.0-g sample of $NH_3$ is mixed with 200 g of $O_2$, a 40.0-g quantity of NO is formed. What is the percent yield?

* **9-41** Given the following equation:

$$3K_2MnO_4 + 4CO_2 + 2H_2O \rightarrow 2KMnO_4 + 4KHCO_3 + MnO_2$$

How many moles of $MnO_2$ will be produced if 9.50 mol of $K_2MnO_4$, $6.02 \times 10^{24}$ molecules of $CO_2$, and 90.0 g of $H_2O$ are mixed?

This planet, earth, is surrounded by a sea of gases. We live at the bottom of this sea.

# Chapter 10
# The Gaseous State of Matter

## Purpose of Chapter 10

In Chapter 10 we develop an understanding of the gaseous state and apply quantitative relationships common to gases.

## Objectives for Chapter 10

After completion of this chapter, you should be able to:

1 Convert between the various units used to express the pressure of a gas. (10-1)
2 Apply the kinetic theory of gases to explain some characteristics of the gaseous state. (10-2)
3 Apply Boyle's law to problems involving the effect of a change of pressure on the volume of a gas. (10-3)
4 Use Charles' law to calculate the effect of a change of temperature on the volume of a gas. (10-4)
5 Calculate the effect of a change of pressure on the temperature of a gas by application of Gay-Lussac's law. (10-4)
6 Apply the kinetic theory of gases to describe the basis for Boyle's, Charles', and Gay-Lussac's laws. (10-2, 10-3, 10-4)
7 Use the combined gas law for a quantity of gas at two sets of conditions. (10-5)
8 Apply Graham's law to problems involving the relationship between average velocities and molar masses of two gases. (10-6)
9 Solve problems involving Dalton's law and Avogadro's laws. (10-7, 10-8)
10 Use the ideal gas law to calculate an unknown property of a sample of gas ($P$, $V$, $T$, or $n$) when the other properties are known. (10-9)
11 Convert between moles of a gas and the volume at STP using the molar volume as a conversion factor. (10-10)
12 Calculate the density of a gas from its molar mass and the molar volume. (10-10)
13 Apply the ideal gas law or the molar volume to stoichiometric calculations involving gases. (10-11)

If there is one thing that held up the development of modern chemistry, it was the ignorance that scientists had of the nature of gases. Even air, the sea of gases in which we live, was only vaguely understood until the experiments of Antoine Lavoisier in France and Joseph Priestley in England in the late 1700s solved much of the mystery. These scientists proved that air is not just one substance but is actually composed mainly of a mixture of two elements, oxygen and nitrogen. Their research not only enlightened science on the nature of air but also laid the foundation for modern quantitative chemistry.

There is a disadvantage and an advantage in the study of gases. The disadvantage is that most gases are invisible, and thus we must infer their presence indirectly from measurements of their properties. When working with the solid or liquid phase, at least you can see them. The advantage is that many of the properties that we will study are the same for all gases. This permits many convenient generalizations or "laws" concerning their behavior. On the other hand, few generalizations apply to all solids or liquids. As a result, they must be studied individually.

Many of the "laws" that we will study in this chapter have been known for centuries. We will cheat a little, however, by first introducing the modern assumptions about the nature of gases called the kinetic theory. This should make the "laws" seem entirely reasonable and even predictable. After discussing the various gas laws, we will expand the general scheme for working stoichiometry problems (introduced in Chapter 9) to include gases.

As background for this chapter you should be familiar with the concept of the mole as discussed in Chapter 8 and the procedure for solving stoichiometry problems discussed in Chapter 9.

## 10-1 The Pressure of a Gas

In 1643, an Italian scientist name Evangelista Torricelli experimented with an apparatus that is now known as a **barometer.** Torricelli filled a long glass tube with the dense, liquid metal, mercury, and inverted the tube into a bowl of mercury so that no air could enter the tube. Torricelli found that the mercury in the tube would stay suspended to a height of about 76 cm no matter how long or wide the tube (see Figure 10-1).

At the time scientists thought that since "nature abhors a vacuum," it was a vacuum in the top of the tube that held up the column of mercury. Torricelli had a different idea. He suggested that it was actually the weight of the air on the outside that pushed up the level of mercury. Otherwise, since tubes 2 and 3 in Figure 10-1 would have "more vacuum" than tube 1, the level would rise higher in those tubes. To prove his point, Torricelli took his barometer up a mountain, where he reasoned that the atmosphere would be less dense and would thus support less mercury in the tube. He was correct; the level of the barometer fell as he ascended the mountain.

*The weight of a quantity of matter pressing on a surface is an example*

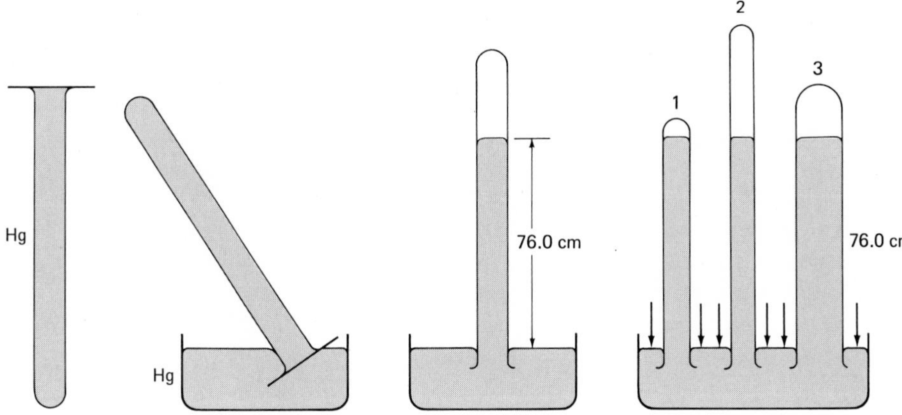


**Figure 10-1** THE BAROMETER. When a long tube is filled with mercury and inverted in a bowl of mercury, the atmosphere supports the column to a height of 76.0 cm.

*of* **force. Pressure** *is defined as the force (in this case the weight) applied per unit area.* This can be expressed mathematically as

$$P \text{ (pressure)} = \frac{F \text{ (force)}}{A \text{ (area)}}$$

In a barometer, the pressure exerted by the atmosphere on the outside is equal to the pressure exerted by the column of mercury on the inside. As the following calculations indicate, the force of a column of mercury depends on the width of the column but the pressure does not. In the barometer,

$$P_{\text{air (outside)}} = P_{\text{Hg (inside)}}$$

$$P_{\text{Hg}} = P_{\text{air}} = \frac{\text{weight of Hg}}{\text{area}}$$

$$\text{Weight of Hg} = \text{volume} \times \text{density}$$

$$\text{Volume} = \text{height} \times \text{area of opening}$$

In tube 1 the area of the opening is 1.00 cm$^2$.

$$\text{Volume} = 76.0 \text{ cm} \times 1.00 \text{ cm}^2 = 76.0 \text{ cm}^3$$

$$\text{Weight} = 76.0 \not{\text{cm}^3} \times 13.6 \text{ g}/\not{\text{cm}^3} = 1030 \text{ g}$$

$$\text{Pressure} = 1030 \text{ g}/1.00 \text{ cm}^2 = 1030 \text{ g/cm}^2$$

Now we can compare the pressure exerted by the mercury in tube 1 with that in tube 3, which has an area of 10 cm$^2$ at the opening. The force in tube 3 is

$$(76.0 \not{\text{cm}} \times 10.0 \not{\text{cm}^2}) \times 13.6 \text{ g}/\not{\text{cm}^3} = 10{,}300 \text{ g}$$

The pressure however, is the same as in tube 1.

$$P = \frac{F}{A} = \frac{10{,}300 \text{ g}}{10.0 \text{ cm}^2} = 1030 \text{ g/cm}^2$$

The human body has roughly 10,000 $cm^2$ of surface area (with, obviously, much individual variation). This means that we feel about $10^7$ g (about 20,000 lb) of force from the air under normal atmospheric conditions. On days when it feels like that whole force is only on our heads, it seems like we are "under a lot of pressure." However, despite how it may feel at times, the force is spread out over the total body area, so we should feel quite comfortable at the bottom of the ocean of gas where we live.

*The normal pressure of the atmosphere at sea level is defined as exactly one atmosphere (1 atm) and is used as the standard.* As we have seen, this is equivalent to the pressure exerted by a column of mercury 76.0 cm (760 mm) high.

$$1.00 \text{ atm} = 76.0 \text{ cmHg} = 760 \text{ mmHg} = 760 \text{ torr}$$

(The unit of mmHg is defined as torr in honor of Evangelista Torricelli.)

In addition to torr, there are several other units of pressure that have special uses. The special use of the unit and its relation to 1 atm is listed in Table 10-1.

**Example 10-1**

What is 485 torr expressed in atmospheres?

**Procedure**

The conversion factors are

$$\frac{760 \text{ torr}}{\text{atm}} \quad \text{and} \quad \frac{1 \text{ atm}}{760 \text{ torr}}$$

**Solution**

**See Problems 10-1 through 10-7.**

$$485 \cancel{\text{torr}} \times \frac{1 \text{ atm}}{760 \cancel{\text{torr}}} = \underline{\underline{0.638 \text{ atm}}}$$

## 10-2 The Kinetic Theory of Gases

In the following sections, we will see how several early investigators were able to present the results of experiments or observations that eventually led to the acceptance of "laws" concerning the behavior of gases. Science does not accept observation as laws without years of research and many experiments. The early investigators would probably be disappointed, however, if they had known how predictable their observations are based on a simple understanding of the nature of gases. Our understanding of gases

**Table 10-1** One Atmosphere

| Unit | Special Use |
|---|---|
| 760 mmHg or 760 torr | Most chemistry laboratory measurements for pressures in the neighborhood of one atmosphere |
| 14.7 lb/in.$^2$ | U.S. pressure gauges |
| 29.9 in. of Hg | U.S. weather reports |
| 101.325 kPa (kilopascals) | The SI unit of pressure (1 newton/m$^2$) |
| 1.013 bars | Used in physics and astronomy mainly for very low pressures (millibars) or very high pressures (kilobars) |

can be summarized by the following assumptions, which taken together are known as the kinetic theory of gases.

1 A gas is composed of very small particles called molecules (atoms in the case of noble gases), which are widely spaced and occupy negligible volume. A gas is thus mostly empty space.

2 The molecules of the gas are in rapid, random motion, colliding with each other and the sides of the container. Pressure is a result of these collisions with the sides of the container. (See Figure 10-2.)

3 All collisions involving gas molecules are elastic. (The total energy of two colliding molecules is conserved. A ball bouncing off the pavement undergoes inelastic collisions, since it does not bounce as high each time.)

4 The gas molecules have negligible attractive (or repulsive) forces between them.

5 The temperature of a gas is related to average kinetic energy of the gas molecules. Also, at the same temperature different gases have the same average kinetic energy. Kinetic energy (K.E.) is given by

$$\text{K.E.} = \tfrac{1}{2}mv^2 \qquad (m = \text{mass},\ v = \text{velocity})$$

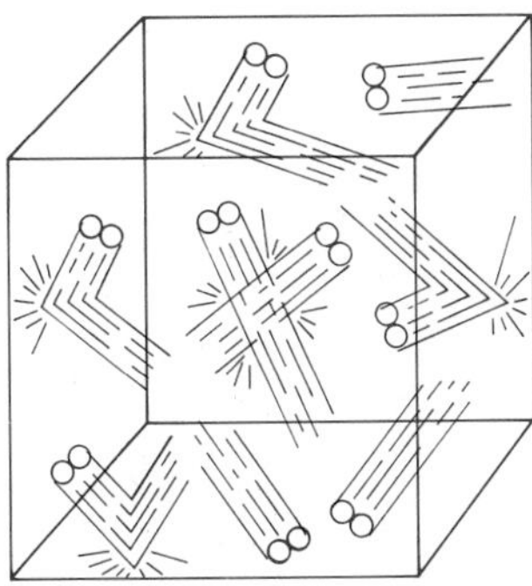

**Figure 10-2** GASES. Gas molecules are in rapid, random motion.

The higher the temperature, the higher the average kinetic energy of the molecules of the gas. *Theoretically, all molecular motion stops at absolute zero.*

Many characteristics of gases that we have all noticed can be expected from an understanding of the kinetic theory of gases.

1 Since gases are mostly empty space, they are highly *compressible.*

2 The small density of a gas can also be explained by the small amount of matter in the gas. For example, 1.0 g of *liquid* water has a volume of 1.0 mL. As *steam* at 100 °C and 1 atm pressure, the 1.0 g of water occupies 1700 mL.

3 Because of the rapid motion of the molecules, gases mix thoroughly. In a chemistry building it doesn't take long for the whole campus to know that hydrogen sulfide experiments are on the schedule. The noxious smell of rotten eggs soon permeates the area.

**See Problems 10-8 through 10-12.**

4 The rapid motion of gas molecules also explains why gases expand to fill the space of any container uniformly.

## 10-3 Boyle's Law

**Review Appendix B on proportionalities and Appendix F on graphs.**

The first quantitative relationship concerning gases was suggested in 1660 by the British chemist, Sir Robert Boyle. **Boyle's law** *states that there is an inverse relationship between the pressure exerted on a quantity of gas and its volume if the temperature is held constant.* (See Appendix B for a discussion of direct and inverse relationships and Appendix F for graphs of these relationships.)

Boyle's law can be illustrated quite simply with an apparatus as shown in Figure 10-3. In experiment 1, a certain quantity of gas ($V_1$ = 10.0 mL) is trapped in a U-shaped tube by some mercury. Since the level of mercury is the same in both sides of the tube and the right side is open to the atmosphere, the pressure on the trapped gas is the same as the atmospheric pressure ($P_1$ = 760 torr).

When mercury is added to the tube, the pressure on the trapped gas is increased to 950 torr (760 torr originally plus 190 torr from the added mercury). Notice in experiment 2 that the *increase in pressure* has caused a *decrease in volume* to 8.0 mL.

The inverse relation of Boyle is represented as

$$V \propto \frac{1}{P}$$

or as an equality with $k$ the constant of proportionality:

$$V = \frac{k}{P} \quad \text{or} \quad PV = k$$

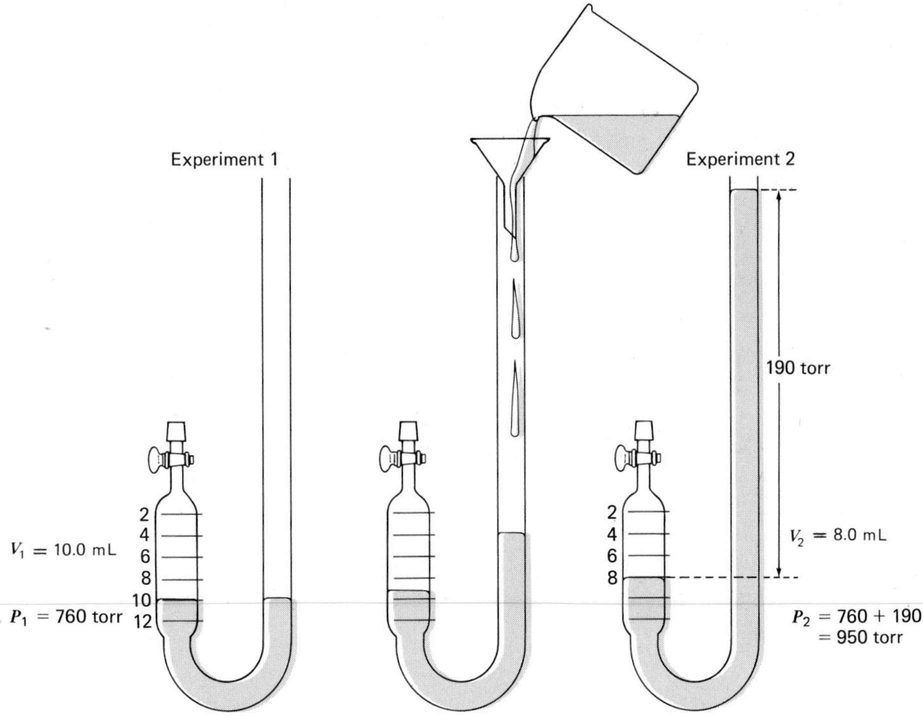


**Figure 10-3** BOYLE'S LAW APPARATUS. Addition of mercury in the apparatus causes an increase in pressure on the trapped gas. This leads to a reduction in the volume.

Notice in experiment 1 in Figure 10-3 that

$$\underbrace{P_1V_1 = 760 \text{ torr} \times 10.0 \text{ mL}}_{\text{Indicates one set of conditions}} = 7600 \text{ torr} \cdot \text{mL} = k$$

In experiment 2,

$$\underbrace{P_2V_2 = 950 \text{ torr} \cdot 8.0 \text{ mL}}_{\text{Indicates a second set of conditions}} = 7600 \text{ torr} \cdot \text{mL} = k$$

As predicted by Boyle's law, notice that in both experiments $PV$ equals the same value. Therefore, for a quantity of gas under two sets of conditions at the same temperature,

$$P_1V_1 = P_2V_2 = k$$

We can use this equation to calculate how a volume of a gas changes when the pressure changes. For example, if $V_2$ is a new volume that is to be found at a certain new pressure, $P_2$, the equation becomes

$$V_2 = V_1 \times \frac{P_1}{P_2}$$

$$\text{Final volume} = \text{initial volume} \times \text{pressure correction factor}$$

In working problems of this nature, first determine whether $V_2$ (the final volume) is smaller or larger than $V_1$ (the initial volume). If $V_2$ is smaller, the pressure correction factor must be less than one, which means that the numerator is smaller than the denominator. If $V_2$ is larger than $V_1$, the pressure factor must be greater than one, which means that the numerator is larger than the denominator. The following two examples illustrate the use of Boyle's law.

---

**Example 10-2**

Inside a certain automobile engine, the volume of a cylinder is 457 mL when the pressure is 1.05 atm. When the gas is compressed, the pressure increases to 5.65 atm at the same temperature. What is the volume of the compressed gas?

**Procedure**

The final volume equals the initial volume times the pressure factor. Since the pressure increases, the volume decreases and the pressure factor must be less than one.

**Solution**

$$V_2 = V_1 \times \frac{P_1}{P_2}$$

$$V_2 = 475 \text{ mL} \times \frac{1.05 \cancel{\text{atm}}}{5.65 \cancel{\text{atm}}} = \underline{\underline{88.3 \text{ mL}}}$$

(Notice that the units of pressure are the same. If the initial and final pressures are given in different units, one must be converted to the other.)

---

**Example 10-3**

If the volume of a gas is 3420 mL at a pressure of 2.17 atm, what is the pressure if the gas expands to 8.75 L?

**Procedure**

The Boyle's law relationship can be solved for $P_2$:

$$P_2 = P_1 \times \frac{V_1}{V_2}$$

Final pressure = initial pressure × volume correction factor

In this case, the volume increases from 3420 mL (3.42 L) to 8.75 L. An increase in volume means a decrease in pressure, so the volume factor is less than one.

**Solution**

$$P_2 = P_1 \times \frac{V_1}{V_2}$$

$$P_2 = 2.17 \text{ atm} \times \frac{3.42 \cancel{L}}{8.75 \cancel{L}} = \underline{\underline{0.848 \text{ atm}}}$$

Boyle's law can now be seen as a direct result of the kinetic theory of gases. In Figure 10-4, the behavior of four average molecules of a certain gas have been illustrated. The distance traveled in a unit of time is represented by the length of the arrow. Since these are average molecules,* they all travel the same distance per unit time at the same temperature. Notice that if the volume is decreased there are more frequent collisions with the walls, which in turn means a higher pressure.

**See Problems 10-13 through 10-19.**

## 10-4 Charles' Law and Gay-Lussac's Law

It was more than a century before any further development in the quantitative understanding of gases occurred. However, in 1787, a French scientist, Jacques Charles, showed that there was a mathematical relationship between the volume of a gas and the temperature if the pressure is held constant. Charles showed that any gas expands by a definite fraction as the

*In fact, individual molecules have a range of velocities at a certain temperature. Temperature relates to the *average* velocity of a large number of molecules of a certain compound or element.

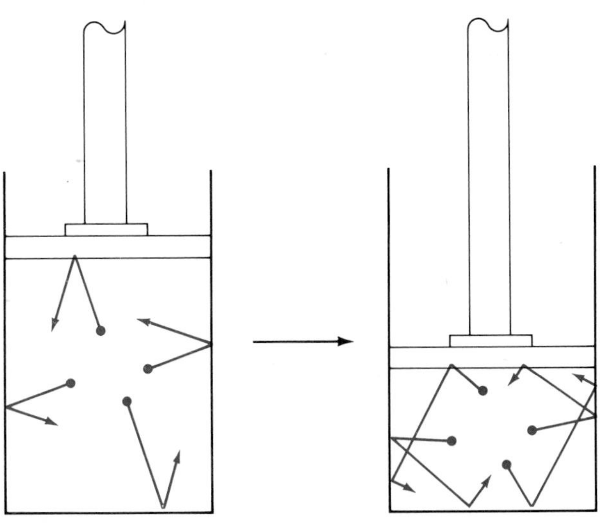

**Figure 10-4** BOYLE'S LAW AND KINETIC THEORY. When the volume decreases, the pressure increases because of the more frequent collisions with the walls of the container.

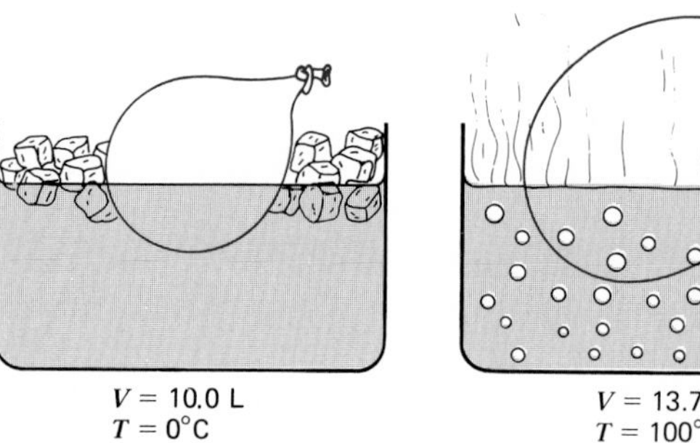


**Figure 10-5** THE EFFECT OF TEMPERATURE ON VOLUME. When the temperature increases, the volume of the balloon increases.

temperature rises. He found that the volume increases by a fraction of $\frac{1}{273}$ for each 1 °C rise in the temperature. (See Figure 10-5.)

In Figure 10-6, the data from the four experiments on the left is plotted in a graph on the right. Notice that this is a linear relationship. (Review Appendix F for a discussion of linear relationships.)

Notice from the graph in Figure 10-6 that the straight line is extrapolated (extended beyond the experimentally measured points) to where the gas would *theoretically* have zero volume. Obviously, it is impossible for matter to have zero volume, but since all gases condense to form solids or liquids at low temperatures anyway, there is no such thing as a "gas law" at very low temperatures. The temperature at zero volume (−273 °C) is significant, however, as this turns out to be the *lowest possible temperature* (−273.15 °C, to be more precise). The behavior of matter at this temperature is discussed shortly. *A temperature scale that assigns zero as the lowest possible temperature is known as the* **absolute** *or* **Kelvin** *scale.* Since the magnitude of the Celsius and Kelvin degrees is the same, we have the following simple relationship between scales where $T$ is the number of kelvins and $t$(C) represents the Celsius temperature.

$$T = [t(\mathrm{C}) + 273]\ \mathrm{K}$$

We can now restate more simply the observations of Charles, which is known as **Charles' law:** *The volume of a gas is directly proportional to the*

| Exp | $t$(C) | $V$(L) |
|---|---|---|
| 1 | 100 | 136.7 |
| 2 | 50 | 118.3 |
| 3 | 10 | 103.7 |
| 4 | 0 | 100.0 |

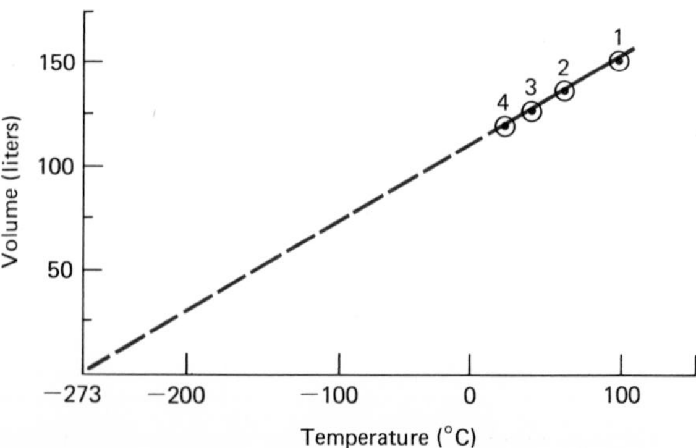


**Figure 10-6** VOLUME AND TEMPERATURE.

*temperature in kelvins (T) at constant pressure.* This can be expressed mathematically as

$$V \propto T$$

As an equality with $k$ as the constant of proportionality,

$$V = kT \quad \text{or} \quad \frac{V}{T} = k$$

For a quantity of gas under two sets of conditions at the same pressure,

$$\frac{V_1}{T_1} = \frac{V_2}{T_2}$$

This equation can be used to calculate how a volume of gas changes when the temperature changes. For example, if $V_2$ is a new volume that we are to find at a certain temperature $T_2$ the equation becomes

$$V_2 = V_1 \times \frac{T_2}{T_1}$$

Final volume = initial volume × temperature correction factor

In this case, remember that if the temperature decreases, the volume must also decrease, which means that the temperature factor is less than one. On the other hand, if the temperature increases, the volume must increase, which means that the temperature factor is more than one.

**Example 10-4**

A certain quantity of gas in a balloon has a volume of 185 mL at a temperature of 50 °C. What is the volume of the balloon if the temperature is lowered to −17 °C? Assume that the pressure remains constant.

**Procedure**

Since the temperature decreases, the volume decreases. The temperature factor must therefore be less than one.

**Solution**

$$V_2 = V_1 \times \frac{T_2}{T_1} \qquad \begin{aligned} T_2 &= (-17 + 273)\text{ K} = 256\text{ K}^* \\ T_1 &= (50 + 273)\text{ K} = 323\text{ K} \end{aligned}$$

$$V_2 = 185\text{ mL} \times \frac{256\ \cancel{\text{K}}}{323\ \cancel{\text{K}}} = \underline{\underline{147\text{ mL}}}$$

*[Notice that temperature *must* be expressed in kelvins. Also, a Celsius reading with two significant figures (e.g., −17 °C) becomes a Kelvin reading with three significant figures (e.g., 256 K).]

**See Problems 10-20 through 10-26.**

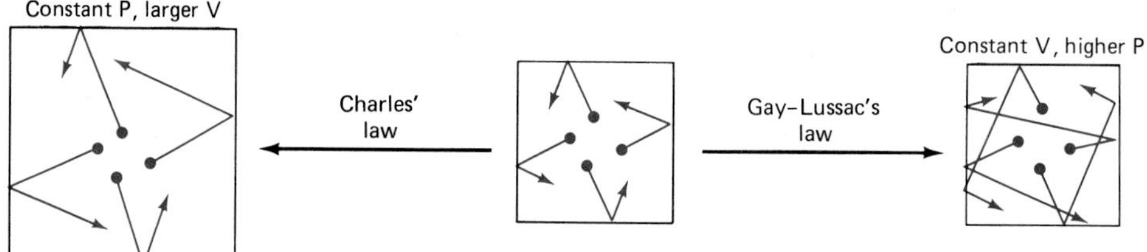


**Figure 10-7** GAY-LUSSAC'S AND CHARLES' LAWS AND KINETIC THEORY. When the temperature increases, the molecules travel faster. If the volume is constant, the pressure increases as a result of the more frequent collisions. If the pressure is constant, the volume increases in order that the frequency of collisions remains the same.

Again, Charles' law is now easily predicted from kinetic theory. In Figure 10-7, the action of four molecules has been illustrated. If the temperature is increased, the molecules travel farther per unit time and their collisions are also more powerful. If the pressure is to remain constant, the volume must increase correspondingly. On the other hand, if the volume is held constant when the temperature is increased, it is obvious that the more frequent and more powerful collisions will lead to an increase in pressure. This relationship between pressure and temperature at constant volume is known as **Gay-Lussac's law.** It was proposed in 1802 by Gay-Lussac as a result of his experiments. Gay-Lussac's law can be stated as follows:

$$P \propto T \qquad P = kT$$

For a sample of gas under two sets of conditions at the same volume:

$$\frac{P_1}{T_1} = \frac{P_2}{T_2} \qquad P_2 = P_1 \times \frac{T_2}{T_1}$$

**Example 10-5**

A quantity of gas in a steel container has a pressure of 760 torr at 25 °C. What is the pressure in the container if the temperature is increased to 50 °C?

**Procedure**

Since the temperature increases, the pressure increases. The temperature correction factor must therefore be greater than one.

**Solution**

$$P_2 = P_1 \times \frac{T_2}{T_1} \qquad \begin{aligned} T_2 &= (50 + 273)\text{ K} = 323\text{ K} \\ T_1 &= (25 + 273)\text{ K} = 298\text{ K} \end{aligned}$$

$$P_2 = 760 \text{ torr} \times \frac{323 \not{K}}{298 \not{K}} = \underline{\underline{824 \text{ torr}}}$$

## 10-5 The Combined Gas Law

See Problems 10-27 through 10-32.

*Boyle's, Charles', and Gay-Lussac's laws can be combined into one relationship, which is appropriately called* **the combined gas law:**

$$\frac{PV}{T} = k$$

It is apparent that the volume of a gas is very much dependent on conditions. This is unlike the volumes of liquids and solids, which are almost incompressible by comparison. When discussing gas volumes, then, it is useful to define certain conditions that are universally accepted as standard. These conditions are called **standard temperature and pressure (STP).** *These conditions are*

*Standard temperature:* *0 °C or 273 K*

*Standard pressure:* *760 torr or 1 atm*

The following examples illustrate the use of the combined gas law and STP.

**Example 10-6**

A 25.8-L quantity of gas has a pressure of 690 torr and a temperature of 17 °C. What is the volume if the pressure is changed to 1.85 atm and the temperature to 345 K?

**Solution**

| *Initial Conditions* | *Final Conditions* |
|---|---|
| $V_1 = 25.8 \text{ L}$ | $V_2 = ?$ |
| $P_1 = 690 \not{\text{torr}} \times \frac{1 \text{ atm}}{760 \not{\text{torr}}} = 0.908 \text{ atm}$ | $P_2 = 1.85 \text{ atm}$ |
| $T_1 = (17 + 273) \text{ K} = 290 \text{ K}$ | $T_2 = 345 \text{ K}$ |

$$\frac{P_1V_1}{T_1} = \frac{P_2V_2}{T_2} \qquad V_2 = V_1 \times \frac{P_1}{P_2} \times \frac{T_2}{T_1}$$

Final volume = initial volume × pressure correction factor × temperature correction factor

In this problem notice that the pressure increases, which decreases the volume. The pressure factor should therefore be less than one. On the other hand, the temperature increases, which increases the volume. The temperature factor is greater than one.

$$V_2 = 25.8 \text{ L} \times \frac{0.908 \text{ atm}}{1.85 \text{ atm}} \times \frac{345 \text{ K}}{290 \text{ K}} = \underline{\underline{15.1 \text{ L}}}$$

**Example 10-7**

A 5850-ft$^3$ quantity of natural gas measured at STP was purchased from the Gas Company. Only 5625 ft$^3$ was received at the house. Assuming that all of the gas was delivered, what was the temperature at the house if the delivery pressure was 1.10 atm?

**Solution**

| *Initial Conditions* | *Final Conditions* |
|---|---|
| $V_1 = 5850 \text{ ft}^3$ | $V_2 = 5625 \text{ ft}^3$ |
| $P_1 = 1.00 \text{ atm}$ | $P_2 = 1.10 \text{ atm}$ |
| $T_1 = 273 \text{ K}$ | $T_2 = ?$ |

$$\frac{P_1V_1}{T_1} = \frac{P_2V_2}{T_2} \qquad T_2 = T_1 \times \frac{P_2}{P_1} \times \frac{V_2}{V_1}$$

In this case, the final temperature is corrected by a pressure and a volume correction factor. Since the final pressure is higher, the pressure factor must be greater than one to increase the temperature. The final volume is lower, so the volume correction factor must be less than one to decrease the temperature.

$$T_2 = 273 \text{ K} \times \frac{1.10 \text{ atm}}{1.00 \text{ atm}} \times \frac{5625 \text{ ft}^3}{5850 \text{ ft}^3} = 289 \text{ K}$$

$$t(\text{C}) = 289 \text{ K} - 273 = \underline{\underline{16 \text{ °C}}}$$

**See Problems 10-33 through 10-41.**

## 10-6 Graham's Law

According to the kinetic theory, the molecules of two different gases at the same temperature have the same average kinetic energy.

Gas 1: K.E. $= \frac{1}{2}m_1v_1^2$ ($m_1$ and $v_1$ = mass and velocity of gas 1)

Gas 2: K.E. $= \frac{1}{2}m_2v_2^2$ ($m_2$ and $v_2$ = mass and velocity of gas 2)

Since $(\text{K.E.})_1 = (\text{K.E.})_2$, we have the following relationship.

$$\tfrac{1}{2}m_1v_1^2 = \tfrac{1}{2}m_2v_2^2$$

$$\frac{v_1}{v_2} = \sqrt{\frac{m_2}{m_1}}$$

This is known as **Graham's law of diffusion.** *It means that at the same temperature, the heavier the gas molecules the slower is its speed or velocity.* Thus lighter gases diffuse or mix faster than heavier gases (see Figure 10-8).

**Example 10-8**
How much faster does a He atom travel relative to an $O_2$ molecule at the same temperature?

**Solution**
At the same temperature, $(K.E.)_{He} = (K.E.)_{O_2}$. Therefore,

$$m_{He} = 4.00 \text{ amu} \qquad m_{O_2} = 32.0 \text{ amu}$$

$$\frac{v_{He}}{v_{O_2}} = \sqrt{\frac{m_{O_2}}{m_{He}}} = \sqrt{\frac{32.0 \cancel{\text{amu}}}{4.00 \cancel{\text{amu}}}} = \sqrt{8.00} = 2.83$$

$$v_{He} = 2.83v_{O_2}$$

On the average, He atoms travel 2.83 times faster than $O_2$ molecules.

**Example 10-9**
Nitrogen dioxide ($NO_2$) diffuses at a rate 1.73 times as fast as an unknown gas. What is the molar mass of the unknown gas?

**Procedure**
The rate of diffusion is a function of the average velocity of the molecules. Also, molar mass (in g/mol) can be used in place of formula weight (in amu). Thus Graham's law is often expressed as

$$\frac{r_1}{r_2} = \sqrt{\frac{MM_2}{MM_1}} \qquad (r = \text{rate of diffusion, } MM = \text{molar mass})$$

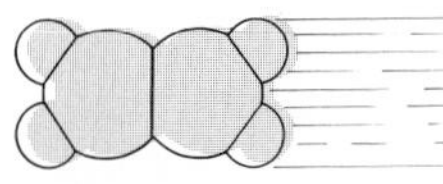

**Figure 10-8** KINETIC ENERGY. To have the same kinetic energy, a light molecule travels at a higher velocity than a heavy molecule.

**Solution**

Let

$$r_1 = \text{rate of diffusion of } NO_2 \ (r_1 = r_{NO_2})$$
$$r_2 = \text{rate of diffusion of the unknown gas}$$

From the problem:

$$r_{NO_2} = 1.73r_2 \qquad MM_{NO_2} = 46.0 \text{ g/mol}$$

$$\frac{r_{NO_2}}{r_2} = \frac{1.73\, \cancel{r_2}}{\cancel{r_2}} = \sqrt{\frac{MM_2}{MM_{NO_2}}}$$

$$1.73 = \sqrt{\frac{MM_2}{46.0 \text{ g/mol}}}$$

Square both sides of the equation to remove the square root sign on the right.

$$(1.73)^2 = \left(\sqrt{\frac{MM_2}{46.0 \text{ g/mol}}}\right)^2 \qquad 2.99 = \frac{MM_2}{46.0 \text{ g/mol}}$$

$$MM_2 = \underline{\underline{138 \text{ g/mol}}}$$

**See Problems 10-42 through 10-49.**

## 10-7 Dalton's Law of Partial Pressures

According to the kinetic theory, the pressure exerted by a quantity of gas molecules at the same temperature is independent of their identity. As a result, pressure is simply a function of the total number of gas molecules present. This is illustrated in Figure 10-9. In this situation, we have two containers of gas of equal volumes at the same temperature. In container 1 we have 0.10 mol ($6.0 \times 10^{22}$ molecules) of $N_2$. In container 2 there is a mixture of three gases totaling 0.10 mol. Notice that as predicted by kinetic theory, the pressures in both containers are the same.

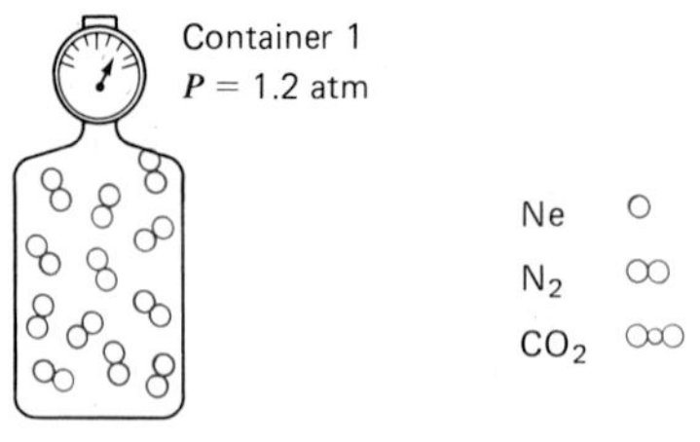


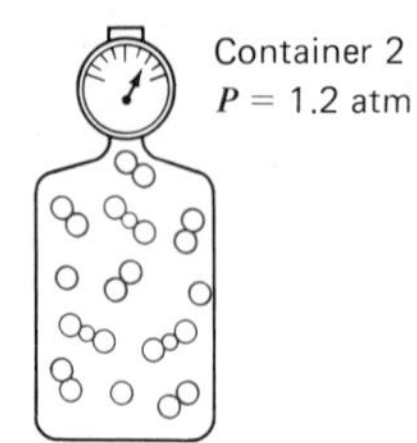


0.050 mol ($3.0 \times 10^{22}$ molecules) $N_2$
0.025 mol ($1.5 \times 10^{22}$ molecules) $CO_2$
0.025 mol ($1.5 \times 10^{22}$ atoms) Ne
0.100 mol ($6.0 \times 10^{22}$ particles) total

**Figure 10-9** PRESSURES OF A PURE GAS AND A MIXTURE OF GASES. Pressure depends only on the number of molecules at a certain temperature and not on their identity.

In container 2 the total pressure is the sum of the individual component gases. Each component contributes to the total pressure proportional to the percent of its molecules to the total. For example, if 50% of the molecules are $N_2$, as in container 2, then 50% of the pressure, or 0.60 atm, is due to $N_2$. This phenomenon was first observed by John Dalton (author of the modern atomic theory) in the early 1800s. *The principle is thus known as* **Dalton's law,** *which states: The total pressure in a system is the sum of the partial pressure of each component gas.*

$$P_T = P_1 + P_2 + P_3, \text{ etc.}$$

**Example 10-10**

Three gases, Ar, $N_2$, and $H_2$, are mixed in a 5.00-L container. The Ar has a pressure of 255 torr, the $N_2$ has a pressure of 228 torr, and the $H_2$ has a pressure of 752 torr. What is the total pressure in the container?

**Solution**

$$P_{Ar} = 255 \text{ torr} \qquad P_{N_2} = 228 \text{ torr} \qquad P_{H_2} = 752 \text{ torr}$$

$$P_T = P_{Ar} + P_{N_2} + P_{H_2} = 255 \text{ torr} + 228 \text{ torr} + 752 \text{ torr} = \underline{\underline{1235 \text{ torr}}}$$

The pressure of our atmosphere is exerted by a mixture of many gases, some necessary ($N_2$, $O_2$, $CO_2$, $H_2O$) and some unhealthy ($CO$, $NO_2$, $SO_2$). The major components, however, are $N_2$ (about 78% of the molecules) and $O_2$ (about 21%). According to Dalton's law, 21% of the pressure of the atmosphere is due to $O_2$. Therefore, the partial pressure of $O_2$ at sea level is

$$P_{O_2} = 0.21 \times 760 \text{ torr} = 160 \text{ torr}$$

On top of the highest mountain, Mt. Everest, the total pressure is 270 torr, so the partial pressure of $O_2$ is only 57 torr or about one-third of normal. A person could not get enough oxygen to survive for long at such a low pressure. At that altitude, even the most conditioned climber must use an oxygen mask, which gives an increased partial pressure of oxygen to the lungs.

In chemistry laboratory experiments, gases are often collected by displacement of water (see Figure 10-10). In this case, the gas would not be pure but would contain some gaseous water molecules (water vapor). To obtain the pressure of the gas that was collected, the pressure due to the $H_2O$ (the vapor pressure of water) must be subtracted from the total pressure. The vapor pressure of water at a certain temperature is always the same and can be obtained from a table, as discussed in Chapter 11.

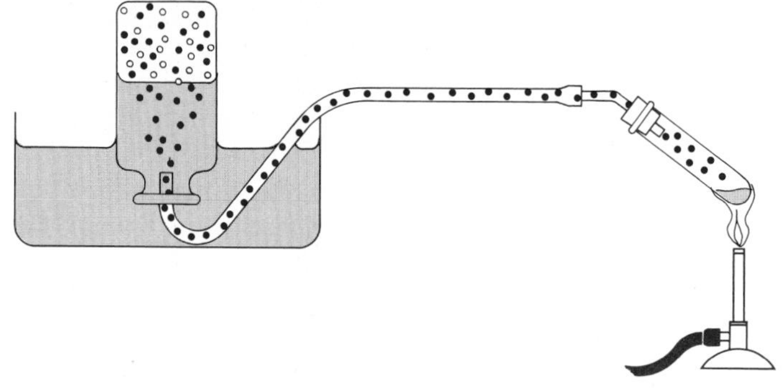

**Figure 10-10** COLLECTION OF A GAS OVER WATER. When a certain gas is collected over water, it is mixed with water vapor.

**Example 10-11**
A sample of hydrogen gas is collected over water. The pressure of the sample is 752 torr. If the vapor pressure of water at the temperature of the experiment is 26 torr, what is the pressure exerted by the pure hydrogen gas?

**Procedure**
Subtract the vapor pressure of $H_2O$ from the total pressure.

**Solution**

$$P_T = P_{H_2} + P_{H_2O}$$

$$P_{H_2} = P_T - P_{H_2O} = 752 \text{ torr} - 26 \text{ torr} = \underline{\underline{726 \text{ torr}}}$$

**Example 10-12**
On a humid day the normal atmosphere ($P = 760$ torr) is composed of 2.10% water molecules. Of the remaining gas, 78.0% is $N_2$ and 21.0% is $O_2$ (the other 1.0% is a mixture of many gases). What is the partial pressure of $H_2O$, $N_2$, and $O_2$?

**Procedure**

1 Find the partial pressure of $H_2O$.

2 Find the pressure of the remaining gas.

3 Find $P_{N_2}$ and $P_{O_2}$.

**Solution**

1 $\quad$ 2.10% of $P_T$ is $0.0210 \times 760 = \underline{\underline{16.0 \text{ torr } (H_2O)}}$

2 $\quad$ $P_T = P_{H_2O} + P_{N_2} + P_{O_2} + P_{\text{other}}$

$$(P_{N_2} + P_{O_2} + P_{\text{other}}) = P_T - P_{H_2O} = 760 - 16 = 744 \text{ torr}$$

3 $P_{N_2} = 0.780 \times 744 \text{ torr} = \underline{\underline{580 \text{ torr } (N_2)}}$

$P_{O_2} = 0.210 \times 744 \text{ torr} = \underline{\underline{156 \text{ torr } (O_2)}}$

**See Problems 10-50 through 10-59.**

## 10-8 Avogadro's Law

**Avogadro's law** *states that equal volumes of gases at the same pressure and temperature contain equal numbers of molecules.* This was originally advanced as a hypothesis to explain how volumes of gases react. Since the original hypothesis, however, kinetic theory has become accepted, and this "hypothesis" is generally considered to be a "law." According to kinetic theory, gas molecules do not interact, thus the total number of molecules is all that is important, and not their identity. Notice that both containers in Figure 10-9 have the same total number of molecules, although container 1 has a pure gas present and container 2 has a mixture. Otherwise, all measurable properties are identical (e.g., $P$, $V$, $T$, and $n_{\text{total}}$).

*A corollary of Avogadro's law is that the volume of a gas is proportional to the total number of molecules (moles) of gas present at constant P and T.* This is expressed as follows, where $n$ equals the number of moles:

$$V \propto n \qquad \text{(as a proportion)}$$

$$V = kn \qquad \text{(as an equality)}$$

$$\frac{V_1}{n_1} = \frac{V_2}{n_2} \qquad \text{(for two conditions)}$$

**Example 10-13**

A balloon that is not inflated but is full of air has a volume of 275 mL and contains 0.0120 mol of air. As shown in Figure 10-11, a piece of Dry Ice (solid $CO_2$) weighing 1.00 g is placed in the balloon and the neck tied. What is the volume of the balloon after the Dry Ice has vaporized? (Assume constant $T$ and $P$.)

**Solution**

| *Initial Conditions* | *Final Conditions* |
|---|---|
| $V_1 = 275 \text{ mL}$ | $V_2 = ?$ |
| $n_1 = 0.0120 \text{ mol}$ | $n_2 = \text{mol air} + \text{mol } CO_2$ |

$$n_2 = 0.0120 + \left(1.00 \cancel{\text{g } CO_2} \times \frac{1 \text{ mol}}{44.0 \cancel{\text{g } CO_2}}\right)$$

$$= 0.0120 + 0.0227$$

$$= 0.0347 \text{ mol}$$

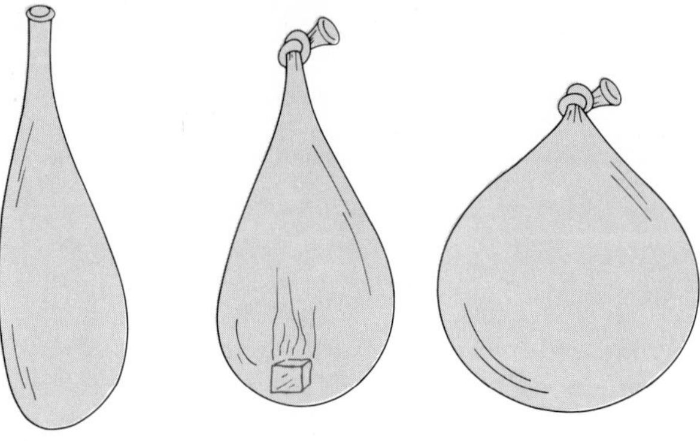

**Figure 10-11** ILLUSTRATION OF AVOGADRO'S LAW. The addition of the carbon dioxide to the balloon increases the number of gas molecules, which increases the volume.

$$\frac{V_1}{n_1} = \frac{V_2}{n_2}$$

$$V_2 = V_1 \times \frac{n_2}{n_1} = 275 \text{ mL} \times \frac{0.0347 \cancel{\text{mol}}}{0.0120 \cancel{\text{mol}}}$$

$$V_2 = \underline{\underline{795 \text{ mL}}}$$

**See Problems 10-60 through 10-63.**

## 10-9 The Ideal Gas Law

So far, the following proportionalities pertaining to the volume of a gas have been established.

Boyle's law: $V \propto \frac{1}{P}$

Charles' law: $V \propto T$

Avogadro's law: $V \propto n$

These three relations can be combined into one proportionality:

$$V \propto \frac{nT}{P}$$

This can be changed to an equality by introducing a constant of proportionality. The constant used in this case is $R$ and is called the **molar gas constant.** Traditionally, the constant is placed between $n$ and $T$.

$$V = \frac{nRT}{P} \quad \text{or} \quad PV = nRT$$

The value for the gas constant is

$$R = 0.0821 \frac{\text{L} \cdot \text{atm}}{\text{K} \cdot \text{mol}} = 62.4 \frac{\text{L} \cdot \text{torr}}{\text{K} \cdot \text{mol}}$$

Notice that the units of $R$ require specific units for $P$, $V$, and $T$. This law is known as the **ideal gas law.** The ideal gas law allows a variety of calculations involving one sample of gas. From the four possible variables ($P$, $V$, $n$, and $T$), one can be calculated if the other three are known (assuming that one knows the value of the gas constant, $R$). The following examples illustrate the use of the ideal gas law.

**Example 10-14**

What is the pressure of a 1.45-mol sample of a gas if the volume is 20.0 L and the temperature is 25 °C?

**Solution**

$$P = ? \qquad V = 20.0 \text{ L} \qquad n = 1.45 \text{ mol} \qquad T = (25 + 273) \text{ K} = 298 \text{ K}$$

$$PV = nRT$$

$$P = \frac{nRT}{V} = \frac{1.45 \cancel{\text{mol}} \times 0.0821 \dfrac{\cancel{\text{L}} \cdot \text{atm}}{\cancel{\text{K}} \cdot \cancel{\text{mol}}} \times 298 \cancel{\text{K}}}{20.0 \cancel{\text{L}}}$$

$$= \underline{\underline{1.77 \text{ atm}}}$$

**Example 10-15**

What is the temperature (in °C) of a 9.65-g quantity of $O_2$ in a 4560-mL container if the pressure is 895 torr?

**Solution**

$$P = 895 \text{ torr} \qquad T = ? \qquad V = 4560 \text{ mL} = 4.56 \text{ L}$$

$$n = 9.65 \cancel{\text{g O}_2} \times \frac{1 \text{ mol}}{32.0 \cancel{\text{g O}_2}} = 0.302 \text{ mol}$$

$$PV = nRT \qquad T = \frac{PV}{nR}$$

$$T = \frac{895 \cancel{\text{torr}} \times 4.56 \cancel{\text{L}}}{0.302 \cancel{\text{mol}} \times 62.4 \dfrac{\cancel{\text{L}} \cdot \cancel{\text{torr}}}{\text{K} \cdot \cancel{\text{mol}}}} = 217 \text{ K}$$

$$t(\text{C}) = 217 - 273 = \underline{\underline{-56 \text{ °C}}}$$

**Example 10-16**

What is the volume of 1.00 mol of a gas at STP?

**Solution**

$P = 1.00 \text{ atm} \qquad V = ? \qquad T = 273 \text{ K} \qquad n = 1.00 \text{ mol}$

$$PV = nRT \qquad V = \frac{nRT}{P}$$

$$V = \frac{1.00 \text{ mol} \times 0.0821 \dfrac{\text{L} \cdot \text{atm}}{\text{K} \cdot \text{mol}} \times 273 \text{ K}}{1.00 \text{ atm}} = \underline{\underline{22.4 \text{ L}}}$$

Notice that the ideal gas law is used when *one* set of conditions is given with one missing variable. The combined gas law is used when *two* sets of conditions are given with one missing variable.

This law is called ideal because it follows from the assumptions of the kinetic theory, which describes an ideal gas. The molecules of an ideal gas have no volume and have no attraction for each other. The molecules of a "real" gas obviously have a volume and there is some interaction between molecules, especially at high pressures and low temperatures where the molecules are pressed close together. Fortunately, at normal temperatures and pressures found on the surface of the earth, gases have close to "ideal" behavior. Therefore, the use of the ideal gas law is justified. If we lived on the planet Jupiter, however, where pressure is measured in thousands of earth atmospheres, the ideal gas law would not provide accurate answers. Other relationships would have to be used that take into account the volume of the molecules and the interaction between molecules.

**See Problems 10-64 through 10-73.**

## 10-10 The Molar Volume and Density of a Gas

The volume of one mole of a gas at STP as calculated in Example 10-16 is known as the **molar volume**. *One mole of any gas (either pure or a mixture) occupies 22.4 liters at STP.* This fact can be included in the definition of a mole as described in Chapter 8. For example, three molar quantities can be described for a gas such as $CO_2$. Notice that only the molar mass depends on the identity of the gas.

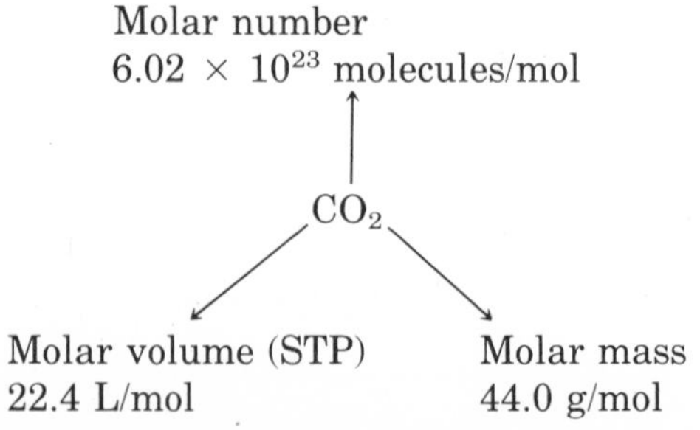

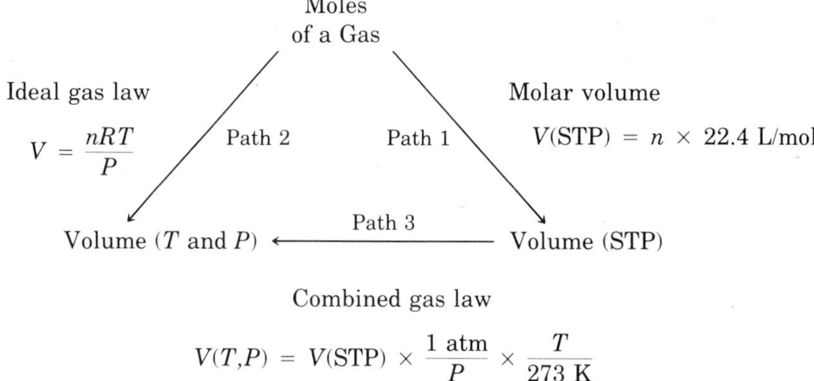


**Figure 10-12** MOLES AND THE VOLUME OF A GAS. The number of moles relates to the volume by two paths.

Before working sample problems illustrating the use of molar volume, let's summarize the relationships discussed between moles and volume of a gas. In Figure 10-12, notice that moles can be converted to volume at STP using the molar volume relationship (path 1). Moles of a gas can also be converted to volume at some other temperature and pressure using the ideal gas law (path 2). Finally, the volume at STP of a quantity of gas and the volume at some other temperature and pressure relate by the combined gas law (path 3).

---

**Example 10-17**

What is the weight of 4.55 L of $O_2$ measured at STP?

**Procedure**

Utilizing path 1 in Figure 10-12, convert volume to moles and then moles to the weight. The two possible conversion factors for converting volume to moles are

$$\frac{1\ \text{mol}}{22.4\ \text{L}} \quad \text{and} \quad \frac{22.4\ \text{L}}{\text{mol}}$$

$$\text{volume (STP)} \rightarrow \text{moles} \rightarrow \text{weight}$$

**Solution**

$$4.55\ \cancel{\text{L}} \times \frac{1\ \cancel{\text{mol}}}{22.4\ \cancel{\text{L}}} \times \frac{32.0\ \text{g}}{\cancel{\text{mol}}} = \underline{\underline{6.50\ \text{g}}}$$

---

**Example 10-18**

A sample of gas that weighs 3.20 g occupies 2.00 L at 17 °C and 380 torr. What is the molar mass of the gas?

**Procedure**

There are two ways that one can work this problem. The first uses paths 1 and 3 in Figure 10-12.

1. Convert given volume to volume at STP.
2. Convert volume at STP to moles.
3. Convert moles and weight to molar mass as follows: From Chapter 8,

$$n \text{ (number of moles)} = \text{weight in g(wt)} \times \frac{1 \text{ mol}}{\text{molar mass (MM)}}$$

or

$$n = \frac{\text{wt}}{MM} \qquad MM = \frac{\text{wt}}{n}$$

**Solution**

| *Initial Conditions* | *Final Conditions* (STP) |
|---|---|
| $V_1 = 2.00 \text{ L}$ | $V_2 = ?$ |
| $P_1 = 380 \text{ torr}$ | $P_2 = 760 \text{ torr}$ |
| $T_1 = (17 + 273)\text{K} = 290 \text{ K}$ | $T_2 = 273 \text{ K}$ |

1
$$\frac{P_1V_1}{T_1} = \frac{P_2V_2}{T_2} \qquad V_2 = V_1 \times \frac{P_1}{P_2} \times \frac{T_2}{T_1}$$

$$V_2 = 2.00 \text{ L} \times \frac{380 \text{ torr}}{760 \text{ torr}} \times \frac{273 \text{ K}}{290 \text{ K}} = \underline{\underline{0.941 \text{ L (STP)}}}$$

2
$$0.941 \text{ L} \times \frac{1 \text{ mol}}{22.4 \text{ L}} = 0.0420 \text{ mol of gas}$$

3
$$\text{Molar mass} = \frac{\text{wt}}{n} = \frac{3.20 \text{ g}}{0.0420 \text{ mol}} = \underline{\underline{76.2 \text{ g/mol}}}$$

The alternative procedure is more direct, since it uses the ideal gas law as shown in path 2 in Figure 10-12. To find the molar mass using the ideal gas law, substitute the relationship for $n$ ($n$ = wt/$MM$) into the gas law.

$$PV = nRT$$

$$n = \frac{\text{wt}}{MM}$$

$$PV = \frac{\text{wt}}{MM} RT$$

Solving for $MM$, we have

$$MM = \frac{(\text{wt})(RT)}{PV}$$

$$= \frac{3.20 \text{ g} \times 62.4 \frac{\cancel{\text{L}} \cdot \cancel{\text{torr}}}{\cancel{\text{K}} \cdot \text{mol}} \times 290 \cancel{\text{K}}}{380 \cancel{\text{torr}} \times 2.00 \cancel{\text{L}}}$$

$$= \underline{\underline{76.2 \text{ g/mol}}}$$

In Chapter 2 the densities of solids and liquids were given in units of g/mL. Since gases are much less dense, units in this case are usually given in g/L (STP). The density of a gas at STP can be calculated by dividing the molar mass by the molar volume.

$$CO_2 \qquad 44.0 \text{ g/mol} \approx 22.4 \text{ L/mol}$$

$$\frac{44.0 \text{ g/}\cancel{\text{mol}}}{22.4 \text{ L/}\cancel{\text{mol}}} = \underline{\underline{1.96 \text{ g/L(STP)}}}$$

The densities of several gases are given in Table 10-2.

## 10-11 Stoichiometry Involving Gases

Gases are formed or consumed in many chemical reactions. Since the volume of a gas relates to the number of moles by the ideal gas law or the molar volume relationship, we can update the general procedure for stoichiometry problems shown in Figure 9-6. In Figure 10-13 the number of individual molecules has been deleted from the scheme, since problems involving actual numbers of molecules are not often encountered. We have also included in this figure the relationships used to convert from one unit to another. Note that the molar volume relationship ($n = V$(STP)/22.4 L can be used to convert from volume to moles when the conditions are at STP.

**Table 10-2** The Densities of Gases

| Gas | Density [g/L(STP)] | Gas | Density [g/L(STP)] |
|---|---|---|---|
| $H_2$ | 0.090 | $O_2$ | 1.43 |
| He | 0.179 | $CO_2$ | 1.96 |
| $N_2$ | 1.25 | $SF_6$ | 6.52 |
| Air (average) | 1.29 | $UF_6$ | 15.7 |

**See Problems 10-74 through 10-85.**

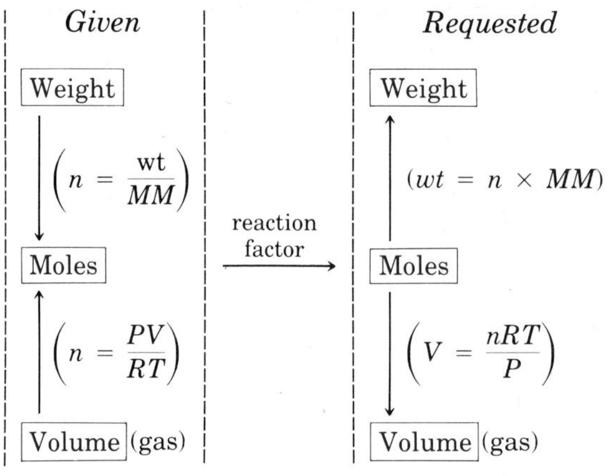


**Figure 10-13** GENERAL PROCEDURE FOR STOICHIOMETRY PROBLEMS.

The following examples illustrate the relationship of gas laws to stoichiometry.

**Example 10-19**

Given the following balanced equation;

$$2\text{Al}(s) + 6\text{HCl}(aq) \rightarrow 2\text{AlCl}_3(aq) + 3\text{H}_2(g)$$

What weight of Al is needed to produce 50.0 L of $H_2$ measured at STP?

**Procedure**

The general procedure is shown below, In this case, it is easier to use the molar volume to convert volume directly to moles.

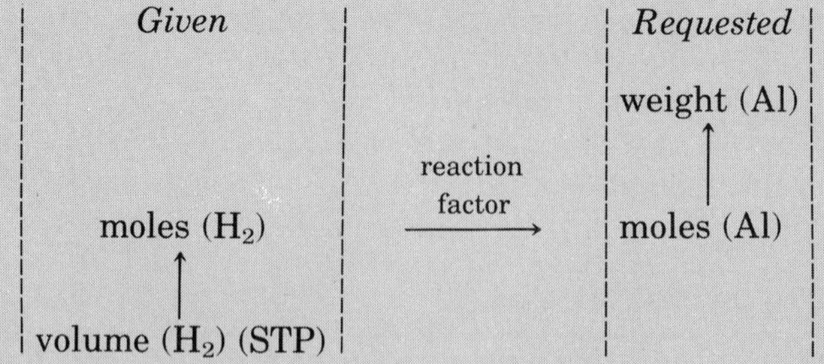


**Solution**

$$50.0 \not{\text{L (STP)}} \times \frac{1 \not{\text{mol H}_2}}{22.4 \not{\text{L (STP)}}} \times \frac{2 \not{\text{mol Al}}}{3 \not{\text{mol H}_2}} \times \frac{27.0 \text{ g}}{\not{\text{mol Al}}} = \underline{\underline{40.2 \text{ g}}}$$

**Example 10-20**
Given the following balanced equation:

$$4NH_3(g) + 5O_2(g) \rightarrow 4NO(g) + 6H_2O(l)$$

What volume of NO gas measured at 550 torr and 25 °C will be produced from 19.5 g of $O_2$?

**Procedure**

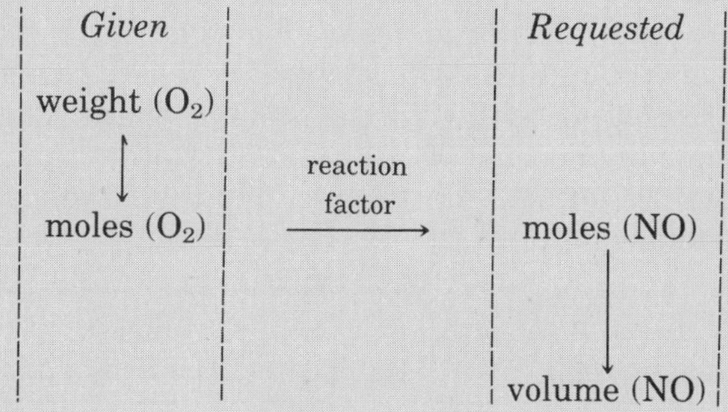


**Solution**

$$19.5 \cancel{g\ O_2} \times \frac{1 \cancel{mol\ O_2}}{32.0 \cancel{g\ O_2}} \times \frac{4 \text{ mol NO}}{5 \cancel{mol\ O_2}} = \underline{\underline{0.488 \text{ mol NO}}}$$

**1** Using the ideal gas law

$$V = \frac{nRT}{P} = \frac{0.488 \cancel{mol} \times 62.4 \frac{L \cdot \cancel{torr}}{\cancel{K} \cdot \cancel{mol}} \times 298 \cancel{K}}{550 \cancel{torr}} = \underline{\underline{16.5 \text{ L}}}$$

**2** Convert moles of NO to volume at STP and then volume at STP to volume at 550 torr and 25 °C.

$$0.488 \cancel{mol} \times 22.4 \text{ L}/\cancel{mol} = 10.9 \text{ L (STP)}$$

$$V_2 = V_1 \times \frac{P_1}{P_2} \times \frac{T_2}{T_1}$$

| *Initial Conditions* | *Final Conditions* |
|---|---|
| $V_1 = 10.9$ L | $V_2 = ?$ |
| $P_1 = 760$ torr | $P_2 = 550$ torr |
| $T_1 = 273$ K | $T_2 = (25 + 273)\text{K} = 298$ K |

$$V_2 = 10.9 \text{ L} \times \frac{760 \cancel{torr}}{550 \cancel{torr}} \times \frac{298 \cancel{K}}{273 \cancel{K}} = \underline{\underline{16.4 \text{ L}}}$$

**See Problems 10-86 through 10-93.**

## 10-12 Chapter Summary

Beginning in the 1660s prominent scientists reported their observations concerning the behavior of air and other gases. The first observation by Torricelli using his barometer involved the fact that gases exert pressure and are therefore legitimate forms of matter. The other observations or "laws" that followed provided the foundation of our modern understanding of gases in particular but all matter in general. We now realize from the kinetic theory of gases that these laws are all reasonable and predictable. Since the kinetic theory tells us, among other things, that gases are composed of moving molecules (or atoms of noble gases), we predict that the identity of gas molecules is not important with regard to most of the properties studied. Exceptions to this are the average velocity of a molecule and the density, which are dependent on the identity of the gas.

The gas laws presented in this chapter and their applications are summarized in the following table.

| Gas law | Relationship | Meaning | Constant Conditions | Application |
|---|---|---|---|---|
| Boyle's | $V \propto \frac{1}{P}$ | $V\uparrow\ P\downarrow$ | $T$, $n$ | Relates $V$ and $P$ of a gas at two different conditions |
| Charles' | $V \propto T$ | $V\uparrow\ T\uparrow$ | $P$, $n$ | Relates $V$ and $T$ of a gas at two different conditions |
| Gay-Lussac's | $P \propto T$ | $P\uparrow\ T\uparrow$ | $V$, $n$ | Relates $P$ and $T$ of a gas at two different conditions |
| Combined | $PV \propto T$ | $PV\uparrow\ T\uparrow$ | $n$ | Relates $P$, $V$, and $T$ of a gas at two different conditions |
| Graham's | $v \propto \frac{1}{\sqrt{\text{MM}}}$ | $v\uparrow\ \text{MM}\downarrow$ | $T$ | Relates MM and $v$ of two different gases at a certain $T$ |
| Dalton's | $P_T = P_1 + P_2$, etc. | $P_T\uparrow\ n_T\uparrow$ | $T$, $V$ | Relates $P_T$ to partial pressures of component gases |
| Avogadro's | $V \propto n_T$ | $V\uparrow\ n_T\uparrow$ | $P$, $T$ | Relates $V$ and $n$ of a gas at two different conditions |

| Gas law | Relationship | Meaning | Constant Conditions | Application |
|---|---|---|---|---|
| Ideal | $PV \propto nT$ | $PV\uparrow \; nT\uparrow$ | — | Relates $P$, $V$, $T$, or $n$ to the other three variables |

$V$ = volume  
$T$ = kelvin temperature  
$P$ = pressure  
$n$ = moles  
$n_T$ = total moles  
$MM$ = molar mass  
$v$ = average velocity  
$\uparrow$ = quantity increases  
$\downarrow$ = quantity decreases

In addition to these relationships another relationship presented was the molar volume. This relationship [22.4 L/mol (STP)] can be used as a conversion factor between moles and volume at STP. To convert between moles and volume at conditions other than STP the use of the ideal gas law is convenient.

$$\text{Volume } (T, P) \xleftarrow{\text{ideal gas law}} \underline{\underline{\text{Moles of gas}}} \xrightarrow{\text{molar volume}} \text{Volume (STP)}$$

The molar volume can also be used to calculate the density of a gas.

Finally, since the volume of a gas relates to moles, its application in stoichiometric calculations can be illustrated as follows:

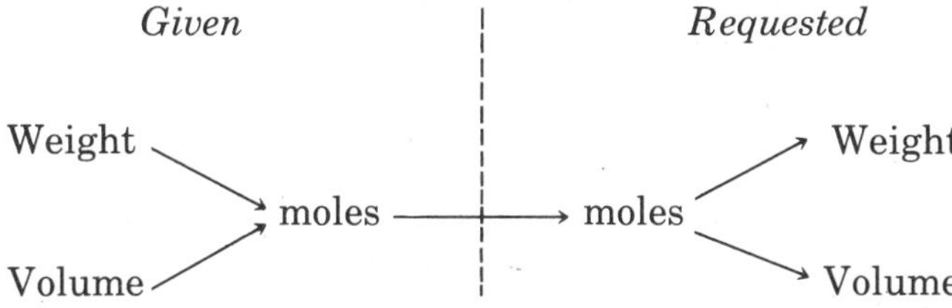


## Exercises

### Units of Pressure

**10-1** Make the following conversions.
(a) 1650 torr to atm
(b) $3.50 \times 10^{-5}$ atm to torr
(c) 185 lb/in.$^2$ to torr
(d) 5.65 kPa to atm
(e) 190 torr to lb/in.$^2$
(f) 85 torr to kPa

**10-2** Make the following conversions.
(a) 30.2 in. of Hg to torr
(b) 25.7 kilobars to atm
(c) 57.9 kPa to lb/in.$^2$
(d) 0.025 atm to torr

**10-3** A column of mercury (density 13.6 g/ml) is 15.0 cm high. A cross section of the column has an area of 12.0 cm$^2$. What is the force of the mercury at the bottom of the tube? What is the pressure in g/cm$^2$ and in atmospheres?

**10-4** The atmospheric pressure on the planet Mars is 10.3 millibars. What is this pressure in earth atmospheres?

**10-5** The atmospheric pressure on the planet Venus is 0.0920 kilobar. What is this pressure in earth atmospheres?

**10-6** The density of water is 1.00 g/mL. If

water is substituted for mercury in the barometer, how high (in feet) would be a column of water supported by 1 atm? A water well is 40 ft deep. Can suction be used to raise the water to ground level?

**10-7** A tube containing an alcohol (density 0.890 g/mL) is 1.00 m high and has a cross section of 15.0 $cm^2$. What is the total force at the bottom of the tube? What is the pressure? How high would be an equivalent amount of mercury assuming the same cross section?

### The Kinetic Theory of Gases

**10-8** It is harder to move your arms in water than in air. Explain based on the kinetic theory.

**10-9** A balloon expands in all directions. Explain.

**10-10** How can more air be added to a tire if it is already filled?

**10-11** A sunbeam forms when light is reflected from dust suspended in the air. Even if the air is still, the dust particles can be seen to bounce around randomly. Explain.

* **10-12** The kinetic theory assumes that the volume of the molecules and their interactions are negligible for gases. Explain why the asumptions may not be true when the pressure is very high and the temperature is very low.

### Boyle's Law

**10-13** A gas has a volume of 6.85 L at a pressure of 0.650 atm. What is the volume of the gas if the pressure is decreased to 0.435 atm?

**10-14** If a gas has a volume of 1560 mL at a pressure of 812 torr, what will be the volume if the pressure is increased to 2.50 atm?

**10-15** A gas has a volume of 125 mL at a pressure of 62.5 torr. What is the pressure if the volume is decreased to 115 mL?

* **10-16** A gas in a piston engine is compressed by a ratio of 15:1. If the pressure before compression is 0.950 atm, what pressure is required to compress the gas? (Assume constant temperature.)

**10-17** A few miles above the surface of the earth the pressure drops to $1.00 \times 10^{-5}$ atm. What would be the volume of a 1.00-L sample of gas at sea level pressure (1.00 atm) if it were taken to that altitude? (Assume constant temperature.)

* **10-18** The volume of a gas is measured as the pressure is varied. The four measurements are reported as follows:

| Experiment | Volume (mL) | Pressure (torr) |
|---|---|---|
| 1 | 125 | 450 |
| 2 | 145 | 385 |
| 3 | 175 | 323 |
| 4 | 220 | 253 |

Make a graph of the volume on the $x$ axis and the pressure on the $y$ axis. What is the average value of the constant of proportionality, $k$?

**10-19** How does the kinetic theory of gases explain Boyle's law?

### Charles' Law

**10-20** A balloon has a volume of 1.55 L at 25 °C. What would be the volume if the balloon is heated to 100 °C? (Assume constant $P$.)

**10-21** A sample of gas has a volume of 677 mL at 23 °C. What is the volume of the gas if the temperature is increased to 46 °C?

**10-22** A balloon has a volume of 325 mL at 17 °C. What is the temperature if the volume increases to 392 mL?

* **10-23** The temperature of a sample of gas is 0 °C. When the temperature is increased, the volume increases by a factor of 1.25 (i.e., $V_2 = 1.25V_1$). What is the final temperature in degrees Celsius?

**10-24** A quantity of gas has a volume of $3.66 \times 10^4$ L. What will be the volume if the temperature is changed from 455 K to 50 °C?

**10-25** The volume of a gas is measured as the temperature is varied. The four measurements are reported as follows:

| Experiment | Volume (L) | Temperature (°C) |
|---|---|---|
| 1 | 1.54 | 20 |
| 2 | 1.65 | 40 |
| 3 | 1.95 | 100 |
| 4 | 2.07 | 120 |

Make a graph of the volume on the $x$ axis and the Kelvin temperature on the $y$ axis. What is the average value of the constant of proportionality, $k$?

**10-26** How does the kinetic theory of gases explain Charles' law?

## Gay-Lussac's Law

**10-27** A confined quantity of gas is at a pressure of 2.50 atm and a temperature of −22 °C. What is the pressure if the temperature increases to 22 °C?

**10-28** A quantity of gas has a volume of 3560 mL at a temperature of 55 °C and a pressure of 850 torr. What is the temperature if the volume remains unchanged but the pressure is decreased to 0.652 atm?

**10-29** An aerosol spray can has gas under a pressure of 1.25 atm at 25 °C. The can explodes when the pressure reaches 2.50 atm. At what temperature will this happen? (Do not throw these cans into a fire!)

**10-30** The pressure in an automobile tire is 28.0 lb/in.$^2$ on a chilly morning of 17 °C. After it is driven for awhile, the temperature of the tire rises to 40 °C. What is the pressure in the tire if the volume remains constant?

**10-31** The pressure of a confined volume of gas is measured as the temperature is raised. The four measurements are reported as follows:

| Experiment | Pressure (torr) | Temperature (K) |
|---|---|---|
| 1 | 550 | 295 |
| 2 | 685 | 372 |
| 3 | 745 | 400 |
| 4 | 822 | 445 |

Make a graph of the pressure on the $x$ axis and the temperature on the $y$ axis. What is the average value of the constant of proportionality, $k$?

**10-32** How does the kinetic theory of gases explain Gay-Lussac's law?

## The Combined Gas Law

**10-33** Which of the following are legitimate expressions of the combined gas law?

(a) $PV = kT$ (d) $\frac{P}{T} \propto \frac{1}{V}$

(b) $PT \propto V$ (e) $VT \propto P$

(c) $\frac{P_1T_1}{V_1} = \frac{P_2T_2}{V_2}$

**10-34** Which of the following are not STP conditions?

(a) $T = 273$ K (e) $t(C) = 273$ °C

(b) $P = 760$ atm (f) $P = 760$ torr

(c) $T = 0$ K (g) $t(C) = 0$ °C

(d) $P = 1$ atm

**10-35** A 5.50-L volume of gas has a pressure of 0.950 atm at 0 °C. What is the pressure if the volume decreases to 4.75 L and the temperature increases to 35 °C?

**10-36** A quantity of gas has a volume of 17.5 L at a pressure of 6.00 atm and a temperature of 100 °C. What is its volume at STP?

**10-37** A quantity of gas has a volume of $4.78 \times 10^{-4}$ mL at a temperature of −50 °C and a pressure of 78.0 torr. If the volume changes to $9.55 \times 10^{-5}$ mL and the pressure to 155 torr, what is the temperature?

**10-38** A gas has a volume of 64.2 L at STP. What is the volume at 77.0 °C and 7.55 atm?

**10-39** A quantity of gas has a volume of $6.55 \times 10^{-5}$ L at STP. What is the pressure if the volume changes to $4.90 \times 10^{-3}$ L and the temperature remains at 273 K?

**10-40** A balloon has a volume of 1.55 L at 25 °C and 1.05 atm pressure. If it is cooled in the freezer, the volume shrinks to 1.38 L and the pressure drops to 1.02 atm. What is the temperature in the freezer?

**10-41** A bubble from a deep-sea diver in the ocean starts with a volume of 35.0 mL at a temperature of 17 °C and a pressure of 11.5 atm. What is the volume of the bubble when it reaches the surface? Assume that the pressure at the surface is 1 atm and the temperature is 22 °C?

## Graham's Law

**10-42** A bowling ball weighs 6.00 kg and a bullet weighs 1.50 g. If the bowling ball is rolled down an alley at 20.0 miles/hr, what is the velocity of a bullet having the same kinetic energy?

**10-43** Order the following gases in order of increasing average speed (rate of diffusion) at the same temperature.
(a) $CO_2$ (b) $SO_2$ (c) $N_2$ (d) $SF_6$ (e) $N_2O$ (f) $H_2$

**10-44** What is the rate of diffusion of $N_2$ molecules relative to Ar atoms?

**10-45** How does the average velocity of $H_2$ molecules compare with $CO_2$ molecules a the same temperature?

**10-46** A certain gas diffuses twice as fast as $SO_3$. What is the molar mass of the unknown gas?

**10-47** Carbon monoxide diffuses 2.13 times faster than an unknown gas. What is the molar mass of the unknown gas?

**10-48** Gas A has a molar mass 1.76 times greater than gas B. How fast does gas A diffuse relative to gas B?

* **10-49** To make enriched uranium for use in nuclear reactors or for weapons, $^{235}U$ must be separated from $^{238}U$. Although $^{235}U$ is the isotope needed for fission, only 0.7% of U atoms are this isotope. Separation is a difficult and expensive process. Since $UF_6$ is a gas, Graham's law can be applied to separate the isotopes. How much faster does a $^{235}UF_6$ molecule travel on the average compared with a $^{238}UF_6$ molecule?

## Dalton's Law

**10-50** Three gases are mixed in a 1.00-L container. The partial pressure of $CO_2$ is 250 torr, $N_2$ is 375 torr, and He is 137 torr. What is the pressure of the mixture of gases?

**10-51** The total pressure in a cylinder containing a mixture of two gases is 1.46 atm. If the partial pressure of one gas is 750 torr, what is the partial pressure of the other gas?

**10-52** Air is about 0.90% Ar. If the barometric pressure is 756 torr, what is the partial pressure of Ar?

**10-53** A sample of oxygen is collected over water in a bottle. If the pressure inside the bottle is made equal to the barometric pressure, which is 752 torr, and the vapor pressure of water at the temperature of the experiment is 24 torr, what is the pressure of pure oxygen?

* **10-54** A mixture of two gases is composed of $CO_2$ and $O_2$. The partial pressure of $O_2$ is 256 torr, and it represents 35% of the molecules of the mixture. What is the total pressure of the mixture?

**10-55** A container holds two gases, A and B. Gas A has a partial pressure of 0.455 atm, and gas B has a partial pressure of 0.175 atm. What percent of the molecules in the mixture is gas A?

**10-56** A volume of gas is composed of $N_2$, $O_2$, and $SO_2$. If the total pressure is 1050 torr, what is the partial pressure of each gas if the gas is 72.0% $N_2$ and 8.00% $O_2$?

* **10-57** A volume of gas has a total pressure of 2.75 atm. If the gas is composed of 0.250 mol of $N_2$ and 0.427 mol of $CO_2$, what is the partial pressure of each gas?

* **10-58** The following gases are all combined into a 2.00-L container: a 2.00-L volume of $N_2$ at 300 torr, a 4.00-L volume of $O_2$ at 85 torr, and a 1.00-L volume of $CO_2$ at 450 torr. What is the total pressure?

* **10-59** The total pressure of a mixture of two gases is 0.850 atm in a 4.00-L container. Before mixing, gas A was in a 2.50-L container and had a pressure of 0.880 atm. What is the partial pressure of gas B in the 4.00-L container?

### Avogadro's Law

**10-60** A 0.112-mol quantity of gas has a volume of 2.54 L at a certain temperature and pressure. What is the volume of 0.0750 mol of gas under the same conditions?

**10-61** A balloon has a volume of 188 L and contains 8.40 mol of gas. How many moles of gas would be needed to expand the balloon to 275 L? Assume the same temperature and pressure in the balloon.

**10-62** A balloon has a volume of 75.0 mL and contains $2.50 \times 10^{-3}$ mol of gas. How many grams of $N_2$ must be added to the balloon in order for the volume to increase to 164 mL at the same temperature and pressure?

**10-63** A 48.0-g quantity of $O_2$ in a container has a pressure of 0.625 atm. What would be the pressure of 48.0 g of $SO_2$ in the same container at the same temperature?

### The Ideal Gas Law

**10-64** What is the temperature (in degrees Celsius) of 4.50 L of a 0.332-mol quantity of gas under a pressure of 2.25 atm?

**10-65** A quantity of gas has a volume of 16.5 L at 325 K and a pressure of 850 torr. How many moles of gas are present?

**10-66** What weight of $NH_3$ gas has a volume of 16,400 mL, a pressure of 0.955 atm, and a temperature of −23 °C?

**10-67** What is the pressure (in torr) exerted by 0.250 g of $O_2$ in a 250-mL container at 29 °C?

**10-68** What weight of Ne is contained in a large neon light if the volume is 3.50 L, the pressure 1.15 atm, and the temperature 23 °C?

**10-69** A sample of $H_2$ is collected in a bottle over water. The volume of the sample is 185 mL at a temperature of 25 °C and a pressure of 760 torr. The vapor pressure of water at this temperature is 24 torr. What is the weight of $H_2$ in the bottle?

**10-70** One liter of a gas weighs 8.37 g. The gas volume is measured at 1.45 atm pressure and 35 °C. What is the molar mass of the gas?

* **10-71** The Goodyear blimp has a volume of about $2.5 \times 10^7$ L. What is the weight of He (in pounds) in the blimp at 27 °C and 780 torr? The average molar mass of air is 29.0 g/mol. What weight of air (in pounds) would the blimp contain? The difference between these two values is the lifting power of the blimp. What weight could the blimp lift? If $H_2$ is substituted for He, what is the lifting power? Why isn't $H_2$ used?

**10-72** A gaseous compound is 85.7% C and 14.3% H. A 6.58-g quantity of this gas occupies 4500 mL at 77.0 °C and a pressure of 1.00 atm. What is the molar mass of the compound, and what is its molecular formula?

* **10-73** A good vacuum pump on earth can produce a vacuum with a pressure as low as $1.00 \times 10^{-8}$ torr. How many molecules are present in each milliliter at a temperature of 27.0 °C?

### Molar Volume and Density

**10-74** What is the volume of 15.0 g of $CO_2$ measured at STP?

**10-75** What is the weight (in kg) of 175 L of CO measured at STP?

**10-76** What is the volume of $3.01 \times 10^{24}$ molecules of $N_2$ at STP?

**10-77** A 6.5-L quantity of a gas measured at

STP weighs 18.6 g. What is the molar mass of the compound?

**10-78** What is the weight of $6.78 \times 10^{-4}$ L of $NO_2$ measured at STP?

**10-79** What is the density in g/L (STP) of $B_2H_6$?

**10-80** What is the density in g/L (STP) of $Cl_2$?

**10-81** A gas has a density of 1.52 g/L (STP). What is the molar mass of the gas?

**10-82** A gas has a density of 6.14 g/L (STP). What is the molar mass of the gas?

* **10-83** A gas has a density of 3.60 g/L at a temperature of 25 °C and a pressure of 1.20 atm. What is its density at STP?

* **10-84** What is the density in g/L of $N_2$ measured at 500 torr and 22 °C?

* **10-85** What is the density in g/L of $SF_6$ measured at 0.370 atm and 37 °C?

### Stoichiometry Involving Gases

**10-86** Limestone is dissolved by $CO_2$ according to the following equation:

$$CaCO_3(s) + H_2O(l) + CO_2(g) \rightarrow Ca(HCO_3)_2(aq)$$

What volume of $CO_2$ measured at STP would dissolve 115 g of $CaCO_3$?

**10-87** Magnesium in flashbulbs burns according to the following equation:

$$2Mg(s) + O_2(g) \rightarrow 2MgO(s)$$

What weight of Mg combines with 5.80 L of $O_2$ measured at STP?

**10-88** Acetylene is produced from calcium carbide as shown by the following reaction:

$$CaC_2(s) + 2H_2O(l) \rightarrow Ca(OH)_2(s) + C_2H_2(g)$$

What volume of acetylene ($C_2H_2$) measured at 25.0 °C and 745 torr would be produced from 5.00 g of $H_2O$?

**10-89** Nitrogen dioxide is an air pollutant. It is produced from NO (from car exhausts) as follows:

$$2NO(g) + O_2(g) \rightarrow 2NO_2(g)$$

What weight of $O_2$ is required to produce 5.00 L of $NO_2$ measured at 0.950 atm and −13 °C?

**10-90** Butane ($C_4H_{10}$) burns according to the following equation:

$$2C_4H_{10}(g) + 13O_2(g) \rightarrow 8CO_2(g) + 10H_2O(l)$$

(a) What volume of $CO_2$ measured at STP would be produced by 85.0 g of $C_4H_{10}$?
(b) What volume of $O_2$ measured at 3.25 atm and a temperature of 127 °C would be required to react with 85.0 g of $C_4H_{10}$?
(c) What volume of $CO_2$ measured at STP would be produced from 45.0 L of $C_4H_{10}$ measured at 25 °C and 0.750 atm pressure?

**10-91** In March 1979 a nuclear reactor overheated, producing a dangerous hydrogen gas bubble at the top of the reactor core. The following reaction occurring at the high temperatures (about 1500 °C) accounted for the hydrogen (Zr alloys hold the uranium pellets in long rods).

$$Zr(s) + 2H_2O(g) \rightarrow ZrO_2(s) + 2H_2(g)$$

If the bubble had a volume of about 28,000 L at 250 °C and 70.0 atm pressure, what weight in kg and tons of Zr had reacted?

**10-92** Nitric acid is produced according to the equation

$$3NO_2(g) + H_2O(l) \rightarrow HNO_3(aq) + 2NO(g)$$

What volume of $NO_2$ measured at −73 °C and $1.56 \times 10^{-2}$ atm would be needed to produce $4.55 \times 10^{-3}$ mol of $HNO_3$?

* **10-93** Natural gas ($CH_4$) burns according to the following equation:

$$CH_4(g) + 2O_2(g) \rightarrow CO_2(g) + 2H_2O(l)$$

What volume of $CO_2$ measured at 27 °C and 1.50 atm pressure is produced from 27.5 L of $O_2$ measured at −23 °C and 825 torr pressure?

# Review Test on Chapters 8–10

*The following multiple choice questions have one correct answer.*

1. A quantity of aluminum weighs 54.0 g. What does the same number of magnesium atoms weigh?
(a) 12.1 g (b) 24.3 g (c) 48.6 g (d) 97.2 g (e) 6.0 g
2. The number $6.02 \times 10^{22}$ is equivalent to:
(a) 0.100 mol (b) 1.00 mol (c) 10.0 mol (d) 0.500 mol (e) no such number exists
3. What is the weight in grams of 0.250 mol of oxygen atoms?
(a) 16.0 g (b) $1.50 \times 10^{23}$ g (c) 8.00 g (d) 4.00 g (e) 32.0 g
4. What is the weight of $3.01 \times 10^{24}$ He atoms?
(a) 20.0 g (b) 4.00 g (c) 200 g (d) $12.0 \times 10^{24}$ g (e) 2.00 g
5. What is the approximate weight of one carbon atom?
(a) 12 g (b) $2.0 \times 10^{-23}$ g (c) $2.0 \times 10^{23}$ g (d) $0.50 \times 10^{-23}$ g (e) 6.0 g
6. What is the mass of one mole of $H_2C_2O_4$?
(a) 90.0 amu (b) 46.0 g (c) 90.0 g (d) 58.0 g (e) 46.0 amu
7. How many moles of oxygen atoms are in 0.50 mol of $Ca(ClO_3)_2$?
(a) 3.0 (b) 1.0 (c) 0.50 (d) 6.0 (e) 1.50
8. How many moles of oxygen atoms are in $3.01 \times 10^{23}$ molecules of $O_2$?
(a) 0.50 (b) $6.02 \times 10^{23}$ (c) 1.0 (d) 0.25 (e) $1.50 \times 10^{23}$
9. A compound is composed of 0.24 mol of Fe, 0.36 mol of S, and 1.44 mol of O. Its empirical formula is:
(a) $FeS_3O_6$ (b) $Fe_4S_6O_{24}$ (c) $Fe_2S_3O_6$ (d) $Fe_2S_3O_{12}$ (e) $Fe_2S_3O_8$
10. A compound has a molar mass of 84.0 g/mol and an empirical formula of $CH_2N$. Its molecular formula is:
(a) $C_3H_4N_2O$ (b) $C_3H_6N_3$ (c) $C_2H_4N_2$ (d) $C_3H_6N_2$ (e) $C_4H_8N_4$
11. Which of the following elements should be represented in an equation as a diatomic molecule?
(a) C (b) He (c) P (d) B (e) Br
12. What type of reaction does the following equation represent?

$$Ba(s) + 2H_2O(l) \rightarrow Ba(OH)_2(aq) + H_2(g)$$

(a) combination
(b) decomposition
(c) single replacement
(d) double displacement
(e) combustion
13. What type of reaction does the following equation represent?

$$SiH_4(g) + 2O_2(g) \rightarrow SiO_2(s) + 2H_2O(l)$$

(a) combination
(b) decomposition
(c) combustion
(d) single replacement
(e) double displacement
14. Determine which of the answers represents the proper balanced equation for the following statement: Calcium reacts with aqueous hydrocholoric acid to produce aqueous calcium chloride and hydrogen.
(a) $Ca(g) + 2HCl(aq) \rightarrow CaCl_2(l) + H_2(g)$
(b) $Ca(s) + 2HCl(aq) \rightarrow CaCl_2(aq) + 2H(g)$
(c) $2Ca(s) + 2HCl(aq) \rightarrow 2CaCl(aq) + H_2(g)$
(d) $Ca(s) + 2HCl(aq) \rightarrow CaCl_2(aq) + H_2(g)$
(e) $2Ca(s) + H_2Cl(aq) \rightarrow CaCl(aq) + H_2(g)$
15. Given the following equation:

$$2C_2H_2 + 5O_2 \rightarrow 4CO_2 + 2H_2O$$

How many moles of $O_2$ are needed to produce 4 mol of $CO_2$?
(a) 4 mol (b) 2 mol (c) 3 mol (d) 1 mol (e) 5 mol
16. Sulfur trioxide is prepared from the following two reactions:

$$S_8(s) + 8O_2(g) \rightarrow 8SO_2(g)$$

$$2SO_2(g) + O_2(g) \rightarrow 2SO_3(g)$$

How many moles of $SO_3$ are produced from 1 mol of $S_8$?
(a) 1 (b) 2 (c) 4 (d) 8 (e) 16

17. In the combustion of a certain hydrocarbon 16.0 g of $CO_2$ is produced, which represents a 75% yield. The theoretical yield is:
(a) 12.0 g (b) 8.0 g (c) 21.3 g (d) 32.0 g
(e) 44.0 g

18. Given the following combustion reaction:

$$CH_4 + 2O_2 \rightarrow CO_2 + 2H_2O$$

When 16.0 g of $CH_4$ is burned with 32.0 g of $O_2$:
(a) $CH_4$ is the limiting reactant.
(b) $CO_2$ is present in excess.
(c) $CH_4$ is present in excess.
(d) $O_2$ is present in excess.
(e) $H_2O$ is the limiting reactant.

19. Which of the following is the SI unit of pressure?
(a) atm (b) torr (c) Pa (d) in. of Hg
(e) lb/in.$^2$

20. Which of the following is *not* an assumption of the kinetic theory of gases?
(a) Molecules have negligible volume.
(b) Molecules of all gases have the same average velocity at a certain temperature.
(c) Gas molecules have negligible interactions.
(d) Temperature is related to the average kinetic energy of the system.
(e) Molecules are in rapid, random motion.

21. Which of the following is a representation of Boyle's law?
(a) $P \propto \dfrac{1}{V}$ (b) $V \propto P$ (c) $V \propto T$ (d) $P \propto \dfrac{1}{T}$
(e) $V \propto \dfrac{1}{T}$

22. The temperature of a volume of gas is increased from 20 °C to 40 °C at constant pressure. Its volume:
(a) doubles
(b) decreases by half
(c) increases by a factor of $\frac{313}{293}$
(d) decreases by a factor of $\frac{20}{273}$
(e) decreases by a factor of $\frac{293}{313}$

23 Which of the following is a representation of Gay-Lussac's law?
(a) $P_1T_1 = P_2T_2$ (b) $P_1V_1 = P_2V_2$
(c) $V_1T_2 = V_2T_1$ (d) $P_1T_1 = P_2T_2$
(e) $P_1T_2 = P_2T_1$

24. Which of the following are known as standard temperature and pressure (STP)?
(a) 0 K and 1 atm
(b) 0 °F and 760 torr
(c) 0 °C and 760 atm
(d) 273 °C and 1 atm
(e) 273 K and 760 torr

25. Which of the following gases has the highest average velocity at a certain temperature?
(a) oxygen
(b) carbon monoxide
(c) neon
(d) sulfur dioxide
(e) hydrogen chloride

26. A mixture of gases has a total pressure of 2.00 atm. If one gas has a partial pressure of 0.50 atm, what part of the mixture is this gas?
(a) 50% (b) 75% (c) 25% (d) 1.00 atm
(e) 1.50 atm

27. A 22.4-L quantity of $O_2$ at STP:
(a) contains 1 mol of oxygen atoms
(b) weighs 16.0 g
(c) contains $1.20 \times 10^{24}$ oxygen atoms
(d) contains 2 mol of $O_2$ molecules
(e) weighs 48.0 g

28. A gas has a density of 2.68 g/L (STP). The gas is:
(a) $CO_2$ (b) $SO_2$ (c) $NO_2$ (d) COS (e) He

29. Which of the following is a value for the gas constant $R$?
(a) $62.4 \dfrac{\text{L} \cdot \text{atm}}{\text{mol} \cdot \text{K}}$
(b) $62.4 \dfrac{\text{L} \cdot \text{torr}}{\text{mol} \cdot \text{K}}$
(c) $82.1 \dfrac{\text{L} \cdot \text{atm}}{\text{mol} \cdot \text{K}}$
(d) $0.0821 \dfrac{\text{L} \cdot \text{atm}}{\text{mol} \cdot {}^\circ\text{C}}$
(e) $0.0821 \dfrac{\text{L} \cdot \text{torr}}{\text{mol} \cdot \text{K}}$

30. Given the following equation:

$$C(s) + H_2O(l) \rightarrow CO(g) + H_2(g)$$

What volume of gas measured at STP would be produced from 24.0 g of carbon?
(a) 22.4 L (b) 89.6 L (c) 44.8 L (d) 11.2 L
(e) 4.0 L

**Problems**

1. Fill in the blanks

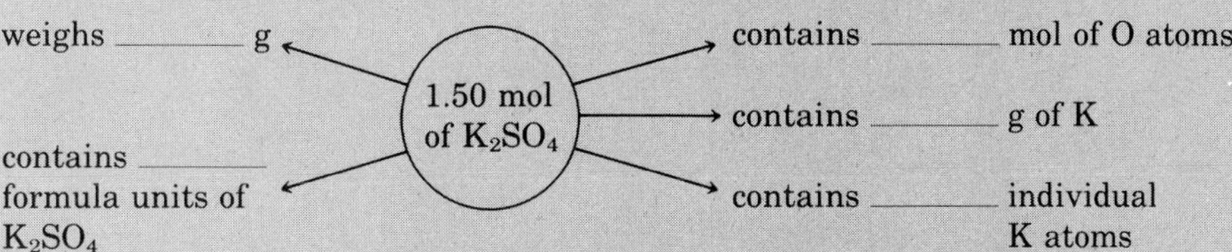


2. Fill in the blanks

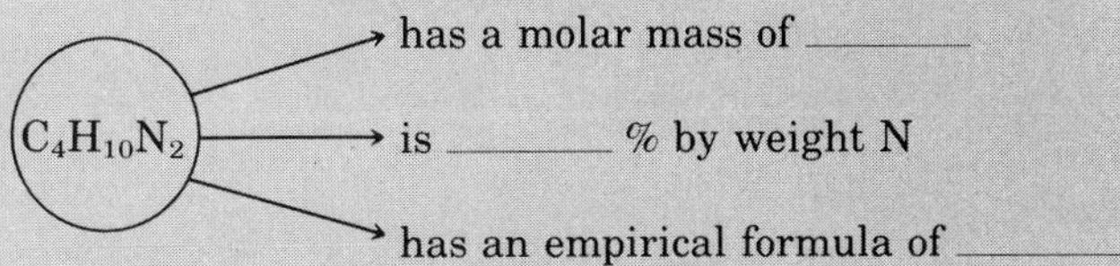


3. A compound is composed of 1.75 g of Fe and 0.667 g of O. What is its empirical formula?
4. A compound is 9.60% H, 16.4% C, and 74.0% B by weight. Its molar mass is 146 g/mol. What is its molecular formula?
5. Complete the following equations. Add subscripts on formulas of elements, coefficients to balance the equations, physical states where needed.

   (a) __K(__) + $Al_2Cl_6(s) \xrightarrow{\Delta}$ __KCl(__) + __Al($s$)

   (b) __$C_6H_6(l)$ + __O__(__) $\rightarrow$ __$CO_2$(__) + __$H_2O(l)$

   (c) __$Cl_2O_3(g)$ + __$H_2O(l) \rightarrow$ __$HClO_2(aq)$

   (d) __$K_2S(aq)$ + __$H_3PO_4(aq) \rightarrow$ __$H_2S$(__) + __$K_3PO_4(aq)$

   (e) __$B_4H_{10}(g)$ + __$H_2O(l) \rightarrow$ __$B(OH)_3(aq)$ + __H__(__)

6. Freon 12 ($CCl_2F_2$) is a gas used as a refrigerant. It is prepared according to the following equation:

$$3CCl_4(l) + 2SbF_3(s) \rightarrow 3CCl_2F_2(g) + 2SbCl_3(s)$$

   (a) How many moles of $SbF_3$ are needed to produce 0.0350 mol of freon?
   (b) How many moles of $SbCl_3$ are produced if 150 g of freon is also produced?

(c) What weight of $CCl_4$ is needed to react with 850 g of $SbF_3$?
(d) How many individual molecules of $CCl_4$ are needed to produce 12.0 kg of $SbCl_3$?
(e) What volume of freon measured at 25 °C and 0.760 atm is produced from 14.9 g of $CCl_4$?

7. Fill in the blanks.

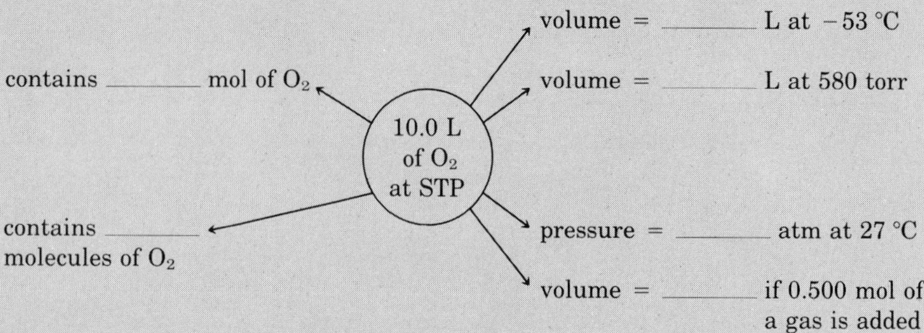


8. A compound can be vaporized at 100 °C. At that temperature it is found that 1.19 g of the compound occupies 250 mL at 756 torr. Analysis of the compound shows that it is 49.0% C, 48.3% Cl, and 2.72% H. What is the molecular formula of the compound?

On earth water exists in all three physical states, liquid, solid, and gas. Clouds are formed as the gas condenses into liquid or solid.

# Chapter 11
# Water: The Liquid and the Solid States

## Purpose of Chapter 11

In Chapter 11 we examine the liquid and solid states of water and the processes involved in changes from one state to another.

## Objectives for Chapter 11

After completion of this chapter, you should be able to:

1. Compare the geometric structure and the polarity of a water molecule to similar molecules. (11-1)
2. Determine whether specified compounds interact by hydrogen bonding or other dipole-dipole interactions. (11-2)
3. Compare the extent of interactions between water molecules in the three states of matter. (11-3)
4. Predict how melting points and boiling points of two specified compounds compare based on the interactions between basic particles of those compounds. (11-4)
5. Describe the changes that occur as water, initially in the form of ice, is heated from below 0 °C to above 100 °C. (11-5)
6. Solve problems involving the heat energy involved in changes of state by application of heats of fusion and vaporization. (11-6)
7. Describe how the interactions between water molecules are reflected by the heats of fusion and vaporization of water. (11-6)
8. Describe the process and effects of the evaporation of a liquid below its boiling point. (11-7)
9. Describe the relationships between the equilibrium vapor pressure of a liquid, its boiling point, and its normal boiling point. (11-7)
10. Write equations illustrating the formation of water and the reactions of water with certain metals and nonmetals. (11-8)
11. Write the formulas of hydrates and determine their composition. (11-8)

So common and simple yet so fascinating and unique—this is the compound known as water. Without it, chemical scientists doubt that even a simple form of life could exist. In fact, several years ago we searched for evidence of life on the planet Mars. Predictably, we looked for a sign of one of the main ingredients for life, which is free water. Water, together with carbon dioxide from the air, minerals from the soil, and energy from the sun, is changed to chemical energy in the carbon compounds of vegetation. When this chemical energy is used for fuel or food, the combustion or metabolism process cycles the water back to the environment. About two-thirds of our planet is covered with water. Without large bodies of water, it would not be habitable by life as we know it. Water has an unusual ability to moderate climate by absorption, storage, and release of heat energy. Water also has a unique ability to act as a solvent and thus serve as a medium for many important chemical reactions. We will save this topic for subsequent chapters. In this chapter we will concentrate on the nature of water itself and compare its properties as a liquid and a solid to other substances.

As background for this chapter you should review:

1 The Lewis structure of water (Section 6-6)

2 Electronegativity, bond polarity, and molecular polarity (Section 6-11,12)

3 The kinetic theory of gases (Section 10-2)

4 Specific heat (Section 2-8)

## 11-1 The Structure and Polarity of Water

The fact that a water molecule is angular rather than linear (see Figure 11-1*a*) has important consequences for its properties. Oxygen and hydrogen have a large difference in electronegativity ($3.5 - 2.1 = 1.4$), and thus the bond between them is polar. If the molecule were linear, water would be a nonpolar molecule since the two individual bond dipoles would cancel. Since it is actually angular, the two bond dipoles do not cancel and the molecule has a net dipole (see Figure 11-1*b*).

The partial charges that result from the polarity of the bonds can be represented as residing on the hydrogens and the oxygen. As we will see shortly, however, the water molecule behaves as if the partial charges are located at the corners of a slightly distorted tetrahedron. If the oxygen is placed in the center of the tetrahedron, two of the corners of the tetrahedron are occupied by hydrogens, which carry a partial positive charge. The other two corners are occupied by the unshared electron pairs, which are the locations of the partial negative charges (see Figure 11-1*c*). The reason for this geometry is that the pairs of electrons either shared in bonds or unshared are as far apart as possible. Tetrahedral geometry allows four pairs of electrons to be in such an arrangement.

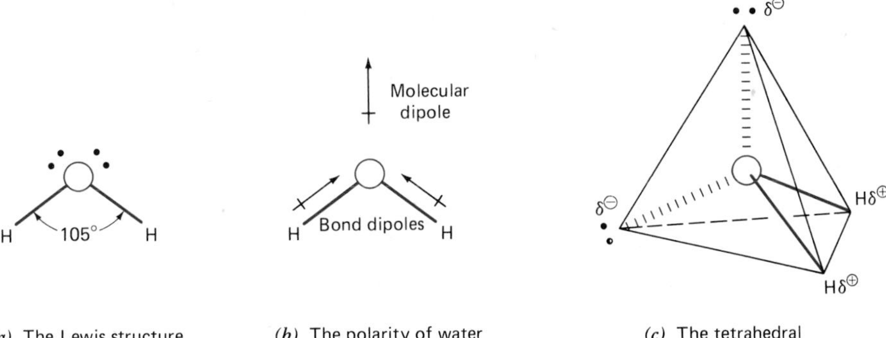


*(a)* The Lewis structure of water

*(b)* The polarity of water

*(c)* The tetrahedral representation of water

**Figure 11-1** THE STRUCTURE OF WATER.

## 11-2 Hydrogen Bonding

*Molecules that are polar (have a dipole) can interact because of electrostatic attractions for each other. This is known as a* **dipole-dipole interaction** (see Figure 11-2). The strength of a particular dipole-dipole interaction depends on the degree of polarity of the molecules. Oxygen is one of the most electronegative elements, so its bonds with hydrogen are especially polar. The same is true for the N—H and H—F bonds. *A dipole-dipole interaction between the hydrogen in an O—H, N—H, or H—F bond and a highly electronegative atom on another molecule is called a* **hydrogen bond.** It is, in fact, not an actual covalent bond, but it is much stronger and more complex than other dipole-dipole interactions. (See Figure 11-2.)

As we will see shortly, these comparatively strong interactions between $H_2O$ molecules have a profound effect on the properties and physical state of water compared with other compounds such as $H_2S$. In $H_2S$ the H—S bond has very low polarity due to the small difference in electronegativity between the H and S. As a result, the molecular polarity is very small, and the interactions between $H_2S$ molecules are very weak.

The phenomenon of hydrogen bonding is not restricted to molecules of the same compound. Interactions between molecules of different biologically

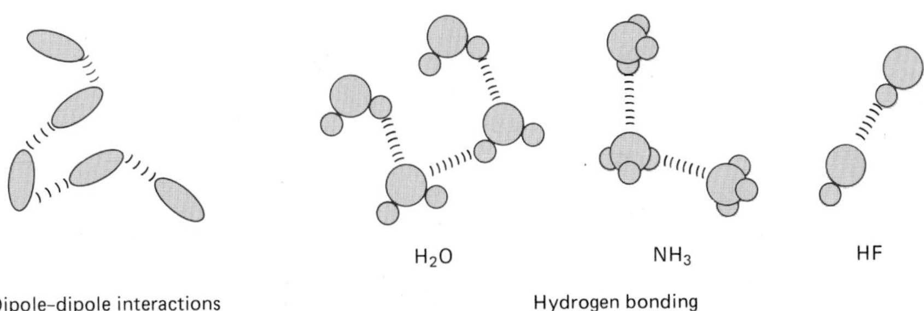


**Figure 11-2** DIPOLE INTERACTIONS AND HYDROGEN BONDING. Hydrogen bonding is a particularly strong form of a dipole-dipole interaction.

See Problems 11-1 through 11-8.

important compounds are especially important. For example, it is the force that holds the two strands of DNA together. For example,

$$\begin{array}{ccccccc} & & \mathrm{H} & & & \mathrm{H} & \\ & & | & & & | & \\ \mathrm{Y} & - & \mathrm{N} & - & \mathrm{H}\cdots & \mathrm{O} & -\ \mathrm{X} \end{array}$$

## 11-3 A Comparison of the Three Physical States of Water

It wouldn't be fair to discuss the gaseous state without some due regard to the other two physical states: solid and liquid. The example we will use is no less than the most common and important chemical, water. It is a convenient compound to use as an example since, under conditions on earth, it exists in all three states: vapor (gas), liquid, and solid (ice).

In the previous chapter, we discussed the kinetic theory of gases. Two of those assumptions about the nature of gases also apply to the molecules (or other particles) in liquids and solids:

1 The particles are in motion and thus have kinetic energy.

2 The average kinetic energy of the particles is related to the temperature of the system.

Two assumptions for gases do not apply to liquids and solids, however. For liquids and solids:

1 The motion of the particles is not completely random.

2 The particles do not behave independently as there are forces of attraction between molecules.

Liquids and solids are known as condensed states of matter. Unlike gases, the molecules are closely packed and occupy a significant part of the total volume. As a result, these states have small compressibility and high densities compared with gases.

In Figure 11-3 the motion of water in the three physical states has been illustrated. As a gas, the water molecules behave independently with random motion. In the solid state, however, the water molecules remain in fixed positions relative to other molecules. Each oxygen in a water molecule has covalent bonds to two hydrogens. In addition, each oxygen has hydrogen bonds to two hydrogens from two other water molecules. All four hydrogens are arranged around the oxygen in a tetrahedral geometry. Movement occurs but is restricted to various types of vibrations within a confined space. Liquids have behavior intermediate between solids and gases. The molecules of liquids can move past one another singly or in groups, but because of the interactions with each other, they remain close together in a condensed state. Since the molecules are not held in fixed positions, however, liquids take the shape of the bottom of the container.

See Problems 11-9, 11-10, and 11-11.

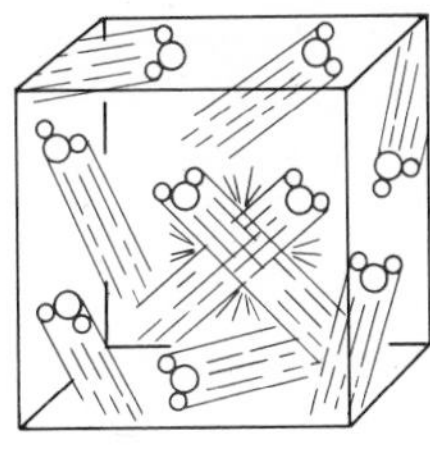

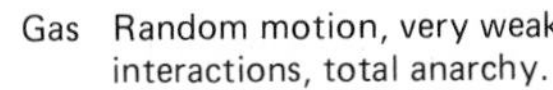

Gas Random motion, very weak interactions, total anarchy.


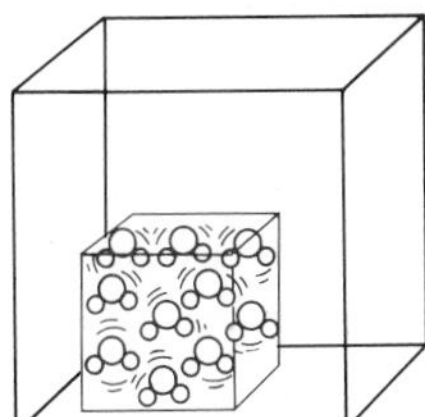

Solid Molecules in fixed positions, motion within a confined volume, highly organized.


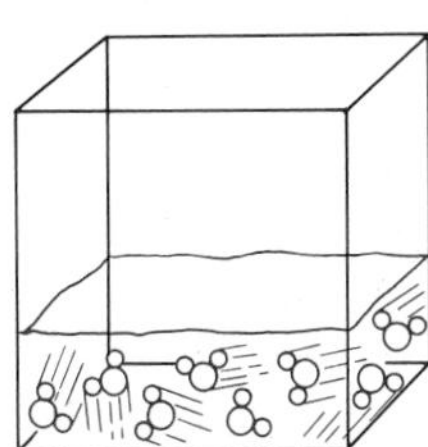

Liquid Molecules are mobile but attractions hold them together in condensed state. More freedom of movement than solid but less than gas.


**Figure 11-3** THE PHYSICAL STATES OF WATER. Interactions between water molecules are different in the three physical states.

## 11-4 Melting Points and Boiling Points

What determines whether a particular substance is a liquid, solid, or a gas? There are two factors that are important: (1) the nature of the interactions between the basic particles of the substance and (2) the temperature and pressure conditions. We will take up these two points individually.

1 Substances whose basic particles (atoms, molecules, or ions) have a high attraction for each other prefer a condensed state at a given temperature. For example, oppositely charged ions in ionic compounds are held together by strong electrostatic interactions. Thus these compounds are solids at room temperature. We also mentioned that $CO_2$, $CH_4$, $N_2$, and $O_2$ are nonpolar molecules. This means that the individual molecules have little attraction for each other and thus are gases a room temperature. There are forces of attraction between nonpolar molecules (actually all molecules) called London forces. Generally, London forces are weak but become stronger for heavier molecules. Thus heavy molecules like carbon tetrachloride (154 g/mol) and octane in gasoline (114 g/mol) are liquids at room temperature despite being nonpolar. Water and alcohol are polar covalent compounds and interact to some extent by hydrogen bonding. These compounds are also in the intermediate or liquid state at room temperature.

2 If the temperature is low enough and the pressure high enough, even molecules or atoms with the weakest interactions can be condensed to the liquid or solid states. On the other hand, if the temperature is high enough, anything can be vaporized to form a gas. The temperature relates to the kinetic energy of the molecules. As the temperature increases, the kinetic energy or motion of all molecules increases regardless of physical state. Eventually, the kinetic energy of the molecules overcomes the forces holding the molecules in fixed positions and the substance melts and, at a higher temperature, boils.

*The temperature at which the solid changes to the liquid state is called the* **melting point.** An operational definition of the **boiling point** *is the temperature at which bubbles of vapor rise through the liquid.* A more formal definition of boiling point will be presented shortly. The temperature at which melting and boiling take place depends to an extent on the strength of the intermolecular forces holding the basic particles of the substance together. Ionic compounds generally have high melting and boiling points because the ion-ion forces holding the ions together are very strong. On the other hand, compounds with the lowest boiling points (known as the most volatile) are generally nonpolar compounds of low molar mass where intermolecular attractions are weak. (Some melting and boiling points are presented in Tables 11-2 and 11-3, respectively, along with other data that will be discussed shortly.) Polar covalent compounds and nonpolar compounds of high molar mass have intermediate interactions and usually have intermediate boiling and melting points. The molecules that have hydrogen bonding in the liquid state (e.g., $H_2O$, $NH_3$, and HF) all have unusually high boiling points compared with other hydrogen compounds in the same group. (See Table 11-1.) Since the other hydrogen compounds in a group have low polarity (e.g., $H_2S$, $H_2Se$, and $H_2Te$), their boiling points show a trend more typical of nonpolar compounds. That is, the higher the molar mass the higher the boiling point.

**See Problems 11-12 through 11-15.**

## 11-5 Changes in State

Let's look at water as an example of the changes in state that occur as the substance is heated. We will start with an ice cube cooled to −10 °C in a freezer and discuss what happens as the ice is heated uniformly at a constant

**Table 11-1** Boiling Points of Some Binary Hydrides

| Group | 2nd Period | 3rd Period | 4th Period | 5th Period |
|---|---|---|---|---|
| VIIA | HF (17 °C) | HCl (−84 °C) | HBr (−70 °C) | HI (−37 °C) |
| VIA | $H_2O$ (100 °C) | $H_2S$ (−61 °C) | $H_2Se$ (−42 °C) | $H_2Te$ (−2.0 °C) |
| VA | $NH_3$ (−33 °C) | $PH_3$ (−88 °C) | $AsH_3$ (−62 °C) | $SbH_3$ (−18 °C) |

rate (see Figure 11-4). As heat is added to the ice, the temperature of the solid slowly rises, meaning that the molecules are vibrating faster and faster about their fixed positions (Figure 11-4*a*). At 0 °C and 1 atm pressure something begins to happen. The motion of an average molecule becomes great enough so that it can overcome some of the forces (hydrogen bonds) holding it in a fixed position. Thus individual molecules and groups of molecules begin to move past one another. The solid ice is melting (Figure 11-4*b*). Despite the breakdown of the rigid structure of the solid, groups of molecules still interact with each other, so they remain in a condensed state. The melting process is endothermic (absorbs heat), and the added heat is consumed by the process of breaking some of the hydrogen bonds between molecules. The breaking of hydrogen bonds increases the potential energy of the system by moving the molecules out of their fixed positions. Since potential energy is *not* related to temperature, notice that the temperature remains constant as long as the melting process is still occurring.

An unusual property of water compared with most substances is that the solid state is less dense than the liquid. As a result, ice floats. Without this property, lakes and oceans would freeze from the bottom up and remain frozen all summer except for the surface. In the case of water, the $H_2O$ molecules have a relatively open arrangement in the solid state. When ice melts, some of the hydrogen bonds holding the molecules in the open structure break and the structure collapses somewhat to form the liquid state.

After all of the ice has melted, the added heat again increases the motion or velocity of the molecules, thus increasing the temperature of the liquid water (Figure 11-4*c*). The velocity of the molecules continues to increase

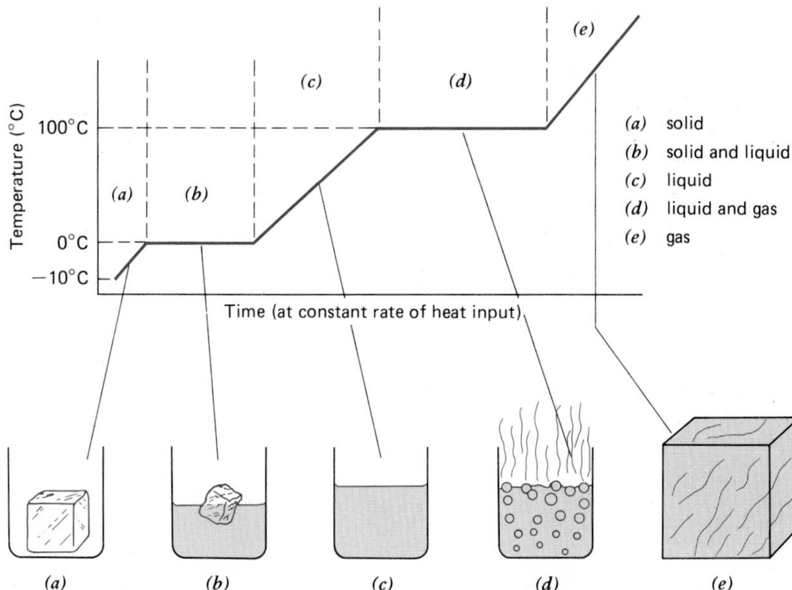


**Figure 11-4** THE HEATING CURVE FOR WATER.

until the temperature reaches 100 °C at 1 atm pressure. At this temperature bubbles of steam form in the liquid, indicating that boiling has begun (see Figure 11-4*d*). In fact, the average velocity of the water molecules has become so great that all remaining attractions to other molecules holding them in the liquid state are overcome and the water turns to gas. Again, the added heat is being consumed to break attractive forces between molecules and move them apart, which increases the potential energy. The temperature remains constant until all of the water has changed to steam. After that, added heat increases the average velocity of the now independent gas molecules and the temperature of the gas rises (Figure 11-4*e*).

The changes in state that a substance undergoes while being heated are somewhat analogous to a dance floor crowded with gyrating, vibrating couples. At first the beat of the music is slow. Everyone just gyrates (moves about a fixed position) and vibrates in one spot, held there, presumably, by the attraction of a partner. This is equivalent to the motion of molecules in the solid state. As the beat of the music increases, however, motion increases (heats up) in response. Eventually, there comes a point where it is impossible for everyone to keep up the beat and still dance in one spot. The attractions are overcome, and the people move around the floor. The dance floor is still crowded, but motion of the people is translational (from position to position). This is similar to the movement of molecules in the liquid state. If we now wished to extend the analogy to the ridiculous, we could forsee a point where the beat and the corresponding motion becomes so intense that the dancers can no longer be confined to the floor and thus move freely and independently throughout the whole room from floor to ceiling. Just as molecules of a gas are spread far apart, the dancers would no longer be crowded since the whole volume of the room is utilized. (See Figure 11-5.)

**See Problems 11-16 through 11-18.**

## 11-6 Energy in Changes of State

In order to break the forces holding the molecules, ions, or atoms together in fixed positions in a solid, heat energy must be supplied. Thus the process of *melting (fusion)* is endothermic. The opposite process, when a liquid solidifies, is known as *freezing*. This process gives off heat energy and is thus

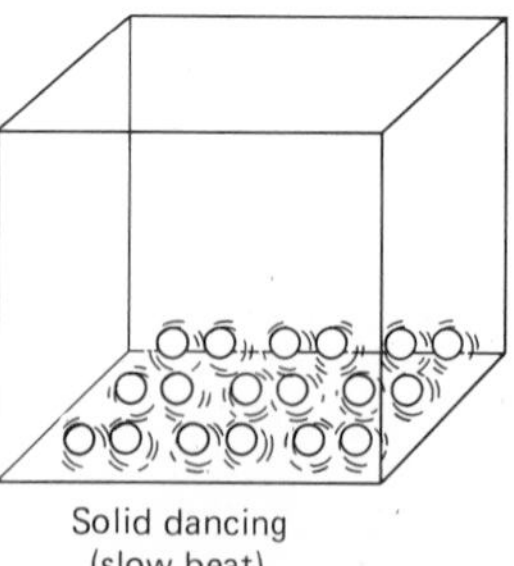


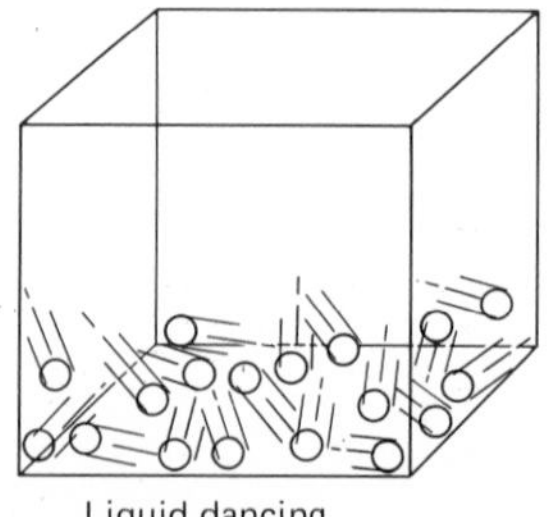


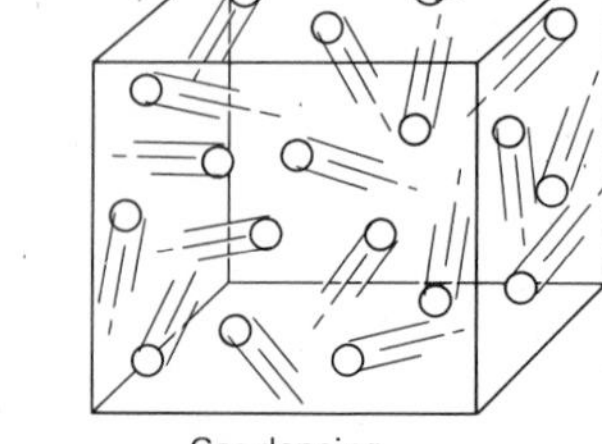


**Figure 11-5** THE EFFECT OF THE BEAT ON DANCING. Temperature is like the beat of music. The faster the beat the more the dancers move.

exothermic. The amount of heat required for melting is known as the heat of fusion and depends on the strength of the interactions between particles in the solid state. *The* **heat of fusion** *of a substance is the amount of heat in calories or joules required to melt one gram of the substance.* Table 11-2 lists the heats of fusion and the melting points of several substances.

Notice that the heat of fusion of water is high compared with other covalent compounds. This fact helps modify weather around the Great Lakes in the United States. When a lake freezes in winter, it releases heat to the environment. In the winter, this heat released by fusion has the same effect as a giant heater and helps keep the thermometer from falling as much as it otherwise would. Since the Great Lakes do not usually completely freeze over, heat is released all through the winter from the lake. On the other hand, spring is delayed because the melting ice absorbs the heat and keeps the thermometer from rising as much as it would without the lake. Siberia in Russia lies about as far north as Minnesota but has few large lakes. As a result, it is much colder there in the winter and also much hotter in the summer.

**Example 11-1**
How many kcal of heat is released when 185 g of water freezes?

**Procedure**
The heat of fusion can be used as a conversion factor relating weight in grams to calories.

**Solution**

$$185 \cancel{g} \times \frac{79.8 \cancel{cal}}{\cancel{g}} \times \frac{1 \text{ kcal}}{10^3 \cancel{cal}} = \underline{\underline{14.8 \text{ kcal}}}$$

**Table 11-2** Heats of Fusion and Melting Points

| Compound | Type of Compound | Heat of Fusion | | Melting Point (°C) |
|---|---|---|---|---|
| | | cal/g | J/g | |
| NaCl | Ionic | 124 | 519 | 801 |
| $H_2O$ | Polar covalent (hydrogen-bonding) | 79.8 | 334 | 0 |
| Ethyl alcohol | Polar covalent (hydrogen-bonding) | 24.9 | 104 | −114 |
| Ethyl ether | Polar covalent | 22.2 | 92.5 | −116 |
| Benzene | Nonpolar covalent | 30.4 | 127 | 5.5 |
| Carbon tetrachloride | Nonpolar covalent | 4.2 | 17.6 | −24 |

The process of boiling a liquid also requires energy to break the forces holding the molecules in a condensed state and separate them to the freedom of the gaseous state. The amount of energy required is known as the heat of vaporization and once again depends on the strength of the intermolecular forces holding the molecules or other particles together in the liquid state. The opposite of boiling is **condensation**. An equal amount of energy is released when a given amount of substance condenses to form a liquid. *The* **heat of vaporization** *is the amount of heat in calories or joules required to vaporize one gram of the substance.* The heats of vaporization and the boiling points of several substances are given in Table 11-3.

Notice again that water has an unusually high heat of vaporization compared with other covalent compounds. The number and strength of the hydrogen bonds between $H_2O$ molecules accounts for its high heat of vaporization.

**Review Section 2-8 on specific heat.**

In summary, we have traced the energy changes involved in converting ice to water and then water to steam. Notice that if we have 1 g of ice at 0 °C, it takes about 80 cal to melt the ice to water at 0 °C, 100 cal to heat the water from 0 °C to 100 °C (see Section 2-8 on specific heat), and 540 cal to vaporize the water at 100 °C to steam at 100 °C. This is a total of 720 cal to convert 1 g of ice at 0 °C to steam at 100 °C.

**Example 11-2**

Steam causes more severe burns than an equal mass of water. Compare the heat released when 3.00 g of steam at 100 °C condenses and then cools to 60 °C to the heat released when the same mass of water at 100 °C cools to 60 °C.

**Table 11-3** Heats of Vaporization and Boiling Points

| Compound | Type of Compound | Heat of Vaporization cal/g | Heat of Vaporization J/g | Normal Boiling Point (°C) |
|---|---|---|---|---|
| NaCl | Ionic | 3130 | 13,100 | 1465 |
| $H_2O$ | Polar covalent (hydrogen-bonding) | 540 | 2,260 | 100 |
| Ethyl alcohol | Polar covalent (hydrogen-bonding) | 204 | 854 | 78.5 |
| Ethyl ether | Polar covalent | 89.6 | 375 | 34.6 |
| Benzene | Nonpolar | 94 | 393 | 80 |
| Carbon tetrachloride | Nonpolar | 46 | 192 | 76 |

**Procedure**
Use the heat of vaporization as a conversion factor between weight in grams and calories. Also use the specific heat of water as a conversion factor between weight, calories, and degrees Celsius.

**Solution**

Steam: Heat released in condensation:

$$3.00\ \cancel{g} \times \frac{540\ \text{cal}}{\cancel{g}} = 1620\ \text{cal}$$

Heat released in cooling:

$$3.00\ \cancel{g} \times 40\ {}^{\circ}\cancel{C} \times \frac{1.00\ \text{cal}}{\cancel{g}\ {}^{\circ}\cancel{C}} = 120\ \text{cal}$$

Total heat released = 1620 cal + 120 cal = <u>1740 cal released</u>

Water: Heat released in cooling = <u>120 cal</u>

Notice that almost 15 times more heat was released by the steam at 100 °C than the water at the same temperature.

---

**Example 11-3**
How many kilocalories of heat are required to convert 250 g of ice at −15 °C to steam at 100 °C?

**Procedure**
The heats of fusion and vaporization as well as the specific heats of ice and water (Table 2-6) are needed.

**Solution**
To heat ice from −15 °C to its melting point (0 °C), or a total of 15 °C:

$$250\ \cancel{g} \times 15\ {}^{\circ}\cancel{C} \times 0.492\ \frac{\cancel{\text{cal}}}{\cancel{g} \cdot {}^{\circ}\cancel{C}} \times \frac{1\ \text{kcal}}{10^3\ \cancel{\text{cal}}} = 1.8\ \text{kcal}$$

To melt the ice at 0 °C:

$$250\ \cancel{g} \times 79.8\ \frac{\cancel{\text{cal}}}{\cancel{g}} \times \frac{1\ \text{kcal}}{10^3\ \cancel{\text{cal}}} = 20.0\ \text{kcal}$$

To heat the water from 0 °C to its boiling point (100 °C), or a total of 100 °C:

$$250\ \cancel{g} \times 100\ {}^{\circ}\cancel{C} \times 1.00\ \frac{\cancel{\text{cal}}}{\cancel{g} \cdot {}^{\circ}\cancel{C}} \times \frac{1\ \text{kcal}}{10^3\ \cancel{\text{cal}}} = 25.0\ \text{kcal}$$

To vaporize the water at 100 °C:

$$250\ \cancel{g} \times 540\ \frac{\cancel{cal}}{\cancel{g}} \times \frac{1\ kcal}{10^3\ \cancel{cal}} = 135\ kcal$$

Total: $1.8 + 20.0 + 25.0 + 135 = \underline{\underline{182\ kcal}}$

**See Problems 11-19 through 11-38.**

## 11-7 Evaporation and Vapor Pressure

We all know that water doesn't have to be boiled for it to change to the vapor. Cold water or even snow changes to vapor if left open to the atmosphere. *The process of vaporization of a liquid below its boiling point is called* **evaporation**. *The vaporization of a solid is called* **sublimation**. How can molecules at temperatures below a boiling point have enough kinetic energy to escape to the vapor? To answer this we must remember that a certain temperature relates to the *average* kinetic energy of the molecules. In fact, at a given temperature individual molecules have a wide range of kinetic energies. In Figure 11-6 the distribution of velocities of certain molecules at two temperatures has been illustrated. Notice that the average is higher for the higher temperature, as expected.

The dotted line represents the minimum kinetic energy necessary for a certain molecule at the surface of the liquid to escape to the gaseous state. Even though the *average* energy is not enough to escape, some molecules with above average energy can escape to form a gas. Notice that at a higher temperature a higher fraction of molecules have the minimum energy needed to escape.

Now we will examine an important effect of this information. In Figure 11-7 a beaker of water has been placed inside a glass container that has a mercury manometer attached. The manometer allows us to measure the pressure inside the container. Initially, the air is dry inside the container and the pressure is the same inside and out. As the more energetic molecules in the liquid find their way to the surface and escape to the vapor, the pressure above the liquid increases (Dalton's law). As more molecules escape, some other molecules return to the liquid state. Eventually, a point is reached

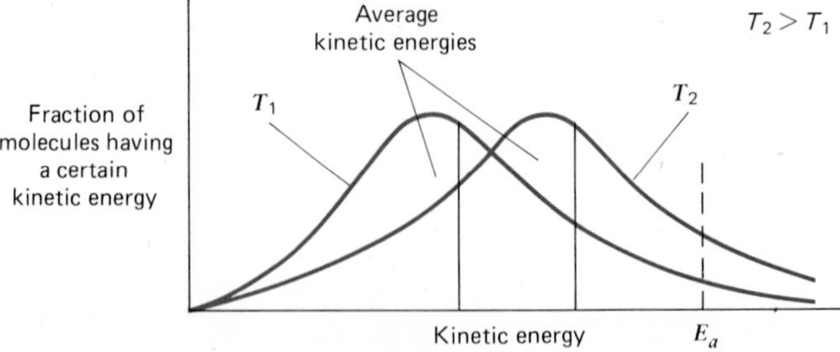


**Figure 11-6** THE DISTRIBUTION OF KINETIC ENERGIES AT TWO TEMPERATURES. The average kinetic energy is higher at $T_2$ than $T_1$.

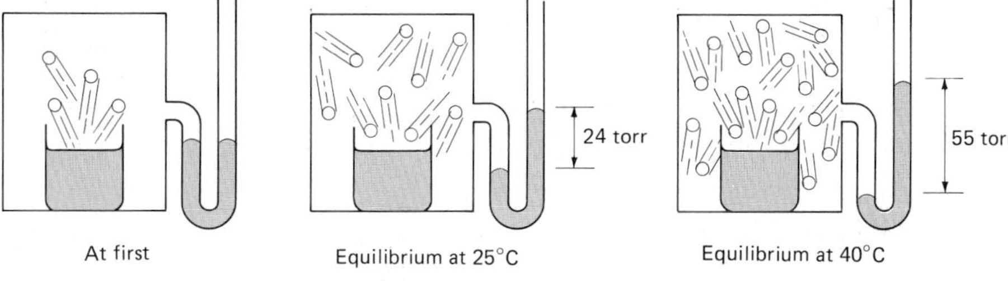


**Figure 11-7** EQUILIBRIUM VAPOR PRESSURE. A certain fraction of molecules escape to the vapor state above a liquid.

where molecules escaping from the surface of the liquid are all replaced by others condensing from the gas. At this point, when the rate of vaporization equals the rate of condensation, the pressure stabilizes and does not change. The system is said to have reached a point of equilibrium.

Equilibrium

Rate of evaporation = rate of condensation

*The pressure exerted by the vapor above a liquid at a certain temperature is called its* **equilibrium vapor pressure**. Notice that at a higher temperature more molecules have the minimum energy to escape and, as a result, the vapor pressure is higher. Solids can also have an equilibrium vapor pressure. Dry Ice has a high vapor pressure and sublimes rapidly. Regular ice has a very low vapor pressure but enough so that snow slowly evaporates or sublimes even though the temperature does not rise above freezing.

The vapor pressure of water and two other liquids are shown as a function of temperature in Figure 11-8. At 100 °C notice that the vapor pressure of water is 760 torr, or 1 atm. *When the vapor pressure of a liquid equals the restraining or the atmospheric pressure, the liquid boils. The* **normal boiling point** *of a liquid is the temperature at which the vapor pressure is equal to 760 torr.* Notice that the vapor pressure of both alcohol and ether reach 1 atm at temperatures below that of water. Thus they boil at around 78 °C and 34 °C, respectively. The actual boiling point of water where you live depends on the elevation of your community. In high elevations, where

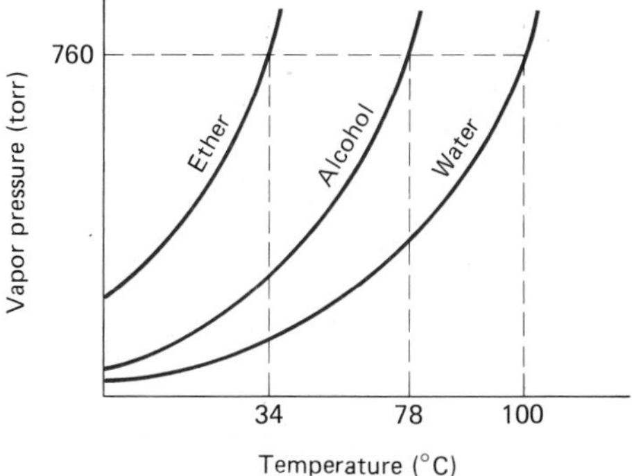


**Figure 11-8** VAPOR PRESSURE AND TEMPERATURE.

the air pressure is significantly lower than that at sea level, the boiling point of water is noticeably lower. On top of Pike's Peak in Colorado, the atmospheric pressure is about 450 torr and the boiling point of water is only 86 °C. On the highest mountain, Mt. Everest, the boiling point is 76 °C. Cooking would obviously take much longer (if it can be done at all) at these elevations. On the other hand, a pressure cooker confines the vapor so that the pressure inside the cooker is greater than atmospheric pressure. The result is a higher boiling point for water and a faster cooking time.

In Figure 11-7, the beaker of water was in a closed container where equilibrium with the vapor was reached. What would happen if the water were left in the open? In that case, the water would eventually all evaporate as long as water molecules could escape (see Figure 11-9).

An important and interesting phenomenon is happening to the water left behind. It is getting cooler (if the water is in an insulated container so that heat does not enter from the outside). The reason is that only those molecules with the highest energy are escaping. This lowers the *average* of the remaining molecules, and thus the temperature (which relates to the average) also decreases. It is like the class average on a chemistry exam. If all grades of those who receive 95% or above are not included, the average would look a lot better (i.e., lower) to many of the students.

Cooling by evaporation is well known. The cool feeling after even a warm shower is not just a feeling; it is actually cooler. When we get out of the water, the evaporation quickly cools the water on our bodies and us along with it. In fact, this is how our bodies are cooled on hot days. Perspiration evaporates and carries off excess heat from the body. On humid days, this cooling mechanism is less efficient. The atmosphere on humid*

*Relative humidity is the percent of saturation of water vapor in the atmosphere. For example, at 25 °C the vapor pressure of water is 24 torr (100% saturation). If the actual pressure of water in the air is 20 torr, the relative humidity is

$$\frac{20 \text{ torr}}{24 \text{ torr}} \times 100\% = 83\% \text{ of saturation}$$

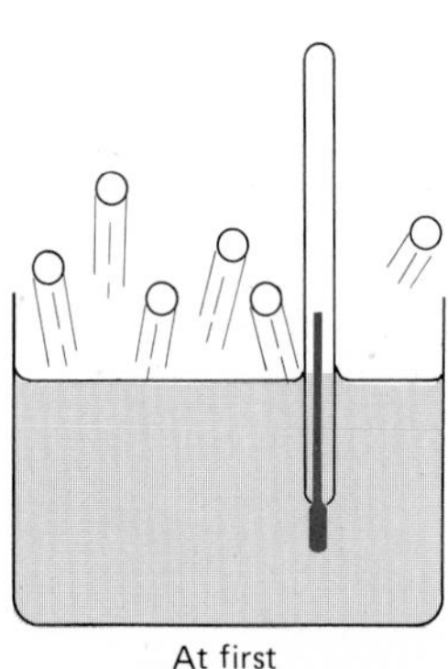
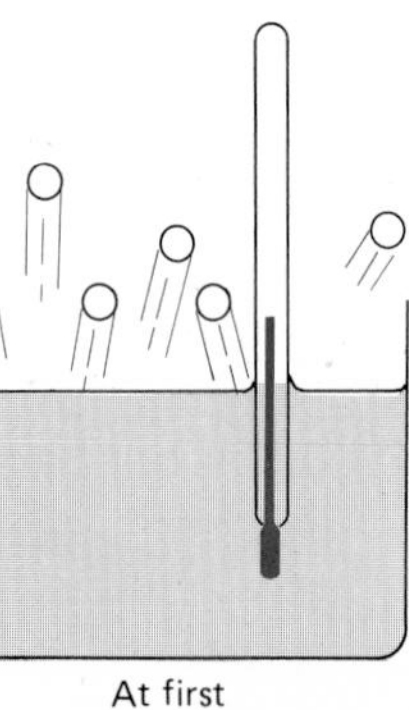


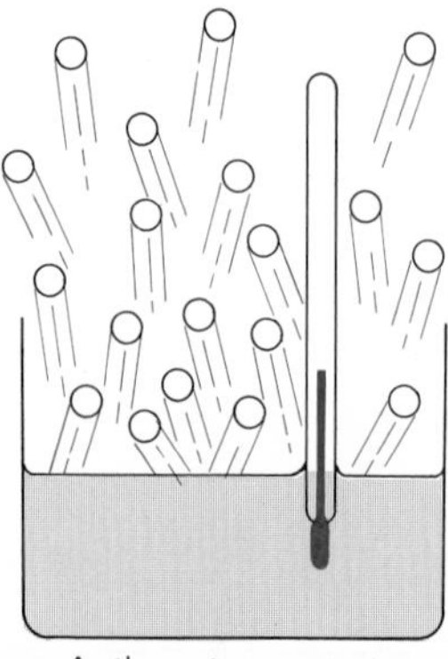


**Figure 11-9** THE EVAPORATION OF WATER. Liquids cool by evaporation.

days already holds a large portion of the water vapor that it can hold (the vapor pressure at that temperature), so evaporation is much slower and the cooling effect is greatly diminished. When they say "It's not the heat, it's the humidity," they're right.

**See Problems 11-39 through 11-52.**

## 11-8 The Formation and Reactions of Water

Water is formed principally from three types of reactions, as illustrated by the following equations:

**Review Section 9-2 on types of reactions.**

Combination: $2H_2(g) + O_2(g) \rightarrow 2H_2O(l)$

Combustion: $CH_4(g) + 2O_2(g) \rightarrow CO_2(g) + 2H_2O(l)$
Methane

$C_6H_{12}O_6(aq) + 6O_2(g) \rightarrow 6CO_2(g) + 6H_2O(l)$
Glucose

Double displacement: $HCl(aq) + NaOH(aq) \rightarrow NaCl(aq) + H_2O(l)$
(Acid-base)

All of these reactions are exothermic, which means that energy is released along with products. Hydrogen burns smoothly unless the hydrogen and oxygen are allowed to mix and then ignited. This, of course, causes a potentially dangerous explosion. This reaction finds use, however, in certain space rockets because of the light weight of hydrogen and the large amount of energy released by the reaction. Methane and glucose are both burned as fuels. Methane (natural gas) is used to heat some of our homes and for cooking, and glucose combustion is the principal source of energy for our bodies. With the help of certain enzymes the combination of glucose and oxygen takes place at a controlled rate, giving us a steady source of energy. This process is called metabolism. The production of water from the reaction of aqueous acids and hydroxides (called bases) is a type of double displacement reaction called a neutralization reaction. This type of reaction is discussed in detail in Chapter 13.

Water is also involved as a reactant in many and varied types of reactions. Three types will be mentioned here: the reaction with certain elements (single replacement reactions), the reaction with oxides (combination reactions), and the formation of hydrates (combination reactions).

Water reacts with both metals and nonmetals. The reactions with several metals is illustrated by the following equations.

$$2Na(s) + 2H_2O(l) \rightarrow 2NaOH(aq) + H_2(g)$$

$$Ca(s) + 2H_2O(l) \rightarrow Ca(OH)_2(aq) + H_2(g)$$

$$Zn(s) + H_2O(\text{steam}) \xrightarrow{\Delta} ZnO(s) + H_2(g)$$

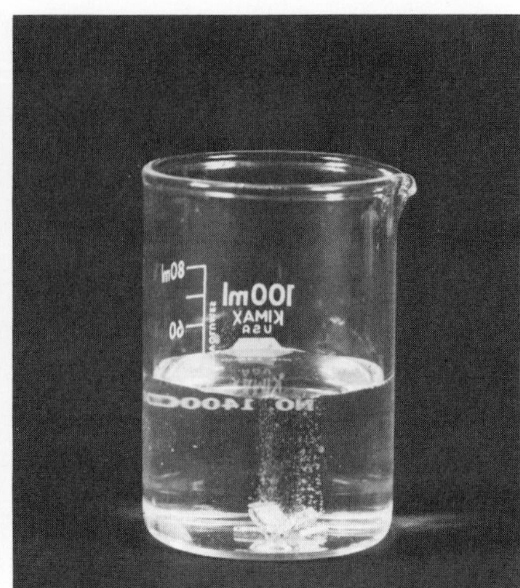


**Figure 11-10** THE REACTIONS OF METALS WITH WATER. Sodium and calcium react with water in a similar manner but at different rates.

Sodium reacts with water so rapidly that the sudden evolution of heat ignites the hydrogen as it is evolved. Other alkali metals react in a similar fashion. Calcium reacts very slowly with water with a steady evolution of hydrogen gas (See Figure 11-10.) Other alkaline earth metals react slowly with water except for beryllium. Beryllium and zinc react with steam at a high temperature. The chemical reactivity of a metal is usually judged by its activity toward water. Thus sodium and other alkali metals are considered very reactive metals. Alkaline earth metals are somewhat less reactive, and other metals such as copper, silver, and gold, are considered comparatively unreactive since they do not react with water under any conditions.

Certain nonmetals also react with water, as illustrated by the following equations:

$$2F_2(g) + 2H_2O(l) \rightarrow 4HF(aq) + O_2(g)$$

$$Cl_2(g) + H_2O(l) \rightleftharpoons HCl(aq) + HOCl(aq)$$

$$C(s) + H_2O(\text{steam}) \rightarrow CO(g) + H_2(g)$$

The reaction of fluorine is violent, as are many reactions involving fluorine, the most reactive element. Chlorine, a yellow-green gas, reacts with water, but the reaction is incomplete (reaches a state of equilibrium). The unreacted $Cl_2$ accounts for the green color when chlorine is used as a disinfectant in swimming pools. The reaction of carbon with steam produces $H_2$ and CO, which together are known as "water gas." It is a method to convert coal (mostly elemental carbon) into a relatively cheap combustible fuel gas.

Water also reacts with certain nonmetal oxides to form acids as follows:

$$SO_2(g) + H_2O(l) \rightarrow H_2SO_3(aq)$$

sulfur dioxide — sulfurous acid

This reaction is one of those that account for what is known as "acid rain."

Water reacts with certain metal oxides to form metal hydroxides, which are known as bases.

$$CaO(s) + H_2O(l) \rightarrow Ca(OH)_2(s)$$

These latter two reactions will be studied in more detail in Chapter 13.

Water reacts with certain compounds to form hydrates as follows:

$$CuSO_4(s) + 5H_2O(l) \rightarrow CuSO_4 \cdot 5H_2O$$

Pale green solid — Dark blue solid

Water in a hydrate is called water of hydration. Hydrates are actual solid compounds that are distinctly different from the anhydrous compound (the form without water of hydration). In the hydrate the water molecules are attached to either the cation by ion-dipole forces or to the anion by hydrogen bonds. Since these attractions are not nearly as strong as covalent bonds, the water molecules can usually be removed without decomposition of the rest of the compound. As a result, many hydrates can be "dehydrated" by heating.

$$CaCl_2 \cdot 2H_2O(s) \xrightarrow{\Delta} CaCl_2(s) + 2H_2O(g)$$

Hydrates are named by adding the word "hydrate" to the regular compound name as discussed in Chapter 7. Greek prefixes are added to designate the number of $H_2O$ molecules. (See Table 7-4.) For example,

| | |
|---|---|
| $CuSO_4 \cdot 5H_2O$ | copper(II) sulfate pentahydrate |
| $CaCl_2 \cdot 2H_2O$ | calcium chloride dihydrate |

In the calculation of molar mass of a hydrate, waters of hydration are included.

**See Problems 11-53 through 11-60.**

## 11-9 Chapter Summary

Perhaps the single most important fact about a water molecule is that it is nonlinear. This, and the fact that the H—O bond is highly polar, leads to a significant net polarity for the entire molecule. As a result, water molecules associate with each other by means of a comparatively strong dipole-dipole interaction known as a hydrogen bond. Hydrogen bonding occurs between a partially positive hydrogen on one molecule and a highly electronegative atom (i.e., O, N, and F) on another atom.

Water exists as a gas, liquid, or a solid under earth conditions. The main difference between each state concerns the degree of association between the individual water molecules. A comparison of the three different physical states of water and the changes between states is summarized as follows:

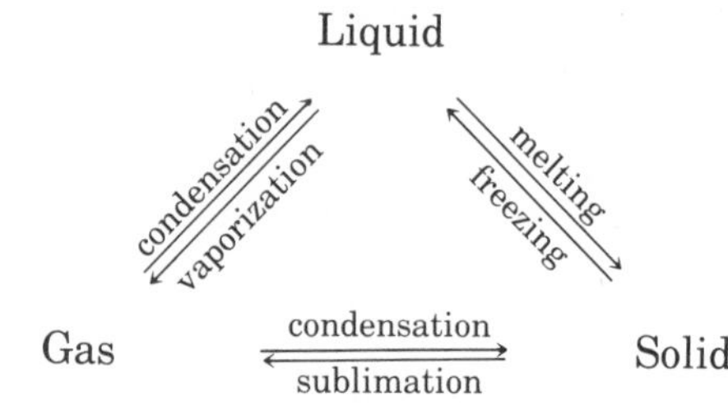


| Gas | Liquid | Solid |
| --- | --- | --- |
| There is negligible association between individual $H_2O$ molecules and random motion. | Molecules are mobile, but significant association holds molecules in condensed state. | Water molecules held in fixed positions by hydrogen bonding; motion restricted to vibrations within a fixed volume. |

The fact that water molecules interact by hydrogen bonding has a profound effect on its melting and boiling points. Hydrogen bonding is not as strong an interaction as the ion-ion forces of ionic compounds but is stronger than the weak forces of most other dipole-dipole interactions. Thus water melts and boils at temperatures well below ionic compounds but much higher than would be expected if it did not have hydrogen bonding.

The melting and boiling process require a specific amount of energy for a substance. These quantities are known respectively as the heat of fusion and the heat of vaporization. Again, the magnitude of these quantities reflects the type and strength of the interaction that must be overcome to effect melting and vaporization. The information discussed can be summarized as follows:

| Type of Compound | Type of Interaction | Melting and Boiling Points, Heats of Fusion, and Vaporization | Examples |
| --- | --- | --- | --- |
| Ionic | Ion-ion | High | NaCl, $CaSO_4$ |
| Polar covalent | Hydrogen bonding | Intermediate | $H_2O$, $NH_3$ |
| Polar covalent | Dipole-dipole | Intermediate to low[a] | $SO_2$, CO, HBr |
| Nonpolar | London forces | Intermediate to low[b] | $CO_2$, $N_2$, $CCl_4$ |

[a]Depends on the extent of molecular polarity and molar mass.
[b]Depends on the molar mass.

When a solid such as ice is heated, it goes through five stages: (a) the heating of the solid, (b) the melting of the solid, (c) the heating of the liquid, (d) the boiling of the liquid, and (e) the heating of the gas. The added heat increases the kinetic energy of the molecules and thus the temperature in steps (a), (c), and (e). In steps (b) and (d) the added heat increases the potential energy of the molecules, which means that the temperature remains constant during the process.

A molecule must have a certain minimum kinetic energy to escape from the interactions in the liquid to the vapor. Since molecules have a range of energies at a certain temperature, some of the molecules with above-average energy may escape from the surface, creating an equilibrium vapor pressure. The higher the temperature, the larger the fraction of molecules having this minimum energy, and thus the higher the vapor pressure. The liquid boils when the vapor pressure equals the atmospheric pressure. The normal boiling point is the temperature at which the vapor pressure equals 1 atm pressure. When evaporation is allowed to occur so that the vapor escapes and no heat enters the liquid from the outside, the liquid cools. This is due to the loss of the molecules with above-average kinetic energy lowering the average kinetic energy (and the temperature) of the remaining molecules.

Water is both a product and a reactant in many important chemical reactions. It is formed mainly from the combustion of hydrogen or hydrogen-containing compounds. It is a reactant with many elements, both metals and nonmetals, and with many oxides. Water is also associated with many ionic compounds in the form of "hydrates."

## Exercises

### Dipole-Dipole Interactions and Hydrogen Bonding

**11-1** If $H_2O$ were a linear molecule, why wouldn't it be polar?

**11-2** The $H_2S$ molecule is also bent similar to $H_2O$, but it has a very small molecular dipole. Why?

**11-3** Write the Lewis structure of $NH_3$ with the electron pair and hydrogens in a tetrahedral arrangement. (See Figure 11-1*c*.) Is $NH_3$ a polar molecule?

**11-4** The $PH_3$ molecule can be represented in a tetrahedral arrangement similar to $NH_3$. Is $PH_3$ a polar molecule?

**11-5** Which of the following molecules have dipole-dipole interactions in the liquid state?

(a) HBr
(b) $SO_2$ (nonlinear)
(c) $CO_2$ (linear)
(d) $BF_3$ (trigonal planar)
(e) $K_2S$
(f) CO

**11-6** Which of the following molecules have dipole-dipole interactions in the liquid state?

(a) $SCl_2$ (nonlinear)
(b) $PH_3$
(c) $CCl_4$ (tetrahedral)
(d) $CaCl_2$
(e) $CS_2$ (linear)
(f) FCl

**11-7** Which of the following molecules have significant hydrogen bonding in the liquid state?

(a) HF
(b) $NCl_3$
(c) $H_2NCl$
(d) $H_2O$
(e) $CH_4$ (tetrahedral)
(f) H—C(=O)—O—H
(g) $CH_3Cl$ (tetrahedral)

**11-8** Which should have stronger hydrogen bonding, $NH_3$ or $H_2O$?

### States of Matter and Changes of State

**11-9** Liquids mix more slowly than gases. Why?

11-10 Why is a gas compressible and a liquid not compressible?

11-11 Describe the forces holding $H_2O$ molecules together in the liquid state.

**11-12** Which compound is predicted to have a higher melting point, $CO_2$ or OCS? Both molecules are linear.

11-13 Which compound is predicted to have a higher melting point, LiCl or HCl?

**11-14** Which compound is predicted to have a higher boiling point, $OF_2$ or $H_2O$? (Both molecules are angular.)

11-15 Explain why $H_2O$ has a higher boiling point than $NH_3$.

**11-16** Which has the higher kinetic energy, $H_2O$ molecules in the form of ice at 0 °C or in the form of water at 0 °C?

11-17 Which has the higher potential energy, $H_2O$ molecules in the form of ice at 0 °C or in the form of water at 0 °C?

**11-18** Which has the higher potential energy, $H_2O$ molecules in the form of steam at 100 °C or in the form of water at 100 °C?

### Energy in Changes of State

**11-19** Which of the following processes are endothermic?
(a) fusion (b) melting (c) boiling (d) condensation

**11-20** If 550 cal of heat is added to solid $H_2O$, NaCl, and benzene at their respective melting points, what weight of each will be changed to a liquid?

**11-21** When 25.0 g of ethyl ether freezes, how many calories are liberated? If 25.0 g of $H_2O$ freezes, how many calories are liberated? Which would be a more effective liquid in a large lake to modify climate? How many joules of heat are required to melt 125 g of ethyl alcohol at its melting point?

11-22 How many kilocalories are required to vaporize 2.50 kg of $H_2O$ at its boiling point?

**11-23** How many calories of heat are released when 18.0 g of benzene condenses at its boiling point?

11-24 What weight of carbon tetrachloride can be vaporized by addition of 1 kcal to the liquid at its boiling point? What weight of water would be vaporized by the same amount of heat?

**11-25** Refrigerators cool when a liquid extracts heat when it is vaporized. Before the invention of freon ($CF_2Cl_2$), ammonia was used. Freon is nontoxic, and $NH_3$ is a pungent and toxic gas. The heat of vaporization of $NH_3$ is 325 cal/g and of freon is 38.6 cal/g. How many calories can be extracted by the vaporization of 450 g of each of these compounds? Which is the better refrigerant on the basis of weight?

11-26 Molten ionic compounds are used as a method to store heat. How many kilojoules of heat are released when 4.50 kg of NaCl crystallizes at its melting point?

**11-27** How many joules of heat are released when 275 g of steam at 100.0 °C is condensed and cooled to room temperature (25.0 °C)?

11-28 How many kilocalories of heat are required to melt 55.6 g of ice and then heat the water to 50.0 °C?

**11-29** How many calories are required to change 132 g of ice at −20.0 °C to steam at 100.0 °C? (The specific heat of ice is 0.492 cal/g · °C.)

11-30 How many joules are released when 450 g of steam at 100.0 °C is condensed, cooled, frozen, and then cooled to −10.0 °C? (The specific heat of ice is 2.06 J/g · °C.)

* **11-31** A sample of steam is condensed at 100.0 °C and then cooled to 75.0 °C. If 6780 cal of heat was released, what is the weight of the sample?

* **11-32** What weight of ice at 0 °C can be changed into steam at 100 °C by 1 kcal of heat?

**11-33** How many calories of heat are needed to heat 120 g of ethyl alcohol from 25.5 °C to its boiling point and then vaporize the alcohol? (The specific heat of ethyl alcohol is 0.590 cal/g · °C.)

* **11-34** A 10.0-g sample of benzene is condensed from the vapor at its boiling point and the liquid allowed to cool. If 1200 calories were released, what is the final temperature of the liquid benzene? (The specific heat of benzene is 0.410 cal/g · °C.)

**11-35** Ethyl ether and ethyl alcohol are both polar covalent molecules. What accounts for the considerable higher heat of vaporization for alcohol?

**11-36** A certain compound has a heat of fusion of about 600 cal/g. Is it likely to have a high or low melting point?

**11-37** A certain compound has a normal boiling point of −75 °C. Is it likely to have a comparatively high or low heat of vaporization?

**11-38** A certain compound has a normal boiling point of 845 °C. Is it likely to have a high or low melting point?

### Vapor Pressure

**11-39** Explain how molecules of a liquid may go into the vapor state if the temperature is below the boiling point.

**11-40** What is meant by the word "equilibrium" in equilibrium vapor pressure?

**11-41** What is the difference between boiling point and *normal* boiling point?

**11-42** A liquid has a vapor pressure of 850 torr at 75 °C. Is the substance a gas or a liquid at 75 °C and 1 atm pressure?

**11-43** On top of Mt. Everest the atmospheric pressure is about 260 torr. What is the boiling point of ethyl alcohol at that pressure? If the temperature is 10 °C, is ethyl ether a gas or a liquid under conditions on Mt. Everest? (Refer to Figure 11-8.)

**11-44** At a certain temperature one liquid has a vapor pressure of 240 torr and another 420 torr. Which liquid probably has the lower boiling point? Which probably has the lower heat of vaporization?

**11-45** On the planet Mars the temperature can reach as high as a comfortable 50 °F (10 °C) on the equator. The atmospheric pressure is about 8 torr on Mars, however. Can liquid water exist on Mars under these conditions? What would the atmospheric pressure have to be before liquid water could exist at this temperature? What would happen to a glass of water if it was set out on the surface of Mars? (Refer to Figure 11-8.)

**11-46** If the atmospheric pressure is 500 torr, what is the approximate boiling point of water, ethyl alcohol, and ethyl ether? (Refer to Figure 11-8.)

* **11-47** On a hot, humid day the relative humidity is 70% of saturation. If the temperature is 34 °C, the vapor pressure of water is 39.0 torr (this would be 100% of saturation). What weight of water is in each 100 L of air under these conditions?

**11-48** How can the boiling point of a pure liquid be raised?

**11-49** Rubbing alcohol feels cool when applied to the skin even if the alcohol is initially at room temperature when first applied. Why does the alcohol feel cool?

**11-50** Ethyl chloride boils at 12 °C. When it is sprayed on the skin, it freezes a small part of the skin and thus serves as a local anesthetic. Explain how it cools the skin.

**11-51** A beaker of a liquid with a vapor pressure of 350 torr at 25 °C is set alongside a beaker of water, and both are allowed to evaporate. Which liquid cools faster? Why?

**11-52** Which cools faster, water at 90 °C or water at 30 °C, when both are allowed to evaporate?

## Reactions of Water and Hydrates

**11-53** Complete and balance the following equations.
(a) $K(s) + H_2O(l) \rightarrow$
(b) $Br_2(l) + H_2O(l) \rightarrow$
(c) $K_2O(s) + H_2O(l) \rightarrow$
(d) $CO_2(g) + H_2O(l) \rightarrow$
(e) $C_2H_4(g) + O_2(g) \rightarrow$

**11-54** Complete and balance the following equations.
(a) $Sr(s) + H_2O(l) \rightarrow$
(b) $C(s) + H_2O(l) \xrightarrow{\Delta}$
(c) $ZnO(s) + H_2O(l) \rightarrow$
(d) $SO_3(g) + H_2O(l) \rightarrow$
(e) $B_4H_{10}(g) + O_2(g) \rightarrow B_2O_3(s) +$

**11-55** Name the following hydrates.
(a) $MgCO_3 \cdot 5H_2O$
(b) $Na_2SO_4 \cdot 10H_2O$
(c) $FeBr_3 \cdot 6H_2O$

**11-56** Name the following hydrates.
(a) $Na_2CO_3 \cdot 7H_2O$
(b) $Sr(NO_3)_2 \cdot 4H_2O$
(c) $CrC_2O_4 \cdot H_2O$

**11-57** Calculate the percent by weight of water in strontium hydroxide octahydrate.

**11-58** Calculate the percent by weight of water in sodium sulfide nonahydrate.

**11-59** Epsom salts is a magnesium sulfate hydrate. If the compound is 51.1% by weight water, what is its formula?

**11-60** A potassium carbonate compound is 20.7% water. What is its formula?

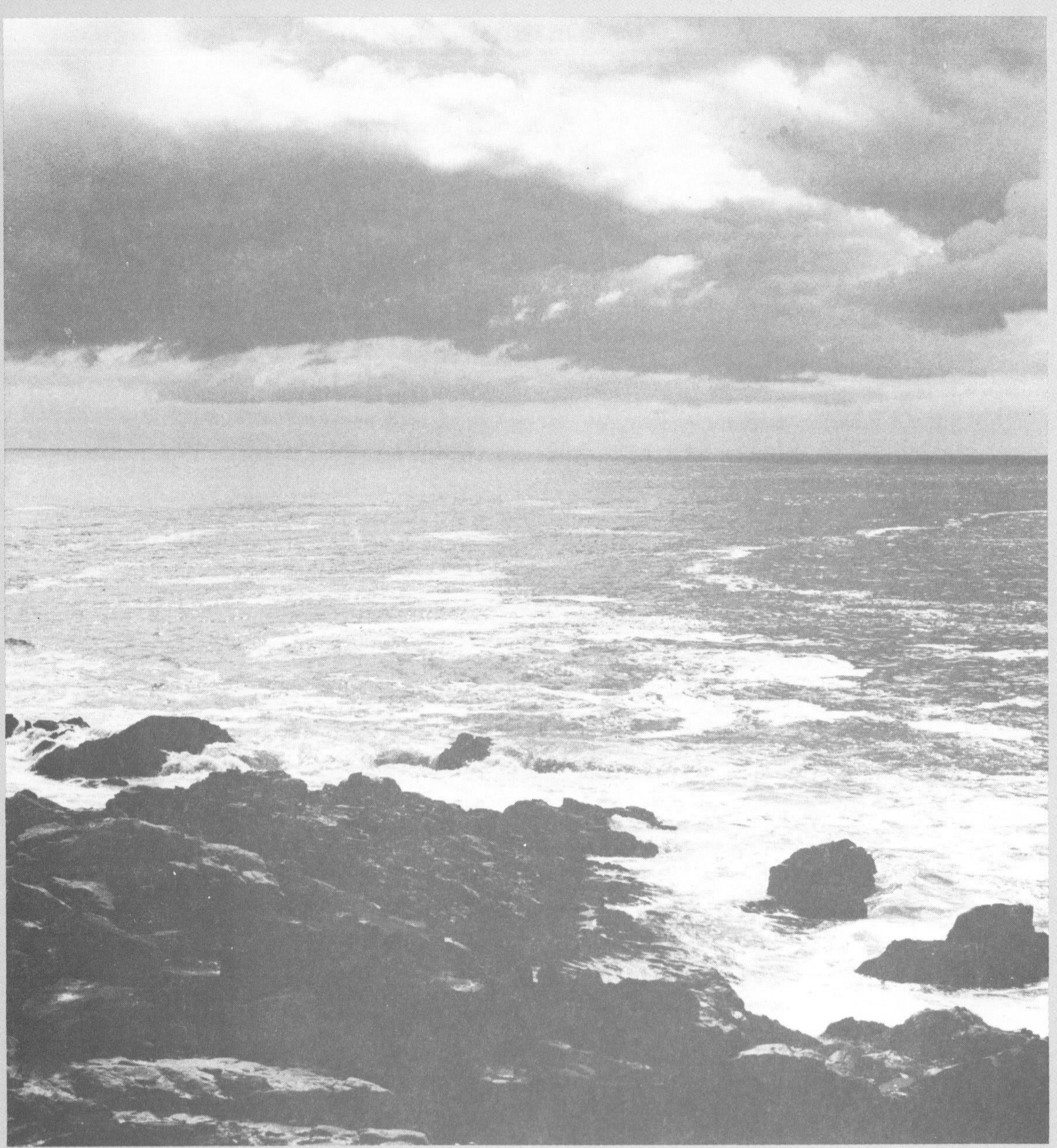

The large oceans that cover most of the surface of the earth are not composed of pure water. They contain many dissolved substances.

# Chapter 12
# Aqueous Solutions

## Purposes of Chapter 12

In Chapter 12 we examine (1) water as a solvent, (2) some reactions that occur in water, (3) concentrations of solutes, and (4) properties of solutions.

## Objectives for Chapter 12

After completion of this chapter, you should be able to:

1 Describe the conductivity properties and compositions of nonelectrolytes, strong electrolytes, and weak electrolytes in water solution. (12-1, 12-2)

2 Describe the process of solution of an ionic compound in water. (12-2)

3 Write equations illustrating the solution of various ionic compounds in water. (12-2)

4 Determine whether a specified ionic compound is soluble in water given a table of solubilities. (12-3)

5 Use a table of solubilities to predict metathesis reactions. (12-4)

6 Write balanced molecular, total ionic, and net ionic equations for specified metathesis reactions. (12-4)

7 Solve problems involving percent composition of a solute. (12-5)

8 Apply the definition of molarity to solve the following types of problems. (12-6)
   (a) Preparation of a specified quantity and concentration of a solution
   (b) Calculation of the quantity of solute in a given quantity of solution
   (c) Dilution of a concentrated solution to make a specified dilute solution
   (d) Calculation of stoichiometric yields involving solutions

9 Describe how the following properties of a solvent are affected by the presence of a solute. (12-7)
   (a) Vapor pressure
   (b) Boiling point
   (c) Melting point
   (d) Osmotic pressure

10 Use the molality of a solution to calculate the freezing point and the boiling point of an aqueous solution. (12-7)

In the previous chapter we discussed some of the properties of pure water. Actually, if we look about us, we rarely see pure water. Perhaps the closest example that we can observe is the water in rain and clouds, but even that water contains various amounts of dissolved gases. Water has a rather unique ability to become contaminated because it dissolves a wide variety of substances. The oceans, for example, contain so much dissolved substances, especially sodium chloride, that the water is unsuitable for drinking or irrigation of crops. The ability of water to dissolve substances is not all bad news by any means. Water in the blood carries dissolved oxygen and nutrients throughout the body and then carries dissolved wastes to organs to be removed. In fact, life itself is thought to have emerged from the ancient oceans that contained the dissolved ingredients.

Since water dissolves many substances, it is reasonable to expect that many of these dissolved compounds react chemically with each other. In this chapter we will see why water is such an effective solvent and thus serves as a medium in which chemical reactions occur. We will also examine how concentration is expressed and how the properties of a solvent are altered by the presence of a solute.

As background to this chapter, you should be familiar with the following topics.

1 Ionic compounds and how they are identified (Section 6-3)

2 The charges, formulas, and names of ions (Table 7-3)

3 Polarity (Sections 6-11 and 6-12)

4 Stoichiometry (Sections 9-3 and 10-11)

5 Water and hydrogen bonding (Sections 11-1 and 11-2)

## 12-1 The Conductivity of Aqueous Solutions

Water supports many chemical reactions because of its extensive ability to act as a solvent. A **solvent** *is a medium, usually a liquid, that dissolves or disperses another substance called a* **solute** *to form a homogeneous mixture of solute and solvent called a* **solution** (see Figure 12-1). The solution has the same physical state as the solvent.

One important property of aqueous solutions concerns their ability to act as conductors of electricity. **Electricity** *is simply a flow of negatively charged electrons through a substance called a* **conductor**. Metals such as copper are useful in electrical wires because they allow the flow of electricity along a length of continuous wire. Glass, on the other hand, does not allow the flow of electricity and is called a **nonconductor** or **insulator**. Pure water is also a nonconductor, as shown in Figure 12-2. In the figure, the wires A and B (called **electrodes**) are separated and immersed in the water. When the battery is connected, the light does not shine. This tells us that

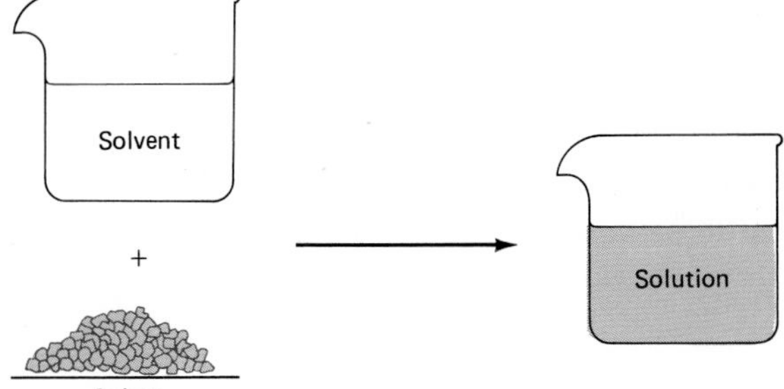


**Figure 12-1** A SOLUTION. A solution is a homogeneous mixture of solute and solvent that has the same physical state as the solvent.

the water prohibits the transfer of electrons between the two wires and the circuit is therefore broken. The presence of certain solutes such as sugar and alcohol have no effect on the ability of water to conduct electricity, since the light remains out when the wires are immersed in these solutions. *Compounds whose aqueous solutions do not conduct electricity are called* **nonelectrolytes**.

There is another class of compounds that have a profoundly different effect. When the electrodes are immersed in aqueous solutions of compounds such as sodium chloride or potassium hydroxide, the light shines brightly.

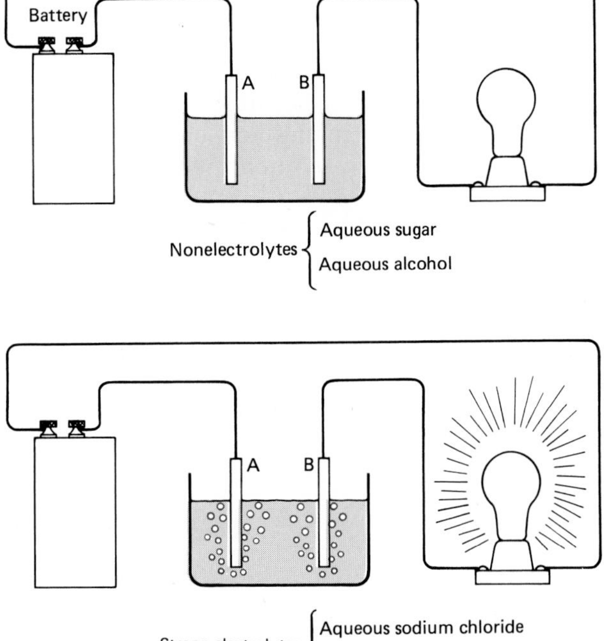


**Figure 12-2** ELECTROLYTES AND NONELECTROLYTES. Solutions of electrolytes conduct electricity, solutions of nonelectrolytes do not.

Obviously, the presence of these compounds allows the conduction of electricity between wire A and wire B. *Compounds whose aqueous solutions conduct electricity are called* **electrolytes**. *If the conduction of electricity is strong, the solute is considered a* **strong electrolyte**. When electrodes are immersed in aqueous solutions of certain other compounds, the light glows but very faintly. *Compounds whose aqueous solutions allow a very limited conduction of electricity are called* **weak electrolytes**. The behavior of weak electrolytes is highlighted in Chapter 13.

## 12-2 Water as a Solvent

In Chapter 11 we discussed the polar nature of the water molecule. Because of this polarity water molecules interact with each other in the liquid and solid states by means of hydrogen bonds. Hydrogen bonding is a particularly strong version of a dipole-dipole interaction. (See Figure 12-3.)

We are now ready to appreciate what occurs when solid ionic compounds are added to water. As illustrated in Figure 12-4, there is an electrostatic interaction between the negative dipoles of water molecules and the positive ions on the surface of the crystal. Likewise, the positive dipoles of water are attracted to the negative ions. *The electrostatic interactions between an ion and the dipoles of polar covalent molecules are known as* **ion-dipole forces**.

The ion-dipole forces attempt to lift the ions away from the crystal surface and suspend them in solution. This attempt is countered in what may be considered a tug-of-war by the ion-ion electrostatic interactions (known as the lattice energy) holding the crystal together. The actual quantitative values for these forces are known, but for now we can appreciate that if the forces are comparable the ions can be lifted from the crystal lattice by the water molecules and be held in the aqueous medium surrounded by an "escort" of water molecules. *In solution the ions are said to be* **hydrated**.

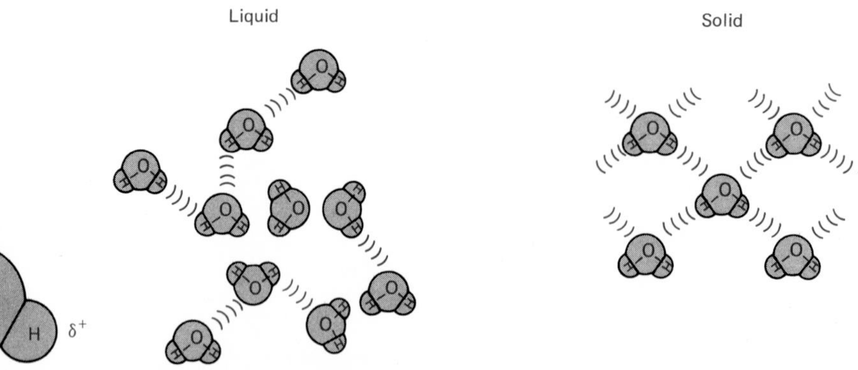

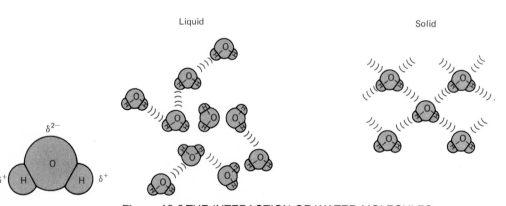

**Figure 12-3** THE INTERACTION OF WATER MOLECULES. Because of the polar nature of water molecules, hydrogen bonding occurs in both the liquid and solid states.

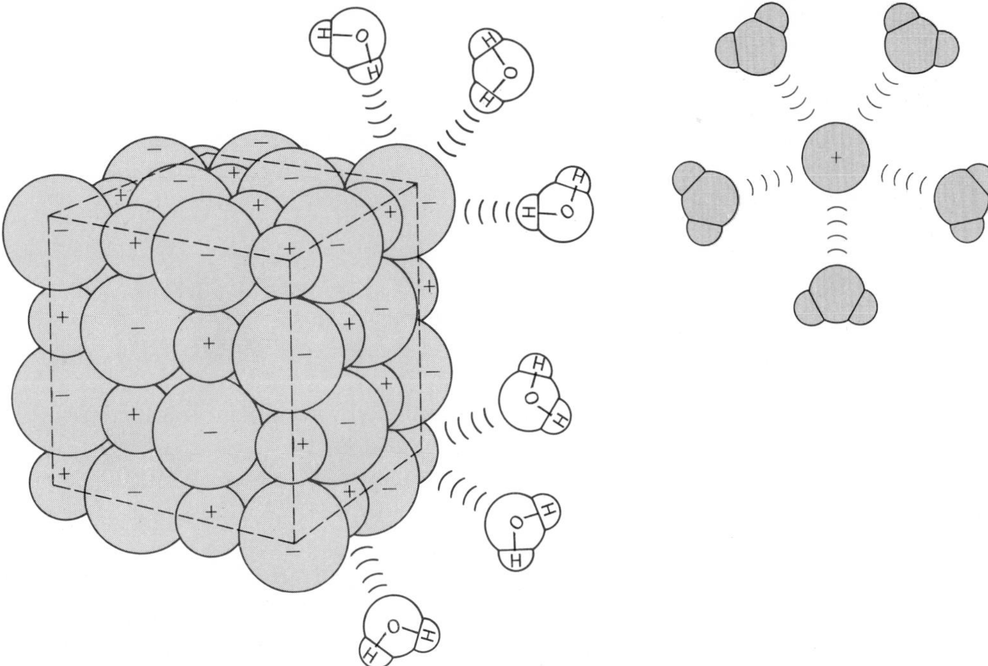

**Figure 12-4** THE INTERACTION OF WATER AND IONIC COMPOUNDS. There is an electrostatic interaction between the polar water molecules and the ions. This is a dipole–ion force.

By virtue of its polar nature, water can thus dissolve a wide variety of polar substances, especially ionic compounds. In some cases, however, the energy that holds a crystal together is strong enough to withstand the forces pulling the crystal apart and little solute dissolves (e.g., AgCl, $CaCO_3$). *Compounds that dissolve are said to be* **soluble,** *and those that do not dissolve appreciably are said to be* **insoluble.**

In summary, the solution process (see Figure 12-5) occurs because the attractive forces between the solute ions and the solvent water molecules (ion-dipole forces) are strong enough to overcome the forces holding the solute together (ion-ion forces) as well as the forces between the water molecules (dipole-dipole forces).*

As an analogy to these interactions, there is the well-known phenomenon of the interaction of a group of young men and a group of young women. The men interact together (boy-boy forces) quite well and are happy, as do the ladies (girl-girl forces). When the two groups come together, however, the boy-girl forces are usually strong enough to overcome the respective

*There are other factors involved in the solution process that can be important, but generally these forces determine the solubility of a compound in water.

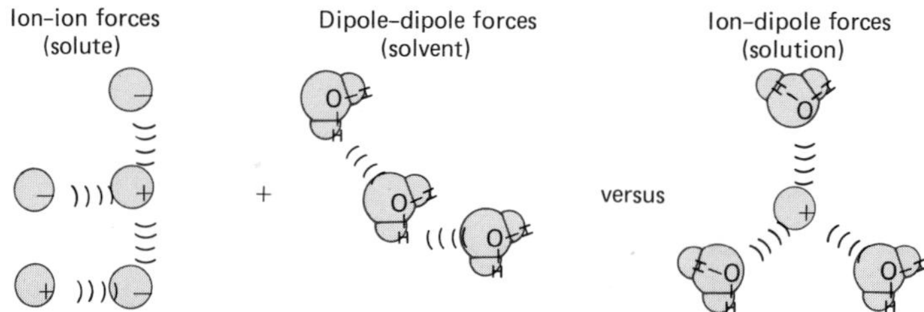


**Figure 12-5** FORCES IN THE SOLUTION PROCESS. When an ionic compound dissolves in water, the ion–dipole forces are comparable to the ion–ion forces holding the crystal together.

boy-boy and girl-girl forces. As a result, a thorough mixing occurs. It is simply a competition of forces: the strongest wins.

The solution process of an ionic compound in aqueous solution can be represented by a chemical equation as follows:

$$NaCl(s) \xrightarrow{xH_2O} Na^+(aq) + Cl^-(aq)$$

The coefficient of $H_2O$, $x$, above the arrow is an undetermined, large number of molecules needed to surround or *hydrate* the ions. Its actual value is not important. The abbreviation $(aq)$ following the symbols for the ions indicates that the ion is in aqueous solution. This indicates that the ions are hydrated and that positive and negative ions are no longer associated with each other in the solution. The water molecules that surround the ion in several layers tend to diminish the effect of the charge by spreading it out.

**Example 12-1**

Write equations illustrating the solution of the following ionic compounds in water: (a) $Na_2CO_3$, (b) $CaCl_2$, (c) 2 mol of $K_3PO_4$, and (d) 3 mol of $(NH_4)_2SO_4$.

**Solution**

Review the charges and formulas of the ions discussed in Chapters 6 and 7. The ions present in water are the same as those present in the solid ionic compound.

(a) $Na_2CO_3(s) \xrightarrow{xH_2O} 2Na^+(aq) + CO_3^{2-}(aq)$

(b) $CaCl_2(s) \xrightarrow{xH_2O} Ca^{2+}(aq) + 2Cl^-(aq)$

(c) $2K_3PO_4(s) \xrightarrow{xH_2O} 6K^+(aq) + 2PO_4^{3-}(aq)$

(d) $3(NH_4)_2SO_4(s) \xrightarrow{xH_2O} 6NH_4^+(aq) + 3SO_4^{2-}(aq)$

**See Problems 12-1 through 12-3.**

The solution of ionic compounds in water form solutions that are electrolytes *because of the presence of ions*. Besides ionic compounds, many polar covalent compounds also dissolve in water. These compounds may form strong electrolytes, weak electrolytes, or nonelectrolytes depending on the degree of ion formation in the solution. This is illustrated by the following three examples.

1 $HCl(g) \xrightarrow{xH_2O} H^+(aq) + Cl^-(aq)$ HCl is a strong electrolyte, since essentially all HCl molecules break into ions in solution.

2 $HF(g) \xrightarrow{xH_2O} HF(aq)$

$HF(aq) \rightleftharpoons H^+(aq) + F^-(aq)$ HF is a weak electrolyte since its ionization is incomplete. It is a reversible reaction where only a small percentage of the dissolved HF molecules are dissociated into ions.

3 $CH_3OH(l) \xrightarrow{xH_2O} CH_3OH(aq)$ $CH_3OH$ (methyl alcohol) is a nonelectrolyte, since no ions are formed when the compound dissolves.

The solution of covalent compounds to form strong and weak electrolytes will be discussed in the next chapter. When covalent compounds dissolve to form nonelectrolytes, it is due to the formation of dipole–dipole forces between solute molecules and water molecules as shown in Figure 12-6.

Nonpolar compounds such as gasoline or carbon tetrachloride do not dissolve in water. Since there are no appreciable solute-solvent forces due to the lack of a charge or a dipole on the nonpolar solute molecules, there are no forces strong enough to overcome the dipole–dipole forces of the water molecules themselves. The water molecules prefer to stay together, which means that mixing or solution does not occur.

## 12-3 Solubility of Ionic Compounds

Each solid compound has a limit as to *the maximum amount of that compound that can dissolve in a given amount of solvent at a certain temperature. This quantity is known as the compound's* **solubility** at a certain temperature. Insoluble compounds such as AgCl have extremely low solubilities in water (but they do dissolve to a very small extent). Soluble compounds

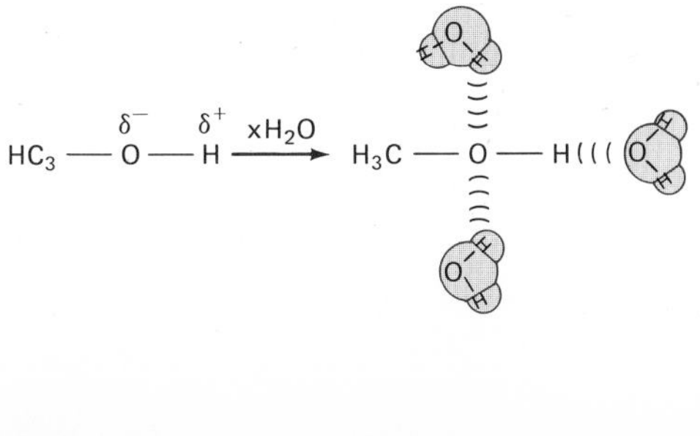

$HC_3 — \overset{\delta^-}{O} — \overset{\delta^+}{H} \xrightarrow{xH_2O} H_3C — O — H$

**Figure 12-6** METHYL ALCOHOL IN WATER. For a nonelectrolyte in water, there are dipole–dipole interactions between solute and water molecules.

such as table sugar (sucrose) have a relatively high degree of solubility. (If you have ever dumped a large amount of sugar in iced tea, you have noticed that even sugar has limits of solubility.) The solubility of a compound in water at a certain temperature can be listed as a definite numerical quantity, but the relative terms "soluble" and "insoluble" will suit our purposes here.

The solubilities of ionic compounds have been determined by experiments in the past. The results of these determinations may be summarized as in Table 12-1 as an aid in predicting whether a compound is soluble or insoluble. This particular table refers to compounds of some of the more common anions that are encountered. If a specified cation is not mentioned as forming an insoluble compound with a certain anion, it is assumed to be soluble. (e.g., $CuSO_4$ is soluble.)

The following examples illustrate the use of Table 12-1 to determine whether an ionic compound is soluble or insoluble in water.

**Example 12-2**

Use Table 12-1 to predict whether the following compounds are soluble or insoluble: (a) NaI, (b) CdS, (c) $Ba(NO_3)_2$, (d) $SrSO_4$.

(a) According to Table 12-1, all alkali metal (IA) compounds of the ions listed are *soluble*. Therefore, NaI is soluble.

(b) All $S^{2-}$ compounds are insoluble except those formed with IA and IIA metals and $NH_4^+$. Since Cd is in group IIB, CdS is insoluble.

(c) All $NO_3^-$ compounds are soluble. Therefore, $Ba(NO_3)_2$ is soluble.

(d) The $Sr^{2+}$ ion forms an insoluble compound with $SO_4^{2-}$. Therefore, $SrSO_4$ is insoluble.

**See Problems 12-4 and 12-5.**

**Table 12-1** Solubility Rules for Some Ionic Compounds

| Anion | Cations Forming Insoluble Compounds | Slightly Soluble Compounds |
|---|---|---|
| $Cl^-$, $Br^-$, $I^-$ | $Ag^+$, $Hg_2^{2+}$, $Pb^{2+}$ | $PbCl_2$, $PbBr_2$ |
| $NO_3^-$, $ClO_3^-$, $ClO_4^-$, $C_2H_3O_2^-$ | None | $AgC_2H_3O_2$ |
| $SO_4^{2-}$ | $Pb^{2+}$, $Ba^{2+}$, $Sr^{2+}$ | $CaSO_4$, $Ag_2SO_4$ |
| $CO_3^{2-}$, $SO_3^{2-}$, $PO_4^{3-}$ | All except IA metals and $NH_4^+$ | |
| $S^{2-}$ | All except IA metals, IIA metals, and $NH_4^+$ | |
| $OH^-$ | All except IA metals, $Ba^{2+}$, $Sr^{2+}$, and $NH_4^+$ | $Ca(OH)_2$ |
| $O^{2-}$ | All except IA metals, $Ca^{2+}$, $Sr^{2+}$, and $Ba^{2+}$ | |

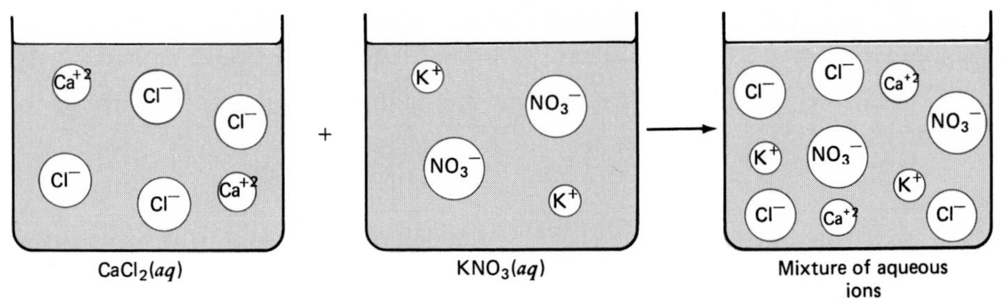


**Figure 12-7** A MIXTURE OF $CaCl_2$ AND $KNO_3$ SOLUTIONS. No reaction occurs when these solutions are mixed.

## 12-4 Ionic Equations and Metathesis Reactions

When an ionic compound dissolves in water, the cation and the anion behave independently. What happens if two different ionic compounds, such as $CaCl_2$ and $KNO_3$, are dissolved in water in separate beakers and then mixed? In this particular case, nothing happens. We simply have a solution containing four different ions as illustrated in Figure 12-7. No cation is associated with a particular anion.

In other cases when solutions of ionic compounds are mixed, a chemical reaction occurs. Although there are several types of reactions that may occur, we are now concerned with a reaction that leads to the formation of a **precipitate** (*an insoluble compound*). To understand this, remember that certain ionic compounds *do not* dissolve to any appreciable extent in water because of the strong interaction between the cation and the anion. Such a compound is AgCl (silver chloride). In fact, *any time $Ag^+$ ions and $Cl^-$ ions are present in the same solution, they are strongly attracted to each other and combine to form solid AgCl.* Thus if solutions of NaCl and $AgNO_3$ are mixed as shown in Figure 12-8, a white, solid precipitate of AgCl is formed, leaving the soluble compound, $NaNO_3$, in solution.

*The formation of a precipitate from mixing solutions of two soluble compounds is known as a* **metathesis reaction.** *It is one of two types of double displacement reactions.* The other is commonly referred to as a neutralization reaction and is discussed in Chapter 13. The example illustrated in Figure 12-8 constitutes a method to prepare two new compounds from two others. The AgCl formed can be removed by filtration, and the $NaNO_3$ can then be recovered by boiling away the solvent water. The $NaNO_3$ remains behind

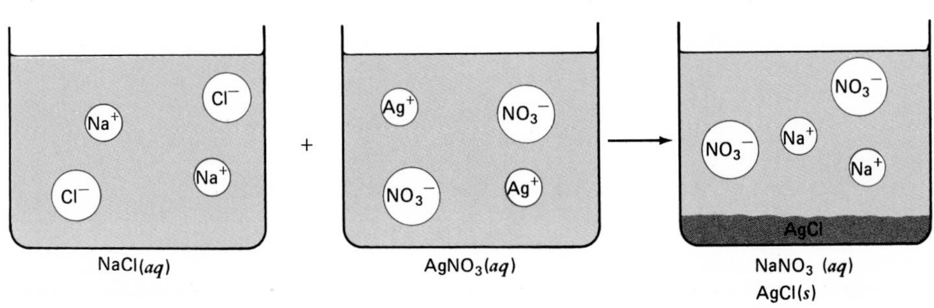


**Figure 12-8** A MIXTURE OF NaCl AND $AgNO_3$ SOLUTIONS. When these solutions are mixed, a precipitate forms.

after the water is boiled away because $NaNO_3$, an ionic compound, is nonvolatile.

The following equation represents the chemical reaction illustrated in Figure 12-8.

$$NaCl(aq) + AgNO_3(aq) \longrightarrow AgCl(s) + NaNO_3(aq)$$

This is the molecular* form of the equation. *In the* **molecular equation,** *all reactants and products are shown as compounds.*

*When all cations and anions in solution are written separately, the resulting equation is known as the* **total ionic equation**:

$$Na^+(aq) + Cl^-(aq) + Ag^+(aq) + NO_3^-(aq) \rightarrow AgCl(s) + Na^+(aq) + NO_3^-(aq)$$

In the total ionic equation, notice that the $Na^+(aq)$ and the $NO_3^-(aq)$ ions are written in an identical state on both sides of the equation. From algebra you are aware that when identical numbers or variables appear on both sides of an equation, they can be subtracted from both sides of the equation as follows:

$$\begin{array}{rcl} x + y &=& x + 17 + z \\ -x &=& -x \\ \hline y &=& 17 + z \end{array}$$

*In the same manner, identical substances that appear on both sides of a chemical equation can be subtracted out to produce the* **net ionic equation.** In our example, the net ionic equation is:

$$Ag^+(aq) + Cl^-(aq) \rightarrow AgCl(s)$$

*Ions that are identical on both sides of an equation are called* **spectator ions,** *since they are not involved directly with the chemical reaction.* The net ionic equation is important, since it focuses on the "main action" of what really occurs in solution.

Metathesis reactions can be predicted by examination of possible products of a reaction and reference to Table 12-1 to see if either of the possible products is insoluble. The following examples illustrate the use of Table 12-1 to predict metathesis reactions.

Write the balanced molecular equation, the total ionic equation, and the net ionic equation for any reaction that occurs when solutions of the following are mixed.

**Example 12-3**

A solution of $Na_2CO_3$ is mixed with a solution of $CaCl_2$.

*"Molecular" is a misnomer in this case, since no real molecules exist in ionic compounds.

**Solution**

To begin, write out the formulas of the possible products resulting from a metathesis reaction. In this case the possible products from the exchange of ions are NaCl and $CaCO_3$.

If both of these compounds are soluble, no reaction occurs. In this case, however, Table 12-1 tells us that $CaCO_3$ is insoluble. Thus, a reaction occurs as illustrated by the following balanced equation written in molecular form:

$$Na_2CO_3(aq) + CaCl_2(aq) \rightarrow CaCO_3(s) + 2NaCl(aq)$$

The equation written in total ionic form is

$$2Na^+(aq) + CO_3^{2-}(aq) + Ca^{2+}(aq) + 2Cl^-(aq) \rightarrow CaCO_3(s) + 2Na^+(aq) + 2Cl^-(aq)$$

Notice that the $Na^+$ and the $Cl^-$ ions are spectator ions. Elimination of the spectator ions on both sides of the equation leaves the net ionic equation

$$Ca^{2+}(aq) + CO_3^{2-}(aq) \rightarrow CaCO_3(s)$$

**Example 12-4**

A solution of KOH is mixed with a solution of $MgI_2$.

**Solution**

The possible metathesis reaction products are KI and $Mg(OH)_2$. The information in Table 12-1 indicates that $Mg(OH)_2$ is insoluble. The balanced molecular equation for this reaction is

$$2KOH(aq) + MgI_2(aq) \rightarrow Mg(OH)_2(s) + 2KI(aq)$$

The total ionic equation is

$$2K^+(aq) + 2OH^-(aq) + Mg^{2+}(aq) + 2I^-(aq) \rightarrow Mg(OH)_2(s) + 2K^+(aq) + 2I^-(aq)$$

Elimination of spectator ions gives the following net ionic equation:

$$Mg^{2+}(aq) + 2OH^-(aq) \rightarrow Mg(OH)_2(s)$$

**Example 12-5**

A solution of $KNO_3$ is mixed with a solution of $CaBr_2$.

**Solution**

The possible metathesis reaction products are KBr and $Ca(NO_3)_2$. Since these compounds are both soluble, no precipitate forms and a reaction does not occur.

**See Problems 12-6 through 12-13.**

## 12-5 Concentration: Percent by Weight

It is necessary at this point to introduce a method of expressing *the amount of a solute that is present in a given amount of solution. This is known as the* **concentration** *of the solution.* The chemist has another convenient but vague way of expressing concentration by terming the solution **dilute** *when not very much solute is present or* **concentrated** *when a lot of solute is present.* Obviously, a more quantitative method is necessary in order to work stoichiometry problems involving solutions. This is fundamental to any chemistry laboratory experience.

There are several ways of expressing concentration quantitatively: percent by weight, mole fraction, molarity, molality, and normality. All of these have their special use, but percent by weight and molarity are used in most laboratory situations and will be discussed here. Molality is used in connection with the physical properties of solutions and will be discussed in that section.

**Percent by weight,** *a rather straightforward method of expressing concentration, simply relates the weight of the solute as a percent of the total weight of the solution.* Therefore, in 100 g of a solution that is 25% by weight HCl there are 25 g of HCl and 75 g of $H_2O$. The formula for percent by weight is

$$\%\text{ by weight (solute)} = \frac{\text{weight of solute}}{\text{weight of solution}} \times 100\%$$

**Example 12-6**

What is the percent by weight of NaCl if 1.75 g of NaCl is dissolved in 5.85 g of $H_2O$?

**Procedure**

Find the total weight of the solution and then the percent of NaCl.

**Solution**

Total weight
1.75 g NaCl (solute)
5.85 g $H_2O$ (solvent)
7.60 g solution

$$\frac{1.75\text{ g NaCl}}{7.60\text{ g solution}} \times 100\% = \underline{\underline{23.0\%\text{ by weight NaCl}}}$$

**Example 12-7**

A solution is 14.0% by weight $H_2SO_4$. How many moles of $H_2SO_4$ are in 155 g of solution?

**Procedure**

1 Find the weight of $H_2SO_4$ in the solution.

2 Convert weight to moles.

**Solution**

1 Multiply the weight of compound by the percent in fraction form to find the weight of $H_2SO_4$.

$$155 \text{ g compound} \times \frac{14.0 \text{ g } H_2SO_4}{100 \text{ g compound}} = 21.7 \text{ g } H_2SO_4$$

2 The molar mass of $H_2SO_4$ is

$$2.0 \text{ g (H)} + 32.1 \text{ g (S)} + 64.0 \text{ g (O)} = 98.1 \text{ g}$$

$$21.7 \text{ g} \times \frac{1 \text{ mol}}{98.1 \text{ g}} = \underline{\underline{0.221 \text{ mol } H_2SO_4}}$$

**See Problems 12-14 through 12-19.**

## 12-6 Concentration: Molarity

Molarity ($M$) is defined as the number of moles of solute ($n$) per liter of solution ($V$):

$$M = \frac{n \text{ (moles of solute)}}{V \text{ (liters of solution)}}$$

Thus a 1.00 $M$ solution of HCl contains 1.00 mol (36.5 g) of HCl dissolved in enough $H_2O$ to make 1.00 L of solution.

**Example 12-8**

What is the molarity of $H_2SO_4$ in a solution made by dissolving 49.0 g of pure $H_2SO_4$ in enough water to make 250 mL of solution?

**Procedure**

Write down the formula for molarity, what you have been given, and then solve for what's requested.

**Solution**

$$M = \frac{n}{V}$$

$$(n) \qquad 49.0 \text{ g } H_2SO_4 \times \frac{1 \text{ mol}}{98.1 \text{ g } H_2SO_4} = 0.499 \text{ mol}$$

$$(V) \qquad 250\ \cancel{\text{mL}} \times \frac{10^{-3}\ \text{L}}{\cancel{\text{mL}}} = 0.250\ \text{L}$$

$$\frac{n}{V} = \frac{0.499\ \text{mol}}{0.250\ \text{L}} = \underline{\underline{2.00\ M}}$$

---

**Example 12-9**
What weight of HCl is present in 155 mL of a 0.540 $M$ solution?

$$M = \frac{n}{V} \qquad n = M \times V$$

$$M = 0.540\ \text{mol/L} \qquad V = 155\ \text{mL} = 0.155\ \text{L}$$

$$n = 0.540\ \text{mol}/\cancel{\text{L}} \times 0.155\ \cancel{\text{L}} = 0.0837\ \text{mol HCl}$$

$$0.0837\ \cancel{\text{mol HCl}} \times \frac{36.5\ \text{g}}{\cancel{\text{mol HCl}}} = \underline{\underline{3.06\ \text{g HCl}}}$$

---

**Example 12-10**
Concentrated laboratory acid is 35.0% by weight HCl and has a density of 1.18 g/mL. What is its molarity?

**Procedure**
Since a volume was not given, you can start with any volume you wish. The molarity will be the *same* for 1 mL as for 25 L. To make the problem as simple as possible, assume that you have exactly 1 L of solution ($V = 1.00$ L) and go from there. The number of moles of HCl ($n$) in 1 L can be obtained as follows:

1 Find the weight of 1 L from the density.

2 Find the weight of HCl in 1 L using the percent by weight and the weight of 1 L.

3 Convert the weight of HCl to moles of HCl.

**Solution**
Assume that $V = 1.00$ L

1 The weight of 1.00 L ($10^3$ mL) is

$$10^3\ \cancel{\text{mL}} \times 1.18\ \text{g}/\cancel{\text{mL}} = 1180\ \text{g solution}$$

2 The weight of HCl in 1.00 L is

$$1180\ \cancel{\text{g solution}} \times \frac{35.0\ \text{g HCl}}{100\ \cancel{\text{g solution}}} = 413\ \text{g HCl}$$

**3** The number of moles of HCl in 1.00 L is

$$413 \text{ g} \times \frac{1 \text{ mol}}{36.5 \text{ g}} = 11.3 \text{ mol HCl}$$

$$\frac{n}{V} = \frac{11.3 \text{ mol}}{1.00 \text{ L}} = \underline{\underline{11.3\ M}}$$

**See Problems 12-20 through 12-29.**

One of the common laboratory exercises for the experienced as well as the beginning chemist is to make a certain volume of a *dilute* solution from a *concentrated* solution. In this case a certain amount (number of moles) of solute is withdrawn from the concentrated solution and diluted with water to form the dilute solution. The quantity of moles needed for the dilute solution is designated $n_d$. It is calculated from the molarity and volume for a specified dilute solution as follows:

$$M_d \times V_d = n_d$$

The moles of solute taken from the concentrated solution is designated $n_c$. The moles of solute in the concentrated solution relate to the volume and molarity of the concentrated solution as follows:

$$M_c \times V_c = n_c$$

Since the moles of solute used in the dilute solution is the same as that taken from the concentrated solution, moles of solute remain constant, and we can write

$$\text{Moles solute} = n_d = n_c$$

Since quantities equal to the same quantity are equal to each other, we have the simple relationship between the dilute solution and the volume and molarity of the concentrated solution that is used.

$$M_c \times V_c = M_d \times V_d$$

**Example 12-11**

What volume of 11.3 *M* HCl is needed to mix with water to make 1.00 L of 0.555 *M* HCl? (See also Figure 12-9.)

**Procedure**

$$M_c \times V_c = M_d \times V_d$$

$$V_c = \frac{M_d \times V_d}{M_c}$$

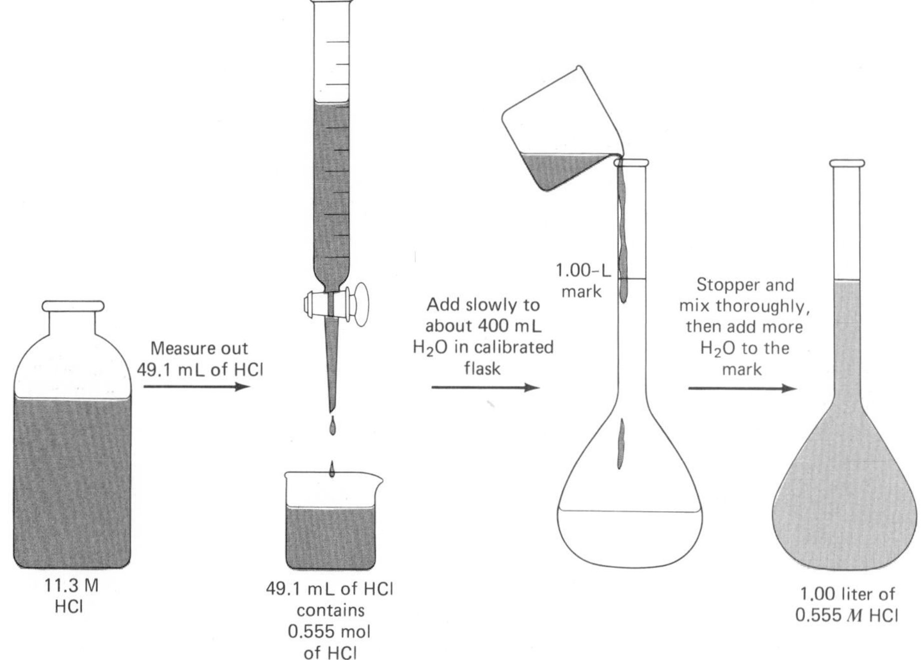
Measure out
49.1 mL of HCl
Add slowly to
about 400 mL
H2O in calibrated
flask
1.00-L
mark
Stopper and
mix thoroughly,
then add more
H2O to the
mark
11.3 M
HCl
49.1 mL of HCl
contains
0.555 mol
of HCl
1.00 liter of
0.555 M HCl

**Figure 12-9** DILUTION OF CONCENTRATED HCl. (*Note*: Water is never added directly to concentrated acid, because it may splatter and cause severe burns.)

**Solution**

$$V_c = \frac{0.555 \cancel{\text{mol/L}} \times 1.00 \text{ L}}{11.3 \cancel{\text{mol/L}}} = 0.0491 \text{ L} = \underline{\underline{49.1 \text{ mL}}}$$

**Example 12-12**

What is the molarity of a solution of KCl that is prepared by dilution of 855 mL of a 0.475 $M$ solution to a volume of 1.25 L?

**Procedure**

$$M_c \times V_c = M_d \times V_d$$

$$M_d = \frac{M_c \times V_c}{V_d}$$

**Solution**

$$M_c = 0.475\ M \qquad M_d = ?$$

$$V_c = 855 \text{ mL} = 0.855 \text{ L} \qquad V_d = 1.25 \text{ L}$$

$$M_d = \frac{0.475 \text{ M} \times 0.855 \cancel{\text{L}}}{1.25 \cancel{\text{L}}} = \underline{\underline{0.325\ M}}$$

See Problems 12-30 through 12-38.

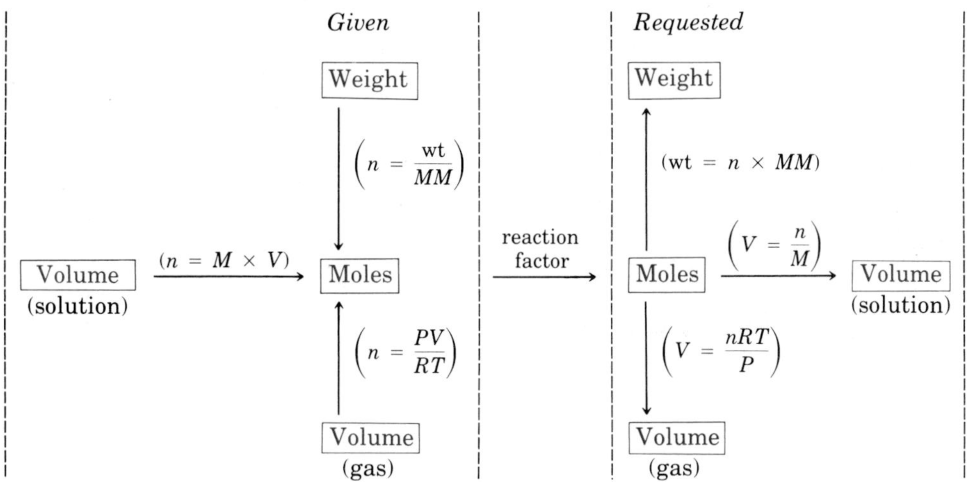


**Figure 12-10** GENERAL PROCEDURE FOR STOICHIOMETRY.

## 12-7 Stoichiometry Involving Solutions

In Chapter 9 a general procedure was presented for working stoichiometry problems. This was expanded in Chapter 10 to include gases. We can now further expand the general procedure to include stoichiometry in aqueous solution, since moles relate to volume of solution by molarity (see Figure 12-10).

The following examples illustrate the role of solutions in stoichiometry.

**Example 12-13**

Given the following balanced equation:

$$3NaOH(aq) + H_3PO_4(aq) \rightarrow Na_3PO_4(aq) + 3H_2O(l)$$

What volume of 0.250 $M$ NaOH is required to react completely with 4.90 g of $H_3PO_4$?

**Procedure**

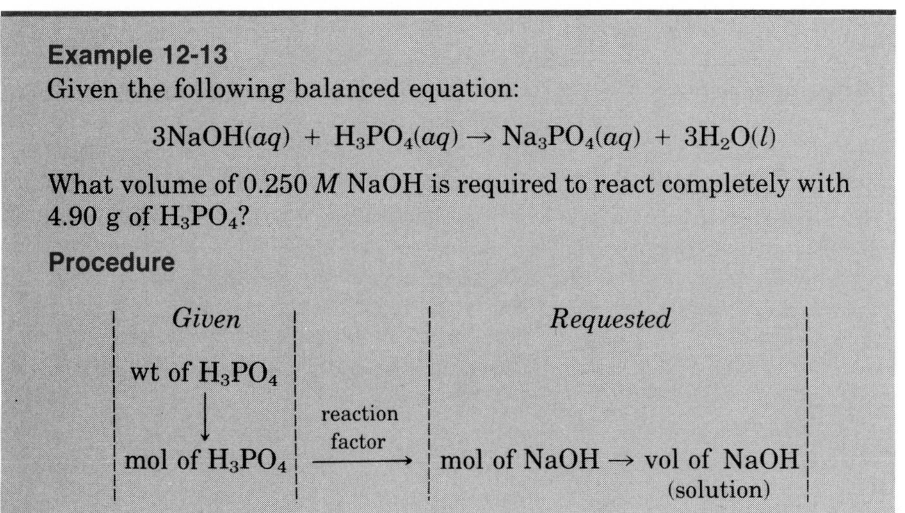

**Solution**

$$4.90 \text{ g } H_3PO_4 \times \frac{1 \text{ mol } H_3PO_4}{98.0 \text{ g } H_3PO_4} \times \frac{3 \text{ mol NaOH}}{1 \text{ mol } H_3PO_4} = 0.150 \text{ mol NaOH}$$

$$\text{vol NaOH} = \frac{\text{mol NaOH}}{M} = \frac{0.150 \text{ mol}}{0.250 \text{ mol/L}} = \underline{\underline{0.600 \text{ L}}}$$

**Example 12-14**

Given the following balanced equation:

$$Cd(NO_3)_2(aq) + K_2S(aq) \rightarrow CdS(s) + 2KNO_3(aq)$$

What weight of CdS would be produced from 15.8 mL of a 0.122 $M$ $Cd(NO_3)_2$ solution with excess of $K_2S$ present?

**Procedure**

| *Given* | reaction factor | *Requested* |
|---|---|---|
| vol of $Cd(NO_3)_2$ (solution) → mol of $Cd(NO_3)_2$ | ⟶ | mol of CdS ↓ wt of CdS |

**Solution**

$$\text{mol of } Cd(NO_3)_2 = V \times M$$
$$= 0.0158 \text{ L} \times 0.122 \text{ mol/L}$$
$$= 0.00193 \text{ mol } Cd(NO_3)_2$$

$$0.00193 \text{ mol } Cd(NO_3)_2 \times \frac{1 \text{ mol CdS}}{1 \text{ mol } Cd(NO_3)_2} \times \frac{144 \text{ g CdS}}{\text{mol CdS}} = \underline{\underline{0.278 \text{ g CdS}}}$$

**Example 12-15**

Given the following balanced equation:

$$2KCl(aq) + Pb(NO_3)_2(aq) \rightarrow PbCl_2(s) + 2KCl(aq)$$

What volume of 0.200 $M$ KCl is needed to react completely with 185 mL of 0.245 $M$ $Pb(NO_3)_2$ solution?

**Procedure**

| *Given* | reaction factor | *Requested* |
|---|---|---|
| vol of $Pb(NO_3)_2$ (solution) → mol of $Pb(NO_3)_2$ | ⟶ | mol of KCl → vol of KCl (solution) |

**Solution**

$$\text{mol Pb(NO}_3)_2 = V[\text{Pb(NO}_3)_2] \times M[\text{Pb(NO}_3)_2]$$

$$0.185\ \cancel{\text{L}} \times 0.245\ \text{mol}/\cancel{\text{L}} = 0.0453\ \text{mol Pb(NO}_3)_2$$

$$0.0453\ \cancel{\text{mol Pb(NO}_3)_2} \times \frac{2\ \text{mol KCl}}{1\ \cancel{\text{mol Pb(NO}_3)_2}} = 0.0906\ \text{mol KCl}$$

$$V\ (\text{KCl}) = \frac{n(\text{KCl})}{M(\text{KCl})} = \frac{0.0906\ \cancel{\text{mol}}}{0.200\ \cancel{\text{mol}}/\text{L}} = 0.453\ \text{L}$$

$$0.453\ \cancel{\text{L}} \times \frac{1\ \text{mL}}{10^{-3}\ \cancel{\text{L}}} = \underline{\underline{453\ \text{mL}}}$$

**Example 12-16**

Given the following balanced equation:

$$2\text{HCl}(aq) + \text{K}_2\text{S}(aq) \rightarrow \text{H}_2\text{S}(g) + 2\text{KCl}(aq)$$

What volume of $H_2S$ measured at STP would be evolved from 1.65 L of a 0.552 *M* HCl solution with excess $K_2S$ present?

**Procedure**

| *Given* | reaction factor | *Requested* |
|---|---|---|
| vol of HCl → mol of HCl (solution) | ⟶ | mol of $H_2S$ ↓ vol of $H_2S$ (gas) |

**Solution**

$$V\ (\text{solution}) \times M(\text{HCl}) = n(\text{HCl})$$

$$1.65\ \cancel{\text{L}} \times 0.552\ \text{mol}/\cancel{\text{L}} = 0.911\ \text{mol HCl}$$

(Since the volume of the gas is at STP, the molar volume relationship can be used rather than the ideal gas law.)

$$0.911\ \cancel{\text{mol HCl}} \times \frac{1\ \cancel{\text{mol H}_2\text{S}}}{2\ \cancel{\text{mol HCl}}} \times \frac{22.4\ \text{L (STP)}}{\cancel{\text{mol H}_2\text{S}}} = \underline{\underline{10.2\ \text{L (STP)}}}$$

**See Problems 12-39 through 12-47.**

## 12-8 The Physical Properties of Solutions

One of the first phenomena discussed in this text was the difference in the physical properties between pure substances and mixtures. Pure substances, such as water, have distinct and unvarying melting points and boiling points,

Figure 12-11 VAPOR PRESSURE LOWERING. A nonvolatile solute reduces the number of solvent molecules escaping to the vapor.

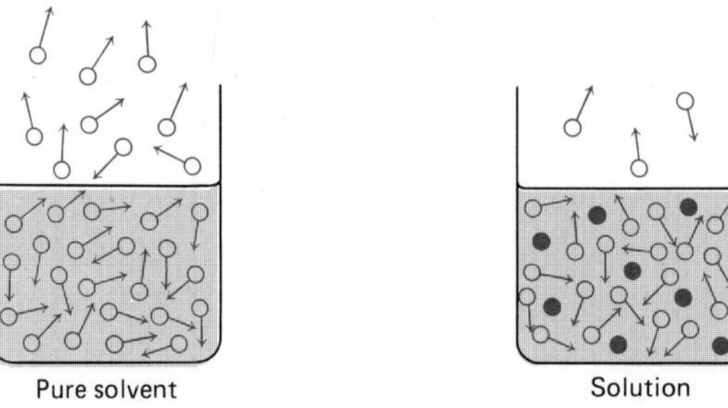


while mixtures, such as an aqueous solution, melt and boil over a temperature range and at temperatures different from that of the pure solvent.

There are four properties of a pure solvent that are affected by the addition of a solute to form the homogeneous mixture we know of as a solution: (1) vapor pressure, (2) boiling point, (3) freezing point, and (4) osmotic pressure. These will be discussed individually.

1 *Vapor pressure. The presence of a nonvolatile solute* in a solvent lowers the equilibrium vapor pressure of the pure solvent.* (See Section 11-7.) A simple explanation of why this happens is illustrated in Figure 12-11. In the solution a certain proportion of the solute particles are near the surface. They do no escape to the vapor and also block off access to the surface to some of the solvent molecules. Thus fewer of the solvent molecules can escape to the vapor, which means that the solution has a lower equilibrium vapor pressure than the pure solvent. As we might predict from this model, the more solute particles present, the greater the effect of vapor pressure lowering. *The effects of a solute on the vapor pressure, boiling point, melting point, and osmotic pressure are known as* **colligative properties.** Colligative properties depend only on the relative amounts of solute and solvent present and not on the identity of the solute.

An example of the effect of vapor pressure lowering is provided by the Dead Sea in Israel. This large body of water has no outlet to the ocean, so dissolved substances have accumulated, forming a very concentrated solution. Even though it exists in an arid region with high summer temperatures, it evaporates very slowly compared to a fresh water lake or even the ocean. It evaporates slowly because of its reduced vapor pressure. If the water evaporated from the Dead Sea at the same rate as fresh water, it would nearly dry up in a matter of a few years.

2 *Boiling point. A direct effect of the lowered vapor pressure of a solution is a higher boiling point compared with the solvent.* Since a liquid boils at the temperature at which its vapor pressure equals the atmospheric pressure, a higher temperature is necessary to cause a solution to reach the same vapor pressure as the solvent.

*A nonvolatile solute is one that has essentially no vapor pressure at the relevant temperatures.

The amount of boiling point elevation is given by the following equation:

$$\Delta T_b = K_b m$$

where

$\Delta T_b$ = the number of Celsius degrees that the boiling point is raised.

$K_b$ = a constant characteristic of the solvent. For water $K_b$ = 0.512 °C·kg/mol. Other values of $K_b$ are given for particular solvents in the exercises.

$m$ = a concentration unit called molality. Molality emphasizes a relationship between moles of solute and weight of solvent rather than the volume, as does molarity. The definition of molality is:

$$\text{Molality } (m) = \frac{\text{moles of solute}}{\text{kg of solvent}}$$

The calculation of molality and its use will be illustrated shortly.

3 *Freezing point. Just as the boiling point of a solution is higher than the pure solvent, the freezing point is lower.* This effect has many applications. When water freezes, it expands, and it will likely break its container. This would occur in the automobile radiator and engine block. Antifreeze (ethylene glycol) is used to lower the freezing point of water in the radiator and prevent freezing and the damage it would cause. (It also raises the boiling point, preventing the fluid from boiling away in the summer.) The amount of freezing point lowering is given by the equation:

$$\Delta T_f = K_f m$$

where

$\Delta T_f$ = the number of Celsius degrees that the freezing point is lowered.

$K_f$ = a constant characteristic of the solvent. For water $K_b$ = 1.86 °C·kg/mol.

$m$ = molality of the solution.

Examples of these calculations are as follows:

**Example 12-17**

What is the molality of methyl alcohol in a solution made by dissolving 18.5 g of methyl alcohol ($CH_3OH$) in 850 g of water?

**Solution**

$$\text{Molality} = \frac{\text{mol solute}}{\text{kg solvent}} \qquad \text{mol solute} = \frac{18.5 \not{g}}{32.0 \not{g}/\text{mol}} = 0.578 \text{ mol}$$

$$\text{kg solvent} = \frac{850 \not{g}}{10^3 \not{g}/\text{kg}} = 0.850 \text{ kg}$$

$$\text{Molality} = \frac{0.578\ \text{mol}}{0.850\ \text{kg}} = \underline{\underline{0.680\ \text{m}}}$$

**Example 12-18**
What is the boiling point of an aqueous solution containing 468 g of sucrose ($C_{12}H_{22}O_{11}$) in 350 g of water?

**Procedure**

$$\Delta T_b = K_b m = 0.512\ °\text{C}\cdot\text{kg/mol} \times \frac{\text{mol solute}}{\text{kg solvent}}$$

**Solution**

$$\text{mol solute} = \frac{468\ \cancel{\text{g}}}{342\ \cancel{\text{g}}/\text{mol}} = 1.37\ \text{mol}$$

$$\text{kg solvent} = \frac{350\ \cancel{\text{g}}}{10^3\cancel{\text{g}}/\text{kg}} = 0.350\ \text{kg}$$

$$\Delta T_b = 0.512\ °\text{C}\cdot\cancel{\text{kg}}/\cancel{\text{mol}} \times \frac{1.37\ \cancel{\text{mol}}}{0.350\ \cancel{\text{kg}}} = 2.0\ °\text{C}$$

The normal boiling point of water = 100.0 °C
The boiling point of the solution = 100.0 °C + $\Delta T_b$ = 100.0 + 2.0
= $\underline{\underline{102.0\ °\text{C}}}$

**Example 12-19**
What is the freezing point of the solution in Example 12-18?

**Procedure**

$$\Delta T_f = K_f m = 1.86\ °\text{C}\cdot\text{kg/mol} \times \frac{\text{mol solute}}{\text{kg solvent}}$$

**Solution**

$$\Delta T_f = 1.86\ °\text{C}\cdot\cancel{\text{kg}}/\cancel{\text{mol}} \times \frac{1.37\ \cancel{\text{mol}}}{0.350\ \cancel{\text{kg}}} = 7.3\ °\text{C}$$

The freezing point of water = 0.0 °C
The freezing point of the solution = 0.0 °C − $\Delta T_f$ = 0.0 °C − 7.3 °C
= $\underline{\underline{-7.3\ °\text{C}}}$

Notice that the liquid range was extended by over 9 °C for the solution.

**4** *Osmotic pressure* An important phenomenon in the life process is known as osmosis. **Osmosis** *is the tendency for a solvent to move through a thin porous membrane from a dilute solution to a more concentrated solution.* The membrane is said to be **semipermeable**, which means small solvent molecules can pass through but solute species cannot. Figure 12-12 illustrates osmosis. On the right there is a pure solvent, and on the left a solution. The two are separated by a semipermeable membrane. Solvent molecules can pass through the membrane in both directions, but the rate at which they diffuse to the right is lower because solute particles block some of the pores of the membrane. As a result, the water level rises on the left and drops on the right. This creates an increased pressure on the left, which eventually counteracts the osmosis, and equilibrium is established. *The extra pressure required to establish this equilibrium is known as the* **osmotic pressure**. Like other colligative properties, it is dependent on the concentration of the solute.

We see an example of the osmosis process whenever we leave our hands in a soapy water or salt water solution. The movement of water molecules from the cells of our skin to the more concentrated solution causes them to be wrinkled. Pickles are wrinkled because the cells of the cucumber have been dehydrated by the salty brine solution. In fact, brine solutions preserve many foods because the concentrated solution of salt removes water from the cells of bacteria thus killing the bacteria. Trees and plants obtain water from the absorption of water through the semipermeable membranes in their roots into the more concentrated solution inside the root cells. Osmosis has many important applications in addition to life processes. In Figure 12-12, if pressure greater than the osmotic pressure is applied on the left, reverse osmosis takes place and solvent molecules move from the solution to the pure solvent. This process is used in desalination plants that convert sea water (a solution) to drinkable water. This is important in areas of the world such as the middle east, where there is a shortage of fresh water.

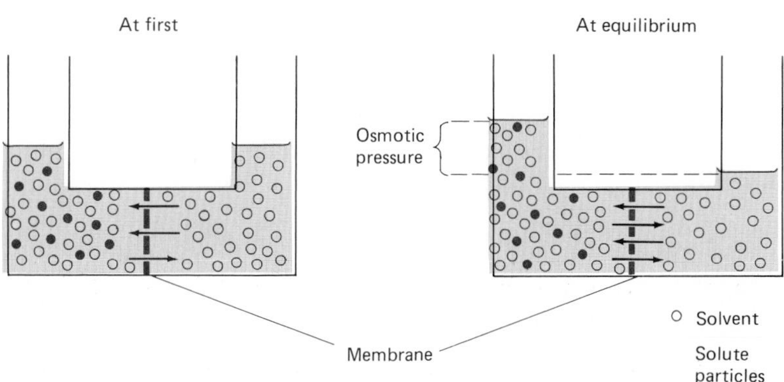


**Figure 12-12** OSMOTIC PRESSURE. Osmosis causes dilution of the more concentrated solution.

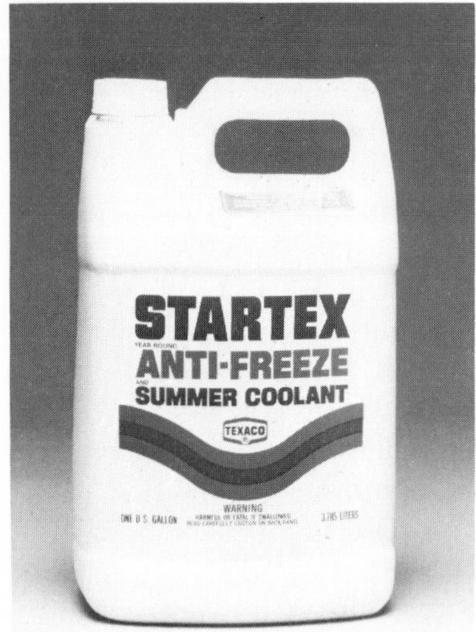

**Figure 12-13** COLLIGATIVE PROPERTIES. Antifreeze not only lowers the freezing point but raises the boiling point of the engine coolant. This salt lake does not dry up because of a low vapor pressure.

Electrolytes have a more pronounced effect on colligative properties than do nonelectrolytes. For example, one mole of NaCl dissolves in water to produce two moles of particles, one mole of $Na^+$ and one mole of $Cl^-$.

$$\mathrm{NaCl}(s) \xrightarrow{\mathrm{H_2O}} \mathrm{Na}^+(aq) + \mathrm{Cl}^-(aq)$$

Thus one mole of NaCl lowers the freezing point approximately twice as much as one mole of a nonelectrolyte. This effect is put to good use in the U.S. snow belt where sodium chloride is spread on snow and ice to cause melting even though the temperature is below freezing. Even more effective in melting ice is calcium chloride ($CaCl_2$). This compound produces three moles of ions ($Ca^{2+} + 2Cl^-$) per mole of solute and is therefore three times as effective per mole in lowering the freezing point as a nonelectrolyte. Calcium chloride is occasionally used on roads when the temperature is too low for sodium chloride to be effective. Electrolytes would be very effective as an antifeeze for the water in a radiator, but, unfortunately, their presence in water causes severe corrosion of most metals (see Figure 12-13).

**See Problems 12-48 through 12-68.**

## 12-9 Chapter Summary

The most common and inexpensive chemical known serves as the most effective medium for many chemical reactions. This is because of the rather extensive ability of water to act as a solvent. Solvents and solutes mix homogeneously to form solutions.

Water itself is essentially a nonconductor of electricity. The presence of certain solutes may or may not affect this property, however. A dissolved substance may be classified as a nonelectrolyte, strong electrolyte, or weak electrolyte as follows:

| Type of Solute | Property in Water | Reason | Examples |
|---|---|---|---|
| Nonelectrolyte | Solution is a nonconductor of electricity. | Ions are not formed in solution. | $C_{12}H_{22}O_{11}$ (sugar) $CH_3OH$ (methyl alcohol) |
| Strong electrolyte | Solution is a good conductor of electricity. | Ions are formed in solution. | NaCl $K_2SO_4$ |
| Weak electrolyte | Solution is a weak conductor of electricity. | Limited amount of ions are formed in solution. | HF $HNO_2$ |

Water is most effective in dissolving ionic or polar covalent compounds. For soluble ionic compounds, the dipole-ion interactions between solute and solvent are strong enough to overcome ion-ion forces holding the ionic crystal together. If the ion-ion forces are too strong to be overcome, the ionic compound remains undissolved and is said to be insoluble. Many polar covalent compounds also dissolve in water to form either nonelectrolytes, strong electrolytes, or weak electrolytes.

The trends in solubility of some common ionic compounds can be classified in a solubility table. This table (Table 12-1) is useful in predicting a type of double displacement reaction called a metathesis reaction. These reactions can be represented by equations in three ways as follows:

Molecular: $AB(aq) + CD(aq) \rightarrow AD(s) + CB(aq)$

Total ionic: $A^+(aq) + B^-(aq) + C^+(aq) + D^-(aq) \rightarrow AD(s) + C^+(aq) + D^-(aq)$

Net ionic: $A^+(aq) + D^-(aq) \rightarrow AD(s)$

Notice that the spectator ions ($C^+$ and $B^-$) are eliminated from the total ionic equation to form the net ionic equation. The net ionic equation highlights the driving force of the reaction.

Three concentration units were used in this chapter: percent by weight, molarity, and molality. Their definitions are as follows:

| Percent by weight | Molarity | Molality |
|---|---|---|
| $\frac{\text{weight of solute}}{\text{weight of solvent}} \times 100\%$ | $M = \frac{\text{mol solute}}{\text{L solution}}$ | $m = \frac{\text{mol solute}}{\text{kg solvent}}$ |

Since molarity relates volume of a solution to moles of solute, it can be incorporated into the general scheme for stoichiometry problems along with weight of a compound (Chapter 9) and volume of a gas (Chapter 10) as follows:

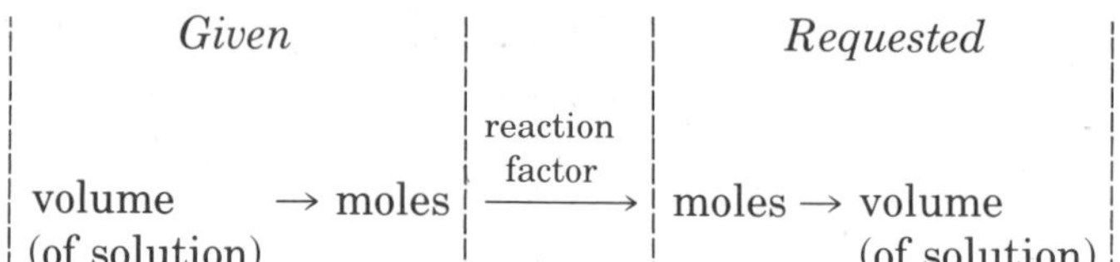


In the final section, the effect of a nonvolatile solute on the properties of a solvent was discussed. Since it is found that these effects relate directly to the number of particles in a given amount of solvent, these effects are known as colligative properties. The four major properties of a solvent that are altered by the presence of a solute are summarized as follows:

| Property | Effect | Result |
|---|---|---|
| Vapor pressure | Lowered | Solutions evaporate slower than pure solvents. |
| Boiling point | Raised | Solutions boil at higher temperatures than pure solvents. |
| Melting point | Lowered | Solutions freeze at lower temperatures than pure solvents. |
| Osmotic pressure | Raised | Solvent from dilute solutions diffuse through a semipermeable membrane into concentrated solutions. |

## Exercises

### Solutions of Electrolytes

**12-1** Three hypothetical binary compounds dissolve in water. The compound AB is a strong electrolyte, AC is a weak electrolyte, and AD is a nonelectrolyte. Describe the extent to which each of these solutions conducts electricity and how each compound exists in solution.

12-2 Describe the process whereby the compound $Na_2S$ dissolves in water and what ions are present in the solution.

**12-3** Write equations illustrating the solution of each of the following strong electrolytes in water.

(a) $K_2Cr_2O_7$ (b) $Li_2SO_4$ (c) $Cs_2SO_3$ (d) $Ca(ClO_3)_2$ (e) 2 mol of $(NH_4)_2S$ (f) 4 mol of $Ba(OH)_2$

### Solubility of Ionic Compounds and Methathesis Reactions

**12-4** Referring to Table 12-1, determine which of the following compounds are insoluble in water.

(a) $Na_2S$ (b) $PbSO_4$ (c) $MgSO_3$ (d) $Ag_2O$ (e) $(NH_4)_2S$ (f) $HgI_2$

12-5 Referring to Table 12-1, determine which of the following compounds are insoluble in water.

(a) NiS (b) $Hg_2Br_2$ (c) $Al(OH)_3$ (d) $Rb_2SO_4$ (e) CaS (f) $BaCO_3$

**12-6** Write the balanced molecular equation for any reaction that occurs when the following solutions are mixed.
(a) KI and $Pb(C_2H_3O_2)_2$
(b) $AgClO_3$ and $KNO_3$
(c) $Sr(ClO_3)_2$ and $Ba(OH)_2$
(d) BaS and $Hg_2(NO_3)_2$
(e) $FeCl_3$ and KOH

**12-7** Write the balanced molecular equation for any reaction that occurs when the following solutions are mixed.
(a) $Ba(C_2H_3O_2)_2$ and $Na_2SO_4$
(b) $NaClO_4$ and $Pb(NO_3)_2$
(c) $Mg(NO_3)_2$ and $Na_3PO_4$
(d) SrS and $NiI_2$

**12-8** Write the total ionic and the net ionic equation for each of the following reactions.
(a) $K_2S(aq) + Pb(NO_3)_2(aq) \rightarrow PbS(s) + 2KNO_3(aq)$
(b) $(NH_4)_2CO_3(aq) + CaCl_2(aq) \rightarrow CaCO_3(s) + 2NH_4Cl(aq)$
(c) $2AgClO_4(aq) + Na_2CrO_4(aq) \rightarrow Ag_2CrO_4(s) + 2NaClO_4(aq)$

**12-9** Write the total ionic equation for any reaction that occurred in Problem 12-6.

**12-10** Write the total ionic equation for any reaction that occurred in Problem 12-7.

**12-11** Write the net ionic equation for any reaction that occurred in Problem 12-9.

**12-12** Write the net ionic equation for any reaction that occurred in Problem 12-10.

* **12-13** Write balanced molecular equations indicating how the following ionic compounds could be prepared by a metathesis reaction using any other ionic compounds. In some cases the desired compound may be soluble and must be recovered by boiling off the solvent water after removal of a precipitate.
(a) $CuCO_3$
(b) $PbSO_3$
(c) $Hg_2I_2$
(d) $NH_4NO_3$
(e) $KC_2H_3O_2$

## Percent by Weight

**12-14** What is the percent by weight of solute in a solution made by dissolving 9.85 g of $Ca(NO_3)_2$ in 650 g of water?

**12-15** What is the percent by weight of solute if 14.15 g of NaI is present in 75.55 g of solution?

**12-16** A solution is 10.0% by weight NaOH. How many moles of NaOH are dissolved in 150 g of solution?

**12-17** A solution is 8.5% by weight $NH_4Cl$. What weight of $NH_4Cl$ is present for each 100 g of water?

**12-18** An aqueous solution contains 150 g of $KNO_3$ and is 23.2% by weight $KNO_3$. What is the weight of the solution?

**12-19** A solution contains 1 mol of NaOH dissolved in 9 mol of ethyl alcohol ($C_2H_5OH$). What is the percent by weight NaOH?

## Molarity

**12-20** What is the molarity of a solution made by dissolving 2.44 mol of NaCl in enough water to make 4.50 L of solution?

**12-21** What is the molarity of a solution made by dissolving 0.345 mol of HBr in enough water to make 750 mL of solution?

**12-22** How many moles of Epsom salts ($MgSO_4 \cdot 7H_2O$) are present in 15.6 L of a 0.0542 *M* solution?

**12-23** What is the volume in liters of a solution containing 2.00 mol of KOH that is 0.250 *M*?

**12-24** Fill in the blanks in the following table.

| Solute | M | Amount of Solute | Volume of Solution |
|---|---|---|---|
| (a) KI | _____ | 2.40 mol | 2.75 L |
| (b) $C_2H_5OH$ | _____ | 26.5 g | 410 mL |
| (c) $NaC_2H_3O_2$ | 0.255 | 3.15 mol | _____ L |
| (d) $LiNO_2$ | 0.625 | _____ g | 1.25 L |
| (e) $BaCl_2$ | _____ | 0.250 mol | 850 mL |
| (f) $Na_2SO_3$ | 0.054 | _____ mol | 0.45 L |
| (g) $K_2CO_3$ | 0.345 | 14.7 g | _____ mL |
| (h) LiOH | 1.24 | _____ g | 1650 mL |
| (i) $H_2SO_4$ | 0.905 | 0.178g | _____ mL |

**12-25** What is the molarity of a solution made by dissolving $2.50 \times 10^{-4}$ g of baking soda ($NaHCO_3$) in enough water to make 2.54 mL of solution?

**12-26** What is the molarity of the hydroxide ion and the barium ion if 13.5 g of $Ba(OH)_2$ is dissolved in enough water to make 475 mL of solution?

**12-27** A solution is 25.0% by weight calcium nitrate and has a density of 1.21 g/mL. What is its molarity?

* **12-28** A solution of concentrated NaOH is 16.4 *M*. If the density of the solution is 1.43 g/mL, what is the percent by weight NaOH?

* **12-29** Concentrated nitric acid is 70.0% $HNO_3$ and is 14.7 *M*. What is the density of the solution?

### Dilution

**12-30** What volume of 4.50 *M* $H_2SO_4$ should be diluted with water to form 2.50 L of 1.50 *M* acid?

**12-31** If 450 mL of a certain solution is diluted to 950 mL with water to form a 0.600 *M* solution, what was the molarity of the original solution?

**12-32** One liter of a 0.250 *M* solution of NaOH is needed. The only available solution of NaOH is a 0.800 *M* solution. Describe how to make the desired solution.

**12-33** What is the volume in liters of a 0.440 *M* solution if it was made by dilution of 250 mL of a 1.25 *M* solution?

**12-34** What is the molarity of a solution made by diluting 3.50 L of a 0.200 *M* solution to a volume of 5.00 L?

**12-35** What volume of water in milliliters should be added to 1.25 L of 0.860 *M* HCl so that its molarity will be 0.545?

* **12-36** What volume in milliliters of *pure* acetic acid should be used to make 250 mL of 0.200 *M* $HC_2H_3O_2$? The density of the pure acid is 1.05 g/mL.

**12-37** What volume of water in milliliters should be *added* to 400 mL of a solution containing 35.0 g of KBr to make a 0.100 *M* KBr solution?

* **12-38** What would be the molarity of a solution made by mixing 150 mL of 0.250 *M* HCl with 450 mL of 0.375 *M* HCl?

### Stoichiometry Involving Solutions

**12-39** Given the following reaction:

$$3KOH(aq) + CrCl_3(aq) \rightarrow Cr(OH)_3(s) + 3KCl(aq)$$

What weight of $Cr(OH)_3$ would be produced if 500 mL of 0.250 *M* KOH were added to a solution containing excess $CrCl_3$?

**12-40** Given the following reaction:

$$2KCl(aq) + Pb(NO_3)_2(aq) \rightarrow PbCl_2(s) + 2KNO_3(aq)$$

What weight of $PbCl_2$ is produced from 1.25 L of 0.450 *M* KCl solution?

**12-41** Given the following reaction:

$$Al_2(SO_4)_3(aq) + 3\ BaCl_2(aq) \rightarrow 3BaSO_4(s) + 2AlCl_3(aq)$$

What weight of $BaSO_4$ is produced from 650 mL of 0.320 *M* $Al_2(SO_4)_3$?

**12-42** Given the following reaction:

$$3Ba(OH)_2(aq) + 2Al(NO_3)_3(aq) \rightarrow 2Al(OH)_3(s) + 3Ba(NO_3)_2(aq)$$

What volume of 1.25 *M* $Ba(OH)_2$ is required to produce 265 g of $Al(OH)_3$?

**12-43** Given the following reaction:

$$2AgClO_4(aq) + Na_2CrO_4(aq) \rightarrow Ag_2CrO_4(s) + 2NaClO_4(aq)$$

What volume of a 0.600 *M* solution of $AgClO_4$ is needed to produce 160 g of $Ag_2CrO_4$?

**12-44** Given the following reaction:

$$3Ca(ClO_3)_2(aq) + 2Na_3PO_4(aq) \rightarrow Ca_3(PO_4)_2(s) + 6NaClO_3(aq)$$

What volume of a 2.22 *M* solution of $Na_3PO_4$ is needed to react with 580 mL of a 3.75 *M* solution of $Ca(ClO_3)_2$?

**12-45** Given the reaction:

$$2HNO_3(aq) + 3H_2S(aq) \rightarrow 2NO(g) + 3S(s) + 4H_2O$$

(a) What volume of 0.350 $M$ $HNO_3$ will completely react with 275 mL of 0.100 $M$ $H_2S$?
(b) What volume of NO gas measured at 27 °C and 720 torr will be produced from 650 mL of 0.100 $M$ $H_2S$ solution?

* **12-46** Given the following reaction:

$$2NaOH(aq) + MgCl_2(aq) \rightarrow Mg(OH)_2(s) + 2NaCl(aq)$$

What weight of $Mg(OH)_2$ would be produced by mixing 250 mL of 0.240 $M$ NaOH with 400 mL of 0.100 $M$ $MgCl_2$?

* **12-47** Given the following reaction:

$$CO_2(g) + Ca(OH)_2(aq) \rightarrow CaCO_3(s) + H_2O$$

What is the molarity of a 1.00-L solution of $Ca(OH)_2$ that would completely react with 10.0 L of $CO_2$ measured at 25.0 °C and a pressure of 0.950 atm?

## Molality and Colligative Properties

**12-48** What is the molality of a solution made by dissolving 25.0 g of NaOH in (a) 250 g of water? (b) 250 g of alcohol ($C_2H_5OH$)?

**12-49** What is the molality of a solution made by dissolving 1.50 kg of KCl in 2.85 kg of water?

* **12-50** What is the molality of an aqueous solution that is 10.0% by weight $CaCl_2$?

**12-51** What weight of NaOH is present in 550 g of water if the concentration is 0.720 m?

**12-52** What weight of water is present in a 0.430 $m$ solution containing 2.58 g of $CH_3OH$?

* **12-53** A 1.00 $m$ KBr solution weighs 1.00 kg. What weight of water is present?

**12-54** What is the freezing point of a 0.20 $m$ aqueous solution of a nonelectrolyte?

**12-55** What is the boiling point of a 0.45 $m$ aqueous solution of a nonelectrolyte?

**12-56** Ethylene glycol ($C_2H_6O_2$) is used as an antifreeze. What weight of ethylene glycol should be added to 5.00 kg of water to lower the freezing point to −5.0 °C? (Ethylene glycol is a nonelectrolyte.)

**12-57** What is the boiling point of the solution in Problem 12-56?

**12-58** Methyl alcohol can also be used as an antifreeze. What weight of methyl alcohol ($CH_3OH$) must be added to 5.00 kg of water to lower its freezing point to −5.0 °C?

**12-59** What is the molality of an aqueous solution that boils at 101.5 °C?

**12-60** What is the boiling point of a 0.15 $m$ solution of a solute in liquid benzene? For benzene, $K_b$ = 2.53 and the boiling point of pure benzene is 80.1 °C.

**12-61** What is the boiling point of a solution of 75.0 g of naphthalene ($C_{10}H_8$) in 250 g of benzene? (See Problem 12-60.)

**12-62** What is the freezing point of a solution of 100 g of $CH_3OH$ in 800 g benzene? For benzene, $K_f$ = 5.12 and the freezing point of pure benzene is 5.5 °C.

**12-63** What is the freezing point of a 10.0% by weight solution of $CH_3OH$ in benzene? (See Problem 12-62.)

* **12-64** What is the freezing point of each of the following in 100 g of water?
(a) 10.0 g of $CH_3OH$ (b) 10.0 g of NaCl
(c) 10.0 g of $CaCl_2$

**12-65** When one is immersed in the salty ocean water for an extended period one gets unusually thirsty. Explain.

**12-66** Dehydrated fruit is wrinkled and shriveled up. When put in water, the fruit expands and becomes smooth again. Explain.

**12-67** Explain how pure water can be obtained from a solution without boiling.

**12-68** In industrial processes it is often necessary to concentrate a dilute solution (much harder than diluting a concentrated solution). Explain how the principle of reverse osmosis can be applied.

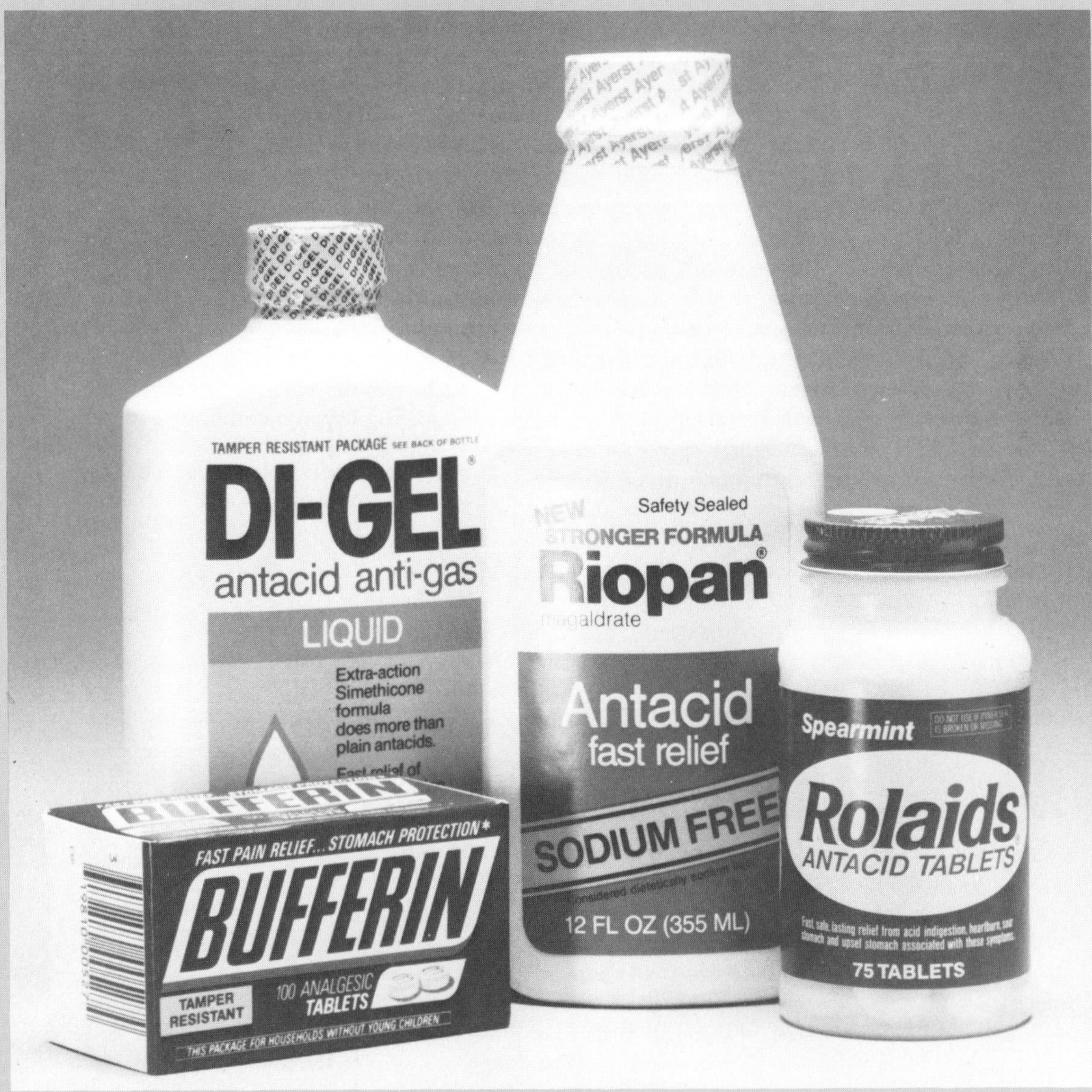


Antacids belong to a class of compounds called bases. The reactions of acids and bases are discussed in this chapter.

## Chapter 13
# Acids, Bases, and Salts

### Purpose of Chapter 13

In Chapter 13 we classify substances as acids, bases, or salts, and describe chemical reactions illustrating this behavior.

### Objectives for Chapter 13

After completion of this chapter, you should be able to:

1. Apply the Arrhenius definition to identify compounds as acids or bases and to write equations illustrating this behavior. (13-1)
2. Give the names and formulas of some common acids and bases derived from specified anions or cations. (13-1)
3. Write equations for neutralization reactions and identify the salts formed. (13-2)
4. Write equations for a stepwise neutralization of a polyprotic acid by a base. (13-2)
5. Distinguish between the behavior of a strong and a weak acid in water. (13-3)
6. Describe the dynamic equilibrium involved in the partial ionization of a weak acid or base in water. (13-3)
7. Calculate $[OH^-]$ from a specified $[H_3O^+]$ and vice versa by use of $K_w$. (13-4)
8. Distinguish between acidic, basic, or neutral solutions in terms of $[H_3O^+]$ and $[OH^-]$. (13-4)
9. Convert $[H_3O^+]$ to pH and vice versa. (13-5)
10. Distinguish between acidic, basic, or neutral solutions in terms of pH. (13-5)
11. Determine the conjugate base of a specified acid or the conjugate acid of a specified base. (13-6)
12. Write specified proton exchange reactions by application of the Brønsted-Lowry definition. (13-6)
13. Determine the extent of ionization (complete, partial, or none) of a specified acid from an appropriate table of acids. (13-7)
14. Determine the extent of hydrolysis (complete, partial, or none) of a specified ion from an appropriate table. (13-7)
15. Identify the solutions of certain salts as acidic, basic, or neutral. (13-7)
16. Determine whether a specified solution acts as a buffer. (13-8)
17. Write equations illustrating the reaction of oxides as acids or bases. (13-9)

It should be obvious by now that chemists have a hopeless habit of classifying or grouping as much information as possible. But with more than 100 elements and millions of compounds and chemical reactions, any time we can detect a common thread in any mass of information it does make sense to take advantage of it. Thus we've put elements into groups and periods, classified chemical reactions into types, and grouped compounds as ionic or covalent, electrolytes or nonelectrolytes, and soluble or insoluble. Classification is as helpful in chemistry as it is in human relations. If we classify a fellow human being as a "nerd" or a "fox," it means that that person's appearance or behavior falls into certain expected patterns (in our minds, anyway). Likewise, if we classify a substance as an "acid" or a "base," it means that the compound has a certain behavior common to other compounds in that particular classification. The classification of substances as acids, bases, and salts is among the oldest classifications in chemistry and is still one of the most important. These compounds and their behavior is the subject of this chapter.

Our first goal will be simply to identify acids and bases and understand the reasons why they are classified as such. We will then do some finer classifying as to acid and base strength and how the pH scale relates to this strength. Finally, we will look at acids and bases through a slightly wider-angle lens so as to expand our understanding of acid-base behavior.

The background for this chapter includes:

1 Ionic and covalent compounds (Sections 6-4 and 6-5)

2 Nomenclature of acids (Section 7-3)

3 Ionic and net ionic equations (Section 12-4)

4 Molarity (Section 12-6)

## 13-1 The Nature of Acids and Bases

Historically, acids were a group of compounds classified as such by their behavior and properties rather than some common elemental composition. Even since the days of the alchemist in the Middle Ages compounds known as *acids* had the following common characteristics.

Acids:

1 Taste sour (don't try this—it's OK for vinegar or carbonated water, but battery acid could ruin your tongue)

2 Cause certain organic dyes to change color (an example is litmus, which turns from blue to red in acids)

3 Dissolve certain metals, such as zinc, with the liberation of a gas (see Figure 13-1)

**Figure 13-1** ZINC AND LIMESTONE IN ACID. Both zinc (right) and limestone ($CaCO_3$) (left) react with acid to liberate a gas.

4 Dissolve limestone ($CaCO_3$), with the liberation of a gas (see Figure 13-1)

5 React with bases to form salts and water

Bases, on the other hand:

1 Taste bitter (again, don't try it)

2 Are slippery or soapy feeling

3 Cause certain organic dyes to change color (red litmus turns blue in basic solution)

4 React with acids to form salts and water

It was easy to classify substances as acid, base, or neither from a few simple laboratory tests. It wasn't until less than 100 years ago, however, that the foundation for a currently accepted model was suggested. *In 1884, Svante Arrhenius suggested that* **acids** *are substances that produce $H^+$ ions and* **bases** *are substances that produce $OH^-$ ions in aqueous solution.*

If we now look at the formulas of some of the well-known acids, we can see how these compounds behave as acids. Notice that they are all hydrogen-containing compounds and all produce $H^+$ as well as an anion (different for each acid) in water.

| | |
|---|---|
| Muriatic or hydrochloric acid | $HCl \rightarrow H^+ + Cl^-$ |
| Oil of vitriol or sulfuric acid | $H_2SO_4 \rightarrow H^+ + HSO_4^-$ |
| Vinegar or acetic acid | $HC_2H_3O_2 \rightleftharpoons H^+ + C_2H_3O_2$ |
| Carbonated water or carbonic acid | $H_2CO_3 \rightleftharpoons H^+ + HCO_3^-$ |

The acids listed above are not ionic compounds when pure. They form ions only when mixed with water. In fact, a chemical reaction occurs between the acid molecules and the $H_2O$ molecules that results in the removal of a $H^+$ ion from the rest of the molecule. As discussed in Chapters 6 and 12, acids (such as HCl) and $H_2O$ are polar covalent compounds. When HCl dissolves in water, there is a dipole-dipole electrostatic attraction between the negative dipole of one molecule and the positive dipole of the other as illustrated in Figure 13-2. (Actually, there are many more $H_2O$ molecules involved with each HCl than those shown in the figure.)

A tug-of-war has developed. On the one side the $H_2O$'s are pulling on the HCl molecule, attempting to split the molecule into a $H^+$ ion and a $Cl^-$ ion. On the other side, the HCl covalent bond is trying mightily to hold the molecule together. In the case of HCl the $H_2O$ molecules are clearly the winners, since the interaction is strong enough to break the bonds of all dissolved HCl molecules. Therefore, HCl is an acid since an $H^+$ is produced in the solution as a result of the **dissociation** (breaking apart) of HCl. (Since dissociation produces ions in this case, the process is also called **ionization**.)

To illustrate the importance of water in the ionization process, the reaction of an acid with water can be represented as

$$HCl + H_2O \rightarrow H_3O^+(aq) + Cl^-(aq)$$

*Instead of $H^+$, the acid species is often represented as $H_3O^+$, which is known as the* **hydronium ion**. The hydronium ion is simply a representation

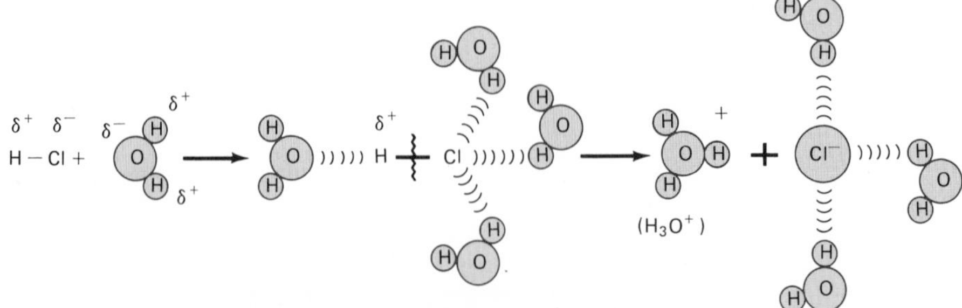


**Figure 13-2** THE INTERACTION OF HCl AND $H_2O$. The dipole–dipole interaction between $H_2O$ and HCl leads to the breaking of the HCl bond.

of the $H^+$ ion in a hydrated form. The acid species is represented as $H_3O^+$ rather than $H^+$ because it is somewhat closer to what is believed to be the actual situation. In fact, the nature of $H^+$ in aqueous solution is even more complex than $H_3O^+$ (i.e., $H_5O_2^+$, $H_7O_3^+$, etc.) In any case, the acid species is represented as $H^+$, $H^+(aq)$, or $H_3O^+$ $(aq)$ depending on the convenience of the particular situation. *Just remember that all refer to the same species in aqueous solution.* If $H^+(aq)$ is used, it should be understood that it is not just a bare proton in aqueous solution but is associated with water molecules (it is hydrated).

*The common property of acids in water is the formation of $H^+(aq)$ by this definition.* We can now see by the following net ionic equations how the $H^+(aq)$ ion accounts for the reactions long known to indicate acid behavior mentioned in the previous section.

1 Acids react with zinc and give off a gas ($H_2$)

$$Zn(s) + \underline{2H^+(aq)} \rightarrow Zn^{2+}(aq) + H_2(g)$$

2 Acids react with limestone ($CaCO_3$) and give off a gas ($CO_2$).

$$CaCO_3(s) + \underline{2H^+(aq)} \rightarrow Ca^{2+}(aq) + H_2O + CO_2(g)$$

3 Acids react with bases.

$$\underline{H^+(aq)} + OH^-(aq) \rightarrow H_2O$$

Other molecules containing hydrogen may not be acidic in $H_2O$ because the hydrogens are (1) too strongly attached to the rest of the molecule to be ionized or (2) not polar enough to create a strong interaction with $H_2O$. Sometimes both reasons are important. The hydrogens in $CH_4$ and $NH_3$, and the three hydrogens on the carbon in acetic acid are examples of hydrogens that do not ionize in aqueous solution to form $H^+$ ion.

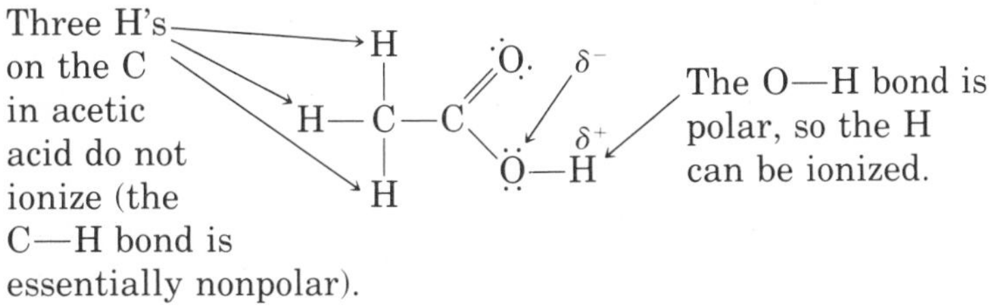

Three H's on the C in acetic acid do not ionize (the C—H bond is essentially nonpolar).

$$H-\underset{H}{\overset{H}{\underset{|}{\overset{|}{C}}}}-C\begin{matrix}\overset{\delta^-}{=\ddot{O}:} \\ -\ddot{O}-\overset{\delta^+}{H}\end{matrix}$$

The O—H bond is polar, so the H can be ionized.

Unlike the acids, which are covalent compounds before being ionized by the water molecules, many common bases consist of ionic metal hydroxides in the solid state. The solution process simply breaks down the solid ionic crystals to the same ions in solution as discussed in Section 12-2. For example,

$$Na^+OH^-(s) \xrightarrow{xH_2O} Na^+(aq) + OH^-(aq)$$

**See Problems 13-1 through 13-5.**

Some commonly known bases are caustic soda or lye (NaOH), caustic potash (KOH), slaked lime [$Ca(OH)_2$], and ammonia ($NH_3$).

## 13-2 Neutralization and Salts

One of the characteristics of acids and bases described earlier in this chapter is that the $H^+(aq)$ from an acid reacts with the $OH^-(aq)$ from a base to form water:

$$H^+(aq) + OH^-(aq) \rightarrow H_2O$$

This is the net ionic equation of the process known as neutralization. **Neutralization** *is the reaction between an acid and a base to form a salt and water. A* **salt** *is an ionic compound composed of the cation from a base and the anion from an acid.* It is amazing that acids such as hydrochloric acid and bases such as sodium hydroxide are both corrosive and dangerous compounds. When mixed in exactly stoichiometric amounts, however, the product is simply a solution of common table salt (NaCl).* The corrosive agents [$H^+(aq)$ and $OH^-(aq)$] annihilate each other, leaving in solution the cation from the base ($Na^+$) and the anion from the acid ($Cl^-$). The salt can be recovered by boiling off the water. The following molecular, total ionic, and net ionic equations illustrate the neutralization discussed:

$$NaOH(aq) + HCl(aq) \rightarrow NaCl(aq) + H_2O$$

$$Na^+(aq) + OH^-(aq) + H^+(aq) + Cl^-(aq) \rightarrow Na^+(aq) + Cl^-(aq) + H_2O$$

$$H^+(aq) + OH^-(aq) \rightarrow H_2O$$

The word "salt" is often thought of as just one substance, sodium chloride, as formed in the previous reaction. Actually, a salt can result from many different combinations of anions and cations from a variety of neutralizations.* The following neutralization reactions illustrate the formation of some other salts.

| Acid | + | Base | → | Salt | + | Water |
|---|---|---|---|---|---|---|
| 1 $HNO_3(aq)$ | + | $KOH(aq)$ | → | $KNO_3(aq)$ | + | $H_2O$ |
| 2 $H_2SO_4(aq)$ | + | $2NaOH(aq)$ | → | $Na_2SO_4(aq)$ | + | $2H_2O$ |
| 3 $2HClO_4(aq)$ | + | $Ca(OH)_2(aq)$ | → | $Ca(ClO_4)_2(aq)$ | + | $2H_2O$ |
| 4 $2H_3PO_4(aq)$ | + | $3Ba(OH)_2(aq)$ | → | $Ba_3(PO_4)_2(aq)$ | + | $6H_2O$ |

*Don't try this without supervision: Much heat evolution, with boiling and splattering, can occur.

*Salts are also produced from metathesis reactions (Chapter 12) or directly from the elements in the case of binary salts (Chapter 9).

Sulfuric acid ($H_2SO_4$) is known as a **polyprotic acid**, *which means that it can produce more than one mole of* $H^+$ *ion per mole of acid*, as follows:

$$H_2SO_4 \rightarrow 2H^+(aq) + SO_4^{2-}(aq)$$

Thus for *complete* neutralization, two moles of NaOH are required for one mole of $H_2SO_4$ as shown in reaction 2 above. Likewise, one mole of $Ca(OH)_2$ produces two moles of $OH^-$ ions, as follows.

$$Ca(OH)_2 \rightarrow Ca^{2+}(aq) + 2OH^-(aq)$$

Thus two moles of $HClO_4$ are required for each mole of $Ca(OH)_2$ as shown in reaction 3 above. To illustrate these two reactions in more detail, the total ionic and net ionic equations are shown in the following equations.

2 $\underline{2H^+(aq)} + \cancel{SO_4^{2-}}(aq) + \cancel{2Na^+(aq)} + \underline{2OH^-(aq)} \rightarrow$

$$\cancel{2Na^+(aq)} + \cancel{SO_4^{2-}} + 2H_2O$$

$$2H^+(aq) + 2OH^-(aq) \rightarrow 2H_2O$$

3 $\underline{2H^+(aq)} + \cancel{2ClO_4}^-(aq) + \cancel{Ca^{2+}(aq)} + \underline{2OH^-(aq)} \rightarrow$

$$\cancel{Ca^{2+}(aq)} + \cancel{2ClO_4}^-(aq) + 2H_2O$$

$$2H^+(aq) + 2OH^-(aq) \rightarrow 2H_2O$$

In the case of polyprotic acids partial neutralizations are possible. The complete ionization of $H_2SO_4$ represented previously actually takes place in two steps.

$$H_2SO_4 \rightarrow H^+(aq) + HSO_4^-(aq) \qquad \text{(first ionization of } H_2SO_4\text{)}$$

$$HSO_4^-(aq) \rightleftarrows H^+(aq) + SO_4^{2-}(aq) \qquad \text{(second ionization of } H_2SO_4\text{)}$$

If *one mole* of NaOH is added to *one mole* of $H_2SO_4$, only the $H^+$ from the first ionization is neutralized, as illustrated by the following equations.

$$H_2SO_4(aq) + NaOH(aq) \rightarrow NaHSO_4(aq) + H_2O$$

$\underline{H^+(aq)} + \cancel{HSO_4^-(aq)} + \cancel{Na^+(aq)} + \underline{OH^-(aq)} \rightarrow$

$$\cancel{Na^+(aq)} + \cancel{HSO_4^-(aq)} + H_2O$$

$$H^+(aq) + OH^-(aq) \rightarrow H_2O$$

Notice that the salt formed in this partial neutralization ($NaHSO_4$) has one acidic hydrogen on the anion. Thus the hydrogen sulfate (or bisulfate) ion can act as an acid as shown by the second ionization of $H_2SO_4$ above. This type of salt is known as an acid salt. **Acid salts** *are ionic compounds containing one or more acidic hydrogens on the anion. They result from the* *partial neutralization of polyprotic acids.* One mole of this acid salt can react with one mole of NaOH to complete the neutralization process.

$$NaHSO_4(aq) + NaOH(aq) \rightarrow Na_2SO_4(aq) + H_2O$$

The stepwise neutralization of a polyprotic acid is further illustrated by the neutralization of $H_3PO_4$ with KOH. If *1 mol* of the KOH is added to *1 mol* of $H_3PO_4$, reaction 1 occurs:

1 $$\text{H—PO}_4\text{(H)(H)}(aq) + KOH(aq) \rightarrow KH_2PO_4(aq) + H_2O$$

If 1 mol of KOH is then added to the $KH_2PO_4$ solution, reaction 2 occurs:

2 $$\text{K(H)(H)PO}_4(aq) + KOH(aq) \rightarrow K_2HPO_4(aq) + H_2O$$

Finally, by adding 1 mol of KOH to the $K_2HPO_4$ solution, reaction 3 occurs:

3 $$K_2\text{H—PO}_4(aq) + KOH \rightarrow K_3PO_4(aq) + H_2O$$

Notice that each additional $OH^-$ removes one $H^+$ from the acid or acid salt.

The total reaction, which is the algebraic sum of all three reactions, is

$$H_3PO_4(aq) + 3KOH(aq) \rightarrow K_3PO_4(aq) + 3H_2O$$

This is the reaction that occurs regardless of whether the KOH is added one mole at a time or all three moles at once. In these reactions, the $H_3PO_4$ is identified as an acid, $KH_2PO_4$ and $K_2HPO_4$ as acid salts, and $K_3PO_4$ as a salt. (See Figure 13-3.)

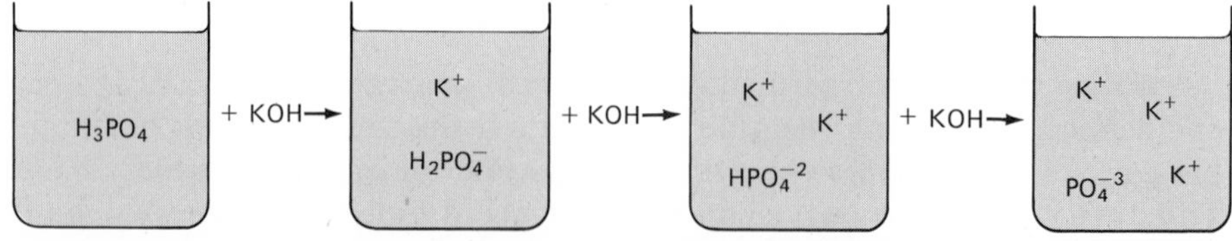


**Figure 13-3** THE NEUTRALIZATION OF $H_3PO_4$. The hydrogens of $H_3PO_4$ can be neutralized one at a time.

**Example 13-1**
Identify each of the following as either an acid, a base, a salt, or an acid salt:

(a) $KClO_4$, (b) $H_2S$, (c) $Ba(HSO_3)_2$, (d) $Al(OH)_3$.

**Answer**

(a) $KClO_4$: salt (b) $H_2S$: acid (c) $Ba(HSO_3)_2$: acid salt (d) $Al(OH)_3$: base

**Example 13-2**
Write the balanced equation in molecular form illustrating the complete neutralization of $Al(OH)_3$ with $H_2SO_4$.

**Procedure**
Each mole of base produces *3 mol* of $OH^-$; each mole of acid produces *2 mol* of $H^+$. Therefore, 2 mol of base (producing 6 mol of $OH^-$) reacts with 3 mol of acid (producing 6 mol of $H^+$) for complete neutralization.

**Answer**

$$2Al(OH)_3 + 3H_2SO_4 \rightarrow Al_2(SO_4)_3 + 6H_2O$$

**Example 13-3**
Write the balanced molecular equation illustrating the reaction of 1 mol of $H_3PO_4$ with 1 mol of $Ca(OH)_2$.

**Procedure**
One mole of base produces 2 mol of $OH^-$, which reacts with 2 mol of $H^+$. The removal of 2 mol of $H^+$ from 1 mol of $H_3PO_4$ leaves the $HPO_4^{2-}$ ion in solution.

**Answer:**

$$H_3PO_4 + Ca(OH)_2 \rightarrow CaHPO_4 + 2H_2O$$

**See Problems 13-6 through 13-15.**

## 13-3 The Strength of Acids and Bases

A 0.10 *M* solution of hydrochloric acid burns holes in your clothes, and you too if not washed off immediately. A 0.10 *M* solution of acetic acid is not particularly corrosive, and when present in vinegar adds a little tang to your salad. Hydrochloric acid and acetic acid at the same concentrations are, at best, distant cousins. The difference lies in the *strength* of their acid behavior, which relates directly to the concentration of $H^+(aq)$ (or $H_3O^+$) produced in the respective solutions.

Hydrochloric acid is known as a strong acid. *Since a* **strong acid** *is completely ionized by water, it behaves as a strong electrolyte.* This is illustrated by the following equation.

$$HCl(g) + H_2O \rightarrow H_3O^+(aq) + Cl^-(aq)$$

Other common strong acids are $HNO_3$, HBr, HI, $HClO_4$, and $H_2SO_4$. Most other common acids behave as weak acids in water. A **weak acid** *is only partially ionized (usually less than 5% at typical molar concentrations)*. Because of the small concentration of ions, weak acids behave as weak electrolytes in water (see Section 12-1). The ionization of a weak acid is incomplete because it is a reversible reaction. A reaction where the reverse reaction also occurs to an appreciable extent is illustrated by double arrows ($\rightleftharpoons$) rather than a single arrow, which implies a reaction that is essentially complete such as is previously shown for HCl. The ionization of the weak acid HF is illustrated as follows:

$$HF(aq) + H_2O \rightleftharpoons H_3O^+(aq) + F^-(aq)$$

The partial ionization of a weak acid is one example of a chemical reaction that reaches a state of equilibrium. *In a reaction at* **equilibrium,** *two reactions are occurring simultaneously.* In the ionization of HF, for example, a forward reaction occurs (to the right), producing ions ($H_3O^+$ and $F^-$), and a reverse reaction occurs (to the left), producing covalent compounds (HF and $H_2O$).

Forward: $HF(aq) + H_2O \rightarrow H_3O^-(aq) + F^-(aq)$

Reverse: $H_3O^+(aq) + F^-(aq) \rightarrow HF(aq) + H_2O$

*At equilibrium, the forward and the reverse reactions occur at the same rate.* For weak acids, the point of equilibrium lies far to the left side of the original ionization equation. This means that most of the fluorine is present in the form of covalent HF molecules rather than fluoride ions. (See Figure 13-4.)

When a system is at equilibrium, the concentration of all species (reactants and products) remains the same but the identity of the individual

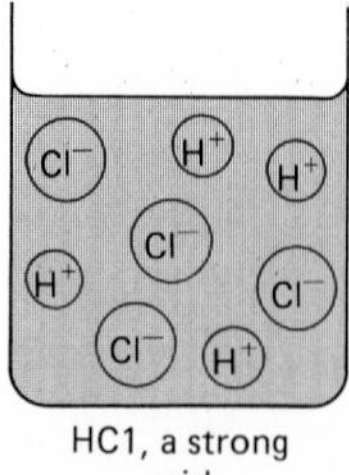


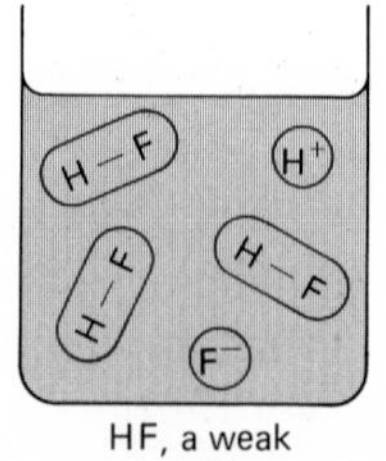


**Figure 13-4** STRONG ACIDS AND WEAK ACIDS. Strong acids are completely ionized in water; weak acids are only partially ionized.

molecules changes. The reaction thus *appears* to have gone so far to the right and then stopped. In fact, at equilibrium, a **dynamic** (*constantly changing*) situation exists where two reactions going in opposite directions at the same rate keep the concentrations of all species constant.

There is a simple analogy to a situation where an amount remains constant as a result of equal rates of change. As shown in Figure 13-5, if the rate at which water enters a lake equals the rate at which it leaves over the spillway, the amount of water in the lake remains constant. In a situation similar to our chemical system the amount of water stays the same but its identity is changing. We will have much more to say about equilibrium in Chapter 15. The subject is introduced now so that we may appreciate the reason for incomplete reactions such as the ionization of a weak acid or base.

The following examples illustrate the difference in acidity (the difference in $H_3O^+$ concentration) between a strong acid and a weak acid. In these examples the appearance of a species in brackets (e.g., $[H_3O^+]$) represents the numerical value of the concentration of that species in moles per liter ($M$).

**Example 13-4**

What is $[H_3O^+]$ in a 0.100 $M$ $HNO_3$ solution?

**Solution**

$HNO_3$ is a strong acid, so the following reaction goes 100% to the right.

$$HNO_3 + H_2O \rightarrow H_3O^+(aq) + NO_3^-(aq)$$

The initial concentration of $HNO_3$ is 0.100 $M$, and all of the $HNO_3$ ionizes to form $H_3O^+$. The reaction stoichiometry tells us that 1 mol of $H_3O^+$ is formed for every mole of $HNO_3$ initially present. Therefore, 0.100 mol of $HNO_3$ forms 0.100 mol of $H_3O^+$ in 1 L. Thus,

$$[H_3O^+] = \underline{\underline{0.100\ M}}$$

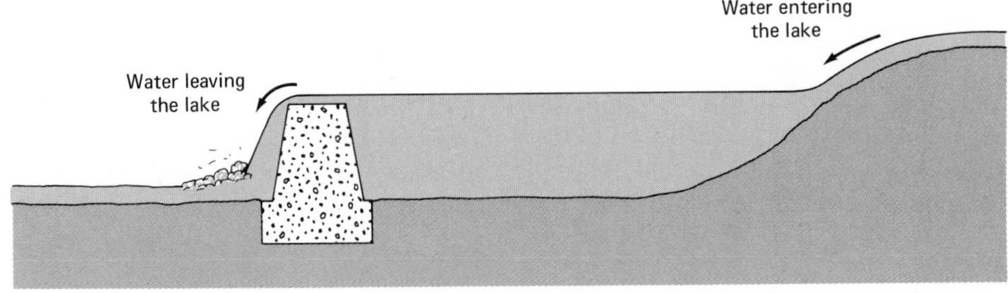


**Figure 13-5** EQUILIBRIUM. The level of the lake is at equilibrium when the flow of water entering the lake equals the flow of water leaving the lake.

**Example 13-5**

What is $[H_3O^+]$ in a 0.100 $M$ $HC_2H_3O_2$ solution that is 1.34% ionized?

**Solution**

Since $HC_2H_3O_2$ is a weak acid, the following ionization reaches equilibrium when 1.34% of the initial $HC_2H_3O_2$ ionizes.

$$HC_2H_3O_2(aq) + H_2O \rightleftharpoons H_3O^+(aq) + C_2H_3O_2^-(aq)$$

The $[H_3O^+]$ is calculated by multiplying the original concentration of acid by the percent **expressed in fraction form**.

$$[H_3O^+] = [\text{original con. of acid}] \times \frac{\%\ \text{ionization}}{100\%}$$

In this case,

$$[H_3O^+] = [0.100] \times \frac{1.34\%}{100\%} = \underline{\underline{1.34 \times 10^{-3}\ M}}$$

(Review Appendix A for similar problems in the use of percent.)

**Review Appendix A on percent problems.**

Notice in the above examples that the concentration of $H_3O^+$ is about 100 times greater in the nitric acid solution compared with the acetic acid solution. Yet both acids were at the same original concentration. *The large difference in acid strength lies in the proportion of the acid molecules originally present that ionize.* We should take note that the concentration of an acid in water does not necessarily correspond to the concentration of $H_3O^+$ ions present in solution.

Caustic soda (NaOH) is a good agent to use to clean drains but not floors. It would ruin the floors and probably the hand that applied it. (Many tragedies could be avoided by keeping caustic soda or lye out of the reach of children.) If you want to clean the floor, a milder caustic agent is needed. Household ammonia is used for this purpose, because it produces a much lower $OH^-$ concentration than lye.

Sodium hydroxide is a **strong base** because 100% of the NaOH that dissolves forms $Na^+$ and $OH^-$ ions:

$$NaOH(s) \rightarrow Na^+(aq) + OH^-(aq)$$

The alkali metal hydroxides and the alkaline earth hydroxides [except $Be(OH)_2$] are all strong bases, although some of the alkaline earth hydroxides dissolve to only a limited extent. These compounds are all ionic in the solid state. As we mentioned before, the solution process simply breaks down the solid lattice of ions into hydrated ions.

Ammonia is a **weak base**, as shown by the following equation.

$$NH_3(aq) + H_2O \rightleftharpoons NH_4^+(aq) + OH^-(aq)$$

Since this equilibrium reaction lies far to the left, only a comparatively small concentration of $OH^-$ ions is produced. Many other neural nitrogen compounds, such as methylamine ($CH_3NH_2$) and pyridine ($C_5H_5N$), behave as weak bases in the same manner as $NH_3$.

**See Problems 13-16 through 13-23.**

## 13-4 The Equilibrium of Water

Before expanding our view of acids and bases in water, let's go back and look at the composition of pure water. In Chapter 12, pure water was classified as a nonconductor of electricity, which meant that ions were not detectable by an ordinary conductivity experiment. With very sensitive instruments, however, we find that there actually is a small concentration of ions present in pure water. The ions arise from the dissociation of water molecules in a reversible reaction that is illustrated as follows:

$$H_2O + H_2O \rightleftharpoons H_3O^+ + OH^-$$

*This process whereby a pure covalent compound dissociates into positive and negative ions is known as* **autoionization**. Although the point of this equilibrium lies very far to the left, there is a small but important concentration of $H_3O^+$ and $OH^-$ that do coexist in pure water. Like other equilibria that we have discussed, this is a dynamic situation, with the forward and reverse reactions occurring simultaneously.

The concentration of each ion at 25 °C is found by experiments to be $1.0 \times 10^{-7}$ mol/L. This means that only about one out of every 500 million water molecules is actually ionized at any one time. Other experimental results tell us that the product of the ion concentrations is a constant. This phenomenon will be explained in more detail in Chapter 15, but for now we will accept it as fact. Therefore, at 25 °C.

$$[H_3O^+][OH^-] = K_w \quad \text{(a constant)}$$

Substituting the actual concentrations of the ions, we can now find the numerical value of the constant.

$$[1.0 \times 10^{-7}][1.0 \times 10^{-7}] = 1.0 \times 10^{-14}$$

*$K_w$ ($1.0 \times 10^{-14}$) is known as the* **ion product** *of water* at 25 °C. The importance of this constant is that it tells us the concentration of $H_3O^+$ and $OH^-$ not only in pure water but in solutions where acids or bases are present. The following example illustrates this relationship.

**Example 13-6**

In a certain solution, $[H_3O^+] = 1.5 \times 10^{-2}\ M$. What is $[OH^-]$ in this solution?

**Procedure**

Use the relationship for $K_w$, $[H_3O^+][OH^-] = 1.0 \times 10^{-14}$, and solve for $[OH^-]$.

**Solution**

$$[H_3O^+][OH^-] = 1.0 \times 10^{-14}$$

$$[OH^-] = \frac{1.0 \times 10^{-14}}{[H_3O^+]} = \frac{1.0 \times 10^{-14}}{1.5 \times 10^{-2}}$$

$$= \underline{\underline{6.7 \times 10^{-13}\ M}}$$

As you can see, there is an inverse relationship between $[H_3O^+]$ and $[OH^-]$. Although inverse relationships have been discussed and illustrated several times previously in this text, the relationship of $[H_3O^+]$ and $[OH^-]$ is illustrated in Figures 13-6 and 13-7. In the figures, concentrations of $[H_3O^+]$ and $[OH^-]$ are listed between $10^{-4}$ and $10^{-10}$ $M$ (higher and lower concentrations exist, however). Both concentrations ($[H_3O^+]$ on the left and $[OH^-]$ on the right) are shown to decrease from top to bottom. In between, there is a rigid pointer with an arrow at each end. In Figure 13-6 the situation in pure water, which is a neutral solution, is illustrated. In a **neutral solution,** the pointer is exactly balanced, indicating that the *concentrations of both ions are equal to $10^{-7}$ M.*

We will now add an *acid* to this neutral solution. By definition, $H_3O^+$ is added to the solution, and the pointer should go up on the left. At the same time, however, the rigid pointer goes *down on the right,* indicating a correspondingly smaller $[OH^-]$. If, instead, a base had been added, the situation would have been just the opposite. Figure 13-7 illustrates the relationship in an acid solution. A calculation such as we made in Example

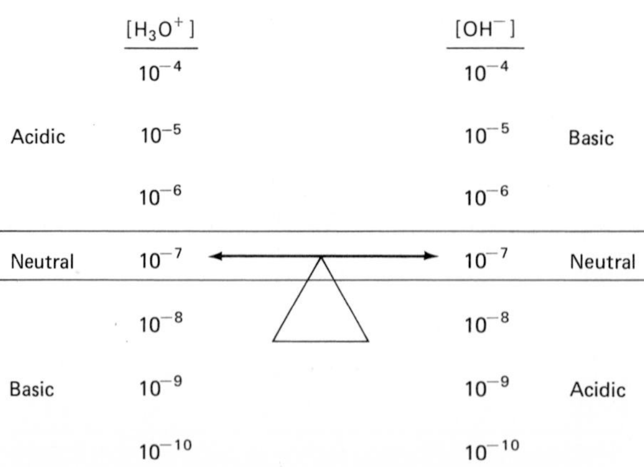


**Figure 13-6** A NEUTRAL SOLUTION. In a neutral solution the $[H_3O^+]$ and the $[OH^-]$ are both equal to $10^{-7}$ $M$.

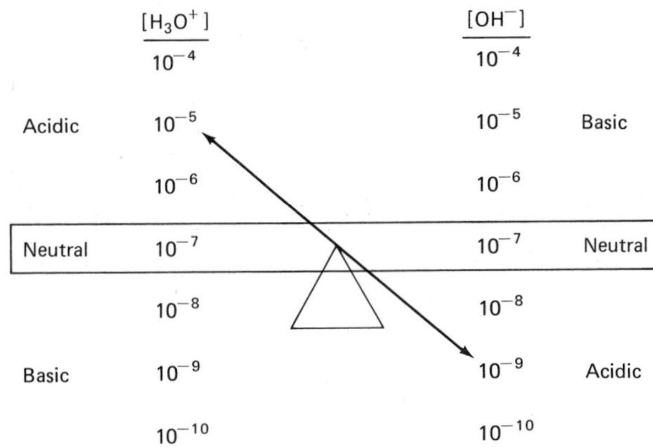


**Figure 13-7** AN ACID SOLUTION. There is an inverse relationship between $[H_3O^+]$ and $[OH^-]$.

13-6 would confirm that when $[H_3O^+]$ is *increased* from $10^{-7}$ to $10^{-5}$ $M$, the $[OH^-]$ is *decreased* from $10^{-7}$ to $10^{-9}$ $M$.

In summary, an acidic, basic, or neutral solution can now be defined in terms of concentrations of ions:

| | |
|---|---|
| Neutral | $[H_3O^+] = [OH^-] = 1.0 \times 10^{-7}$ |
| Acidic | $[H_3O^+] > 1.0 \times 10^{-7}$ and $[OH^-] < 1.0 \times 10^{-7}$ |
| Basic | $[H_3O^+] < 1.0 \times 10^{-7}$ and $[OH^-] > 1.0 \times 10^{-7}$ |

We should now accomodate this information into our understanding of acids and bases. A modern version of the Arrhenius definition of an acid is a substance that produces $H_3O^+$ ions in aqueous solution. But now we see that $H_3O^+$ is present in neutral and even basic solutions as well. A slight modification of the definitions solves this problem. *An* **acid** *is any substance that increases* $[H_3O^+]$ *in water, and a* **base** *is any substance that increases* $[OH^-]$ *in water.* With our new understanding of the equilibrium, notice that a substance can be an acid by directly donating $H^+(aq)$ to the solution (e.g., HCl, $H_2S$), or a substance can be an acid by reacting with $OH^-$ ions, thus removing them from the solution. As illustrated in Figure 13-7, lowering $[OH^-]$ on the right has the effect of raising $[H_3O^+]$ on the left.

**See Problems 13-24 through 13-33.**

**Figure 13-8** pH AND PRODUCTS. Most of us are quite familiar with the use of pH in commercial products.

## 13-5 The pH Scale

It is difficult to watch commercial television for long without hearing some reference to pH. Presumably, the whole population understands what is meant when a hair rinse has a high, low, or controlled pH. (See Figure 138.) In fact, pH is an important and convenient method for expressing $[H_3O^+]$ in aqueous solution. In a previous section, we found that the $[H_3O^+]$ of a 0.100 $M$ acetic acid solution is $1.34 \times 10^{-3}$ $M$. Expressing concentrations such as this in scientific notation is certainly easier than using many zeros,

**Refer to Appendix E for discussion of logarithms.**

but it is still awkward. The solution to this problem is the use of a logarithmic scale called pH. **pH** *is a mathematical definition of* $[H_3O^+]$ *as follows*:

$$pH = -\log[H_3O^+]$$

Much less popular but a valid way to express the $[OH^-]$ is called **pOH**, which is defined as follows:

$$pOH = -\log[OH^-]$$

A simple relationship between pH and pOH can be derived from the ion product of water:

$$[H_3O^+][OH^-] = 1.0 \times 10^{-14}$$

If we now take $-\log$ of both sides of the equation, we have

$$-\log[H_3O^+][OH^-] = -\log(1.0 \times 10^{-14})$$

Since $\log(A \times B) = \log A + \log B$, the equation can be written as

$$-\log[H_3O^+] - \log[OH^-] = -\log 1.0 - \log 10^{-14}$$

Since $\log 1.0 = 0$ and $\log 10^{-14} = -14$, the equation is

$$pH + pOH = 14$$

As you can see from this relationship and the following examples, if pOH is known, so is pH. Therefore, only pH is needed to express both the $[H_3O^+]$ and the $[OH^-]$.

**Example 13-7**
What is the pH of a neutral solution? What is the pOH?

**Solution**
In a neutral solution, $[H_3O^+] = 1.0 \times 10^{-7}$.

$$pH = -\log[H_3O^+] = -\log(1.0 \times 10^{-7})$$
$$= -\log 1.0 - \log 10^{-7}$$
$$= 0.00 - (-7.00) = \underline{\underline{7.00}}*$$

Also in a neutral solution, $[OH^-] = 1.0 \times 10^{-7}$.

$$pOH = \underline{\underline{7.00}}$$

*At first glance it appears that the pH is expressed to three significant figures but the $[H_3O^+]$ to only two. In fact, only the number to the right of the decimal (the mantissa) in the pH should be expressed to two significant figures. The number to the left of the decimal (the characteristic) in the pH refers to the exponent of 10.

**Example 13-8**
What is the pH of a 0.015 *M* solution of $HClO_4$? What is the pOH of this solution?

**Solution**

$$HClO_4 + H_2O \longrightarrow H_3O^+ + ClO_4^-$$

Since this is a strong acid, all of the $HClO_4$ is ionized to produce $H_3O^+$. Therefore, $[H_3O^+] = 0.015$.

$$\begin{aligned} pH &= -\log[H_3O^+] = -\log(1.5 \times 10^{-2}) \\ &= -\log 1.5 - \log 10^{-2} \\ &= -0.18 + 2.00 = \underline{\underline{1.82}} \end{aligned}$$

$$\begin{aligned} pH + pOH &= 14.00 \\ pOH &= 14.00 - pH \\ &= 14.00 - 1.82 = \underline{\underline{12.18}} \end{aligned}$$

**Example 13-9**
In a certain weakly basic solution, pH = 9.62. What is the $[H_3O^+]$?

**Solution**

$$pH = 9.62 = -\log[H_3O^+]$$

$$\log[H_3O^+] = -9.62 = 0.38 - 10^*$$

Since antilog of 0.38 = 2.4 and antilog of $-10 = 10^{-10}$,

$$[H_3O^+] = \underline{\underline{2.4 \times 10^{-10}\ M}}$$

*See Appendix E-4.

Acidic, basic, and neutral solutions were previously defined in terms of concentrations. We can now do the same thing in terms of pH and pOH.

| | |
|---|---|
| Neutral | $pH = pOH = 7.00$ |
| Acidic | $pH < 7.00$ and $pOH > 7.00$ |
| Basic | $pH > 7.00$ and $pOH < 7.00$ |

Remember, then, that a high pH (greater than 7) means basic and a low pH (less than 7) means acidic.

You've probably heard of the Richter scale for measuring earthquake intensity, especially if you're from California. If so, you are aware that an

| | pH | |
|---|---|---|
| Strong acid | 1 | ← 0.1 *M* HCl (stomach acid) |
| | 2 | |
| | | ← Lemon juice, vinegar |
| | 3 | |
| | | ← Orange juice |
| Weak acid | 4 | |
| | 5 | |
| | 6 | |
| | | ← Milk |
| Neutral | 7 | ← Pure water |
| | | ← Blood |
| | 8 | ← Sea water |
| | 9 | ← Baking soda solution |
| Weak base | 10 | |
| | | ← Milk of magnesia |
| | 11 | |
| | | ← Household ammonia |
| | 12 | |
| | 13 | ← 0.1 *M* NaOH (dilute lye) |
| Strong base | 14 | |

**Figure 13-9** THE pH OF SOME COMMMON SUBSTANCES.

earthquake with a reading of 8.0 on the Richter scale is 10 times more powerful than one with a reading of 7.0. This is because the Richter scale, like pH, is a logarithmic scale. In such a scale, a difference of one integer actually represents a 10-fold change. For example, a solution of pH = 4 is 10 times more acidic than a solution of pH = 5. A pH = 1 solution is actually 100,000 times more acidic than a pH = 6 solution, although both solutions are labeled acidic. The pH for some common substances is given in Figure 13-9.

**See Problems 13-34 through 13-44.**

## 13-6 Brønsted-Lowry Acids and Bases

At this point, our understanding of acids and bases is simply that of substances that change the pH of an aqueous solution. We have illustrated how many compounds such as HCl, $HC_2H_3O_2$, $NH_3$, and NaOH can act as either an acid or a base. In addition to compounds, however, many ions such as $NH_4^+$, $CN^-$, and $C_2H_3O_2^-$ also act as acids and bases in water. To better describe the action of ions in water, it is helpful to employ a more general definition of acid and base behavior. The Brønsted-Lowry definition is useful not only for this purpose but to extend the concept of acids and bases to certain other solvents. *In the Brønsted-Lowry definition, an* **acid** *is defined as a proton ($H^+$) donor and a* **base** *as a proton acceptor.* To illustrate this definition, we will again look at the equilibrium involving the common weak base, ammonia ($NH_3$).

$$NH_3 + H_2O \rightleftharpoons NH_4^+ + OH^-$$

Ammonia is a base in the Arrhenius definition because by removing a proton from an $H_2O$ molecule *it produces $OH^-$*. It is also a base in the

Brønsted-Lowry definition because *it accepts an* $H^+$ from $H_2O$ illustrated as follows:

$$\text{H—}\ddot{\text{N}}\text{(—H)—H} + \text{(H)—}\ddot{\text{O}}\text{:(—H)} \rightleftharpoons \text{H—N(—H)(—H)—H}^+ + \text{:}\ddot{\underset{\cdot\cdot}{\text{O}}}\text{—H}^-$$

Although $NH_3$ is a base in both definitions, the $H_2O$ molecule plays the role of an acid in the Brønsted-Lowry sense because it donates the $H^+$ to $NH_3$.

As mentioned previously, this reaction is a reversible reaction that reaches a point of equilibrium. This means that the reverse reaction is also occurring whereby an $NH_4{}^+$ ion donates an $H^+$ to an $OH^-$ to form the original reactants. *Thus in the reverse reaction,* $NH_4{}^+$ *acts as an acid and* $OH^-$ *as a base.*

In the Brønsted-Lowry sense the reaction can be viewed as simply an exchange of an $H^+$. When the base ($NH_3$) reacts, it adds $H^+$ to form an acid ($NH_4{}^+$). When the acid ($H_2O$) reacts, it loses an $H^+$ to form a base ($OH^-$). The $NH_3$,$NH_4{}^+$ and the $H_2O$,$OH^-$ pairs are known as **conjugate acid-base pairs**. Thus a reaction of an acid and a base is said to produce a conjugate base and acid as illustrated in the following equation (A = Brønsted-Lowry acid, B = Brønsted-Lowry base).

$$\underset{B_1}{NH_3} + \underset{A_2}{H_2O} \rightleftharpoons \underset{A_1}{NH_4{}^+} + \underset{B_2}{OH^-}$$

*The conjugate base of a substance can be determined by removing an* $H^+$. *The conjugate acid of a substance is determined by adding an* $H^+$.

$$\text{conjugate acid} \underset{+H^+}{\overset{-H^+}{\rightleftharpoons}} \text{conjugate base}$$

For example

$$H_3PO_4 \underset{+H^+}{\overset{-H^+}{\rightleftharpoons}} H_2PO_4{}^-$$

We have seen that bases by the Arrhenius definition, such as $NH_3$, are also bases in the Brønsted-Lowry sense. Arrhenius acids such as HCl also maintain their status as acids when they are considered as $H^+$ donors. Consider the following equation, which illustrates the acid behavior of HCl.

$$\underset{A_1}{HCl} + \underset{B_2}{H_2O} \rightarrow \underset{A_2}{H_3O^+} + \underset{B_1}{Cl^-}$$

In this reaction $H_2O$ is a base, since it accepts a $H^+$ to form $H_3O^+$. (Remember that $H_2O$ is an acid when $NH_3$ is present.) *A compound that can act as either an acid or base, depending on what other substance is present, is called* **amphoteric.** One must determine whether an amphoteric substance is an acid or a base from the particular reaction. For example, we cannot determine whether $H_2O$ is a Brønsted-Lowry acid or base unless we know what other substance is present (HCl or $NH_3$).

**Example 13-10**
What are the conjugate bases of (a) $H_2SO_3$ and (b) $H_2PO_4^-$?

**Procedure**

$$\text{Acid} - H^+ = \text{conjugate base}$$

**Answer**

(a) $H_2SO_3 - H^+ = \underline{\underline{HSO_3^-}}$ (b) $H_2PO_4^- - H^+ = \underline{\underline{HPO_4^{2-}}}$

**Example 13-11**
What are the conjugate acids of (a) $CN^-$ and (b) $H_2PO_4^-$?

**Procedure**

$$\text{Base} + H^+ = \text{conjugate acid}$$

**Answer**

(a) $CN^- + H^+ = \underline{\underline{HCN}}$ (b) $H_2PO_4^- + H^+ = \underline{\underline{H_3PO_4}}$

**Example 13-12**
Write the equations illustrating the following Brønsted-Lowry acid-base reactions.
(a) $H_2S$ as an acid with $H_2O$
(b) $H_2PO_4^-$ as an acid with $OH^-$
(c) $H_2PO_4^-$ as a base with $H_3O^+$
(d) $CN^-$ as a base with $H_2O$

**Procedure**
A Brønsted-Lowry acid-base reaction produces a conjugate acid and base.

**Answer**

| | Acid | | Base | → | Acid | | Base |
|---|---|---|---|---|---|---|---|
| (a) | $H_2S$ | + | $H_2O$ | → | $H_3O^+$ | + | $HS^-$ |
| (b) | $H_2PO_4^-$ | + | $OH^-$ | → | $HPO_4^{2-}$ | + | $H_2O$ |

(c) $H_2PO_4^- + H_3O^+ \rightarrow H_3PO_4 + H_2O$
Notice that equations (b) and (c) indicate that the $H_2PO_4^-$ ion is amphoteric.
(d) $CN^- + H_2O \rightarrow HCN + OH^-$

See Problems 13-45 through 13-50.

## 13-7 Predicting Acid and Base Reactions in Water

The strength of an acid in water is determined by the concentration of $H_3O^+$ produced by a specified concentration of the acid. There are three categories that can be use to classify acids according to their strength in aqueous solutions. Group 1 acids are all very strong acids that are 100% ionized in water. Group 2 acids are intermediate to weak acids but are all partially ionized in water. Finally, group 3 acids are very weak and actually have negligible ionization or acid behavior in water. The presence of any of these compounds in water has no detectable effect on the pH. A list of some acids in each classification is shown in Table 13-1 in decreasing order of acid strength.

Since group 1 acids are all completely ionized in water, we can conclude that the reverse reaction to the ionization (base behavior of anion) is negligible. Since the conjugate bases of the very strong acids do not exhibit $H^+$ acceptor ability in water, it is apparent that at least in water they are bases

**Table 13-1** Relative Strengths of Some Acids and Bases[a]

| Increasing acid strength (↑) | Acid | $\xrightarrow{-H^+}$ / $\xleftarrow{+H^+}$ | Base | Increasing base strength (↓) |
|---|---|---|---|---|
| 1 Very strong | $HClO_4$ | | $ClO_4^-$ | Very weak (negligible) |
| | $HBr$ | | $Br^-$ | |
| | $HCl$ | | $Cl^-$ | |
| | $HNO_3$ | | $NO_3^-$ | |
| 2 Intermediate or weak | $H_2SO_3$ | | $HSO_3^-$ | Intermediate or weak |
| | $HNO_2$ | | $NO_2^-$ | |
| | $HF$ | | $F^-$ | |
| | $HC_2H_3O_2$ | | $C_2H_3O_2^-$ | |
| | $H_2S$ | | $HS^-$ | |
| | $H_2CO_3$ | | $HCO_3^-$ | |
| | $HOCl$ | | $OCl^-$ | |
| | $HCN$ | | $CN^-$ | |
| 3 Very weak (negligible) | $H_2$ | | $H^-$ | Very strong |
| | $CH_4$ | | $CH_3^-$ | |

[a]The species in the shaded region are all capable of existence in appreciable concentrations in aqueous solutions.

"in name only." On the other hand, the conjugate bases of group 2 acids do show base behavior in water since the reverse reaction of the ionization takes place to an appreciable extent. This suggests that there is an inverse relationship between the strength of an acid and the strength of its conjugate base. That is, if the acid is very strong, its conjugate base is very weak; if the acid is very weak, its conjugate base is very strong. In between, both acid and conjugate base exhibit intermediate behavior. This relationship is analogous to a "seesaw." When one side is up, the other is down.

As a result of the inverse relationship between an acid and its conjugate base, the conjugate bases in Table 13-1 are listed in order of increasing strength (the ability to accept an $H^+$) on the right. Let us now examine each group of acids and their conjugate bases individually so as to predict their behavior in water.

| **Group 1** | Acid | $\longleftrightarrow$ | Conjugate base |
|---|---|---|---|
| | Very strong in water | | Negligible in water |

These strong acids all act identically in water since they are essentially 100% ionized. *This means that the molecular form of the acid on the right does not exist in any appreciable concentration in water.* The ionization is represented as follows:

$$HNO_3 + H_2O \xrightarrow{100\%} H_3O^+ + NO_3^-$$

Since the reaction above is essentially irreversible, we can conclude that the $NO_3^-$ ion has negligible base behavior in water. This is indeed the case.

A similar situation exists for cations of strong bases (e.g., $K^+$, $Na^+$, $Ca^{2+}$). These cations do not show any acid behavior in water. As a result, *aqueous solutions of most salts formed from strong acids and strong bases* [e.g., NaCl, KBr, $Ba(ClO_4)_2$] *are neutral* ($pH = 7.0$). Neither ion shows acid or base behavior in water.

| **Group 2** | Acid | $\longleftrightarrow$ | Conjugate base |
|---|---|---|---|
| | Intermediate or weak in water | | Intermediate or weak in water |

Aqueous solutions of acids in this group are partially ionized. The molecular form of the acid exists in equilibrium with its conjugate base, as the following equation illustrates.

$$HNO_2 + H_2O \rightleftharpoons H_3O^+ + NO_2^-$$

The extent of ionization and thus the $H_3O^+$ concentration for equal molar concentrations of acids decreases down this group. The actual quantitative relationships concerning the extent of ionization are known as equilibrium constants and are discussed in Chapter 15.

In this group, it is important to note that the existence of the equilibrium, especially the reverse reaction, indicates that the anion does exhibit base behavior in water. In fact, when certain salts containing these anions dissolve in water, the resulting solution has an increased $OH^-$ concentration. The pH of these basic solutions is thus greater than 7.0. The basic nature of these anions is illustrated by the following example.

$$NO_2^- + H_2O \rightleftharpoons HNO_2 + OH^-$$

*The reaction of an anion with water to produce $OH^-$ or the reaction of a cation with water to produce $H_3O^+$ is known as a* **hydrolysis reaction.** The extent of hydrolysis of the anions in this group is generally small but increases from top to bottom in Table 13-1. (Notice this is the inverse of the acid behavior of their conjugate acids.) In any case, *aqueous solutions of salts such as $NaC_2H_3O_2$, LiCN, and $Ca(ClO)_2$ are weakly basic due to anion hydrolysis.* (The cations in these examples do not hydrolyze since they are all cations of strong bases.)

In a similar manner *salts containing conjugate acids (cations) of weak bases undergo cation hydrolysis to produce an acidic solution.* For example, solutions of $NH_4Cl$ are slightly acidic because of the following reaction.

$$NH_4^+ + H_2O \rightleftharpoons NH_3 + H_3O^+$$

(Notice in this case that the $Cl^-$ does not undergo hydrolysis since it is the anion of a strong acid.)

If both cation and anion of a salt undergo hydrolysis (e.g., $NH_4^+CN^-$), then the reaction that predominates (goes farther to the right) determines whether the solution is acidic or basic. Equilibrium constants are needed to judge the extent of each reaction. In the case of $NH_4CN$, the constants indicate that the anion hydrolysis is more extensive so the solution is slightly basic.

The case of some acid salts (e.g., $NaHCO_3$) is complicated by the anion that may be either acidic or basic with water illustrated as follows:

$$HCO_3^- + H_2O \rightleftharpoons H_2CO_3 + OH^-$$

$$HCO_3^- + H_2O \rightleftharpoons CO_3^{2-} + H_3O^+$$

Again, the reaction that lies farther to the right predominates and determines whether the pH is greater or less than 7.0. In the case of sodium bicarbonate solutions, the hydrolysis reaction is more extensive and aqueous solutions of this salt are basic. This is why bicarbonate of soda is used as an antacid to neutralize stomach acid. To make the prediction, however, the specific equilibrium constants are needed for the two reactions to tell us the extent of each reaction.

| **Group 3** | Acid | $\longleftrightarrow$ | Conjugate base |
|---|---|---|---|
| | Negligible in water | | Very strong in water |

In this group we have the opposite situation as in group 1. The acids are so weak that they have negligible ionizations or acid properties in water. Their conjugate bases do exist, however, in certain ionic compounds (e.g., $Li^+CH_3^-$). These ions are 100% hydrolyzed in water, which means that they are very strong bases in water. The behavior of $CH_4$ and its conjugate base in water is summarized by the following equations.

$$CH_4 + H_2O \rightarrow \text{no reaction}$$

$$CH_3^- + H_2O \xrightarrow{100\%} CH_4 + OH^-$$

In this case, *only the molecular form of the acid ($CH_4$) exists in water.* The conjugate base ($CH_3^-$) when present in an ionic compound reacts completely with water.

In summary, the species capable of existence in aqueous solution are shaded in Table 13-1. The species not shaded react completely to form either their conjugate base (group 1) or conjugate acid (group 3). In further development of this topic, Table 13-1 could be expanded to include many more compounds. Expanded versions of the table also include many ions that behave as acids in water (e.g., $HSO_4^-$, $NH_4^+$).

**Example 13-13**

Referring to Table 13-1, write an equation illustrating the reaction of each of the following as an acid or base in water. If a partial reaction exists, indicate with double arrows. If the species does not exhibit acid or base nature in water, write "no reaction."

(a) HBr (b) $H^-$ (c) $OCl^-$ (d) $ClO_4^-$ (e) $H_2$ (f) HCN

**Answer**

(a) HBr is a very strong acid: $HBr + H_2O \rightarrow H_3O^+ + Br^-$.
(b) $H^-$ is a very strong base: $H^- + H_2O \rightarrow H_2 + OH^-$.
(c) $OCl^-$ is a weak base (hydrolyzes): $OCl^- + H_2O \rightleftharpoons HOCl + OH^-$.
(d) $ClO_4^-$ is a very weak base: no reaction with water.
(e) $H_2$ is a very weak acid: no reaction with water.
(f) HCN is a weak acid: $HCN + H_2O \rightleftharpoons H_3O^+ + CN^-$.

In the following examples, indicate whether a solution of the salt is acidic, basic, or neutral, and write the reaction illustrating this behavior. To do this, both the cation and anion must be examined for hydrolysis behavior.

**Example 13-14: $KNO_2$**

**Solution**
$K^+$ is the cation of the strong base KOH and does not hydrolyze. The $NO_2^-$ ion, however, is the conjugate base of the weak acid $HNO_2$ and hydrolyzes as follows:

$$NO_2^- + H_2O \rightleftharpoons HNO_2 + OH^-$$

Since $OH^-$ is formed in this solution, the solution is basic.

**Example 13-15: $Ca(NO_3)_2$**

**Solution**
$Ca^{2+}$ is the cation of the strong base, $Ca(OH)_2$, and does not hydrolyze. $NO_3^-$ is the conjugate base of the strong acid $HNO_3$ and does not hydrolyze either. Since neither ion hydrolyzes, the solution is neutral.

**Example 13-16: $H_2N(CH_3)_2{}^+Br^-$** [$HN(CH_3)_2$ is a weak base like $NH_3$.]

**Solution**
$H_2N(CH_3)_2{}^+$ is the conjugate acid of the weak base $HN(CH_3)_2$. It undergoes hydrolysis according to the equation

$$H_2N(CH_3)_2{}^+ + H_2O \rightleftharpoons HN(CH_3)_2 + H_3O^+$$

The $Br^-$ ion is the conjugate base of the strong acid HBr and does not hydrolyze. Since only the cation undergoes hydrolysis, the solution is acidic.

**See Problems 13-51 through 13-63.**

## 13-8 Buffer Solutions

The addition of small amounts of a strong acid or base to pure water causes drastic changes in the pH. For example, if 0.10 mol of HCl is added to 1 L of pure water, the pH changes from 7.0 to 1.0. On the other hand, if 0.10 mol of KOH is added to 1 L of pure water, the pH changes from 7.0 to 13.0. In contrast, our bloodstream is a water solution with almost the same pH as pure water (the pH of blood = 7.4), but unlike pure water, the pH of blood changes very little despite all of the acids and bases we ingest in our foods. The blood resists changes in pH because it is a buffer solution. A **buffer solution** *is a solution that can resist changes in pH from the addition of limited amounts of a strong acid or base.*

To understand how a buffer solution works, let's again consider the dissociation of a weak acid:

$$HCN + H_2O \rightleftharpoons H_3O^+ + CN^-$$

In a solution of HCN, the point of equilibrium lies far to the left, which means that most of the HCN is present as undissociated molecules. Now let's examine what happens if a small amount of the strong base $K^+OH^-$ is added to a solution of HCN. Remember that even a small amount of KOH would make the pH of pure water change to a much higher value. In this case, however, the added $OH^-$ reacts with the $H_3O^+$ present in a neutralization reaction.

$$H_3O^+ + OH^- \rightarrow 2H_2O$$

In a dynamic equilibrium situation, the point of equilibrium can change to make up for a loss. In this case, the $H_3O^+$ lost by neutralization can be replaced by further dissociation of HCN.

$$HCN + H_2O \rightarrow H_3O^+ + CN^-$$

If we add these two reactions, the overall reaction is obvious.

$$HCN + OH^- \rightarrow CN^- + H_2O$$

In summary, the addition of small amounts of a strong base to a weak acid solution does not effect the pH to a large extent because the added $OH^-$ is essentially removed by the undissociated HCN. The undissociated HCN serves as a reservoir to replace any $H_3O^+$ lost by reaction with $OH^-$.

This situation is much like the case of a young man who likes to keep an equilibrium concentration of $10 in his pocket. He can be "buffered" against going broke by having a savings account in the bank. If he should spend $5, he can replace the spent money by withdrawing $5 from his reservoir in the bank. (See Figure 13-10.)

Could this solution also resist a change in pH if additional $H_3O^+$ is added to the solution? No. A solution of a weak acid *alone* could not resist a change in pH in this direction. However, notice that the added $H_3O^+$ could be removed if a reservoir of the conjugate base of the acid (i.e., $CN^-$) is also present in the same solution. The $CN^-$ reacts with the added $H_3O^+$ in the reverse reaction of the dissociation:

$$H_3O^+ + CN^- \rightarrow HCN + H_2O$$

Thus a buffer solution resists changes in pH because it contains *both* a concentration of a weak acid (e.g., HCN) and a salt providing its conjugate base (e.g., NaCN) *in the same solution*. A buffer solution may also be compound of a solution of a weak base (e.g., $NH_3$) and a salt providing its conjugate acid (e.g., $NH_4Cl$).

In Figure 13-10 we discussed the analogy of a fat bank account serving as a "buffer" against going broke. What if that young man should earn some extra money? He could still maintain his equilibrium concentration of $10, but he would have to remove the extra $5 by depositing it in the bank.

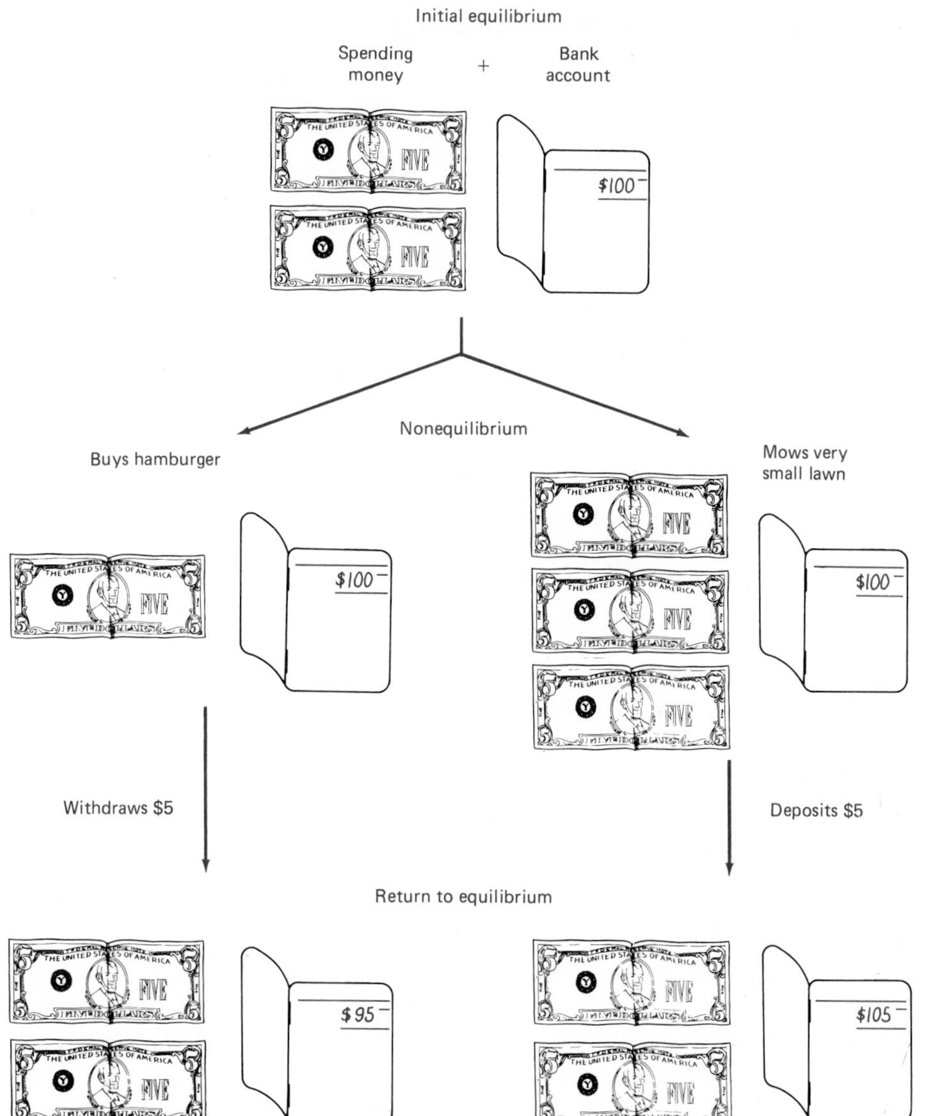


**Figure 13-10** AN ANALOGY TO A BUFFER. The $10 spending money is analogous to the equilibrium concentration of $H_3O^+$ in a buffer solution.

Likewise, any added $H_3O^+$ can be removed from the buffer solution (by reaction with $CN^-$) to form additional undissociated HCN.

As mentioned, a common buffered solution is blood, which must maintain a nearly constant pH for metabolism. This is accomplished by a combination of the following two buffer systems:

$$H_2CO_3 + H_2O \rightleftharpoons H_3O^+ + HCO_3^-$$

$$H_2PO_4^- + H_2O \rightleftharpoons H_3O^+ + HPO_4^{2-}$$

If for some reason the buffer breaks down and the pH of the blood drops lower than 7.2 (acidosis) or rises higher than 7.8 (alkalosis), serious difficulties are encountered—such as dizziness, coma, and muscle spasms. Since we ingest many acidic and basic substances, a healthy buffering system is necessary to maintain a controlled pH.

**See Problems 13-64 through 13-67.**

## 13-9 Oxides as Acids and Bases

Carbonated water is tangy because the carbon dioxide gas dissolves in water to produce the weak carbonic acid:

$$CO_2(g) + H_2O \rightleftharpoons H_2CO_3(aq) + H_2O \rightleftharpoons H_3O^+(aq) + HCO_3^-(aq)$$

Since $CO_2$ can be obtained algebraically by subtracting out the elements of water from the formula of carbonic acid, *the oxide is called an* **acid anhydride** (*acid without water*).

$$\begin{array}{r} H_2CO_3 \\ -H_2O \\ \hline CO_2 \end{array}$$

Most nonmetal oxides form acids when dissolved in water. Some other examples of the more common acid anhydrides and their reactions with water to produce the acid are the following:

$$SO_2(g) + H_2O \rightarrow H_2SO_3(aq)$$

$$N_2O_5(l) + H_2O \rightarrow 2HNO_3(aq)$$

$$B_2O_3(s) + 3H_2O \rightarrow 2H_3BO_3(aq)$$

$$SO_3(g) + H_2O \rightarrow H_2SO_4(aq)$$

Notice that the central nonmetal has the *same oxidation state in both the oxide and the acid.* This is a convenient way to tell in some cases which oxide goes with which acid. For example, is $N_2O_3$ the anhydride for $HNO_3$ or $HNO_2$? The answer is $HNO_2$, since both compounds ($HNO_2$ and $N_2O_3$) have N in a + 3 oxidation state.

The fact that gaseous oxides dissolve in water to form acids is currently one of the most debated and urgent problems with the environment. When high sulfur coal is burned, $SO_2$ is released as a gas. By itself, $SO_2$ is an irritating and pungent gas that combines with water to form $H_2SO_3$, which is a weak acid. Unfortunately, however, some of the $H_2SO_3$ reacts with oxygen in the air to form $H_2SO_4$, which is a strong acid. As a result, water containing dissolved $SO_2$ can become appreciably acidic.* In Canada, northeastern United States, and other parts of the industrialized world the acidity

*On April 10, 1974 a rain fell on Pilochry, Scotland, that had a pH of 2.4, which is about the same as vinegar. This is the most acidic rain ever recorded. Nitrogen oxides formed in automobile engines also contribute to acid rain by forming nitric acid.

of the rain is causing increased environmental problems. Acid rain leaches minerals from the soil, corrodes metals, dissolves limestone ($CaCO_3$) in buildings, and kills fish in lakes. In southern Scandinavia, for example, the fish cannot live in many lakes because of the high acidity. This area is downwind from northern Europe, in which a large amount of industry is concentrated. The Canadian and United States governments currently have a joint commission studying the problem. They have made recommendations toward a solution. The probable solution is to remove $SO_2$ from the smoke of plants that use coal or oil. Carbon dioxide, which is normally present in the air, also makes the rain acidic, but $H_2CO_3$ is such a weak acid that it does not cause such dramatic effects. (See Figure 13-11.)

*Ionic metal oxides dissolve in water to form bases and thus are known as* **base anhydrides.** Some examples of these reactions are

$$Na_2O(s) + H_2O \rightarrow 2NaOH(aq)$$

$$CaO(s) + H_2O \rightarrow Ca(OH)_2(aq)$$

Salt formation occurs in a reaction between an acid anhydride and a base anhydride. For example, the following reaction occurs, forming the same salt as is formed in the neutralization of $H_2SO_3$ with $Ca(OH)_2$ in aqueous solution.

$$SO_2(g) + CaO(s) \rightarrow CaSO_3(s)$$

$$H_2SO_3(aq) + Ca(OH)_2(aq) \rightarrow CaSO_3(s) + 2H_2O$$

**Figure 13-11** THE EFFECT OF ACID RAIN. The deterioration of this ancient statute is blamed on acid rain and air pollution.

See Problems 13-68 through 13-75.

The former reaction is typical of reactions that are being studied as a possible way to remove $SO_2$ from the combustion products of an industrial plant so that some of our abundant high-sulfur coal can be used without harming the environment.

## 13-10 Chapter Summary

Compounds have been classified as acids or bases for hundreds of years based on common sets of chemical characteristics. In this century, however, this acid character is known to be due to formation of $H^+$(aq) [also represented as the hydronium ion ($H_3O^+$)] in aqueous solution. The well-known base character is due to the formation of $OH^-$ (aq) in solution. When acid and base solutions are mixed, the two ions combine to form water in what is known as a neutralization reaction. Complete neutralizations also form ionic compounds called salts, as in reaction 1. Incomplete neutralizations of polyprotic acids produce acid salts, as in reaction 2.

| | Acid | + Base | → Salt | + Water |
|---|---|---|---|---|
| 1 | $HX(aq)$ | $+ M^+OH^-(aq)$ | $\rightarrow M^+X^-(aq)$ | $+ H_2O$ |
| 2 | $H_2Y(aq)$ | $+ M^+OH^-(aq)$ | $\rightarrow M^+HY^-(aq)$ | $+ H_2O$ |

Acids and bases can also be classified as to strength. Strong acids and bases are 100% ionized whereas weak acids and bases are only partially ionized. Partial ionization results because of a reaction that reaches a point of equilibrium in which both molecules and ions are present. For weak acids and bases the point of equilibrium favors the left or molecular side of the equation. Therefore, the $H_3O^+$ concentration in a weak acid solution is considerably lower than in a strong acid solution at the same initial concentration of acid.

Even in pure water there is a very small equilibrium concentration of $H_3O^+$ and $OH^-$, which is found to be $10^{-7}$ *M*. The product of these concentrations is a constant known as the ion product of water. The ion product can be used to calculate the concentration of one ion from that of the other in any aqueous solution.

A convenient method to express the $H_3O^+$ or $OH^-$ concentrations of solutions involves the use of the logarithmic definition, pH and pOH. Examples of solutions of various acidities in terms of $[H_3O^+]$ and pH are as follows:

| Solution | $[H_3O^+]$ | $[OH^-]$ | pH | pOH |
|---|---|---|---|---|
| Strongly acidic | $>10^{-1}$ | $<10^{-13}$ | $<1.0$ | $>13.0$ |
| Weakly acidic | $10^{-4}$ | $10^{-10}$ | 4.0 | 10.0 |
| Neutral | $10^{-7}$ | $10^{-7}$ | 7.0 | 7.0 |
| Weakly basic | $10^{-10}$ | $10^{-4}$ | 10.0 | 4.0 |
| Strongly basic | $<10^{-13}$ | $>10^{-1}$ | $>13.0$ | $<1.0$ |

Our understanding of acid-base behavior can be broadened somewhat by use of the Brønsted-Lowry definition. In this case, acids are $H^+$ donors and bases are $H^+$ acceptors. Acids and bases react to form their conjugate bases and acids respectively. We can use this definition to correlate the strength of the acids and their conjugate bases. In this situation acids and their conjugate bases can be placed into three categories as follows:

| Group | Acid in Water | Conjugate Base in Water | Comments |
|---|---|---|---|
| 1 | $HCl + H_2O \xrightarrow{100\%} Cl^- + H_3O^+$ | $Cl^- + H_2O \rightarrow$ no reaction | 100% ionization of acid<br>No hydrolysis of anion |
| 2 | $HF + H_2O \rightleftharpoons F^- + H_3O^+$ | $F^- + H_2O \rightleftharpoons HF + OH^-$ | Partial ionization of acid<br>Partial hydrolysis of anion |
| 3 | $H_2 + H_2O \rightarrow$ no reaction | $H^- + H_2O \xrightarrow{100\%} H_2 + OH^-$ | No ionization of acid<br>100% hydrolysis of anion |

The reaction of a group 2 or 3 anion as a base in water or the reaction of certain cations in water as acids is known as hydrolysis. The possibility of hydrolysis reactions means that solutions of salts may not be neutral. To predict the acidity of the solution of a salt, possible hydrolysis reactions of both cation and anion must be examined. Four possible combinations are as follows:

| Salt | Solution | Comments |
|---|---|---|
| 1 $NaClO_4$ | Neutral | Neither cation nor anion hydrolyzes. |
| 2 $K_2S$ | Basic | Anion ($S^{2-}$) undergoes hydrolysis.<br>$S^{2-} + H_2O \rightleftharpoons HS^- + OH^-$ |
| 3 $NH_4Br$ | Acidic | Cation ($NH_4^+$) undergoes hydrolysis.<br>$NH_4^+ + H_2O \rightleftharpoons NH_3 + H_3O^+$ |
| 4 $(NH_4)_2S$ | Unable to predict | Both ions undergo hydrolysis, but equilibrium constants are needed to determine which hydrolysis is more extensive. |

When solutions of weak acids are mixed with solutions of salts providing their conjugate bases, a buffer solution is formed. Buffer solutions resist changes in pH from addition of small amounts of a strong acid or base. In a buffer the reservoir of un-ionized acid (e.g., HCN) absorbs added $OH^-$ while the reservoir of the conjugate base (e.g., $CN^-$) absorbs added $H_3O^+$. Weak bases and salts providing their conjugate acids also serve as buffers (e.g., $NH_3$ and $NH_4^+Cl^-$).

Finally, the list of acids and bases was expanded to include oxides. The total list of acids and bases studied is summarized as follows:

| Type | Example | Reactions |
|---|---|---|
| | ACIDS | |
| 1 Molecular hydrogen compounds ($H^+$ + ion) | $HClO_4$ | $HClO_4 + H_2O \rightarrow H_3O^+ + ClO_4^-$ |
| 2 Cations (conjugate acids of weak bases) | $NH_4^+$ | $NH_4^+ + H_2O \rightleftharpoons H_3O^+ + NH_3$ |
| 3 Nonmetal oxides | $SO_3$ | $SO_3 + H_2O \rightarrow H_2SO_4$<br>$H_2SO_4 + H_2O \rightarrow H_3O^+ + HSO_4^-$ |
| | BASES | |
| 1 Ionic hydroxides | $Ca(OH)_2$ | $Ca(OH)_2 \xrightarrow{H_2O} Ca^{2+} + 2OH^-$ |
| 2 Molecular nitrogen compounds | $NH(CH_3)_2$ | $NH(CH_3)_2 + H_2O \rightleftharpoons HNH(CH_3)_2^+ + OH^-$ |
| 3 Anions (conjugate bases of weak acids) | $CN^-$ | $CN^- + H_2O \rightleftharpoons HCN + OH^-$ |
| 4 Metal oxides | $K_2O$ | $K_2O + H_2O \rightarrow 2KOH$<br>$KOH \xrightarrow{H_2O} K^+ + OH^-$ |

## Exercises

### Acids and Bases

**13-1** Give the formulas and names of the acid compounds derived from the following anions.
(a) $NO_3^-$ (b) $NO_2^-$ (c) $ClO_3^-$
(d) $SO_3^{2-}$

**13-2** Give the formulas and names of the base compounds derived from the following cations.
(a) $Cs^+$ (b) $Sr^{2+}$ (c) $Al^{3+}$ (d) $Mn^{3+}$

**13-3** Give the formulas and names of the acid or base compounds derived from the following ions.
(a) $S^{2-}$ (d) $HCO_3^-$
(b) $Li^+$ (e) $BrO_4^-$
(c) $Fe^{2+}$

**13-4** Write equations illustrating the reaction of the acids formed in Problem 13-1 with water.

**13-5** Write equations illustrating the reaction of the acids and bases formed in Problem 13-3 with water.

### Neutralization and Salts

**13-6** Identify each of the following as an acid, base, salt, or acid salt.
(a) $H_2S$ (d) $Ba(HSO_4)_2$
(b) $BaCl_2$ (e) $K_2SO_4$
(c) $H_3AsO_4$ (f) $LiOH$

**13-7** Write the balanced equation showing the complete neutralization of:
(a) $KOH$ by $HC_2H_3O_2$
(b) $Ca(OH)_2$ by $HI$
(c) $H_2SO_4$ by $Ca(OH)_2$

**13-8** Write the balanced equation showing the complete neutralization of:
(a) $HNO_2$ by $NaOH$
(b) $H_2S$ by $CsOH$
(c) $Al(OH)_3$ by $H_2SO_4$

**13-9** Write the total ionic equations and the

net ionic equations for reactions (b) and (c) in Problem 13-7.

**13-10** Write balanced acid-base neutralization reactions that would lead to formation of the following salts or acid salts.
(a) $CaBr_2$ (c) $Ba(HS)_2$
(b) $Sr(ClO_2)_2$ (d) $Li_2S$

**13-11** Write balanced acid-base neutralization reactions that would lead to formation of the following salts or acid salts.
(a) $Na_2SO_3$ (c) $Mg_3(PO_4)_2$
(b) $AlI_3$ (d) $NaHCO_3$

**13-12** Write two equations illustrating the stepwise neutralization of $H_2S$ with LiOH.

**13-13** Write three equations illustrating the stepwise neutralization of $H_3AsO_4$ with NaOH. Write the total reaction.

**13-14** Write the equation illustrating the reaction of 1 mol of $H_2S$ with 1 mol of NaOH.

* **13-15** Write the equation illustrating the reaction between 1 mol of $Ca(OH)_2$ and 2 mol of $H_3PO_4$.

### Strengths of Acids and Base

**13-16** Describe the difference between a strong acid and a weak acid.

**13-17** Dimethylamine [$NH(CH_3)_2$] is a weak base that reacts in water like ammonia ($NH_3$). Write the equilibrium illustrating this reaction.

**13-18** The concentration of a weak monoprotic acid (HX) in water is 0.10 *M*. The concentration of $H_3O^+$ ion in this solution is 0.010 *M*. Is HX a weak or strong acid? What percent of the acid is ionized?

**13-19** A 0.50 mol quantity of an acid is dissolved in 2.0 L of water. In the solution, $[H_3O^+] = 0.25$ *M*. Is this a strong or weak acid? Explain.

**13-20** What is $[H_3O^+]$ in a 0.55 *M* $HClO_4$ solution?

**13-21** What is $[H_3O^+]$ in a 0.55 *M* solution of a weak acid, HX, that is 3.0% ionized?

**13-22** What is $[OH^-]$ in a 1.45 *M* solution of $NH_3$ if the $NH_3$ is 0.95% ionized?

* **13-23** What is $[H_3O^+]$ in a 0.354 *M* solution of $H_2SO_4$? Assume that the first ionization is complete but that the second is only 25% complete.

### Equilibrium of Water and $K_w$

**13-24** If some ions are present in pure $H_2O$, why isn't pure water considered to be an electrolyte?

**13-25** Why can't $[H_3O^+] = [OH^-] = 1.00 \times 10^{-2}$ *M* in water? What would happen if we tried to make such a solution by mixing $10^{-2}$ mol/L of KOH with $10^{-2}$ mol/L of HCl?

**13-26** (a) What is $[H_3O^+]$ when $[OH^-] = 10^{-12}$ *M*?
(b) What is $[H_3O^+]$ when $[OH^-] = 10$ *M*?
(c) What is $[OH^-]$ when $[H_3O^+] = 2.0 \times 10^{-5}$ *M*?

**13-27** (a) What is $[OH^-]$ when $[H_3O^+] = 1.50 \times 10^{-3}$*M*?
(b) What is $[H_3O^+]$ when $[OH^-] = 2.58 \times 10^{-7}$ *M*?
(c) What is $[H_3O^+]$ when $[OH^-] = 56.9 \times 10^{-9}$ *M*?

**13-28** When 0.250 mol of the strong acid $HClO_4$ is dissolved in 10.0 L of water, what is $[H_3O^+]$? What is $[OH^-]$?

**13-29** Lye is a very strong base. What is $[H_3O^+]$ in a 2.55 *M* solution of NaOH? In the weakly basic household ammonia, $[OH^-] = 4.0 \times 10^{-3}$ *M*. What is $[H_3O^+]$?

**13-30** Identify the solutions in Problem 13-26 as acidic, basic, or neutral.

**13-31** Identify the solutions in Problem 13-27 as acidic, basic, or neutral.

**13-32** Identify each of the following as an acidic, basic, or neutral solution.
(a) $[H_3O^+] = 6.5 \times 10^{-3}$ *M*
(b) $[H_3O^+] = 5.5 \times 10^{-10}$ *M*
(c) $[OH^-] = 4.5 \times 10^{-8}$ *M*
(d) $[OH^-] = 50 \times 10^{-8}$ *M*

**13-33** Identify each of the following as an acidic, basic, or neutral solution.

(a) $[OH^-] = 10.0 \times 10^{-8}\,M$
(b) $[OH^-] = 8.1 \times 10^{-4}\,M$
(c) $[H_3O^+] = 518 \times 10^{-8}\,M$

### pH and pOH

**13-34** What is the pH of each of the following solutions?
(a) $[H_3O^+] = 1.0 \times 10^{-6}\,M$
(b) $[H_3O^+] = 1.0 \times 10^{-9}\,M$
(c) $[OH^-] = 1.0 \times 10^{-2}\,M$
(d) $[OH^-] = 2.5 \times 10^{-5}\,M$
(e) $[H_3O^+] = 6.5 \times 10^{-11}\,M$

**13-35** What is the pH of each of the following solutions?
(a) $[H_3O^+] = 1.0 \times 10^{-4}\,M$
(b) $[H_3O^+] = 1.0 \times 10^{-9}\,M$
(c) $[H_3O^+] = 1.5\,M$
(d) $[OH^-] = 0.058 \times 10^{-3}\,M$
(e) $[OH^-] = 460 \times 10^{-14}\,M$

**13-36** What is $[H_3O^+]$ when:
(a) pH = 3.00 (d) pOH = 6.38
(b) pH = 3.54 (e) pH = 12.70
(c) pOH = 8.00

**13-37** What is $[H_3O^+]$ when:
(a) pH = 9.0 (c) pH = 2.30
(b) pOH = 9.0 (d) pH = 8.90

**13-38** Identify each of the solutions in Problems 13-34 and 13-36 as acidic, basic, or neutral.

**13-39** Identify each of the solutions in Problems 13-35 and 13-37 as acidic, basic, or neutral.

**13-40** What is the pH of a 0.075 *M* solution of the strong acid $HNO_3$?

**13-41** What is the pH of a 0.86 *M* solution of the strong base KOH?

**13-42** What is the pH of a 0.018 *M* solution of $Ca(OH)_2$?

**13-43** Identify each of the following solutions as strongly basic, weakly basic, neutral, weakly acidic, or strongly acidic.
(a) pH = 1.5 (e) pOH = 7.0
(b) pOH = 13.0 (f) pH = 8.5
(c) pH = 5.8 (g) pOH = 7.5
(d) pH = 13.0 (h) pH = −1.00

**13-44** Order the following substances in order of *increasing* acidity.
(a) household ammonia, pH = 11.4
(b) vinegar, pH = 2.6
(c) grape juice, pH = 4.0
(d) sulfuric acid, pH = 0.40
(e) eggs, pH = 7.8
(f) rain water, pH = 5.6

### Brønsted-Lowry Acids and Bases

**13-45** What is the conjugate base of each of the following?
(a) $HNO_3$ (d) $CH_4$
(b) $H_2SO_4$ (e) $H_2O$
(c) $HPO_4^{2-}$ (f) $NH_3$

**13-46** What is the conjugate acid of each of the following?
(a) $NH_2CH_3$ (d) $O^{2-}$
(b) $HPO_4^{2-}$ (e) HCN
(c) $NO_3^-$ (f) $H_2O$

**13-47** Write reactions indicating Brønsted-Lowry acid behavior with $H_2O$ for each of the following. Indicate conjugate acid-base pairs.
(a) $H_2SO_3$ (c) HBr (e) $H_2S$
(b) HClO (d) $HSO_3^-$ (f) $NH_4^+$

**13-48** Write reactions indicating Brønsted-Lowry base behavior with $H_2O$ for each of the following. Indicate conjugate acid-base pairs.
(a) $NH_3$ (c) $HS^-$ (e) $F^-$
(b) $N_2H_4$ (d) $H^-$

**13-49** Most acid salts are amphoteric, as illustrated in Example 13-11. Write equations showing how $HS^-$ can act as a Brønsted-Lowry base with $H_3O^+$ and as a Brønsted-Lowry acid with $OH^-$.

**13-50** Calcium carbonate is the principal ingredient in Rolaids, Tums, and other products used to neutralize the $H_3O^+$ from excess stomach acid (HCl). Show an equation illustrating how the $CO_3^{2-}$ ion acts as a Brønsted-Lowry base with $H_3O^+$. Bicarbonate of soda ($NaHCO_3$) also acts as an antacid (base) in water. Show an equation illustrating how the $HCO_3^-$ ion reacts with $H_3O^+$.

### Predicting Acid and Base Reactions: Hydrolysis

**13-51** Referring to Table 13-1, complete the following equations. When no reaction occurs, write N.R. When a partial reac-

tion exists, indicate it by double arrows.
(a) $HNO_3 + H_2O$ (d) $HOCl + H_2O$
(b) $NO_2^- + H_3O^+$ (e) $NO_3^- + H_3O^+$
(c) $CH_4 + H_2O$ (f) $HClO_4 + H_2O$

**13-52** Referring to Table 13-1, complete the following equations. When no reaction occurs, write N.R. When a partial reaction occurs, indicate with double arrows.
(a) $Br^- + H_3O^+$ (c) $OCl^- + H_2O$
(b) $H_2S + H_2O$ (d) $HBr + H_2O$

**13-53** Complete the following hydrolysis equilibria:
(a) $S^{2-} + H_2O \rightleftharpoons$ ____ $+ OH^-$
(b) $N_2H_5^+ + H_2O \rightleftharpoons N_2H_4 +$ ____
(c) $HPO_4^{2-} + H_2O \rightleftharpoons H_2PO_4^- +$ ____
(d) $HNH(CH_3)_2^+ + H_2O \rightleftharpoons$ ____ $+ H_3O^+$

**13-54** Complete the following hydrolysis equilibria.
(a) $CN^- + H_2O \rightleftharpoons HCN +$ ____
(b) $NH_4^+ + H_2O \rightleftharpoons$ ____ $+ H_3O^+$
(c) $B(OH)_4^- + H_2O \rightleftharpoons H_3BO_3 +$ ____
(d) $Al(H_2O)_6^{3+} + H_2O \rightleftharpoons Al(H_2O)_5(OH)^{2+} +$ ____

**13-55** Write the hydrolysis equilibrium (if any) for each of the following ions.
(a) $F^-$ (d) $HPO_4^{2-}$
(b) $SO_3^{2-}$ (e) $CN^-$
(c) $H_2N(CH_3)_2^+$ (f) $Li^+$

**13-56** Write hydrolysis reactions, if any occur, for each of the following ions.
(a) $Br^-$ (c) $ClO_4^-$ (e) $Ca^{2+}$
(b) $HS^-$ (d) $H^-$

* **13-57** Both $C_2^{2-}$ and its conjugate acid $HC_2^-$ hydrolyze 100% in water. From this information complete the following equation.

$$CaC_2(s) + 2H_2O \rightarrow \underline{\quad\quad}(g) + Ca^{2+}(aq) + 2\underline{\quad\quad}(aq)$$

[The gas formed (acetylene) can be burned as it is produced. This reaction was once important for this purpose as a source of light in old miners' lamps.]

**13-58** Sodium hypochlorite (NaOCl) is the principal ingredient in commercial bleaches. When dissolved in water, it makes a slightly basic solution. Write the equation illustrating this basic behavior.

**13-59** Aqueous NaF solutions are slightly basic whereas aqueous NaCl solutions are neutral. Write the appropriate equation that illustrates this. Why aren't NaCl solutions also basic?

**13-60** Predict whether aqueous solutions of the following salts are acidic, neutral, or basic.
(a) $Ba(ClO_4)_2$ (d) KBr
(b) $N_2H_5^+NO_3^-$ ($N_2H_4$ is a weak base) (e) $NH_4Cl$
(f) $BaF_2$
(c) KHS

**13-61** Predict whether aqueous solutions of the following salts are acidic, neutral, or basic.
(a) $Na_2CO_3$ (c) $NH_4ClO_4$
(b) $K_3PO_4$ (d) $SrI_2$

**13-62** Aqueous solutions of $NH_4CN$ are basic. Write the two hydrolysis reactions and indicate which takes place to the greater extent.

* **13-63** Aqueous solutions of $NaHSO_3$ are acidic. Write the two equations (one hydrolysis and one ionization) and indicate which takes place to the greater extent.

## Buffers

**13-64** Identify which of the following form buffer solutions when 0.50 mol of both compounds are dissolved in 1 L of water.
(a) $HNO_2$ and $KNO_2$
(b) $NH_4Cl$ and $NH_3$
(c) $HNO_3$ and $KNO_2$
(d) $HNO_3$ and $KNO_3$
(e) HOBr and $Ca(OBr)_2$
(f) HCN and KClO
(g) $NH_3$ and $BaBr_2$
(h) $H_2S$ and LiHS
(i) $KH_2PO_4$ and $K_2HPO_4$

**13-65** A certain solution contains dissolved HCl and NaCl. Why can't this solution act as a buffer?

* **13-66** If 0.5 mol of KOH is added to a solu-

tion containing 1.0 mol of $HC_2H_3O_2$, the resulting solution is a buffer. Explain.

13-67 A solution contains 0.50 mol each of HClO and NaClO. If 0.60 mol of KOH is added, will the buffer prevent a significant change in pH? Explain.

## Oxides as Acids and Bases and General Problems

**13-68** Write the formula of the acid or base formed when each of the following anhydrides is dissolved in water.
(a) SrO (c) $P_4O_{10}$ (e) $N_2O_3$
(b) $SeO_3$ (d) $Cs_2O$ (f) $Cl_2O_5$

**13-69** Write the formula of the acid or base formed when each of the following anhydrides is dissolved in water.
(a) BaO (c) $Cl_2O$ (e) $K_2O$
(b) $SeO_2$ (d) $Br_2O$

**13-70** Carbon dioxide is removed from manned space capsules by bubbling the air through a LiOH solution. Show the reaction and the product formed.

* **13-71** Complete the following reaction.

$$Li_2O(s) + N_2O_5(g) \rightarrow \_\_\_\_(s)$$

13-72 The sulfite ion ($SO_3^{2-}$) and sulfur trioxide ($SO_3$) look similar, but one forms a strongly acid solution whereas the other is weakly basic. Write equations illustrating this behavior.

**13-73** Tell whether each of the following compounds forms acidic, basic or neutral solutions when added to pure water. Write the reaction illustrating the acidic or basic behavior where appropriate.
(a) $H_2S$ (f) $Ba(OH)_2$
(b) KClO (g) $Sr(NO_3)_2$
(c) NaI (h) $LiNO_2$
(d) $NH_3$ (i) $H_2SO_3$
(e) $HN_2H_4^+Br^-$ (j) $Cl_2O_3$

**13-74** Tell whether each of the following compounds forms acidic, basic, or neutral solutions when added to pure water. Write the reaction illustrating the acidic or basic behavior where appropriate.
(a) HOBr (e) $SO_2$
(b) CaO (f) $Ba(C_2H_3O_2)_2$
(c) $NH_4ClO_4$ (g) RbBr
(d) $N_2H_4$

* **13-75** In a lab there are five different solutions with pH's of 1.0, 5.2, 7.0, 10.2 and 13.0. The solutions are LiOH, $SrBr_2$, KOCl, $NH_4Cl$, and HI, all at the same concentration. Which pH corresponds to which compound? What is the concentration of these compounds?

Batteries convert chemical energy into electrical energy in a spontaneous chemical reaction. The interconversion of chemical and electrical energy is the topic of this chapter.

# Chapter 14
# Oxidation-Reduction Reactions

## Purposes of Chapter 14

In Chapter 14 we define oxidation-reduction reactions, balance their equations, describe some practical applications, and predict whether or not certain reactions will occur.

## Objectives for Chapter 14

After completion of this chapter, you should be able to:

1 Apply the appropriate rules to calculate the oxidation state of the atoms of each element in a compound or ion. (14-1)

2 Identify the substance oxidized (reducing agent) and the substance reduced (oxidizing agent) in a specified redox reaction. (14-1)

3 Balance redox equations by either the oxidation state method or the ion-electron method in acidic or basic solutions. (14-2, 14-3).

4 Describe how the relationship of the potential energy of reactants and products relates to the reactions occurring in a voltaic cell. (14-4)

5 Write the anode, cathode, and overall reactions occurring in specified voltaic cells, including the Daniell cell, dry cell, and the lead-acid battery. (14-4)

6 Give some practical applications of electrolytic cells. (14-5)

7 Rank the strength of two competing oxidizing agents and reducing agents from observance of a single spontaneous redox reaction. (14-6)

8 Describe the inverse relationship between the oxidizing strength of a metal ion and the reducing strength of the corresponding metal. (14-6)

9 Predict the occurrence and write the equations for spontaneous redox reactions by use of an appropriate table. (14-6)

Some of the most familiar and important chemical reactions have something in common. Reactions such as combustion, metabolism, corrosion or rusting, decay of organic matter, and the generation of power from a battery all involve the exchange of electrons between two reactants. In the previous chapter we discussed reactions that involve an exchange of an $H^+$ ion. In this chapter, we will discuss reactions involving an exchange of electrons that are known as oxidation-reduction or simply redox reactions. This classification is somewhat broader than the types of reactions discussed in Chapter 9. In fact, redox reactions account for many of the reactions of all types listed in Chapter 9 except for double displacement reactions, which were discussed in detail in Chapters 12 and 13.

It is well worth the time to study redox reactions as a separate classification, however. From a knowledge and understanding of the electron exchange process we will do several things. These include (1) balancing complicated equations, (2) understanding how chemical energy is converted into electrical energy in batteries, (3) learning how electrical energy is used to free elements from their compounds, and, finally, (4) seeing how we can predict the occurrence of many chemical reactions from the observance of a few.

As background, review the following:

**1** The formation of ions and binary compounds (Sections 6-3 and 6-4)

**2** The names and formulas of ions (Section 7-3)

**3** Ionic equations (Section 12-4)

## 14-1 The Nature of Oxidation and Reduction

Sodium metal burns in the presence of chlorine gas to form table salt, NaCl (see Figure 14-1). This is represented by the equation

$$2\text{Na}(s) + \text{Cl}_2(g) \rightarrow 2\text{Na}^+\text{Cl}^-(s)$$

The active metal and poisonous gas combine to form a compound used for our food.

Let's take a close look at each element involved in this reaction to identify the changes that have occurred. First, the Na has lost an electron:

$$\text{Na} \rightarrow \text{Na}^+ + e^-$$

Notice that the oxidation state of the Na has changed from zero for the element to +1 for the ion. *A substance whose oxidation state has increased (by losing electrons) in a chemical reaction is said to be* **oxidized.**

The $Cl_2$ has undergone the following change:

$$2e^- + \text{Cl}_2 \rightarrow 2\text{Cl}^-$$

**Figure 14-1** THE FORMATION OF NaCl FROM ITS ELEMENTS. An active metal reacts with a poisonous gas to form sodium chloride.

Notice that the oxidation state of each Cl in $Cl_2$ has decreased from zero to $-1$. *A substance whose oxidation state has decreased (by gaining electrons) in a chemical reaction is said to be* **reduced.**

*Reactions involving an exchange of electrons are known as* **oxidation-reduction** *or simply* **redox reactions.**

*The substance that accepts the electrons from the substance oxidized is called the* **oxidizing agent.** Notice that this is the same as the substance reduced. *The substance that gives up the electrons to the substance reduced is called the* **reducing agent.** The reducing agent is thus the substance oxidized.

The reaction of Na and $Cl_2$ is summarized as follows:

| Reactant | Change | Product | Agent |
|---|---|---|---|
| Na | Oxidation | $Na^+$ | Reducing |
| $Cl_2$ | Reduction | $Cl^-$ | Oxidizing |

Notice that oxidation-reduction reactions are the sum of two half-reactions. *A* **half-reaction** *illustrates either the oxidation or the reduction reaction separately.* Two half-reactions add to make a total reaction in such a way that *electrons lost* in the oxidation half-reaction equal the *electrons gained* in the reduction half-reaction. Therefore, 2 mol of Na react to produce 2 mol of electrons needed for the 1 mol of $Cl_2$ as follows:

$$\begin{array}{ll} \text{Oxidation half-reaction:} & 2Na \rightarrow 2Na^+ + \cancel{2e^-} \\ \text{Reduction half-reaction:} & \cancel{2e^-} + Cl_2 \rightarrow 2Cl^- \\ \hline \text{Total reaction:} & 2Na + Cl_2 \rightarrow 2Na^+Cl^- \end{array}$$

We will have more to say about half-reactions and the balancing of equations later. First, let's concentrate on the identification of substances oxidized and reduced and of oxidizing and reducing agents. To do this, it is necessary to calculate oxidation states so as to recognize the species that have undergone changes. The rules for calculation of oxidation states were given in Section 7-1 but are summarized again here. Sample calculations are given in Chapter 7.

1 For any free or uncombined element the oxidation state is zero.

2 The oxidation state of a monatomic ion is the same as the charge on the ion. (Group IA is +1, group IIA is +2, Group IIIA is usually +3.)

3 The halogens (VIIA) have a −1 oxidation state in binary (two-element) compounds when bound to a less electronegative element.

4 Oxygen is usually −2. Peroxides and superoxides contain oxygen in −1 or $-\frac{1}{2}$ oxidation states, respectively. When bound to F, oxygen has a positive oxidation state.

5 Hydrogen is usually +1. When combined with a less electronegative element (i.e., a metal), H has a −1 oxidation state.

6 The oxidation states of all of the elements in a neutral compound add to zero. In the case of polyatomic ions, the sum of the oxidation states add to the charge on the ion.

**See Problems 14-1 and 14-2.**

---

**Example 14-1**

In the following unbalanced equations, indicate the substance oxidized, the substance reduced, the oxidizing agent, and the reducing agent. Indicate the products that contain the elements that were oxidized or reduced.

(a) $Al + HCl \rightarrow AlCl_3 + H_2$
(b) $CH_4 + O_2 \rightarrow CO_2 + H_2O$
(c) $MnO_2 + HCl \rightarrow MnCl_2 + Cl_2 + H_2O$
(d) $K_2Cr_2O_7 + SnCl_2 + HCl \rightarrow CrCl_3 + SnCl_4 + KCl + H_2O$

**Procedure**

In the equations, we wish to identify the species that contains atoms of an element undergoing a change in oxidation state. At first, it may be necessary to calculate the oxidation state of every atom in the equation until you can recognize the changes by inspection. You will notice that any substance present as a free element is involved in either the oxidation or reduction process and that hydrogen and oxygen are generally not oxidized or reduced unless they are present as a free element.

**Solution**

(a) Oxidation state of element $\quad 0 \quad +1 \; -1 \quad +3 \; -1 \quad 0$

$$Al + HCl \rightarrow AlCl_3 + H_2$$

| Reactant | Change | Product | Agent |
|---|---|---|---|
| Al | Oxidation | $AlCl_3$ | Reducing |
| HCl | Reduction | $H_2$ | Oxidizing |

(b) Oxidation state of element $\quad -4 \; +1 \quad 0 \quad +4 \; -2 \quad +1 \; -2$

$$CH_4 + O_2 \rightarrow CO_2 + H_2O$$

| Reactant | Change | Product | Agent |
|---|---|---|---|
| $CH_4$ | Oxidation | $CO_2$ | Reducing |
| $O_2$ | Reduction | $CO_2$, $H_2O$ | Oxidizing |

(c) Oxidation state of element $\quad +4 \; -2 \quad +1 \; -1 \quad +2 \; -1 \quad 0 \quad +1 \; -2$

$$MnO_2 + HCl \rightarrow MnCl_2 + Cl_2 + H_2O$$

| Reactant | Change | Product | Agent |
|---|---|---|---|
| HCl | Oxidation | $Cl_2$ | Reducing |
| $MnO_2$ | Reduction | $MnCl_2$ | Oxidizing |

(d) Oxidation state of element

$+1 \; +6 \; -2 \quad +2 \; -1 \quad +1 \; -1 \quad +3 \; -1 \quad +4 \; -1 \quad +1 \; -2$

$$K_2Cr_2O_7 + SnCl_2 + HCl \rightarrow CrCl_3 + SnCl_4 + H_2O$$

| Reactant | Change | Product | Agent |
|---|---|---|---|
| $SnCl_2$ | Oxidation | $SnCl_4$ | Reducing |
| $K_2Cr_2O_7$ | Reduction | $CrCl_3$ | Oxidizing |

**See Problems 14-3 through 14-7.**

## 14-2 Balancing Redox Equations: Oxidation State Method

There are two widely used procedures for balancing redox equations. The **bridge** or **oxidation state** *method discussed in this section focuses on* *the atoms of the elements undergoing a change in oxidation state.* This method serves as a helpful introduction to the concepts and is effective in balancing uncomplicated redox equations. *The second procedure is the* **ion-electron** *method, which focuses on the entire molecule or ion containing the atom undergoing a change in oxidation state.* This latter method is generally used in general and subsequent chemistry courses and is discussed in the next section.

The following reaction will be used to illustrate the procedures for balancing equations by the oxidation state method:

$$HNO_3(aq) + H_2S(aq) \rightarrow NO(g) + S(s) + H_2O$$

**1** Identify the atoms whose oxidation states have changed.

$$H\overset{+5}{\underline{\underline{N}}}O_3 + H_2\overset{-2}{\underline{\underline{S}}} \rightarrow \overset{+2}{\underline{\underline{N}}}O + \overset{0}{\underline{\underline{S}}} + H_2O$$

**2** Draw a bridge between the same atoms whose oxidation states have changed, indicating the electrons gained or lost. This is the change in oxidation state. *Be sure that the atoms in question are balanced on both sides of the equation if they are not the same.*

$$\overbrace{\overset{+5}{HNO_3} + H_2S \rightarrow \overset{+2}{NO}}^{+3e} + S + H_2O$$

$$HNO_3 + \underbrace{\underset{-2}{H_2S} \rightarrow NO + \underset{0}{S}}_{-2e^-} + H_2O$$

**3** Multiply the two numbers (+3 and −2) by whole numbers that produce a common number. For 3 and 2 the common number is 6. (For example, $+3 \times \underline{2} = +6$; $-2 \times \underline{3} = -6$.) Use these multipliers as coefficients of the respective compounds or elements.

$$+3e \times (2) = -6e^-$$

$$\underline{2}HNO_3 + \underline{3}H_2S \rightarrow \underline{2}NO + \underline{3}S + H_2O$$

$$-2e^- \times (3) = -6e^-$$

Notice that six electrons are lost (bottom) and six are gained (top).

4 Balance the rest of the equation by inspection. Notice that there are eight H's on the left, so *four* $H_2O$'s are needed on the right. If the equation has been balanced correctly, the O's should balance. Notice that they do.

$$2HNO_3 + 3H_2S \rightarrow 2NO + 3S + 4H_2O$$

**Example 14-2**

Balance the following equations by the oxidation state method.

(a) $Zn + AgNO_3 \rightarrow Zn(NO_3)_2 + Ag$

$-2e^-$

0 +2

$$Zn + AgNO_3 \rightarrow Zn(NO_3)_2 + Ag$$

+1 0

$+1e^-$

The oxidation process (top) should be multiplied by 1 and the reduction process (bottom) should be multiplied by 2.

$-2e^- \times 1 = -2e^-$

$$Zn + 2AgNO_3 \rightarrow Zn(NO_3)_2 + 2Ag$$

$+1e \times 2 = +2e$

The final balanced equation is

$$Zn + 2AgNO_3 \rightarrow Zn(NO_3)_2 + 2Ag$$

(b) $Cu + HNO_3 \rightarrow Cu(NO_3)_2 + H_2O + NO_2$

$-2e^- \times 1 = -2e^-$

0 +2

$$Cu + HNO_3 \rightarrow Cu(NO_3)_2 + H_2O + NO_2$$

+5 +4

$+1e^- \times 2 = +2e^-$

The equation, so far, appears as follows:

$$Cu + 2HNO_3 \rightarrow Cu(NO_3)_2 + H_2O + 2NO_2$$

Notice, however, that four N's are present on the right but only two on the left. The addition of two more $HNO_3$'s balances the N's, and the equation is completely balanced with two $H_2O$'s on the right:

$$Cu + 4HNO_3 \rightarrow Cu(NO_3)_2 + 2H_2O + 2NO_2$$

(In this aqueous reaction, $HNO_3$ serves two functions. Two $HNO_3$'s are reduced to two $NO_2$'s and the other two $HNO_3$'s provide anions for the $Cu^{2+}$ ion. These later $NO_3^-$ ions are present in the solution as spectator ions. Spectator ions are not oxidized, reduced, or otherwise changed during the reaction.)

(c) $Al + H_2SO_4 \rightarrow Al_2(SO_4)_3 + H_2$

The atoms undergoing a change in oxidation state are Al and H. Before you calculate electrons gained or lost, both atoms in question must be balanced on both sides of the equation. In this case, two Al's lose a total of six electrons and two H's gain a total of two electrons.

$$2(-3e^-) \times 1 = -6e^-$$

$$\overset{0}{2Al} + \overset{+2}{H_2}SO_4 \rightarrow \overset{+6}{Al_2}(SO_4)_3 + \overset{0}{H_2}$$

$$2(+1e^-) \times 3 = +6e^-$$

See Problems 14-8 and 14-9.

$$2Al + 3H_2SO_4 \rightarrow Al_2(SO_4)_3 + 3H_2$$

## 14-3 Balancing Redox Equations: The Ion-Electron Method

In the ion-electron method (also known as the half-reaction method), the total reaction is separated into half-reactions, which are then balanced separately and added. Although this method is somewhat more involved than the oxidation state method, it is more realistic for redox reactions in aqueous solutions. The ion-electron method recognizes that the entire molecule or ion, not just one atom, undergoes a change. This method also provides the proper background for the study of electrochemistry, which involves the applications of balanced half-reactions. This will be apparent later in this chapter.

The rules for balancing equations are somewhat different in acidic solution [containing $H^+(aq)$ ion] than in basic solution [containing $OH^-(aq)$ ion]. The two cases are taken up separately, with acid solution reactions discussed first. To simplify the equations, only the net ionic equations are balanced.

The balancing of an equation in aqueous acid solution is illustrated by the following unbalanced equation.

$$Cr_2O_7^{2-}(aq) + Cl^-(aq) + H^+(aq) \rightarrow Cr^{3+}(aq) + Cl_2(g) + H_2O$$

1 Separate the molecule or ion that contains an atom that has been oxidized or reduced and the product containing the atom that changed. If necessary, calculate the oxidation states of individual atoms until you are able to recognize the species that changed. *It is actually not necessary to know the oxidation state.* The reduction process is

$$Cr_2O_7^{2-} \rightarrow Cr^{3+}$$

2 If necessary, balance the atom undergoing a change in oxidation state. In this case it is Cr.

$$Cr_2O_7^{2-} \rightarrow 2Cr^{3+}$$

3 Balance the oxygens by adding $H_2O$ on the side needing the oxygens (one $H_2O$ for each O needed).

$$Cr_2O_7^{2-} \rightarrow 2Cr^{3+} + 7H_2O$$

4 Balance the hydrogens by adding $H^+$ on the other side of the equation from the $H_2O$'s ($2H^+$ for each $H_2O$ added). Notice that the H and O have not undergone a change in oxidation state.

$$14H^+ + Cr_2O_7^{2-} \rightarrow 2Cr^{3+} + 7H_2O$$

5 The atoms in the half-reaction are now balanced. Check to make sure. The charge on both sides of the reaction must now be balanced. To do this, add the appropriate number of electrons to the *more positive* side. The total charge on the left is $(14 \times +1) + (-2) = +12$. The total charge on the right is $(2 \times +3) = +6$. By adding $6e^-$ on the left, the charges balance on both sides and the half-reaction is balanced.

$$6e^- + 14H^+ + Cr_2O_7^{2-} \rightarrow 2Cr^{3+} + 7H_2O$$

6 Repeat the same procedure for the other half-reaction.

$$Cl^- \rightarrow Cl_2$$

$$2Cl^- \rightarrow Cl_2$$

$$2Cl^- \rightarrow Cl_2 + 2e^-$$

7 Before the two half-reactions are added, we must make sure that electrons gained equal electrons lost. Sometimes, the half-reactions must be multiplied through by factors that give the same number of electrons. In this case, if the oxidation process is multiplied through by 3 and the

reduction process by 1, there will be an exchange of $6e^-$. When these two half-reactions are added, the $6e^-$ are canceled from both sides of the equation.

$$3[2Cl^- \rightarrow Cl_2 + 2e^-]$$

$$6Cl^- \rightarrow 3Cl_2 + 6e^-$$

Addition produces the balanced net ionic equation:

$$\cancel{6e^-} + 14H^+ + Cr_2O_7^{2-} \rightarrow 2Cr^{3+} + 7H_2O$$
$$6Cl^- \rightarrow 3Cl_2 + \cancel{6e^-}$$
$$14H^+(aq) + 6Cl^-(aq) + Cr_2O_7^{2-}(aq) \rightarrow 2Cr^{3+}(aq) + 3Cl_2(g) + 7H_2O$$

**Example 14-3**

Balance the following equations for reactions occurring in acid solution by the ion-electron method.

(a) $MnO_4^-(aq) + SO_2(g) + H_2O \rightarrow Mn^{2+}(aq) + SO_4^{2-}(aq) + H^+(aq)$

Reduction: $MnO_4^- \rightarrow Mn^{2+}$

$H_2O$: $MnO_4^- \rightarrow Mn^{2+} + 4H_2O$

$H^+$: $8H^+ + MnO_4^- \rightarrow Mn^{2+} + 4H_2O$

$e^-$: $5e^- + 8H^+ + MnO_4^- \rightarrow Mn^{2+} + 4H_2O$

Oxidation: $SO_2 \rightarrow SO_4^{2-}$

$H_2O$: $2H_2O + SO_2 \rightarrow SO_4^{2-}$

$H^+$: $2H_2O + SO_2 \rightarrow SO_4^{2-} + 4H^+$

$e^-$: $2H_2O + SO_2 \rightarrow SO_4^{2-} + 4H^+ + 2e^-$

The reduction reaction is multiplied by 2 and the oxidation by 5 to produce 10 electrons for each process as shown below.

$$2(5e^- + 8H^+ + MnO_4^- \rightarrow Mn^{2+} + 4H_2O)$$
$$5(2H_2O + SO_2 \rightarrow SO_4^{2-} + 4H^+ + 2e^-)$$
$$\cancel{10e^-} + 16H^+ + 2MnO_4^- \rightarrow 2Mn^{2+} + 8H_2O$$
$$10H_2O + 5SO_2 \rightarrow 5SO_4^{2-} + 20H^+ + \cancel{10e^-}$$
$$10H_2O + 16H^+ + 5SO_2 + 2MnO_4^- \rightarrow 5SO_4^{2-} + 2Mn^{2+} + 8H_2O + 20H^+$$

Notice that $H_2O$ and $H^+$ are present on both sides of the equation. Therefore, $8H_2O$ and $16H^+$ can be subtracted from *both sides*, leaving the final balanced net ionic equation:

$$2MnO_4^-(aq) + 5SO_2(g) + 2H_2O \rightarrow 2Mn^{2+}(aq) + 5SO_4^{2-}(aq) + 4H^+(aq)$$

(b) $Cu(s) + NO_3^-(aq) \rightarrow Cu^{2+}(aq) + H_2O + NO(g)$

Reduction: $NO_3^- \rightarrow NO$

$H_2O$: $NO_3^- \rightarrow NO + 2H_2O$

$H^+$: $4H^+ + NO_3^- \rightarrow NO + 2H_2O$

$e^-$: $3e^- + 4H^+ + NO_3^- \rightarrow NO + 2H_2O$

Oxidation: $Cu \rightarrow Cu^{2+}$

$e^-$: $Cu \rightarrow Cu^{2+} + 2e^-$

Multiply the reduction half-reaction by 2 and the oxidation half-reaction by 3, and then add the two half-reactions:

$$\cancel{6e^-} + 8H^+ + 2NO_3^- \rightarrow 2NO + 4H_2O$$

$$3Cu \rightarrow 3Cu^{2+} + \cancel{6e^-}$$

$$8H^+(aq) + 2NO_3^-(aq) + 3Cu(s) \rightarrow 3Cu^{2+}(aq) + 2NO(g) + 4H_2O$$

**See Problems 14-10 through 14-13.**

In a basic solution, $OH^-$ ion is predominant rather than $H^+$. Therefore, the procedure is adjusted to allow the half-reactions to be balanced with $OH^-$ ions and $H_2O$ molecules in the basic solution.

The balancing of an equation in aqueous base solution is illustrated by the following unbalanced equation.

$$MnO_4^-(aq) + C_2O_4^{2-}(aq) + OH^-(aq) \rightarrow MnO_2(s) + CO_3^{2-}(aq) + H_2O$$

1 Separate the molecule or ion that contains an atom that has been oxidized or reduced and the product containing that atom. The reduction process is

$$MnO_4^- \rightarrow MnO_2$$

2 For every oxygen needed on the oxygen-deficient side, add *two* $OH^-$ ions. This provides one O and one $H_2O$:

$$O\boxed{H + OH} = O + H_2O$$

$$MnO_4^- \rightarrow MnO_2 + 4OH^-$$

3 For every *two* $OH^-$ ions added on the one side, add *one* $H_2O$ to the other side:

$$2H_2O + MnO_4^- \rightarrow MnO_2 + 4OH^-$$

4 Balance the charge by adding electrons to the more positive side as before:

$$3e^- + 2H_2O + MnO_4^- \rightarrow MnO_2 + 4OH^-$$

5 Repeat the same procedure for the other half-reaction:

$$C_2O_4^{2-} \rightarrow CO_3^{2-}$$

(balance C) $$C_2O_4^{2-} \rightarrow 2CO_3^{2-}$$

(balance O) $$4OH^- + C_2O_4^{2-} \rightarrow 2CO_3^{2-} + 2H_2O$$

(balance charge) $$4OH^- + C_2O_4^{2-} \rightarrow 2CO_3^{2-} + 2H_2O + 2e^-$$

6 Multiply the half-reactions so that electrons gained equal electrons lost. In this example the oxidation half-reaction is multiplied by 3 and the reduction half-reaction is multiplied by 2:

$$\cancel{6e^-} + 4H_2O + 2MnO_4^- \rightarrow 2MnO_2 + 8OH^-$$
$$12OH^- + 3C_2O_4^{2-} \rightarrow 6CO_3^{2-} + 6H_2O + \cancel{6e^-}$$

$$\cancel{4H_2O} + 2MnO_4^- + \overset{4}{\cancel{12}}OH^- + 3C_2O_4^{2-} \rightarrow 6CO_3^{2-} + \overset{2}{\cancel{6}}H_2O + 2MnO_2 + \cancel{8OH^-}$$

$$2MnO_4^-(aq) + 3C_2O_4^{2-}(aq) + 4OH^-(aq) \rightarrow 6CO_3^{2-}(aq) + 2MnO_2(s) + 2H_2O$$

When hydrogen is present in a substance oxidized or reduced in basic solution, the procedure requires an additional step. Hydrogens in a compound or ion are balanced by adding an $OH^-$ for each hydrogen *on the same side* and an $H_2O$ on the other side. For example, consider the following process in basic solution:

$$NO_3^- \rightarrow NH_3$$

(balance O) $$3H_2O + NO_3^- \rightarrow NH_3 + 6OH^-$$

Now balance the H's in $NH_3$ by adding an additional three $OH^-$ on the same side and an additional three $H_2O$ on the left.

$$6H_2O + NO_3^- \rightarrow NH_3 + 9OH^-$$

(balance charge) $$8e^- + 6H_2O + NO_3^- \rightarrow NH_3 + 9OH^-$$

**Example 14-4**
Balance the following equation in basic solution by the ion-electron method.

$$Bi_2O_3(s) + OCl^-(aq) + OH^-(aq) \rightarrow, BiO_3^-(aq) + Cl^-(aq) + H_2O$$

Reduction: $OCl^- \rightarrow Cl^-$

$OH^-$: $OCl^- \rightarrow Cl^- + 2OH^-$

$H_2O$: $H_2O + OCl^- \rightarrow Cl^- + 2OH^-$

$e^-$: $2e^- + H_2O + OCl^- \rightarrow Cl^- + 2OH^-$

Oxidation: $Bi_2O_3 \rightarrow BiO_3^-$

Bi: $Bi_2O_3 \rightarrow 2BiO_3^-$

$OH^-$: $6OH^- + Bi_2O_3 \rightarrow 2BiO_3^-$

$H_2O$: $6OH^- + Bi_2O_3 \rightarrow 2BiO_3^- + 3H_2O$

$e^-$: $6OH^- + Bi_2O_3 \rightarrow 2BiO_3^- + 3H_2O + 4e^-$

Multiply the reduction half-reaction by 2 and add to the oxidation half-reaction:

$$\cancel{4e^-} + 2H_2O + 2OCl^- \rightarrow 2Cl^- + 4OH^-$$
$$6OH^- + Bi_2O_3 \rightarrow 2BiO_3^- + 3H_2O + \cancel{4e^-}$$
$$2H_2O + 6OH^- + Bi_2O_3 + 2OCl^- \rightarrow 2Cl^- + 4OH^- + 2BiO_3^- + 3H_2O$$

By eliminating $H_2O$ and $OH^-$ duplications, we have the balanced net ionic equation.

$$Bi_2O_3(s) + 2OCl^-(aq) + 2OH^-(aq) \rightarrow 2BiO_3^-(aq) + 2Cl^-(aq) + H_2O$$

**See Problems 14-14 through 14-17.**

## 14-4 Voltaic Cells

If a car is halfway down the side of a hill and it begins to roll, does it spontaneously roll up or down hill? A silly question perhaps, but the concept is very similar to why chemical reactions proceed in one direction and not the other. The car rolls down the hill because the bottom of the hill represents a lower potential energy state than the top of the hill. As the car rolls down the hill, the potential energy is converted into kinetic energy in the form of motion and heat from friction. Likewise, chemical reactions are spontaneous (occur naturally without external help or stimulus) in one direction because the reactants are at a higher potential energy state (in the form of chemical energy) than the products. (See Figure 14-2.) When the reaction proceeds, the difference in chemical energy between the reactants and the products can be released as heat energy (exothermic reaction) or converted directly to electrical energy. *The* **voltaic cell** (*also called the* **galvanic cell**) *uses a spontaneous redox reaction to generate electrical energy through an*

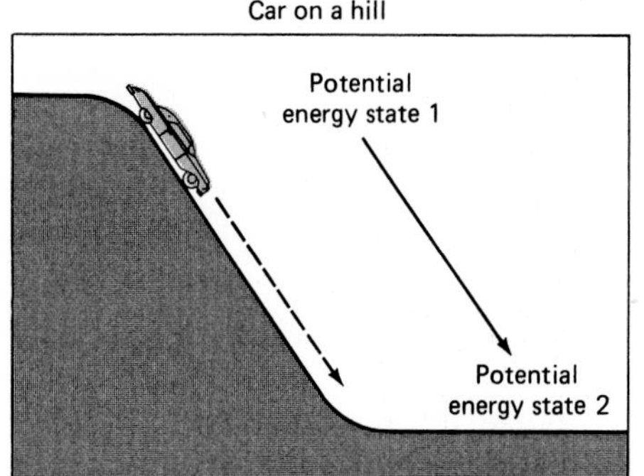


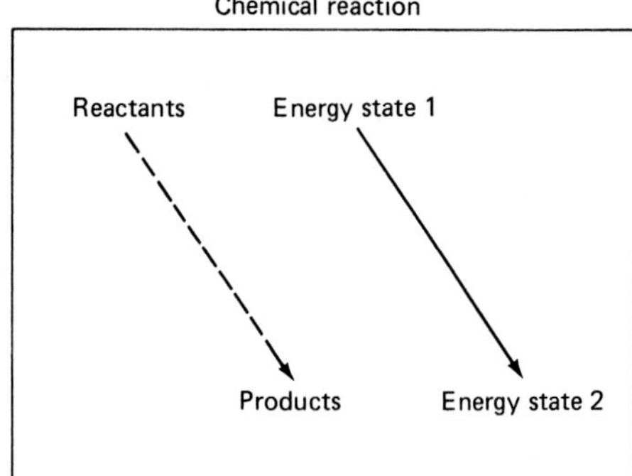


**Figure 14-2** ENERGY STATES. A chemical reaction proceeds in a certain direction for the same reason that a car rolls *down* a hill.

*external circuit.* The following spontaneous reaction is a redox reaction that can be used to form a voltaic cell.

$$Zn(s) + CuSO_4(aq) \rightarrow ZnSO_4(aq) + Cu(s)$$

This is called the **Daniell cell;** it was used to generate electricity for the early telegraph.

If a piece of zinc is placed in an aqueous $CuSO_4$ solution, a reaction takes place directly on the surface of the zinc. This is obvious since a dull, brown coating of copper forms on the zinc. In a voltaic cell, however, the oxidation and reduction process must be separated so that the electrons are not exchanged on one surface but are detoured through an external wire. The Daniell cell is illustrated in Figure 14-3. A zinc strip is immersed in a $Zn^{2+}$ solution, and, in a separate compartment, a copper strip is immersed in a $Cu^{2+}$ solution. A wire connects the two metal strips. The two metal strips are called electrodes. *The* **electrodes** *are the surfaces in a cell at which the reactions takes place. The electrode at which oxidation takes place is called the* **anode**. *Reduction takes place at the* **cathode**.

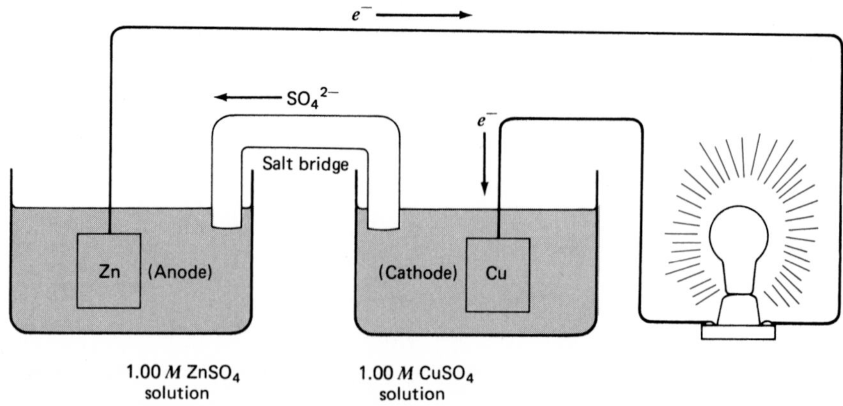


**Figure 14-3** THE DANIELL CELL. This chemical reaction produced electricity for the first telegraphs.

In the compartment on the left, the strip of Zn serves as the anode, since the following reaction occurs when the circuit is connected.

$$Zn \rightarrow Zn^{2+} + 2e^-$$

When the circuit is complete, the two electrons travel in the external wire to the Cu electrode, which serves as the cathode. The reduction reaction occurs at the cathode:

$$2e^- + Cu^{2+} \rightarrow Cu$$

To maintain neutrality in the solution, some means must be provided for the movement of a $SO_4^{2-}$ ion (or some other negative ion) from the right compartment where a $Cu^{2+}$ has been removed to the left compartment where a $Zn^{2+}$ has been produced. The *salt bridge* is an aqueous gel that allows ions to migrate between compartments but does not allow the mixing of solutions. (If $Cu^{2+}$ ions wandered into the left compartment, they would form a coating of Cu on the Zn electrode, thus short-circuiting the cell.)

As the cell discharges (generates electrical energy), the Zn electrode gets smaller and the Cu electrode gets larger. If the external circuit is open (switch off), the reaction stops. Thus the cell holds its charge until needed, and the electrical energy is stored.

Two of the most common voltaic cells in use today are the dry cell (flashlight battery) and the lead-acid cell (car battery). (*The word* **battery** *means a collection of one or more separate cells joined together in one unit.*)

One cell of a lead-acid storage battery is illustrated in Figure 14-4. Each cell is composed of two grids separated by an inert spacer. One grid of a fully charged battery contains metallic lead. The other contains $PbO_2$, which is insoluble in $H_2O$. Both grids are immersed in a sulfuric acid solution (battery acid). When the battery is discharged by connecting the electrodes, the following half-reactions take place spontaneously.

$$\begin{array}{ll} \text{Anode:} & Pb(s) + H_2SO_4(aq) \rightarrow PbSO_4(s) + 2H^+(aq) + 2e^- \\ \text{Cathode:} & \underline{2e^- + 2H^+(aq) + PbO_2(s) + H_2SO_4(aq) \rightarrow PbSO_4(s) + 2H_2O} \\ \text{Total reaction:} & Pb(s) + PbO_2(s) + 2H_2SO_4(aq) \rightarrow 2PbSO_4(s) + 2H_2O \end{array}$$

The electrons released at the Pb anode travel through the external circuit to run lights, starters, radios, or whatever is needed. The electrons return to the $PbO_2$ cathode to complete the circuit. As the reaction proceeds, both electrodes are converted to $PbSO_4$ and the $H_2SO_4$ is depleted. Since $PbSO_4$ is also insoluble, it remains attached to the grids as it forms. The degree of discharge of a battery can be determined by the density of the battery acid. Since the density of a fully discharged battery is 1.05 g/mL, the difference in density between this value and the density of a fully charged battery (1.35 g/mL) gives the amount of charge remaining in the battery. The hydrometer discussed in Chapter 2 is used to determine the density of

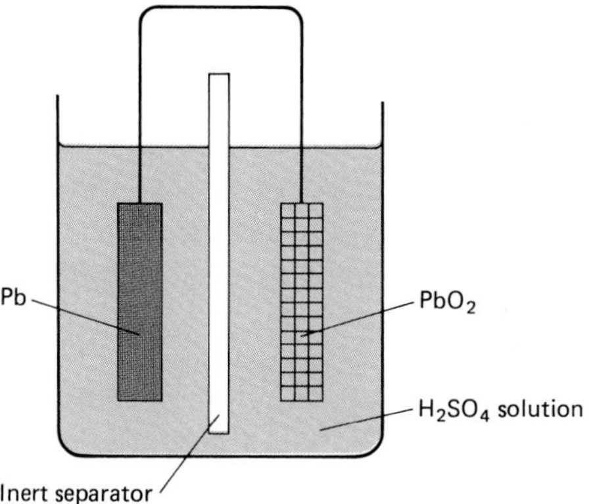


**Figure 14-4** THE LEAD STORAGE BATTERY. The lead storage battery produces electricity to start your car.

the acid. As the electrodes convert to $PbSO_4$, the battery loses power and eventually becomes "dead."

The convenience of a car battery is that it can be recharged. After the engine starts, an alternator or generator is engaged to push electrons back into the cell in the opposite direction they came during discharge. This forces the reverse, nonspontaneous reaction to proceed:

$$2PbSO_4(s) + 2H_2O \rightarrow Pb(s) + PbO_2(s) + 2H_2SO_4(aq)$$

When the battery is fully recharged, the alternator shuts off, the circuit is open, and the battery is ready for the next start.

The dry cell (invented by Leclanché in 1866) is not rechargeable to any extent but is comparatively inexpensive and easily portable. (In contrast, the lead-acid battery is heavy, is expensive, and must be kept upright.) The dry cell illustrated in Figure 14-5 consists of a zinc anode, which is the outer rim, and an inert graphite electrode. (An inert electrode provides a reaction surface but does not itself react.) In between is an aqueous paste containing $NH_4Cl$, $MnO_2$, and carbon. The reactions are as follows:

Anode: $$Zn(s) \rightarrow Zn^{2+}(aq) + 2e^-$$

Cathode: $$2NH_4{}^+(aq) + 2MnO_2(s) + 2e^- \rightarrow Mn_2O_3(s) + 2NH_3(aq) + H_2O$$

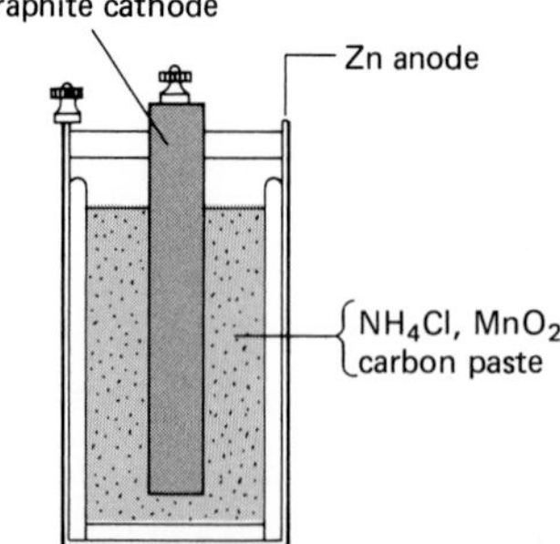


**Figure 14-5** THE DRY CELL. The dry cell is comparatively inexpensive, light, and portable.

Many other batteries have their special advantages. Nickel-cadmium batteries are popular as replacements for dry cells. Although initially more expensive, they are rechargeable and so will last longer. For space travel the fuel cell, which converts some of the energy from the reaction between $H_2$ and $O_2$ (to form water) directly to electrical energy, is used. Fuel cells are very expensive but generate a continual flow of electricity without recharging. Currently, government and industry are conducting a great amount of research for a replacement for the lead-acid battery for use in electric cars. A small electric car needs at least 18 lead-acid batteries, which have to be completely replaced every few months depending on use. Also, much of the power of the lead-acid batteries must be used just to move the batteries around, let alone the passengers. Some encouraging results have been announced with zinc–nickel oxide batteries that are lighter, last longer, and carry more charge than lead-acid batteries. Production of automobiles using new types of batteries is planned for the later 1980s.

**See Problems 14-18 through 14-22.**

## 14-5 Electrolytic Cells

In the previous section we mentioned that a car spontaneously rolls down a hill, releasing energy as it goes. The car can roll up the hill, but obviously energy must be supplied from some source before this can happen. Redox reactions can be forced in the nonspontaneous direction, but, like the car on the hill, energy must be supplied. This is what happened in the previous section when the lead-acid battery was recharged with energy supplied from the engine. In this case the electrical energy is converted back into chemical energy. *Cells that convert electrical energy into chemical energy are called* **electrolytic cells.**

An example of an electrolytic cell is shown in Figure 14-6. When suf-

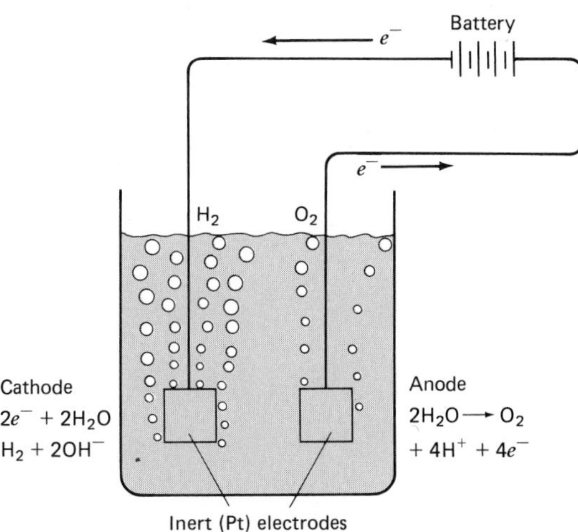


**Figure 14-6** THE ELECTROLYSIS OF WATER. With an input of electrical energy, water can be decomposed to its elements.

ficient electrical energy is supplied to the electrodes from an outside source, the following nonspontaneous reaction occurs.

$$2H_2O \rightarrow 2H_2(g) + O_2(g)$$

In order for this electrolysis to occur, an electrolyte such as $Na_2SO_4$ must be present in solution. Pure water alone does not have a sufficient concentration of ions to allow conduction of electricity.

Electrolysis has many useful applications. For example, silver or gold can be electroplated onto cheaper metals. In Figure 14-7 the metal spoon is the cathode, with the silver bar serving as the anode. When electricity is supplied, the Ag anode produces $Ag^+$ ions and the spoon cathode reduces $Ag^+$ ions to give a layer of Ag. The silver-plated spoon can be polished and made to look as good as sterling silver. (See Figure 14-8.)

Electrolytic cells are used commercially to produce many metals such as sodium, magnesium, and aluminum. The electrolysis of molten (melted) salts, where no water is present, serves this purpose. For example, electrolysis of molten NaCl is the prime source of commercial quantities of elemental chlorine and sodium. As shown in Figure 14-9, at the high temperature of the process, the Na forms in the molten state and is subsequently drained from the top of the cell.

## 14-6 Predicting Spontaneous Redox Reactions

In a previous section, we discussed the following net ionic equation, which represents the spontaneous reaction of the Daniell cell.

$$Zn(s) + Cu^{2+}(aq) \rightarrow Zn^{2+}(aq) + Cu(s)$$

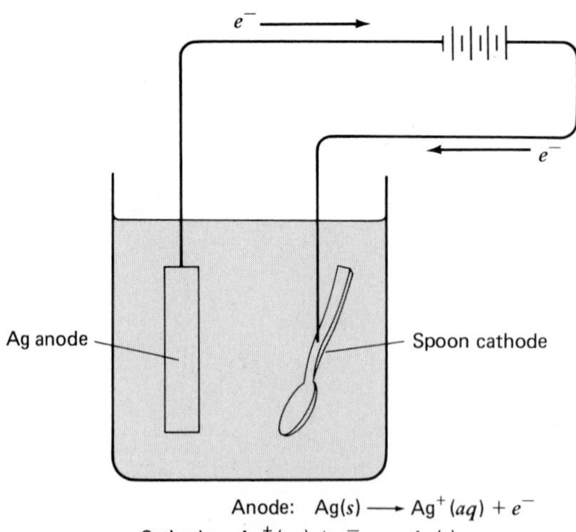


**Figure 14-7** ELECTROPLATING. With an input of energy, a spoon can be coated with silver.

**Figure 14-8** ELECTROPLATING. This service set has been electroplated with a thin coating of silver.

As we will see in this section, this and many other redox reactions could be predicted as being spontaneous if we had the proper information. Predictions are accomplished by observing the results of a few experiments and drawing some conclusions that can lead to a generalization or model. The model explains the experiments and predicts the results of other experiments before they are performed in the laboratory. If the predictions turn out to be correct, the model is generally accepted. If not, the model must be modified

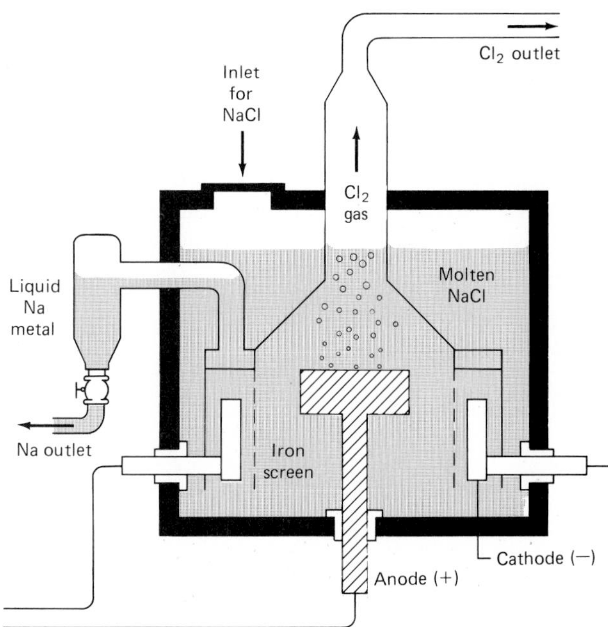


**Figure 14-9** THE ELECTROLYSIS OF NaCl. Elemental sodium and chlorine are prepared by the electrolysis of ordinary table salt in the molten state.

or discarded. This sequence is what is known as the **scientific method**. We will follow this procedure in this section.

Let's begin with two experiments as illustrated in Figure 14-10. When a strip of nickel is placed in an aqueous $ZnCl_2$ solution, no reaction is apparent (experiment 1 in Figure 14-10). Conversely, if a strip of Zn is placed in an aqueous $NiCl_2$ solution, a reaction takes place as shown by the coating of Ni formed on the Zn strip (experiment 2 in Figure 14-10). We have shown that the following net ionic equation represents the spontaneous reaction that occurs.

$$Zn(s) + Ni^{2+}(aq) \rightarrow Zn^{2+}(aq) + Ni(s)$$

In Chapter 13, we indicated that an acid-base reaction (in the Brønsted-Lowry definition) can be considered as a competition for an $H^+$ ion. Likewise, we can propose that an oxidation-reduction reaction is simply a competition for electrons (in this case between $Zn^{2+}$ and $Ni^{2+}$). From the direction of the spontaneous reaction it seems safe to make the following statement.

$$Ni^{2+} + 2e^- \rightarrow Ni$$

occurs more readily than

$$Zn^{2+} + 2e^- \rightarrow Zn$$

The total reaction can also be considered analogous to the results of a tug-of-war between two competitors. In Figure 14-11 two people of unequal strengths are pulling on a rope looped around a post. Obviously, the stronger prevails and moves in the desired direction while the other is forced to move in the opposite direction. Likewise, since $Ni^{2+}$ has a stronger tendency to be reduced (a stronger oxidizing agent) than $Zn^{2+}$, it will do so, but the $Zn^{2+}$-Zn half-reaction is forced to go in the opposite direction as follows:

$$Zn \rightarrow Zn^{2+} + 2e^-$$

We've concluded from experiments that $Ni^{2+}$ is a stronger oxidizing agent than $Zn^{2+}$ by head-on competition. We will now match the winner

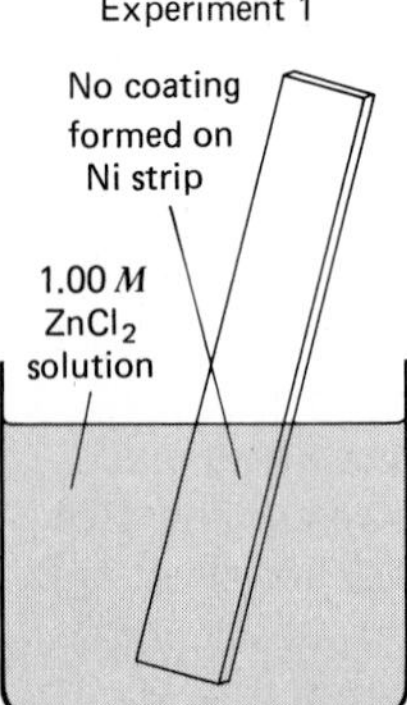


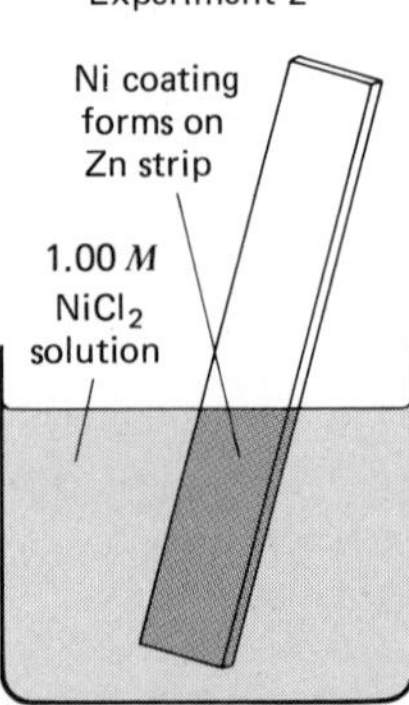


**Figure 14-10** THE REACTION OF Zn AND $Ni^{2+}$. Zinc replaces the $Ni^{2+}$ in a $NiCl_2$ solution.

**Figure 14-11** A TUG-OF-WAR FOR ELECTRONS. The stronger of the two moves in the desired direction. The other moves in the opposite direction.

($Ni^{2+}$) with another challenger ($Cu^{2+}$) to test its strength. Experiments to provide the answer to the challenge of $Cu^{2+}$ to $Ni^{2+}$ are illustrated in Figure 14-12. In this case, a strip of Ni immersed in a $CuCl_2$ solution forms a coat of Cu (experiment 3 in Figure 14-12), but the opposite reaction does not occur (experiment 4). The following net ionic equation illustrates the spontaneous reaction that is observed.

$$Ni(s) + Cu^{2+}(aq) \rightarrow Cu(s) + Ni^{2+}(aq)$$

The results of this contest can be summarized as follows:

$$Cu^{2+} + 2e^- \rightarrow Cu$$

occurs more readily than

$$Ni^{2+} + 2e^- \rightarrow Ni$$

From the results of the two sets of experiments we can make the following ranking as to oxidizing strength of the three ions.

(1) $Cu^{2+}$ (2) $Ni^{2+}$ (3) $Zn^{2+}$

When an ion that is a *strong* oxidizing agent (e.g., $Cu^{2+}$) is reduced, the metal that is formed (e.g., Cu) is a *weak* reducing agent. This phenomenon is analogous to the acid–conjugate base relationship discussed in Section 13-7. There we found that a strong acid ionizes to produce a weak conjugate base and vice versa. Here we find that a strong oxidizing agent reacts to

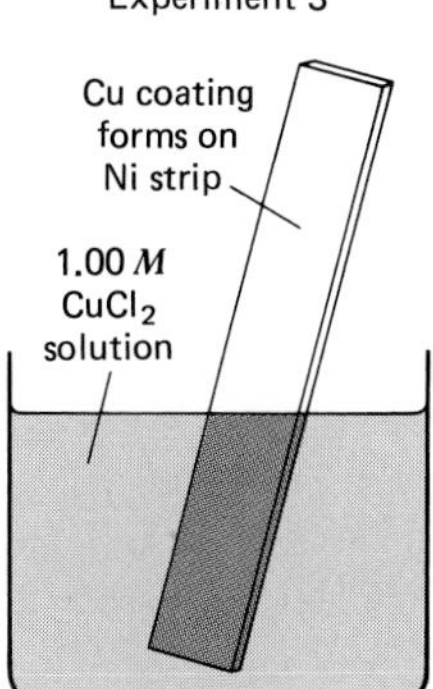


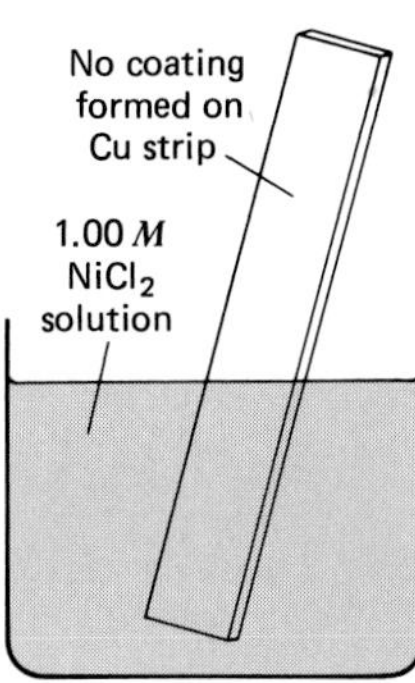


**Figure 14-12** THE REACTION OF Ni AND $Cu^{2+}$. Nickel replaces the $Cu^{2+}$ in a $CuCl_2$ solution.

produce a weak reducing agent and vice versa. This inverse relationship is illustrated by the "seesaw" in Figure 14-13.

Just as the ions were arranged in order of decreasing strength as oxidizing agents, the metals can also be arranged in order of decreasing strength as reducing agents.

$$Zn \rightarrow Zn^{2+} + 2e^-$$

occurs more readily than

$$Ni \rightarrow Ni^{2+} + 2e^-$$

which occurs more readily than

$$Cu \rightarrow Cu^{2+} + 2e^-$$

*Thus $Cu^{2+}$ is the strongest oxidizing agent and Zn is the strongest reducing agent.*

This information is summarized in Table 14-1. The ions (on the left) are listed in order of decreasing strength as oxidizing agents as you go down. The oxidation reaction is indicated by the arrow to the right (→). The metals on the right are thus listed in order of increasing strength as reducing agents as you go down the table. The reduction reaction is indicated by the arrow to the left (←).

The results of the experiments studied so far suggest a generalization (or model). *A spontaneous redox reaction occurs whereby the stronger oxidizing agent reacts with the stronger reducing agent to produce a weaker oxidizing and a weaker reducing agent.*

We can now use the generalization and Table 14-1 to predict the results of an experiment before we try it in the lab. Which of the following two reactions occurs spontaneously?

**1** $$Cu(s) + Zn^{2+}(aq) \rightarrow Cu^{2+}(aq) + Zn(s)$$

or the reverse reaction

**2** $$Zn(s) + Cu^{2+}(aq) \rightarrow Zn^{2+}(aq) + Cu(s)$$

In Table 14-1, notice that $Cu^{2+}$ is a stronger oxidizing agent than $Zn^{2+}$ and that Zn is a stronger reducing agent than Cu. Since the stronger react,

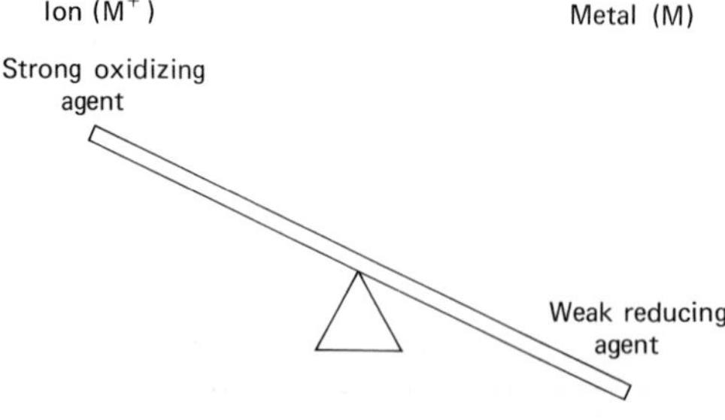


**Figure 14-13** AN ION AND ITS METAL. There is an inverse relationship between the oxidizing strength of an ion and the reducing strength of the metal.

**Table 14-1** Zn, Ni, and Cu

| | | |
|---|---|---|
| Strongest oxidizing agent ⟶ | $Cu^{2+} + 2e^- \rightleftarrows Cu$ | ⟵ Weakest reducing agent |
| | $Ni^{2+} + 2e^- \rightleftarrows Ni$ | |
| Weakest oxidizing agent ⟶ | $Zn^{2+} + 2e^- \rightleftarrows Zn$ | ⟵ Strongest reducing agent |

reaction (2) above is predicted to be the spontaneous reaction. Our prediction is correct since that is the spontaneous reaction utilized in the Daniell cell.

We can now expand Table 14-1 by including more ions and some molecular oxidizing agents in their proper positions in the table. This has been accomplished in Table 14-2.

In some cases more elaborate experiments than we have illustrated here are required to locate precisely certain half-reactions in Table 14-2. Positions in the table are subject to certain restrictions on the concentration of ions present and the pressure of any gas present. Specifically, it is assumed that all ions are present at 1.00 *M* and gas at 1.00 atm pressure.

This table can be put to use in predicting the occurrence of many spontaneous redox reactions. To predict a reaction either the strengths of the two oxidizing agents or the strengths of the two reducing agents can be compared in the table. The oxidizing agent higher in the table prevails and reacts to the right. The one lower in the table is forced to the left. This automatically makes the reactants the stronger two agents and the products the weaker two agents. As illustrated in Figure 14-14, the spontaneous reaction can be visualized as taking place in a clockwise direction (the dashed line).

**Table 14-2** Reducing and Oxidizing Agents

| | | |
|---|---|---|
| Strongest oxidizing agent | $F_2 + 2e^- \rightleftarrows 2F^-$ | Weakest reducing agent |
| ↑ | $Cl_2 + 2e^- \rightleftarrows 2Cl^-$ | ↓ |
| | $Br_2 + 2e^- \rightleftarrows 2Br^-$ | |
| | $Ag^+ + e^- \rightleftarrows Ag$ | |
| | $Cu^{2+} + 2e^- \rightleftarrows Cu$ | |
| | $2H^+ + 2e^- \rightleftarrows H_2$ | |
| Increasing strength of oxidizing agent | $Pb^{2+} + 2e^- \rightleftarrows Pb$ | Increasing strength of reducing agent |
| | $Sn^{2+} + 2e^- \rightleftarrows Sn$ | |
| | $Ni^{2+} + 2e^- \rightleftarrows Ni$ | |
| | $Fe^{2+} + 2e^- \rightleftarrows Fe$ | |
| | $Zn^{2+} + 2e^- \rightleftarrows Zn$ | |
| | $2H_2O + 2e^- \rightleftarrows H_2 + 2OH^-$ | |
| | $Al^{3+} + 3e^- \rightleftarrows Al$ | |
| Weakest oxidizing agent | $Mg^{2+} + 2e^- \rightleftarrows Mg$ | Strongest reducing agent |
| | $Na^+ + e^- \rightleftarrows Na$ | |

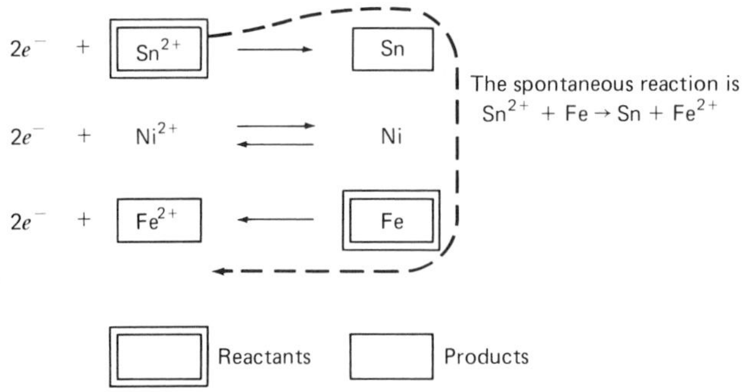

**Figure 14-14** A SPONTANEOUS REACTION. The stronger oxidizing agent reacts with the stronger reducing agent.

In which direction will the following reactions be spontaneous?

**Example 14-5**

$$Pb^{2+}(aq) + 2Cl^-(aq) \overset{?}{\leftrightarrow} Pb(s) + Cl_2(g)$$

In Table 14-2, notice that $Cl_2$ is a stronger oxidizing agent than $Pb^{2+}$ and that Pb is a stronger reducing agent than $Cl^-$. Therefore, the reaction is spontaneous to the left. Metallic lead will dissolve in an aqueous solution of $Cl_2$. ($Cl_2$ in swimming pools is hard on many metals, as you can see.)

$$\underline{\underline{Pb(s) + Cl_2(g) \rightarrow PbCl_2(aq)}}$$

**Example 14-6**

$$Mg(s) + 2H_2O \overset{?}{\leftrightarrow} Mg^{2+}(aq) + 2OH^-(aq) + H_2(g)$$

$H_2O$ is a stronger agent than $Mg^{2+}$, so the reaction proceeds to the right. Notice that all metals listed below the $H_2O$ half-reaction react with water. These are considered the *active* metals because of their high chemical reactivity. Aluminum metal should dissolve in water, but metallic aluminum is covered by a protective layer of aluminum oxide that keeps the metal from coming in contact with water. If it wasn't for this fortunate property, aluminum would not be a useful metal.

**Example 14-7**

$$Ni(s) + 2H^+(aq) \overset{?}{\leftrightarrow} Ni^{2+}(aq) + H_2(g)$$

$H^+$ is a stronger oxidizing agent than $Ni^{2+}$. This reaction also proceeds to the right. Notice that all metals below $H^+$ react with aqueous acid. Those between $H^+$ and $H_2O$ react in acid solution but not in pure $H_2O$. On the other hand, Cu and Ag do not react with $H^+$, since $Cu^{2+}$ is a better oxidizing agent than $H^+$. This is one good reason why copper and silver can be used in coins and jewelry. They are not generally affected by aqueous acid solutions.

**See Problems 14-23 through 14-35.**

## 14-7 Chapter Summary

A common characteristic of a large number of chemical reactions involves an exchange of electrons between reactants. This extensive classification of reactions is known as oxidation-reduction or simply redox reactions. In such a reaction, the substance that gives up or loses the electrons is oxidized and the substance that gains the electrons is reduced. The substances are also classified as oxidizing and reducing agents as follows:

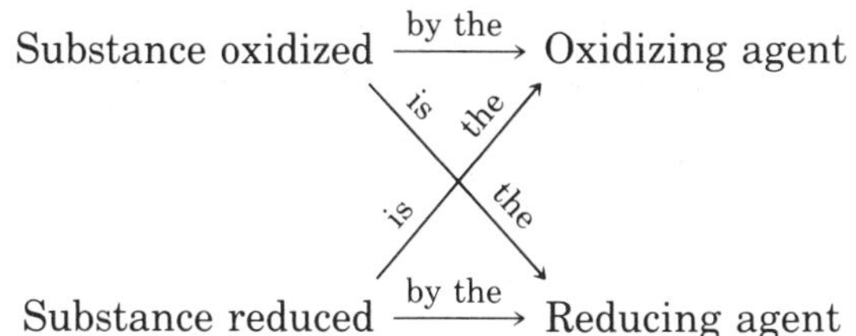


A redox reaction can be broken down into two half-reactions: an oxidation and a reduction. These two processes take place so that all electrons lost in the oxidation process are gained in the reduction process. This fact is useful in balancing oxidation-reduction reactions. Most of these reactions would, at best, be very difficult to balance by inspection methods as described in Chapter 9.

Balancing equations by the oxidation state method requires that the atoms undergoing a change in oxidation state be identified. By equalizing electron gain and electron loss with proper coefficients, equations can be balanced. The more useful method is the ion-electron or half-reaction method. In this method, the entire molecule or ion containing the atom undergoing a change is balanced in two half-reactions. The rules for balancing equations by this method are summarized as follows:

1 Identify the ion or molecule containing the atom that has a change in oxidation state and the product species containing the same atom.

2 Balance the atom undergoing the change.

3 Balance oxygens [acidic: add one $H_2O$ for each O needed.<br>basic: add two $OH^-$ ions for each O needed.

4 Balance H [acidic: add $H^+$ for each H needed.<br>basic: add one $H_2O$ for each two H's needed.

5 Add electrons on the more positive side to balance the charge.

6 Multiply each half-reaction through by a whole number so that electrons cancel when the half-reactions are added.

7 Subtract out any species common to both sides of the equation.

Spontaneous chemical reactions occur because reactants are higher in chemical energy than products. We are familiar with many exothermic reactions where the difference in energy is released as heat. The energy can also be released as electrical energy in a voltaic cell. In a voltaic cell the two half-reactions are physically separated so that electrons travel in an external circuit or wire. The Daniell cell, the car battery, and the dry cell all involve spontaneous chemical reactions where the chemical energy is converted directly into electrical energy. Nonspontaneous chemical reactions can be made to occur. In this case, electrical energy is converted into chemical energy. These reactions take place in electrolytic cells and require an outside source of energy.

Each substance has its own inherent strength as either an oxidizing agent or a reducing agent. This allows the construction of a table where oxidizing and reducing agents are ranked by strength. The rankings are arrived at by observations or measurements with electrical instruments. From this table a great many other spontaneous reactions can be predicted. In Table 14-2 stronger oxidizing agents are ranked higher on the left and stronger reducing agents are ranked lower on the right. Reactions can thus be predicted as follows:

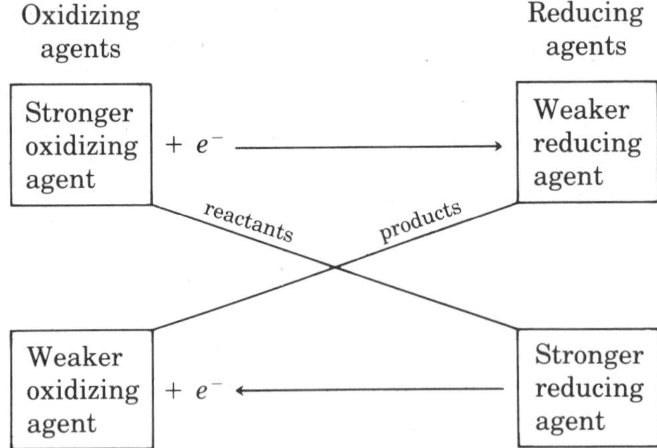

## Exercises

### Oxidation-Reduction

**14-1** What is the oxidation state of the specified atom in each of the following?
(a) the S in $SO_3$ (d) the N in $HNO_3$
(b) the Co in $Co_2O_3$ (e) the Cr in $K_2CrO_4$
(c) the U in $UF_6$ (f) the Mn in $CaMnO_4$

**14-2** What is the oxidation state of the specified atom in each of the following?
(a) the Si in $SiO_2$ (c) the As in $H_3AsO_4$
(b) the H in $CaH_2$ (d) the O in $Na_2O_2$

**14-3** Which of the following reactions are oxidation-reduction reactions?
(a) $2H_2 + O_2 \rightarrow 2H_2O$
(b) $CaCO_3 \rightarrow CaO + CO_2$
(c) $2Na + 2H_2O \rightarrow 2NaOH + H_2$
(d) $2HNO_3 + Ca(OH)_2 \rightarrow Ca(NO_3)_2 + 2H_2O$
(e) $AgNO_3 + KCl \rightarrow AgCl + KNO_3$
(f) $Zn + CuCl_2 \rightarrow ZnCl_2 + Cu$

**14-4** Identify each of the following half-reactions as either oxidation or reduction.
(a) $Na \rightarrow Na^+ + e^-$
(b) $Zn^{2+} + 2e^- \rightarrow Zn$
(c) $Fe^{2+} \rightarrow Fe^{3+} + e^-$
(d) $O_2 + 4H^+ + 4e^- \rightarrow 2H_2O$
(e) $S_2O_8^{2-} + 2e^- \rightarrow 2SO_4^{2-}$

**14-5** Identify each of the following changes as either oxidation or reduction.
(a) $P_4 \rightarrow H_3PO_4$ (d) $Al \rightarrow Al(OH)_4^-$
(b) $NO_3^- \rightarrow NH_4^+$ (e) $S^{2-} \rightarrow SO_4^{2-}$
(c) $Fe_2O_3 \rightarrow Fe^{2+}$

**14-6** For each of the following unbalanced equations, complete the table below.
(a) $MnO_2 + H^+ + Br^- \rightarrow Mn^{2+} + Br_2 + H_2O$
(b) $CH_4 + O_2 \rightarrow CO_2 + H_2O$
(c) $Fe^{2+} + MnO_4^- + H^+ \rightarrow Fe^{3+} + Mn^{2+} + H_2O$

**14-7** For the following two unbalanced equations, complete a table as in Problem 14-6.
(a) $Al + H_2O \rightarrow AlO_2^- + H_2 + H^+$
(b) $Mn^{2+} + Cr_2O_7^{2-} + H^+ \rightarrow MnO_4^- + Cr^{3+} + H_2O$

### Balancing Equations by the Oxidation State Method

**14-8** Balance each of the following equations by the oxidation state method.
(a) $NH_3 + O_2 \rightarrow NO + H_2O$
(b) $Sn + HNO_3 \rightarrow SnO_2 + NO_2 + H_2O$
(c) $Cr_2O_3 + Na_2CO_3 + KNO_3 \rightarrow CO_2 + Na_2CrO_4 + KNO_2$
(d) $Se + BrO_3^- + H_2O \rightarrow H_2SeO_3 + Br^-$

**14-9** Balance the equations in Problem 14-6 by the oxidation state method.

### Balancing Equations by the Ion-Electron Method

**14-10** Balance each of the following half-reactions in acid solution.
(a) $Sn^{2+} \rightarrow SnO_2$
(b) $CH_4 \rightarrow CO_2$
(c) $Fe^{3+} \rightarrow Fe^{2+}$

| Reaction | Species Oxidized[a] | Product of Oxidation | Species Reduced | Product of Reduction | Oxidizing Agent | Reducing Agent |
|---|---|---|---|---|---|---|
| (a) | | | | | | |
| (b) | | | | | | |
| (c) | | | | | | |

[a]Element, molecule, or ion.

(d) $I_2 \rightarrow IO_3^-$
(e) $NO_3^- \rightarrow NO_2$

**14-11** Balance the following half-reactions in acid solution.
(a) $P_4 \rightarrow H_3PO_4$ (b) $NO_3^- \rightarrow NH_4^+$
(c) $Fe_2O_3 \rightarrow Fe^{2+}$

**14-12** Balance each of the following by the ion-electron method. All are in acid solution.
(a) $S^{2-} + NO_3^- + H^+ \rightarrow S + NO + H_2O$
(b) $I_2 + S_2O_3^{2-} \rightarrow S_4O_6^{2-} + I^-$
(c) $SO_3^{2-} + ClO_3^- \rightarrow Cl^- + SO_4^{2-}$
(d) $Fe^{2+} + H_2O_2 + H^+ \rightarrow Fe^{3+} + H_2O$
(e) $AsO_4^{3-} + I^- + H^+ \rightarrow I_2 + AsO_3^{3-} + H_2O$
(f) $Zn + H^+ + NO_3^- \rightarrow Zn^{2+} + NH_4^+ + H_2O$

**14-13** Balance each of the following by the ion-electron method. All occur in acid solution.
(a) $Mn^{2+} + BiO_3^- + H^+ \rightarrow MnO_4^- + Bi^{3+} + H_2O$
(b) $IO_3^- + SO_2 + H_2O \rightarrow I_2 + SO_4^{2-} + H^+$
(c) $Se + BrO_3^- + H_2O \rightarrow H_2SeO_3 + Br^-$
(d) $P_4 + HOCl + H_2O \rightarrow H_3PO_4 + Cl^- + H^+$
(e) $Al + Cr_2O_7^{2-} + H^+ \rightarrow Al^{3+} + Cr^{3+} + H_2O$

**14-14** Balance each of the following half-reactions in basic solution.
(a) $SnO_2^{2-} \rightarrow SnO_3^{2-}$
(b) $ClO_2^- \rightarrow Cl_2$
(c) $Si \rightarrow SiO_3^{2-}$
(d) $NO_3^- \rightarrow NH_3$

**14-15** Balance the following half-reactions in basic solution.
(a) $Al \rightarrow Al(OH)_4^-$ (b) $S^{2-} \rightarrow SO_4^{2-}$
(c) $N_2H_4 \rightarrow NO_3^-$

**14-16** Balance each of the following by the ion-electron method. All occur in basic solution.
(a) $S^{2-} + OH^- + I_2 \rightarrow SO_4^{2-} + I^- + H_2O$
(b) $MnO_4^- + OH^- + I^- \rightarrow MnO_4^{2-} + IO_4^- + H_2O$
(c) $BiO_3^- + SnO_2^{2-} + H_2O \rightarrow SnO_3^{2-} + OH^- + Bi(OH)_3$
*(d) $CrI_3 + OH^- + Cl_2 \rightarrow CrO_4^{2-} + IO_4^- + Cl^- + H_2O$
(*Hint*: Two ions are oxidized; include both in one half-reaction.)

**14-17** Balance each of the following by the ion-electron method. Both are in basic solution.
(a) $ClO_2 + OH^- \rightarrow ClO_2^- + ClO_3^- + H_2O$
(b) $OH^- + Cr_2O_3 + NO_3^- \rightarrow CrO_4^{2-} + NO_2^- + H_2O$

## Voltaic Cells

**14-18** What is the function of the salt bridge in the voltaic cell?

**14-19** In the alkaline battery the following two half-reactions occur.

$$Zn(s) + 2OH^-(aq) \rightarrow Zn(OH)_2(s) + 2e^-$$

$$2MnO_2(s) + 2H_2O + 2e^- \rightarrow 2MnO(OH)(s) + 2OH^-(aq)$$

Which reaction takes place at the anode and which at the cathode? What is the total reaction?

**14-20** The following overall reaction takes place in the silver oxide battery.

$$Ag_2O(s) + H_2O + Zn(s) \rightarrow Zn(OH)_2(s) + 2Ag(s)$$

The reaction takes place in basic solution. Write the half-reaction that takes place at the anode and the half-reaction that takes place at the cathode.

**14-21** In the fuel cell used on the space shuttle, the following half-reactions occur.

$$O_2(g) + 2H_2O + 4e^- \rightarrow 4OH^-(aq)$$
$$H_2(g) + 2OH^-(aq) \rightarrow 2H_2O + 2e^-$$

Which reaction takes place at the anode and which at the cathode? Write the net overall reaction.

**14-22** Sketch a galvanic cell in which the following overall reaction occurs.

$$Ni^{2+}(aq) + Fe(s) \rightarrow Fe^{2+}(aq) + Ni(s)$$

(a) What reaction takes place at the anode and the cathode?
(b) In what direction do the electrons flow in the wire?
(c) In what direction do the anions flow in the salt bridge?

## Predicting Redox Reactions

**14-23** Given the following information concerning metal strips immersed in certain solutions:

| Metal Strip | Solution | Reaction |
|---|---|---|
| Cd | $NiCl_2$ | Ni coating formed |
| Cd | $FeCl_2$ | N.R. (no reaction) |
| Zn | $CdCl_2$ | Cd coating formed |
| Fe | $CdCl_2$ | N.R. |

Write the net ionic equations representing the reactions that occur. Where does $Cd^{2+}$ rank as a reducing agent in Table 14-2?

**14-24** A hypothetical metal (M) forms a coating of Sn when placed in a $SnCl_2$ solution. However, when a strip of Ni is placed in an $MCl_2$ solution, a coating of the metal M forms on the nickel. Write the net ionic equations representing the reactions that occur. Where does $M^{2+}$ rank as a reducing agent in Table 14-2?

**14-25** Gold does not dissolve in water or in a normal acid solution. Fluorine, however, will cause gold to dissolve to form aqueous $Au^{3+}$ solutions. Where does $Au^{3+}$ rank as an oxidizing agent in Table 14-2?

**14-26** Using Table 14-2, predict whether the following reactions occur in aqueous solution. If not, write N.R. (no reaction).
(a) $2Na + 2H_2O \rightarrow H_2 + 2NaOH$
(b) $Pb + Zn^{2+} \rightarrow Pb^{2+} + Zn$
(c) $Fe + 2H^+ \rightarrow Fe^{2+} + H_2$
(d) $Fe + 2H_2O \rightarrow Fe^{2+} + 2OH^- + H_2$
(e) $Cu + 2Ag^+ \rightarrow 2Ag + Cu^{2+}$

**14-27** Using Table 14-2, predict whether the following reactions occur in aqueous solution. If not, write N.R.
(a) $Sn^{2+} + Pb \rightarrow Pb^{2+} + Sn$
(b) $Ni^{2+} + H_2 \rightarrow 2H^+ + Ni$
(c) $Cu + F_2 \rightarrow CuF_2$
(d) $Ni^{2+} + 2Br^- \rightarrow Ni + Br_2$

**14-28** Tell whether a reaction occurs in each of the following cases. Write a balanced equation indicating the spontaneous reaction if one occurs.
(a) Some iron nails are placed in a $CuCl_2$ solution.
(b) Silver coins are dropped into an acid solution [$H^+(aq)$].
(c) A copper penny is placed in a $Pb(NO_3)_2$ solution.
(d) Ni metal is placed in a $Pb(NO_3)_2$ solution.
(e) Sodium metal is placed in water.
(f) Zinc metal is placed in water.

**14-29** Tell whether a reaction occurs in each of the following cases. Write balanced equations illustrating the spontaneous reaction if one occurs.
(a) Aluminum metal makes contact with water.
(b) Iron nails are placed in a $ZnBr_2$ solution.
(c) $H_2$ gas is bubbled into an acid solution of $Cu^{2+}$.
(d) Copper metal is placed in an acid [$H^+(aq)$] solution.
(e) Tin metal is placed in an acid solution.

**14-30** In Chapter 13, we mentioned the corrosiveness of acid rain. Why does rain containing a high $H^+(aq)$ concentration cause more damage to iron exposed in bridges and buildings than pure $H_2O$? Write the reaction between Fe and $H^+(aq)$.

**14-31** $Br_2$ can be prepared from the reaction of $Cl_2$ with NaBr dissolved in sea water. Explain. Write the reaction. Can $Cl_2$ be used to prepare $F_2$ from NaF solutions?

**14-32** Describe how a voltaic cell could be constructed from a strip of iron, a strip of lead, a $Fe(NO_3)_2$ solution, and a

$Pb(NO_3)_2$ solution. Write the anode reaction, the cathode reaction, and the total reaction.

* **14-33** Judging from the relative difference in the strength of the oxidizing agents ($Fe^{2+}$ vs. $Pb^{2+}$) and ($Zn^{2+}$ vs. $Cu^{2+}$), which do you think would be the more powerful cell, the one in Problem 14-32 or the Daniell cell? Why?

**14-34** Why can't elemental Na be formed in the electrolysis of aqueous NaCl? Write the reaction that occurs at the cathode.

**14-35** Certain metals can be purified by electrolysis. For example, a mixture of Ag, Zn, and Fe can be dissolved so that their metal ions are present in aqueous solution. If a solution containing these ions is electrolyzed, which metal ion would be reduced to the metal the easiest?

# Review Test on Chapters 11–14

*The following multiple-choice questions have one correct answer.*

1. In which of the following cases would there be dipole-dipole interactions in the liquid state?
(a) NaCl (b) $H_2$ (c) $CCl_4$ (d) COS (e) $CO_2$
2. Which of the following has hydrogen bonding in the liquid state?
(a) $H_2S$ (b) NaH (c) $CH_4$ (d) $CH_3OH$
(e) $NCl_3$
3. Which of the following nonpolar molecules should have the highest boiling point?
(a) $H_2$ (b) $SF_6$ (c) $CO_2$ (d) $CH_4$ (e) $N_2$
4. Based on interactions in the liquid state, which of the following has the highest heat of vaporization?
(a) $H_2O$ (b) $H_2S$ (c) $CH_4$ (d) $H_2$ (e) $NH_3$
5. When heat is applied to a liquid at its boiling point:
(a) The heat energy is converted into potential energy.
(b) The heat energy is converted into kinetic energy.
(c) The heat energy is converted into both potential and kinetic energy.
(d) It makes the molecules of the vapor move faster than those in the liquid.
6. The normal boiling point of a liquid is defined as the temperature at which:
(a) Bubbles form in the liquid.
(b) The vapor pressure equals the atmospheric pressure.
(c) The vapor pressure equals one atmosphere.
(d) The vapor and the liquid exist in equilibrium.
7. A compound has a melting point of 950 °C. Which of the following statements is most likely true?
(a) Its molecules are nonpolar.
(b) The compound has a low heat of fusion.
(c) Its molecules are polar covalent.
(d) The compound has a low boiling point.
(e) The compound has a high heat of vaporization.
8. When a liquid in an insulated container is allowed to evaporate,
(a) The temperature of the liquid rises.
(b) The temperature of the liquid does not change.
(c) No liquid evaporates unless heat is supplied from the outside.
(d) The vapor that escapes is warmer than the liquid.
9. Which of the following equations illustrates the solution of $Na_2SO_4$ in water?

$$Na_2SO_4(s) \xrightarrow{H_2O}$$

(a) $Na_2^{2+}(aq) + SO_4^{2-}(aq)$
(b) $2Na^+(aq) + SO_4^{2-}(aq)$
(c) $Na^{2+}(aq) + SO_4^-(aq)$
(d) $2Na^{2+}(aq) + SO_4^-(aq)$
10. Which of the following equations represents a metathesis reaction?
(a) $CaCO_3 \rightarrow CaO + CO_2$
(b) $HCl + KOH \rightarrow H_2O + KCl$
(c) $Na_2S + Ni(NO_3)_2 \rightarrow NiS + 2NaNO_3$
(d) $Zn + CuCl_2 \rightarrow ZnCl_2 + Cu$
(e) $CH_4 + 2O_2 \rightarrow CO_2 + 2H_2O$
11. Which of the following is a net ionic equation?
(a) $CaCl_2 + Na_2CO_3 \rightarrow CaCO_3 + 2NaCl$
(b) $Ca^{2+} + 2Cl^- + 2Na^+ + CO_3^{2-} \rightarrow CaCO_3 + 2Cl^- + 2Na^+$
(c) $2Cl^- + 2Na^+ \rightarrow 2Na^+ + 2Cl^-$
(d) $Ca^{2+} + CO_3^{2-} \rightarrow CaCO_3$
12. A 2.0-L quantity of 0.10 *M* solution of HCl contains:
(a) 1.0 mol of HCl (d) 0.05 mol of $H_2O$
(b) 0.20 mol of $H_2O$ (e) 0.20 mol of HCl
(c) 20 mol of HCl
13. Which of the following is not a colligative property?
(a) the boiling point of the solvent (c) vapor pressure lowering
(b) freezing point depression (d) osmotic pressure
14. What is the freezing point of a 0.100 *m* aqueous solution of a nonelectrolyte? For water $K_f = 1.86$.

(a) −1.86 °C
(b) 0.186 °C
(c) −0.186 °C
(d) 18.6 °C
(e) −18.6 °C

15. If solutions of each of the following have the same concentration, which has the lowest freezing point?
(a) $CH_3OH$ (a nonelectrolyte)
(b) NaCl
(c) $Na_2SO_4$
(d) CaS
(e) all are the same

16. Which of the following is the formula for perchloric acid?
(a) HCl
(b) HClO
(c) $HClO_2$
(d) $HClO_3$
(e) $HClO_4$

17. Which of the following is a balanced neutralization reaction that leads to the formation of an acid salt?
(a). $HCl + NaOH \rightarrow NaCl + H_2O$
(b) $H_3PO_4 + KOH \rightarrow KHPO_4 + H_2O$
(c) $H_2SO_4 + 2KOH \rightarrow K_2SO_4 + H_2O$
(d) $NaOH + H_2SO_3 \rightarrow NaHSO_3 + H_2O$

18. Which of the following is a strong acid?
(a) $HNO_2$
(b) $HNO_3$
(c) $H_2SO_3$
(d) HClO
(e) KOH

19. The concentration of an aqueous HCl solution is 0.01 *M*. The pH is:
(a) 1 (b) 2 (c) 3 (d) 12 (e) 7

20. A solution has pH = 8.5. The solution is
(a) strongly basic
(b) weakly basic
(c) neutral
(d) weakly acidic
(e) strongly acidic

21. Which of the following is the conjugate acid of the $HS^-$ ion?
(a) $H_2S^-$
(b) $S^{2-}$
(c) $S^-$
(d) $H_3O^+$
(e) $H_2S$

22. Which of the following conjugate bases does not exhibit basic nature in water?
(a) $S^{2-}$
(b) $NO_2^-$
(c) $ClO_4^-$
(d) $PO_4^{3-}$
(e) $HS^-$

23. A 0.10 *M* solution of a compound has pH = 5.4. The compound is
(a) HCl
(b) KOH
(c) $NH_4Cl$
(d) $K_2S$
(e) NaCl

24. Which of the following pairs of compounds does not form a buffer when aqueous solutions of the two are mixed?
(a) $NaC_2H_3O_2$ and $HC_2H_3O_2$
(b) KBr and HBr
(c) $LiNO_2$ and $HNO_2$
(d) $NH_3$ and $NH_4Cl$

25. Which of the following oxides is the anhydride of $HClO_4$?
(a) $CO_2$
(b) ClO
(c) $Cl_2O_7$
(d) $ClO_3$
(e) $ClO_4^-$

26. Which of the following is an oxidation process?
(a) $H_2CO_3 \rightarrow H_2O + CO_2$
(b) $2Br^- \rightarrow Br_2$
(c) $Ca^{2+} + SO_4^{2-} \rightarrow CaSO_4$
(d) $H^+ + OH^- \rightarrow H_2O$
(e) $SO_4^{2-} \rightarrow S^{2-}$

27. What is the oxidation state of the Br in $Ca(BrO_3)_2$?
(a) +10
(b) +6
(c) +5
(d) +12
(e) −1

28. In the Daniell cell, what allows for for the migration of anions between compartments?
(a) a salt bridge
(b) the electrodes
(c) the external wire
(d) the electrons

29. In the following electrolytic cell, ________ is ________ at the anode.

$$2Cl^-(aq) + Fe^{2+}(aq) \rightarrow Fe(s) + Cl_2(g)$$

(a) $Fe^{2+}$, reduced
(b) $Cl^-$, reduced
(c) $Cl_2$, reduced
(d) $Cl^-$, oxidized
(e) Fe, oxidized

30. The following equation represents a spontaneous reaction:

$$Br_2(aq) + Ni(s) \rightarrow Ni^{2+}(aq) + 2Br^-(aq)$$

Which of the following statements is correct?
(a) $Br_2$ is a stronger reducing agent than $Ni^{2+}$.
(b) $Br^-$ is a stronger reducing agent than Ni.
(c) Ni is an oxidizing agent.
(d) $Br_2$ is oxidized.
(e) $Br_2$ is a stronger oxidizing agent than $Ni^{2+}$.

## Problems

1. How many calories are required to change 25.0 g of ice at −20.0 °C to steam at 100 °C? (The specific heat of ice is 0.492 cal/g · °C, the heat of fusion is 79.8 cal/g, and the heat of vaporization is 540 cal/g.)
2. An aqueous solution is made by dissolving 40.0 g of $H_2SO_4$ in enough water to make 850 mL of solution.
   (a) What is the molarity of the $H_2SO_4$ solution?
   (b) If 100 mL of this solution is diluted to 550 mL, what is the resulting molarity?
   (c) What volume of the original solution is needed to form 2.00 L of a 0.150 *M* solution?
3. (a) When aqueous solutions of lead(II) nitrate and sodium iodide are mixed, a precipitate of lead(II) iodide forms. Write the balanced equation representing this reaction.
   (b) What volume of 0.40 *M* lead nitrate is needed to react with 2.50 L of 0.55 *M* sodium iodide?
4. (a) Write the balanced equation representing the complete neutralization of sulfuric acid with lithium hydroxide.
   (b) What volume (in mL) of 0.25 *M* sulfuric acid is needed to completely react with 1.80 g of lithium hydroxide?
5. (a) When elemental tin is placed in a nitric acid solution, a spontaneous redox reaction occurs to produce $SnO_2$ and $NO_2$. Write the balanced equation representing this reaction.
   (b) What weight of $NO_2$ is produced from the complete reaction of 350 mL of 0.20 *M* nitric acid?
6. (a) Write the total ionic and net ionic equations for the reaction in Problem 3.
   (b) Write the total ionic and net ionic equations for the reaction in Problem 4.
7. What is the freezing point of an aqueous solution containing 10.0 g of the nonelectrolyte ethyl alcohol ($C_2H_5OH$) in 500 g of water?
8. From the equation in Problem 5, answer the following.
   (a) What is the species oxidized?
   (b) What is the species reduced?
   (c) What is the product of oxidation?
   (d) What is the product of reduction?
   (e) What is the oxidizing agent?
   (f) What is the reducing agent?
9. Complete the following equations.
   (a) $HOCl + H_2O \rightleftharpoons$
   (b) $NH_3 + H_2O \rightleftharpoons$
   (c) $KOH + HNO_2 \rightarrow$
   (d) $H_3AsO_3 + 2NaOH \rightarrow$
10. Write equations representing reactions for the following. If no reaction occurs, write N.R.
    (a) $K^+ + H_2O$
    (b) $CaO + H_2O$
    (c) $NH_4^+ + H_2O$
    (d) $CO_2 + H_2O$
    (e) $ClO^- + H_2O$
    (f) $Br^- + H_2O$
11. In a solution of a weak base, $[OH^-] = 6.5 \times 10^{-4}$. What is the value of
    (a) $[H_3O^+]$ (b) pH (c) pOH
12. Given the following table of oxidizing and reducing agents.

    Strongest oxidizing agent → $I_2 + 2e^- \rightleftarrows 2I^-$
    $Cr^{3+} + 3e^- \rightleftarrows Cr$
    $Mn^{2+} + 2e^- \rightleftarrows Mn$
    $Ca^{2+} + 2e^- \rightleftarrows Ca$ ← Strongest reducing agent

    (a) Is the following reaction spontaneous or nonspontaneous?

    $$Mn + Ca^{2+} \rightarrow Ca + Mn^{2+}$$

    (b) Write a balanced, spontaneous equation representing a reaction involving the $Cr,Cr^{3+}$ and the $Mn,Mn^{2+}$ half-reactions.
    (c) If an aqueous solution of $I_2$ (an antiseptic) is spilled on a chromium-coated bumper on an automobile, will a reaction occur? If so, write the balanced equation for the reaction.

Like all chemical reactions the spoiling of food is much slower at low temperatures. Thus food keeps much longer when stored in a refrigerator. The rate at which chemical reactions take place is a topic of this chapter.

# Chapter 15
# Chemical Kinetics and Equilibrium

## Purpose of Chapter 15

In Chapter 15 we discuss chemical reactions at equilibrium, factors that affect their rates, and quantitative aspects of reactions at equilibrium.

## Objectives for Chapter 15

After completion of this chapter, you should be able to:

1. Describe the conditions necessary for colliding reactant molecules to transform into products. (15-1)
2. Describe dynamic equilibrium in terms of reaction rates. (15-1)
3. Explain how a heterogeneous mixture of a salt in a saturated solution of the salt is an example of dynamic equilibrium. (15-2)
4. Describe how the following factors affect the rate of a reaction. (15-3)
   (a) The activation energy
   (b) The temperature
   (c) The concentration of reactants
5. Write the law of mass action for any homogeneous (one-phase) reaction. (15-4)
6. Demonstrate how the law of mass action relates to the rate laws for the hypothetical reaction discussed. (15-4)
7. Calculate the value of an equilibrium constant given appropriate experimental data. (15-4)
8. Use the value of the equilibrium constant to calculate the concentration of a species in the mass action expression. (15-4)
9. Describe how the values of $K_a$ and $K_b$ relate to the strength of a weak acid or base. (15-5)
10. Use the value of $K_a$ or $K_b$ and the initial concentration to calculate the pH of a weak acid, weak base, or a buffer solution. (15-5)
11. Apply the principle of Le Châtelier to show how changes in concentrations, temperature, and volume affect a system at equilibrium. (15-6)
12. Describe the role of a catalyst in a chemical reaction. (15-6)

Thank goodness for the refrigerator! By storing food in this appliance, we can drastically slow the chemical reactions that lead to decay and spoiling. In fact, the reactions can be stopped almost entirely by storing food in the freezer. This is an everyday example of the fact that the rates of all chemical reactions decrease as the temperature drops. Another example of this phenomenon concerns reptiles. Snakes and turtles are animals that become lethargic on chilly mornings. This is because these reptiles are cold-blooded animals whose body temperatures fluctuate with the temperature of the air. Thus, as the temperature drops, their rate of metabolism slows. Metabolism is a series of chemical reactions that generate energy from food. As a result, cold-blooded animals must seek shelter in holes as winter approaches or they won't be able to move at all. Warm-blooded animals such as humans continuously generate enough heat internally to keep the body temperature regulated at around 98 °F (37 °C). This constant internal temperature allows metabolism to continue at a constant rate even in cold weather.

To understand why temperature affects the rate of a chemical reaction we must first step back and examine how a typical reaction takes place. Once we know basically how reactants change into products, the factors that affect the rate of this phenomenon can then be discussed. This we will do in this chapter. We will also concentrate, both qualitatively and quantitatively, on certain reactions that reach a state of equilibrium.

As background for this chapter, review the kinetic theory of gases discussed in Section 10-2, the distribution of molecular kinetic energies as discussed in Section 11-7, and weak acids and bases discussed in Sections 13-3 and 13-6.

## 15-1 Mechanisms and Equilibrium

So far, we have just accepted that reactants change into products. Somehow, atoms in the compounds of reactants get shuffled around to form the new compounds of the products. In this section we wish to examine *how* this may occur. *The path or method whereby reactant molecules transform into product molecules is known as the* **mechanism** *of the reaction. The study of reaction rates and their relation to the mechanism of a reaction is known as* **chemical kinetics**. The mechanisms of reactions can be simple or complicated. We will confine this discussion, however, to a case of a simple, hypothetical reaction. We chose a hypothetical reaction rather than a real one because it can be "tailor-made" to illustrate clearly the principles involved. It can also be kept free of some complications that always seem to occur in real situations and that require lengthy explanations. From an understanding of how this simple reaction takes place we will see how factors such as concentration of reactants and temperature affect the rate and the extent of a reaction.

Our reaction involves the reaction of two hypothetical diatomic molecules $A_2$ and $B_2$. This particular reaction does not continue completely to the right. It is a reversible reaction that appears to stop when appreciable quantities of reactants are still present. Such reactions are indicated by double arrows ($\rightleftharpoons$).

$$A_2(g) + B_2(g) \rightleftharpoons 2AB(g)$$

First, let's concentrate on the forward reaction, which produces the gaseous molecule AB. In Chapter 10 we mentioned that gas molecules are in constant motion, undergoing collisions with each other and the walls of the container. *The* **collision theory** *of chemical reactions assumes that reactions take place through the collisions of molecules.* If the collision between two reacting molecules takes place (1) in exactly the right orientation and (2) with enough energy, a reshuffling of electrons takes place, with new bonds forming and old bonds breaking. If these conditions are not met, the molecules simply recoil from each other unchanged. This is illustrated in Figure 15-1.

In the hypothetical reaction, we assume that at least some of the collisions have the right orientation and sufficient energy. Thus the concentration of the product AB begins to increase as the concentrations of $A_2$ and $B_2$ decrease. Eventually, a point is reached where the concentrations of

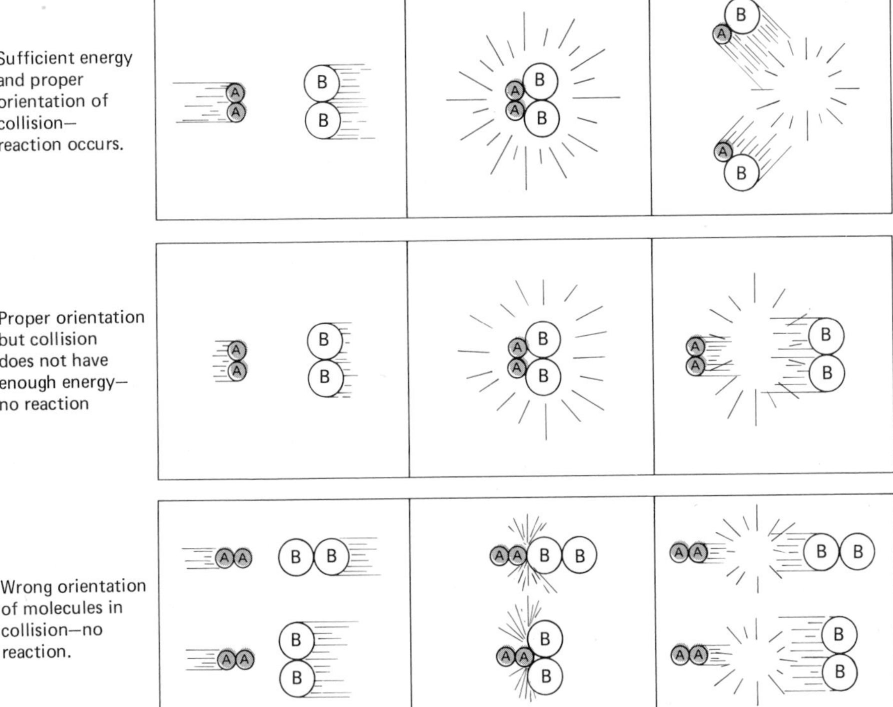
Sufficient energy and proper orientation of collision—reaction occurs.
Proper orientation but collision does not have enough energy—no reaction
Wrong orientation of molecules in collision—no reaction.
A
B

**Figure 15-1** THE REACTION OF $A_2$ AND $B_2$. Reactions occur only when colliding molecules have the minimum amount of energy and the right orientation.

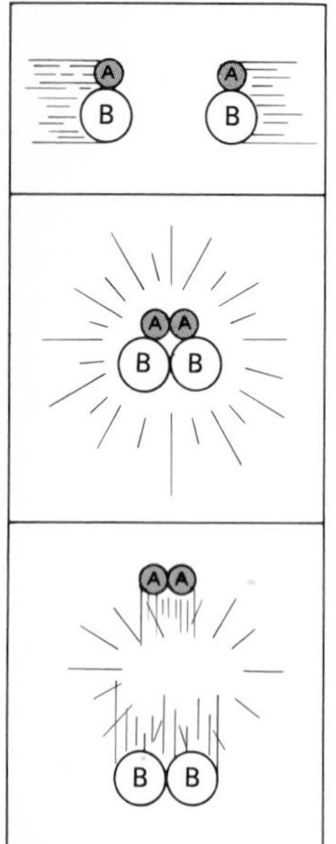


**Figure 15-2** THE REACTION OF 2AB. Collisions between AB molecules having the minimum energy and correct orientation lead to the formation of reactants.

reactants remains constant. To understand how this happens we will conduct an experiment starting with pure AB. In this case, we find that AB molecules can react to form $A_2$ and $B_2$ if collisions between AB molecules have proper orientation and sufficient energy. (See Figure 15-2.) This, of course, is simply the reverse of the original reaction.

Now we will put these two reactions together. If we again start with pure $A_2$ and $B_2$, at first the only reaction leads to the formation of AB. As the concentration of AB increases, however, the reverse reaction becomes important. Eventually, the rate of formation of products (the forward reaction) is exactly offset by the rate of formation of reactants (the reverse reaction). At this point the concentrations of all species ($A_2$, $B_2$, and AB) remain constant. This phenomenon is referred to as a **dynamic equilibrium**, *which emphasizes the changing identity of reactants and products despite the fact that the total amounts of each does not change.* In Figure 15-3 the reaction and the point of equilibrium are illustrated graphically.

If we could shrink ourselves to the molecular level, we would immediately become aware of this dynamic (changing) situation. If one A atom in an $A_2$ molecule was marked so that it could be traced, we would notice that

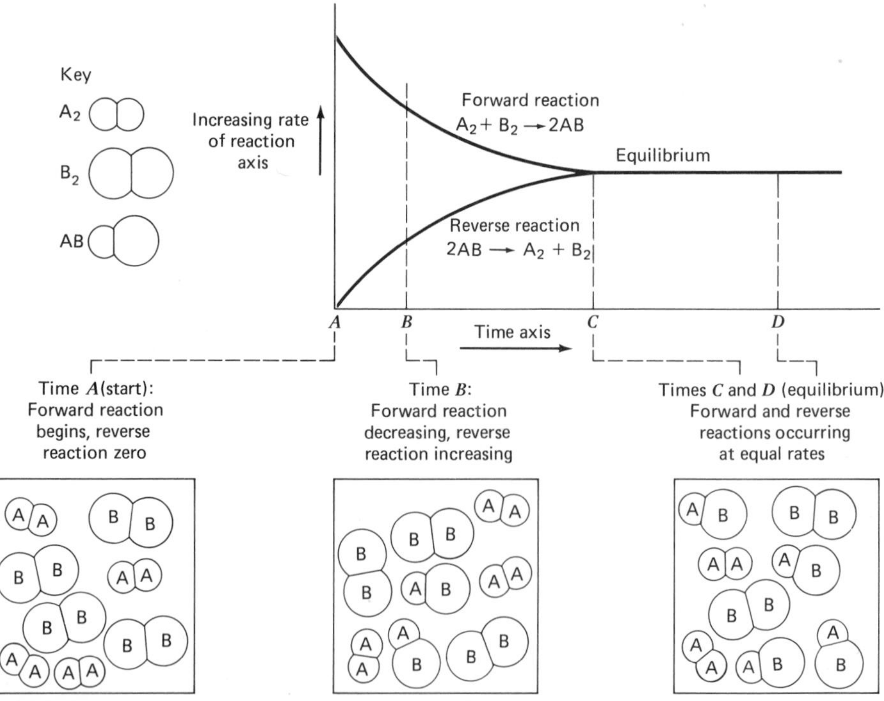


**Figure 15-3** $A_2$, $B_2$, AB EQUILIBRIUM. Equilibrium is achieved when the rates of the forward and reverse reactions are equal.

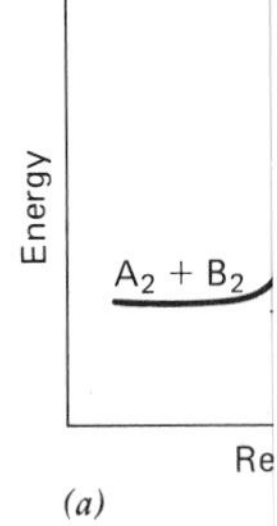

at one moment it is present in an $A_2$ molecule, later it is part of an AB molecule, and still later part of another $A_2$ molecule.

$$\begin{array}{ccccccc} *\mathrm{A} & & *\mathrm{A} & & *\mathrm{A} & & \\ | & & | & & | & & \\ \mathrm{A} & \xrightarrow{+\mathrm{B}_2} & \mathrm{B} & \xrightarrow{+\mathrm{AB}} & \mathrm{A} & \xrightarrow{+\mathrm{B}_2} & \text{etc.} \end{array}$$

## 15-2 Examples of Reactions at Equilibrium

We have previously discussed several equilibrium reactions. In Chapter 9, the following reaction was used to illustrate the calculation of percent yield of an incomplete reaction.

$$N_2(g) + 3H_2(g) \rightleftharpoons 2NH_3(g)$$

The reason a 100% yield was not obtained was because the reaction reached a state of equilibrium. When gaseous $N_2$ and $H_2$ are mixed, gaseous $NH_3$ is formed. Eventually, the reaction reaches a state of equilibrium, when the $NH_3$ decomposes to $N_2$ and $H_2$ as fast as it is being formed.

In Chapter 12, the solubility of ionic compounds was mentioned. When the solution is saturated (the limit of solubility is reached), an equilibrium is established between the undissolved solid phase and the dissolved ions:

$$NaCl(s) \rightleftharpoons Na^+(aq) + Cl^-(aq)$$

In this case a dynamic situation also exists, as the identity of the ions in solution is changing although the concentration is not (see Figure 15-4).

This phenomenon can be demonstrated by the art of growing large crystals from small ones. If a comparatively large crystal of a water-soluble substance such as $CuSO_4$ is suspended in a saturated solution of that substance, the small crystals in the bottom of the beaker get smaller and the large crystal gets larger. The concentration of the substance in solution does not change, nor does the total weight of crystals. It is obvious, however, by

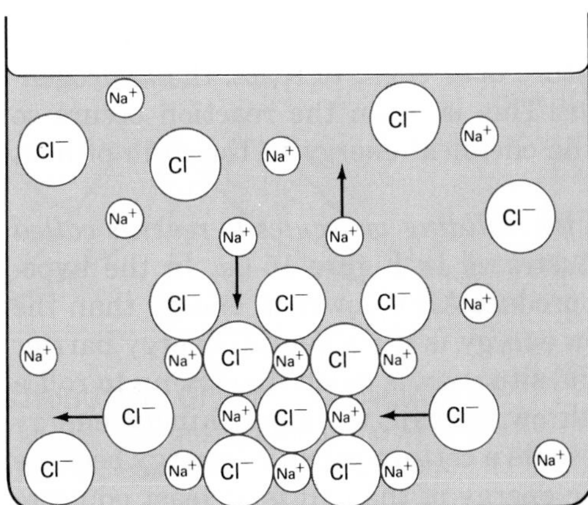


**Figure 15-4** SOLUBILITY AND EQUILIBRIUM. At equilibrium the rate of solution of ions equals the rate at which ions are forming the solid.

Figure 15-5
TAL GROW
size of the c
changes, bu
tal weight d



the higher temperature ($T_2$) the curve intersects the dashed line at a higher point than $T_1$, meaning that a greater fraction of molecules can overcome the activation energy in a collision. As a result, the rates of *all* chemical reactions increase as the temperature increases.

There is another reason why an increase in temperature increases the rate of a reaction. In addition to the energy of the collisions, the rate of a reaction depends on the frequency of collisions (the number of collisions per second):

$$r \text{ (rate)} \propto \text{frequency of collisions}$$

Recall that the kinetic energy (K.E.) of a moving object is given by the relationship

$$\text{K.E.} = \tfrac{1}{2}mv^2 \qquad (m = \text{mass},\ v = \text{velocity})$$

This indicates that the more kinetic energy a moving object has the faster it is moving.

As the temperature increases, the average velocity of the molecules increases. This means that the frequency of collisions also increases. As an analogy, imagine a box containing Ping Pong balls (four white and four dark balls) as shown in Figure 15-8. If we jiggle the box, the balls move around and collide. If we jiggle the box faster (analogous to a higher temperature), the balls move around faster and there are more frequent collisions. The increased noise we hear tells us that, indeed, collisions are not only more energetic but more frequent.

In summary, a rise in temperature increases the rate of a reaction for two reasons: the collisions have more energy, and the collisions occur more frequently.

**The Concentration of Reactants** Let's look again at the box of Ping Pong balls. The frequency of collisions depends not only on the velocity but on the number of Ping Pong balls in the box. Shaking a box containing 20 balls at the same rate as one with 10 balls would obviously create a lot more

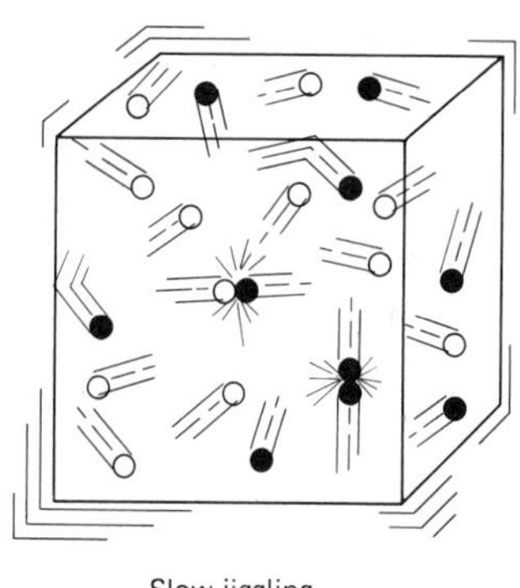


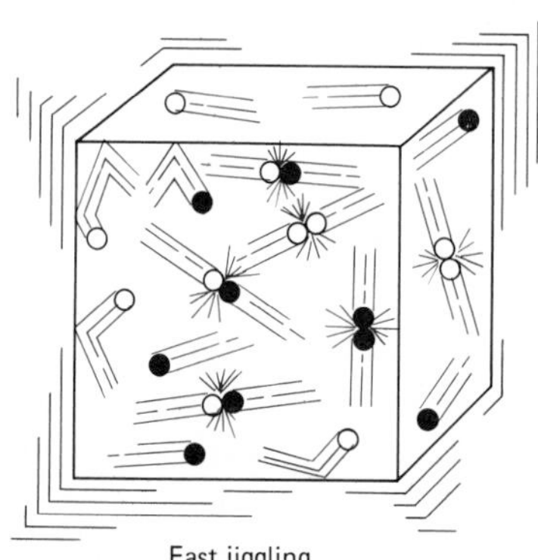


**Figure 15-8** THE EFFECT OF VELOCITY ON COLLISIONS. Collisions occur more frequently (and with more force) as the velocity increases.

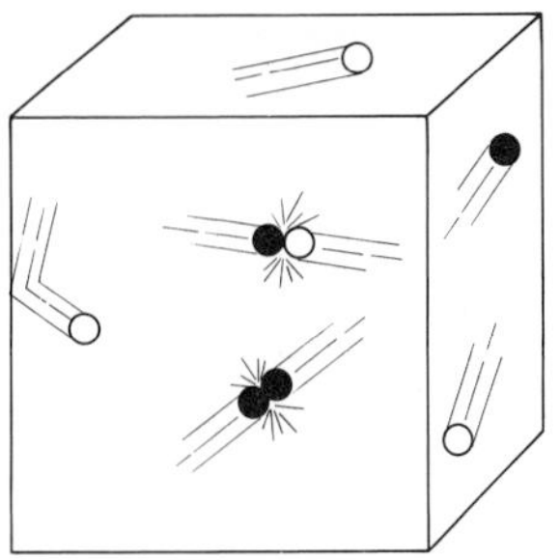
Four white and four dark balls

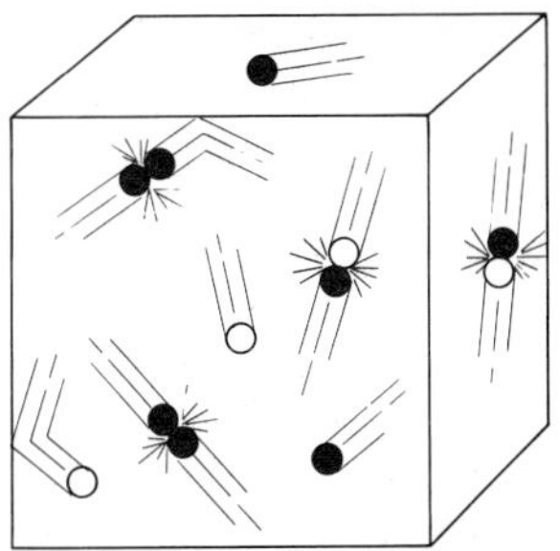
Four white and eight dark balls

**Figure 15-9** THE EFFECT OF CONCENTRATION ON COLLISIONS. Collisions occur more frequently when there are more balls in the box.

racket. In Figure 15-9 we have doubled the number of dark balls but still have only four white balls. This doubles the chances for a dark-white collision for any one white ball. Similarly, if we tripled the number of dark balls, the rate of dark-white collisions is tripled. This relationship can be written mathematically as:

$$\text{Frequency of dark-white collisions} \propto \text{concentration of dark balls}$$

and conversely,

$$\text{Frequency of dark-white collisions} \propto \text{concentration of white balls}$$

In the case of a chemical reaction, the frequency of collisions (and thus the rate of the reaction) is also dependent on the number of reactant molecules in a fixed volume (which is the concentration). In the reaction of $A_2$ and $B_2$, the rate of formation of AB doubles if the concentration of $A_2$ doubles (and $[B_2]$* is held constant). Mathematically, this can be expressed as

$$r_f \propto [A_2] \qquad [B_2] \text{ constant} \qquad (r_f = \text{rate of forward reaction})$$

The result is similar if $[B_2]$ doubles and $[A_2]$ is held constant:

$$r_f \propto [B_2] \qquad [A_2] \text{ constant}$$

These two relations can be combined into one relation known as the rate law for the forward reaction:

$$r_f \propto [A_2][B_2]$$

or, as an equality,

$$r_f = k_f[A_2][B_2]$$

where $k_f$ is called the **rate constant** for the forward reaction. *The* **rate law** *for a reaction expresses the effect that changes in concentration of reactants have on the rate of the reaction.*

Notice in this rate law that if both $[A_2]$ and $[B_2]$ are doubled, the rate doubles twice, or fourfold. In the reverse reaction, both colliding molecules

*$[B_2]$ symbolizes the numerical value of the concentration in units of moles per liter (mol/L).

are identical. Therefore, the doubling of [AB] in effect doubles the concentration of both colliding molecules, and the rate increases fourfold. The rate law for the reverse reaction ($r_r$) is thus

$$r_r = k_r[\text{AB}][\text{AB}] = k_r[\text{AB}]^2$$

where $k_r$ is the rate constant for the reverse reaction.

In our example, we assumed a mechanism for the reaction and then rationalized a rate law from this proposed mechanism. In fact, the sequence is just the opposite. Chemists first determine a rate law from experiments that study the effect of concentrations on the rate of reaction. From this rate law they then propose a mechanism that is consistent with the experiments. In our hypothetical example, we proposed that products form from a simple one-step collision of reactant molecules. Other reactions have rate laws that indicate mechanisms that are more complex. For example, many reactions involve mechanisms in which products form after two or more steps, where each step involves a collision. In any case, the rate of all reactions are dependent in some manner on the concentration of one or more of the reactants.

**See Problems 15-1 through 15-8.**

## 15-4 The Law of Mass Action

Long before much was known about rate laws and mechanisms, experimental results indicated a relationship governing reactions at equilibrium. It was found that the relative proportions of reactants and products were distributed in a predictable manner at a specified temperature. This is illustrated with the following general reaction at equilibrium.

$$a\text{A} + b\text{B} \rightleftharpoons c\text{C} + d\text{D}$$

For a reaction at equilibrium, the following relationship exists and is known as the **law of mass action.**

$$K_{\text{eq}} = \frac{[\text{C}]^c[\text{D}]^d}{[\text{A}]^a[\text{B}]^b}$$

*Notice that the coefficients (a, b, c, and d) of the compounds (A, B, C, and D) become exponents of the compounds, which are expressed as molar concentrations (mol/L).* The products are written in the numerator, and the reactants are written in the denominator. The law of mass action for a particular reaction consists of two parts. *$K_{\text{eq}}$ is called the equilibrium constant and has a definite numerical value at a certain temperature. The ratio to the right of the equality is called the* **mass action expression**.

In Table 15-1, the results of three experiments are listed. In each experiment we start with a different distribution of reactants and products. At equilibrium, the concentrations of all species are distributed so that solution of the mass action expression yields approximately the same constant.

**Table 15-1** The Value of $K_{eq}$

For the reaction $H_2(g) + I_2(g) \rightleftharpoons 2HI(g)$,

$$K_{eq} = \frac{[HI]^2}{[H_2][I_2]} \qquad [X] = \text{concentration in mol/L}$$

| Exp. | Initial Concentration | | | Equilibrium Concentration | | | |
|---|---|---|---|---|---|---|---|
| No. | $[H_2]$ | $[I_2]$ | $[HI]$ | $[H_2]$ | $[I_2]$ | $[HI]$ | $K_{eq}$ |
| 1 | 2.000 | 2.000 | 0 | 0.428 | 0.428 | 3.144 | 53.9 |
| 2 | 0 | 0 | 2.000 | 0.214 | 0.214 | 1.572 | 54.0 |
| 3 | 1.000 | 1.000 | 1.000 | 0.321 | 0.321 | 2.358 | 54.0 |

We can now see that the mass action expression is consistent with our previous discussion of rate laws. In the discussion in the previous section we indicated (Figure 15-3) that the point of equilibrium is reached (time $C$) when the rates of the forward and reverse reactions are equal. This can be expressed mathematically as:

$$r_f = r_r$$

For the hypothetical reaction the two rate laws can now be set equal to each other as follows:

$$k_f[A_2][B_2] = k_r[AB]^2$$

$$\frac{k_f}{k_r} = \frac{[AB]^2}{[A_2][B_2]}$$

Since the ratio of two constants is another constant, the expression can be simplified as follows:

$$\frac{k_f}{k_r} = K_{eq} = \frac{[AB]^2}{[A_2][B_2]}$$

We derived this law of mass action by assuming a mechanism, and from that a rate law. Fortunately, an expression can be written as indicated earlier for any reaction at equilibrium regardless of the details of how the reaction takes place.

We are now ready to write the laws of mass action for some gas phase reactions at equilibrium.

**Example 15-1**

Write the law of mass action for each of the following reactions.

(a) $N_2(g) + 3H_2(g) \rightleftharpoons 2NH_3(g)$
(b) $4NH_3(g) + 5O_2(g) \rightleftharpoons 4NO(g) + 6H_2O(g)$
(c) $PCl_3(g) + Cl_2(g) \rightleftharpoons PCl_5(g)$

**Solutions**

(a) $$K_{eq} = \frac{[NH_3]^2}{[N_2][H_2]^3}$$

(b) $$K_{eq} = \frac{[NO]^4[H_2O]^6}{[NH_3]^4[O_2]^5}$$

(c) $$K_{eq} = \frac{[PCl_5]}{[PCl_3][Cl_2]}$$

**See Problems 15-9 and 15-10.**

We can now illustrate with some examples how the equilibrium constant is calculated from experimental data and then how it is used in calculating equilibrium concentrations.

**Example 15-2**

What is the value for $K_{eq}$ for the following system at equilibrium?

$$2NO(g) + O_2(g) \rightleftharpoons 2NO_2(g)$$

At a certain temperature, the equilibrium concentrations of the gases are $[NO] = 0.890$, $[O_2] = 0.250$, and $[NO_2] = 0.0320$.

**Solution**

For this reaction,

$$K_{eq} = \frac{[NO_2]^2}{[NO]^2[O_2]}$$

$$= \frac{[0.0320]^2}{[0.890]^2[0.250]} = \frac{1.024 \times 10^{-3}}{0.198}$$

$$= \underline{\underline{5.17 \times 10^{-3}}}$$

**Example 15-3**

For the equilibrium

$$N_2(g) + 3H_2(g) \rightleftharpoons 2NH_3(g)$$

complete the table and compute the value of the equilibrium constant.

| Initial Concentration | | | Equilibrium Concentration | | |
|---|---|---|---|---|---|
| $[H_2]$ | $[N_2]$ | $[NH_3]$ | $[H_2]$ | $[N_2]$ | $[NH_3]$ |
| 0.200 | 0.200 | 0 | ____ | ____ | 0.0450 |

**Procedure**

1 As in other stoichiometry problems, find the $[H_2]$ and $[N_2]$ that reacted to form the 0.0450 mol/L of $NH_3$ (Section 9-3).

2 Find the $[H_2]$ and $[N_2]$ remaining at equilibrium by subtracting the concentration that reacted from the initial concentration; that is,

$$[N_2]_{eq} = [N_2]_{initial} - [N_2]_{reacted}$$

3 Substitute the concentrations of all compounds present at equilibrium into the mass action expression and solve to find the value of $K_{eq}$.

**Solution**

1 If 0.0450 mol of $NH_3$ is formed, calculate the number of moles of $N_2$ that reacted (per liter).

$$0.0450 \, \cancel{\text{mol NH}_3} \times \frac{1 \text{ mol N}_2}{2 \, \cancel{\text{mol NH}_3}} = 0.0225 \text{ mol N}_2 \text{ reacted}$$

The number of moles of $H_2$ that reacted are

$$0.0450 \, \cancel{\text{mol NH}_3} \times \frac{3 \text{ mol H}_2}{2 \, \cancel{\text{mol NH}_3}} = 0.0675 \text{ mol H}_2 \text{ reacted}$$

2 At equilibrium,

$$[N_2]_{eq} = 0.200 - 0.0225 = 0.178$$

$$[H_2]_{eq} = 0.200 - 0.0675 = 0.132$$

3 $$K_{eq} = \frac{[NH_3]^2}{[N_2][H_2]^3} = \frac{(0.0450)^2}{(0.178)(0.132)^3} = \underline{\underline{4.95}}$$

---

**Example 15-4**
In the preceding equilibrium, what is the concentration of $NH_3$ at equilibrium if the equilibrium concentration of $N_2$ and $H_2$ is 0.22 mol/L and 0.14 mol/L, respectively?

**Procedure**

In this problem, use the value of $K_{eq}$ found in the previous problem. The concentration of $NH_3$ can be found by substituting the concentrations of the species given and solving for the one unknown.

**Solution**

$$K_{eq} = \frac{[NH_3]^2}{[N_2][H_2]^3} \quad [N_2] = 0.22 \quad [H_2] = 0.14 \qquad K_{eq} = 4.95$$

$$= \frac{[NH_3]^2}{[0.22][0.14]^3} = 4.95$$

$$[NH_3]^2 = 2.99 \times 10^{-3}$$

$$[NH_3] = 5.5 \times 10^{-2} = 0.055$$

Thus the concentration of $NH_3$ = 0.055 mol/L.

**See Problems 15-11 through 15-27.**

The numerical value of $K_{eq}$ is a measure of the extent of the reaction to the right, so it tells us about the *position* of the equilibrium. A large value for $K_{eq}$ (e.g., $10^3$) tells us that the numerator is much larger than the denominator in the mass action expression. This means that the concentrations of at least one of the products in the numerator is considerably larger than the reactants in the denominator. Thus the position of equilibrium lies to the right. Conversely, a small value for $K_{eq}$ (e.g., $10^{-3}$) indicates that the reactants are favored and the position of equilibrium lies significantly to the left. Values for $K_{eq}$ in between indicate appreciable concentrations of both reactants and products present at equilibrium.

## 15-5 Equilibria of Weak Acids and Bases in Water

The following equilibrium illustrates the dissociation of a typical weak acid in water.

$$HOCl(aq) + H_2O \rightleftharpoons H_3O^+(aq) + OCl^-(aq)$$

The law of mass action can be written as:

$$K_{eq} = \frac{[H_3O^+][OCl^-]}{[H_2O][HOCl]}$$

In this equilibrium, the concentration of $H_3O^+$, $OCl^-$, and HOCl can all be varied. The $H_2O$ is the solvent, however, and is present in a very large excess compared with the other species. Since the amount of $H_2O$ actually reacting is very small compared with the total amount present, the concentration of $H_2O$ is essentially a constant. The $[H_2O]$ can therefore be included with the other constant, $K_{eq}$, to product another constant labeled $K_a$. *$K_a$ is known as an* **acid ionization constant**.

$$K_{eq}[H_2O] = K_a = \frac{[H_3O^+][OCl^-]}{[HOCl]}$$

The ionization of the weak base $NH_3$ is illustrated as follows:

$$NH_3(aq) + H_2O \rightleftharpoons NH_4^+(aq) + OH^-(aq)$$

In this case, the *equilibrium constant is labeled $K_b$, which is the* **base ionization constant**.

$$K_b = \frac{[NH_4^+][OH^-]}{[NH_3]}$$

The values for the equilibrium constants for weak acids and bases can be determined experimentally as illustrated in the following examples.

**Example 15-5**

In a 0.20 *M* solution of $HNO_2$ it is found that 0.009 mol/L of the $HNO_2$ dissociates. What is the concentration of $H_3O^+$, $NO_2^-$, and $HNO_2$ at equilibrium, and what is the value of $K_a$?

**Procedure**

1 Write the equilibrium equation.

2 Calculate the concentration of undissociated $HNO_2$ present at equilibrium. Remember that the initial concentration given (0.20 *M*) represents the concentration of undissociated $HNO_2$ present *plus* the concentration that dissociates. Therefore, $[HNO_2]$ at equilibrium is the initial concentration minus the concentration that dissociates.

$$[HNO_2]_{eq} = [HNO_2]_{initial} - [HNO_2]_{dissociated}$$

3 Notice that one $H_3O^+$ and $NO_2^-$ are formed for each $HNO_2$ that dissociates (from the equation stoichiometry).

$$[H_3O^+]_{eq} = [NO_2^-]_{eq} = [HNO_2]_{dissociated}$$

4 Calculate $K_a$.

**Solution**

1 $$HNO_2 + H_2O \rightleftharpoons H_3O^+ + NO_2^-$$

2 $$[HNO_2]_{initial} = 0.20 \qquad [HNO_2]_{dissociated} = 0.009$$

$$[HNO_2]_{eq} = 0.20 - 0.009 = \underline{\underline{0.19}}$$

3 $$\underline{\underline{[H_3O^+] = [NO_2^-] = 0.009}}$$

4 $$K_a = \frac{[H_3O^+][NO_2^-]}{[HNO_2]} = \frac{[0.009][0.009]}{[0.19]}$$

$$= \underline{\underline{4 \times 10^{-4}}}$$

**Similar problems are illustrated in Appendix B (Examples B-13, B-14.)**

**Example 15-6**

A 0.25 *M* solution of HCN has pH = 5.00. What is $K_a$?

**Procedure**

1 Write the equilibrium reaction.

2 Convert pH to $[H_3O^+]$.

3 Notice that $[CN^-] = [H_3O^+]$ at equilibrium.

4 Calculate [HCN] at equilibrium, which is

$$[HCN]_{eq} = [HCN]_{initial} - [HCN]_{dissociated}$$

$[HCN]_{dissociated} = [H_3O^+]_{eq}$, since one $H_3O^+$ is produced at equilibrium for every HCN that dissociates.

5 Use these values to calculate $K_a$.

**Solution**

$$HCN + H_2O \rightleftharpoons H_3O^+ + CN^-$$

$$[H_3O^+] = \text{antilog } 5.00$$

$$[H_3O^+] = 1.0 \times 10^{-5} = [CN^-]$$

$$[HCN]_{eq} = 0.25 - (1.0 \times 10^{-5}) \approx 0.25$$

($\approx$ means approximately equal)

Notice that the amount of HCN that dissociates ($10^{-5}$ *M*) is negligible compared with the initial concentration of HCN (0.25 *M*).* therefore, $[HCN]_{eq} = [HCN]_{initial}$.

$$K_a = \frac{[H_3O^+][CN^-]}{[HCN]}$$

$$= \frac{(1.0 \times 10^{-5})(1.0 \times 10^{-5})}{(0.25)}$$

$$= \underline{\underline{4.0 \times 10^{-10}}}$$

*For the purposes of these calculations (two significant figures), a number will be considered negligible compared with another if it is less than 10% of the larger number.

**Example 15-7**

A 0.10 *M* solution of $NH_3$ is 1.34% ionized. What is the value of $K_b$ (the base ionization constant)?

**Procedure**

Find the concentration of all species in the mass action expression at equilibrium. Substitute these values into the expression and solve for $K_b$.

**Solution**

$$NH_3 + H_2O \rightleftharpoons NH_4^+ + OH^-$$

$$K_b = \frac{[NH_4^+][OH^-]}{[NH_3]}$$

At equilibrium, 1.34% of the $NH_3$ is ionized, or

$$0.0134 \times 0.10 = 1.34 \times 10^{-3} \text{ mol/L}$$

According to the equation, for every $NH_3$ ionized, one $NH_4^+$ and one $OH^-$ will form. Therefore, if $1.34 \times 10^{-3}$ mol/L ionize, at equilibrium

$$[NH_4^+] = [OH^-] = 1.34 \times 10^{-3}$$

The concentration of $NH_3$ at equilibrium is the initial concentration (0.10) minus the concentration that ionizes.

$$[NH_3] = 0.10 - (1.34 \times 10^{-3}) \approx 0.10$$

Substituting these values into the expression,

$$K_b = \frac{(1.34 \times 10^{-3})(1.34 \times 10^{-3})}{(0.10)}$$

$$= 1.8 \times 10^{-5}$$

**See Problems 15-28 through 15-37.**

The values of $K_a$ for some common weak acids are shown in Table 15-2 and of $K_b$ for some weak bases in Table 15-3. Notice that the smaller

**Table 15-2** $K_a$ for Some Weak Acids

$$K_a = \frac{[H_3O^+][X^-]}{[HX]}$$ (HX symbolizes a weak acid, $X^-$ its conjugate base)

| Acid | Formula | $K_a$ | |
|---|---|---|---|
| Hydrofluoric | HF | $6.7 \times 10^{-4}$ | ↓ Decreasing acid strength |
| Nitrous | $HNO_2$ | $4.5 \times 10^{-4}$ | |
| Formic | $HCHO_2$ | $1.8 \times 10^{-4}$ | |
| Acetic | $HC_2H_3O_2$ | $1.8 \times 10^{-5}$ | |
| Hypochlorous | HOCl | $3.2 \times 10^{-8}$ | |
| Hypobromous | HOBr | $2.1 \times 10^{-9}$ | |
| Hydrocyanic | HCN | $4.0 \times 10^{-10}$ | |

**Table 15-3** $K_b$ for Some Weak Bases

$$K_b = \frac{[HB^+][OH^-]}{[B]}$$ (B symbolizes a weak base, $HB^+$ its conjugate acid)

| Base | Formula | $K_b$ | |
|---|---|---|---|
| Dimethylamine | $NH(CH_3)_2$ | $7.4 \times 10^{-4}$ | Decreasing ↓ |
| Ammonia | $NH_3$ | $1.8 \times 10^{-5}$ | base |
| Hydrazine | $N_2H_4$ | $9.8 \times 10^{-7}$ | strength |

the value of $K_a$, the weaker the acid. For bases, the smaller the value of $K_b$, the weaker the base.

The values for $K_a$ or $K_b$ can now be used to calculate the pH from a known concentration of weak acid or base, as illustrated in the following examples.

**Example 15-8**
What is the pH of a 0.155 *M* solution of HOCl?

**Procedure**

1 Write the equilibrium involved.

2 Write the appropriate mass action expression.

3 Let $x = [H_3O^+]$ at equilibrium; since $[H_3O^+] = [OCl^-]$, $x = [OCl^-]$.

4 At equilibrium, $[HOCl]_{eq} = [HOCl]_{initial} - [HOCl]_{dissociated}$.

5 Using the value for $K_a$, solve for $x$.

6 Convert $x$ to pH.

In summary:

| | [HOCl] | $[H_3O^+]$ | $[OCl^-]$ |
|---|---|---|---|
| Initial | 0.155 | 0 | 0 |
| Equilibrium | $0.155 - x$ | $x$ | $x$ |

**Solution**

1 $$HOCl + H_2O \rightleftharpoons H_3O^+ + OCl^-$$

2 $$K_a = \frac{[H_3O^+][OCl^-]}{[HOCl]} = 3.2 \times 10^{-8}$$

3 At equilibrium, $[H_3O^+] = [OCl^-] = x$.

4 At equilibrium, $[HOCl] = 0.155 - x$.

5
$$\frac{[x][x]}{[0.155 - x]} = 3.2 \times 10^{-8}$$

The solution of this equation appears to require the quadratic equation. However, a simplification of this calculation is possible. Notice that $K_a$ is a small number, indicating that the degree of dissociation is small (the equilibrium lies far to the left). This means that $x$ is a very small number. Since very small numbers make little or no difference when added to or subtracted from large numbers, they can be ignored with little or no error. (Refer to the example of the large crowd in Section 2-1.)

$$0.155 - x \approx 0.155$$

Therefore, the expression can now be simplified as follows:

$$\frac{[x][x]}{[0.155 - x]} = \frac{x^2}{0.155} = 3.2 \times 10^{-8}$$

$$x^2 = 5.0 \times 10^{-9}$$

To solve for $x$, take the square root of both sides of the equation:

$$\sqrt{x^2} = \sqrt{5.0 \times 10^{-9}} = \sqrt{50 \times 10^{-10}}$$

$$x = [H_3O^+] = 7.1 \times 10^{-5}$$

(Notice that the $x$ is indeed much smaller than 0.155.)

6
$$\text{pH} = -\log[H_3O^+] = \underline{\underline{4.15}}$$

---

**Example 15-9**

What is the pH of a 0.245 *M* solution of $N_2H_4$?

**Solution**

1
$$N_2H_4 + H_2O \rightleftharpoons HN_2H_4^+ + OH^-$$

2
$$K_b = \frac{[HN_2H_4^+][OH^-]}{[N_2H_4]} = 9.8 \times 10^{-7}$$

3 Let $x = [OH^-] = [HN_2H_4^+]$ (at equilibrium).

4 $[N_2H_4] = 0.245 - x$ (at equilibrium). Since $K_b$ is very small, $x$ is very small. Therefore,

$$[0.245 - x] \approx [0.245]$$

5 $$\frac{[x][x]}{[0.245]} = 9.8 \times 10^{-7}$$

$$x^2 = 2.4 \times 10^{-7}$$

$$x = 4.9 \times 10^{-4} = [OH^-]$$

6 $$pOH = -\log[OH^-] = -\log(4.9 \times 10^{-4}) = 3.31$$

$$pH = 14 - pOH = 14 - 3.31 = \underline{\underline{10.69}}$$

---

**Example 15-10**

What is the pH of a buffer solution that is made by dissolving 0.50 mol of $HC_2H_3O_2$ and 0.50 mol of $NaC_2H_3O_2$ in enough water to make 1.00 L of solution?

**Procedure**

**1** and **2** as in previous examples.

**3** Let $x = [H_3O^+]$ at equilibrium. In this case, $[C_2H_3O_2^-] = 0.50 + x$ (the concentration from the dissolved salt plus that from the dissociation of the acid).

**4** $$[HC_2H_3O_3] = 0.50 - x$$

**5** and **6** as in previous examples.

**Solution**

**1** $HC_2H_3O_2(aq) + H_2O \rightleftharpoons H_3O^+(aq) + C_2H_3O_2^-(aq)$

**2** $$K_a = \frac{[H_3O^+][C_2H_3O_2^-]}{[HC_2H_3O_2]}$$

**3, 4** Let $x = [H_3O^+]$, $[C_2H_3O_2^-] = 0.50 + x$, and $[HC_2H_3O_2] = 0.50 - x$.
Since $K_a$ is small, $x$ is small. Therefore,

$$[C_2H_3O_2^-] = 0.50 + x \approx 0.50$$

$$[HC_2H_3O_2] = 0.50 - x \approx 0.50$$

**5** $$\frac{[x]\cancel{[0.50]}}{\cancel{[0.50]}} = 1.8 \times 10^{-5}\ (K_a)$$

$$x = [H_3O^+] = 1.8 \times 10^{-5}$$

**6** $$pH = -\log(1.8 \times 10^{-5}) = \underline{\underline{4.74}}$$

Notice that for buffers made up of equal molar concentrations of the acid and the salt the pH of the solution is simply the negative log of the constant $K_a$. This is defined as the p$K_a$.

When [HX] = [$X^-$], pH = p$K_a$ and p$K_a$ = $-\log K_a$. p$K_a$ values are useful in determining the pH of various buffer solutions.

**See Problems 15-38 through 15-47.**

## 15-6 The Effect of Stress and Catalysts on Equilibrium

When a chemical system is at equilibrium, any change in conditions may affect the point of equilibrium. Such changes include a change in temperature, pressure, or concentration of a reactant or product. How a system at equilibrium reacts to a change in conditions is summarized by **Le Châtelier's principle,** which states: *When stress is applied to a system at equilibrium, the system reacts in such a way to counteract the stress.*

This principle can be illustrated by a long-distance jogger. When jogging it is desirable to maintain a steady pace, which is analogous to equilibrium. At first the body maintains the pace rather easily, with little change in respiration or heartbeat. After a couple of miles the body begins to tire, which is analogous to stress on the system. If nothing changed, the jogger would slow and finally stop. The body counteracts the stress, however, by faster breathing, faster heartbeat, and increased perspiration. Thus the human body counteracts the stress so that the steady pace (equilibrium) is maintained (see Figure 15-10).

We will illustrate Le Châtelier's principle with the following important industrial process that reaches an equilibrium.

$$2SO_2(g) + O_2(g) \rightleftharpoons 2SO_3(g)$$

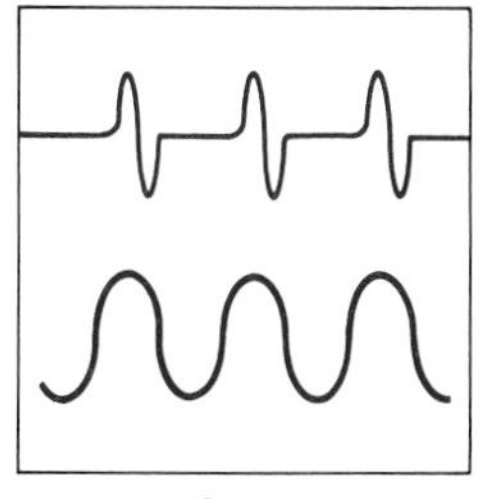


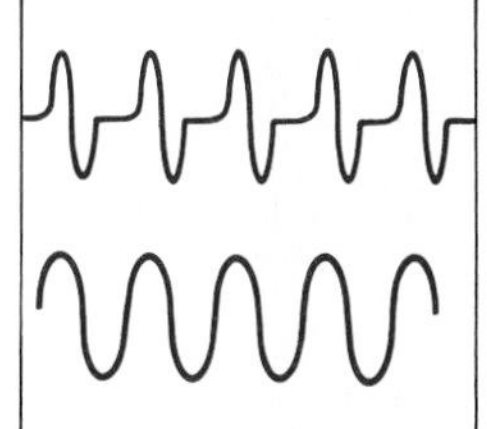

**Figure 15-10** STRESS AND THE JOGGER. To compensate for the increased rate of metabolism of a jogger, the heartbeat and respiration increase.

The $SO_2$ is produced by the combustion of sulfur or sulfur compounds. The $SO_3$ produced from this reaction is then hydrated to form sulfuric acid ($H_2SO_4$):

$$SO_3(g) + H_2O \rightarrow H_2SO_4(aq)$$

In 1980 *40 million tons* of $H_2SO_4$ with a value of \$3.2 billion was produced by this process (see Figure 15-11) in the United States alone. Because of the recession, recent production has been somewhat lower.

It is important to convert the maximum amount of $SO_2$ to $SO_3$ by forcing the point of equilibrium to the right as far as possible.

**1** If the system is at equilibrium and additional $O_2$ is introduced, the system counteracts that stress by removing some of the added $O_2$. It can do this by reaction with $SO_2$. An increase in the concentration of compounds on one side of a reaction at equilibrium ultimately leads to an increase on the other side as the system shifts to remove the stress of the added compound.

The selective removal of a compound from a reaction mixture, such as formation of a precipitate or evolution of a gas, has a similar effect. The system shifts in such a way so as to replace the loss. If the loss is complete, the reaction in the direction of that loss is complete (an irreversible reaction).

The changes in the point of equilibrium can be illustrated by a simple

**Figure 15-11**
A SULFURIC ACID PLANT. This large plant converts sulfur into some of the millions of tons of sulfuric acid manufactured in this country each year.

example. In the following hypothetical reaction, equilibrium is established when [A] = 1.00 mol/L and [B] = 2.00 mol/L.

$$A \rightleftharpoons B \qquad K_{eq} = \frac{[B]}{[A]} = \frac{2.00}{1.00} = 2.00$$

If an additional 1.00 mol/L of A is now introduced into the reaction container, notice that the system is no longer at equilibrium.

$$\frac{[B]}{[A]} = \frac{2.00}{(1.00 + 1.00)} = 1.00 \neq K_{eq}$$

For the system to return to equilibrium, the numerator becomes larger (more B is formed) and the denominator smaller (some A must react). Thus we see mathematically that introducing additional components present on one side of an equation leads to a shift to the other side. In the example above, equilibrium is reestablished when

$$[A] = 1.33$$

$$[B] = 2.67$$

$$\frac{2.67}{1.33} = 2.00 = K_{eq}$$

This indicates that the added A increases the concentration of B by 0.67 mol/L.

In our industrial example, the conversion of $SO_2$ to $SO_3$ is aided by a large excess of $O_2$ in the reaction container.

**2** A change in the volume of the container also affects the point of equilibrium. Notice that the reaction produces a volume reduction when it goes to the right. Three moles of gas ($2SO_2 + 1O_2$ or 67.2 L at STP) react to form two moles of gas ($2SO_3$ or 44.8 L at STP). Recall from Boyle's law that when the volume of a container of gas is decreased this corresponds to an increase in pressure. *Thus, when a mixture of gases at equilibrium is compressed, the volume is lowered and the pressure raised.* The chemical system can counteract this stress by going to a lower volume. *Therefore, when a mixture of gases is compressed, the equilibrium shifts in the direction of the fewest number of molecules of gas.* In the reaction discussed, compression of the gaseous mixture results in the formation of a larger proportion of $SO_3$. When gases are not involved in a reaction at equilibrium, pressure changes have little effect since solids and liquids are virtually incompressible.

**3** The reaction of $SO_2$ with $O_2$ not only produces $SO_3$ as a product but also heat energy since this is an exothermic reaction.

$$2SO_2(g) + O_2(g) \rightleftharpoons 2SO_3(g) + \text{heat}$$

The heat is considered a component of the reaction and is produced in amounts directly proportional to the amount of $SO_3$ produced. If the reaction mixture at equilibrium is cooled, this removes heat from the system. According to Le Châtelier, the system shifts to the right to replace the lost heat. Since $SO_3$ is formed along with the heat, cooling the mixture produces more $SO_3$. On the other hand, heating the mixture forces the equilibrium to shift to the left. By shifting to the left, the system attempts to remove the added heat. This phenomenon is reflected in the value for $K_{eq}$. *For exothermic reactions, $K_{eq}$ becomes larger as the temperature decreases.* This means that, for exothermic reactions, formation of products (on the right) becomes more favorable at lower temperatures.

In all reactions, however, the rate of reaction decreases as the temperature decreases (i.e., the refrigerator effect). Since time is money in industry, the process of maximizing the yield of $SO_3$ must be compromised to an extent. The reaction must be run at a high enough temperature so that it proceeds to equilibrium in a reasonable time. This reaction is actually carried out at about 400 °C.

There is one other factor that can affect the rates at which reactions occur, and this is the presence of a catalyst. *A* **catalyst** *is a substance that is not consumed in a reaction but whose presence increases the rate of both the forward and reverse reactions.*

Catalysts work in different ways in different reactions. They may be intimately mixed with the reaction mixture (homogeneous catalysts), or they may simply provide a surface on which the reactions take place (heterogeneous catalysts). Catalysts do their work by providing an alternate mechanism with a lower activation energy for the reaction. As stated previously, a lower activation energy means that the rate of the reaction is increased. If it is a reaction that reaches equilibrium, both the rates of the forward and reverse reactions are increased and equilibrium is established faster. A catalyst does not affect the value of $K_{eq}$, however, so the eventual distribution of reactants and products is not altered. The normal activation energy for a reaction is analogous to having a mountain pass between two cities. The normal journey is difficult and long. A catalyst is analogous to having access to a tunnel through the mountains. It is much easier and thus faster to drive through the tunnel than going over the mountain pass. (See Figure 15-12.)

In the reaction of $SO_2$ and $O_2$ surface catalysts are used to achieve equilibrium. Most surface catalysts are transition metals such as platinum or palladium or their compounds such as $V_2O_5$. In many processes the exact catalyst used is a closely guarded industrial secret. In any case, the presence of a catalyst allows the reaction to be run at a lower temperature than would otherwise be possible. Without a catalyst the reaction would have to be run at a considerably higher temperature, with a correspondingly lower yield of $SO_3$.

In summary, the production of $SO_3$ from $SO_2$ is favored by:

1 A large excess concentration of $O_2$

**Figure 15-12**
A TUNNEL THROUGH THE MOUNTAIN. The tunnel through the mountain provides an alternate path that is easier and faster than going over a high pass. A catalyst has the same effect on a chemical reaction.

2 Compression of the gaseous mixture

3 A low temperature (but this slows the reaction)

4 The presence of a catalyst (which speeds up the reaction)

While on the topic of catalysts, we should mention their use in helping to reduce air pollution. Automobile exhaust contains poisonous CO and unburned gasoline (mainly $C_8H_{18}$), which contribute to smog. The catalytic converter on the automobile contains finely divided platinum and palladium in an attachment on the exhaust pipe. These metals provide a surface for the following reactions, which occur only at very high temperature without a catalyst:

$$2CO(g) + O_2(g) \rightarrow 2CO_2(g)$$

$$2C_8H_{18}(g) + 25O_2(g) \rightarrow 16CO_2(g) + 18H_2O(g)$$

Both of these reactions are exothermic, which explains why the catalytic converter becomes quite hot when the engine is running.

One drawback to the catalytic converter in an automobile is that it also is effective in speeding the conversion of $SO_2$ to $SO_3$. The combustion of sulfur impurities in gasoline produces $SO_2$, which is then converted to $SO_3$

See Problems 15-48 through 15-51.

in the converter. The $SO_3$ reacts with $H_2O$ to form $H_2SO_4$. Although $H_2SO_4$ is a major component of acid rain, the contribution from the automobile has not yet been determined.

## 15-7 Chapter Summary

Everyday experiences tell us that the rates of chemical reactions are affected by the temperature. Actually, as we have shown in this chapter, there are other factors as well as temperature that affect reaction rates. To grasp how certain factors affect rate, it is helpful to understand the mechanism of a simple reaction. In this case, a hypothetical reaction was chosen that could be kept free of complications yet still be used to describe how reactions are thought to occur.

Reactions occur because of collisions involving reactant molecules. In the case of reactions that reach a point of equilibrium a forward and reverse reaction occurs through collisions of appropriate energy and molecular orientation. Examples of chemical equilibria that have previously been discussed include gaseous reactions in Chapter 9, solution of ionic compounds in Chapter 11, and ionization of weak acids and bases in Chapter 13.

If we accept that collisions of a certain minimum energy are important in the formation of products, it is obvious that the rate of a reaction increases with both the energy and the frequency of collisions. The following three factors will then affect the rate.

| | | |
|---|---|---|
| 1 | The activation energy | All reactions have a certain minimum energy for reactants to form products called the activation energy. Activation energy varies for each set of potential reactants. |
| 2 | The temperature | Higher temperatures increase both the energy of colliding molecules and also the frequency of collisions. |
| 3 | Concentration of reactants | Increased concentration of reactants increases the frequency of collisions. The mathematical relationship between the rate and concentration is known as the rate law. The constant of proportionality in the rate law is called the rate constant. |

The distribution of concentrations of reactants and products can be predicted from the law of mass action. The mass action expression is constructed from a balanced equation and is equal to a constant, $K_{eq}$, at a certain temperature. Examples for the two types of equilibria discussed are as follows:

| *Gaseous* | *Ionic* |
|---|---|
| $A_2(g) + B_2(g) \rightleftharpoons 2AB(g)$ | $HX(aq) + H_2O \rightleftharpoons H_3O^+(aq) + X^-(aq)$ |
| $K_{eq} = \frac{[AB]^2}{[A_2][B_2]}$ | $K_a = \frac{[H_3O^+][X^-]}{[HX]}$ |

The value of the equilibrium constant at a certain temperature is obtained by experimental measurements of the distribution of reactants and products present at equilibrium. The constant and initial concentrations can then be used to calculate the concentration of a certain gas or ion present at equilibrium. The ionization constants for weak acids ($K_a$) and weak bases ($K_b$) are useful for the calculation of the pH of these solutions.

Finally, there was discussion of how the point of equilibrium could be altered. The factors that can affect a system at equilbrium are summarized as follows:

1 Concentration — An increase in concentration on one side of the reaction results in an increase in the concentrations on the other. A decrease on one side leads to an increase on the same side.

2 Volume — For gas phase reactions, compression of the mixture leads to a lower volume and a higher pressure. The equilibrium shifts to the side with the fewer total number of molecules of gas.

3 Temperature — For exothermic reactions an increase in temperature leads to a decrease in the proportion of products. Just the opposite occurs for an endothermic reaction.

Many important reactions do not take place readily or even at all without a catalyst present. A catalyst provides a reaction mechanism with a lower activation energy and thus speeds the reaction. There is no net loss of a catalyst during the course of a reaction, and the point of equilibrium is not shifted.

## Exercises

### The Collision Theory and Reaction Rates

**15-1** What two conditions are necessary before colliding molecules can react to form products?

**15-2** When is a chemical reaction at equilibrium?

**15-3** In the hypothetical reaction between $A_2$ and $B_2$, when was the rate of the forward reaction at a maximum? When was the rate of the reverse reaction at a maximum? (See Figure 15-3.)

**15-4** Besides reaching a point of equilibrium, what is another explanation for certain reactions that do not proceed completely to the right?

**15-5** Explain the following facts from your knowledge of collision theory and the factors that affect the rates of reactions.
(a) The rates of chemical reactions approximately double for each 10 °C rise in temperature.
(b) Eggs cook slower at higher altitudes, where water boils at temperatures less than 100 °C.
(c) $H_2$ and $O_2$ do not start to react to form $H_2O$ except at very high temperatures (unless initiated by a spark).
(d) Wood burns explosively in pure $O_2$ but slowly in air, which is about 20% $O_2$.
(e) Coal dust burns faster than a single lump of coal.
(f) Milk sours if left out for a day or two but will last for two weeks in the refrigerator.

**15-6** Explain the following facts from your knowledge of collision theory and the factors that affect the rates of reactions.
(a) A wasp is lethargic at temperatures below 65 °F.
(b) Charcoal burns faster if you blow on the coals.
(c) A 0.10 *M* boric acid solution ($[H_3O^+] = 7.8 \times 10^{-6}$) can be used for eyewash, but a 0.10 *M* hydrochloric acid solution ($[H_3O^+] = 0.10$) would cause severe damage.
(d) To keep apple juice from fermenting to apple cider, the apple juice must be kept cold.

**15-7** Compare the activation energy for the reverse reaction with that of the forward reaction in Figure 15-6. Which are easier to form, reactants from products or products from reactants? Which system would come to equilibrium faster, a reaction starting with pure products or starting with pure reactants?

**15-8** Write the energy diagram for an endothermic reaction similar to that shown in Figure 15-6. In this case, which is greater, the activation energy for the forward or the reverse reaction? Which system would come to equilibrium faster, a reaction starting with pure products or starting with pure reactants?

### The Law of Mass Action

**15-9** Write the law of mass action for each of the following equilibria.
(a) $CO(g) + Cl_2(g) \rightleftharpoons COCl_2(g)$
(b) $CH_4(g) + 2H_2O(g) \rightleftharpoons CO_2(g) + 4H_2(g)$
(c) $4HCl(g) + O_2(g) \rightleftharpoons 2Cl_2(g) + 2H_2O(g)$
(d) $CH_4(g) + Cl_2(g) \rightleftharpoons CH_3Cl(g) + HCl(g)$

**15-10** Write the law of mass action for each of the following equilibria.
(a) $3O_2(g) \rightleftharpoons 2O_3(g)$
(b) $N_2(g) + 2O_2(g) \rightleftharpoons 2NO_2(g)$
(c) $C_2H_2(g) + 2H_2(g) \rightleftharpoons C_2H_6(g)$

### The Value of $K_{eq}$

**15-11** For the hypothetical reaction $A_2 + B_2 \rightleftharpoons 2AB$, $K_{eq} = 1.0 \times 10^8$. Will reactants or products be favored at equilibrium?

**15-12** For a hypothetical reaction $2C + 3B \rightleftharpoons 2D + F$, $K_{eq} = 5 \times 10^{-7}$. Will reactants or products be favored?

**15-13** For the reaction $H_2(g) + I_2(g) \rightleftharpoons 2HI(g)$, $K_{eq} = 45$ at a certain temperature. Are reactants or products favored at equilibrium?

**15-14** Given the following system:

$$3O_2(g) \rightleftharpoons 2O_3(g)$$

At equilibrium, $[O_2] = 0.35$ and $[O_3] = 0.12$. What is $K_{eq}$ for the reaction at this temperature?

**15-15** Given the following system:

$$N_2(g) + 2O_2(g) \rightleftharpoons 2NO_2(g)$$

At a certain temperature there are $1.25 \times 10^{-3}$ mol of $N_2$, $2.50 \times 10^{-3}$ mol of $O_2$, and $6.20 \times 10^{-4}$ mol of $NO_2$ in a 1.00-L container. What is $K_{eq}$ for this reaction at this temperature?

**15-16** Given the following system:

$$CH_4(g) + 2H_2O(g) \rightleftharpoons CO_2(g) + 4H_2(g)$$

At equilibrium we find 2.20 mol of $CO_2$, 4.00 mol of $H_2$, 6.20 mol of $CH_4$, and 3.00 mol of $H_2O$ in a 30.0-L container. What is $K_{eq}$ for the reaction?

**15-17** Given the following system:

$$C_2H_2(g) + 2H_2(g) \rightleftharpoons C_2H_6(g)$$

At equilibrium it is found that there are 296 g of $C_2H_6$ present along with 3.50 g of $H_2$ and 21.0 g of $C_2H_2$ in a 400 mL container. What is the value of $K_{eq}$?

**15-18** Given the following system:

$$2HI(g) \rightleftharpoons H_2(g) + I_2(g)$$

(a) If we start with [HI] = 0.60, what are the $[H_2]$ and $[I_2]$ that would be present if all of the HI reacts?
(b) If we start with [HI] = 0.60 and $[I_2]$ = 0.20, what are the $[H_2]$ and $[I_2]$ that would be present if all of the HI reacts?
(c) If we start with only [HI] = 0.60 and 0.20 mol/L of HI reacts, what are [HI], $[I_2]$, and $[H_2]$ at equilibrium?
(d) From the information in (c) calculate the value of $K_{eq}$.
(e) What is the value of $K_{eq}$ for the reverse reaction? How does this value differ from the value given in Table 15-1?

**15-19** Given the following system:

$$N_2(g) + 3H_2(g) \rightleftharpoons 2NH_3(g)$$

Initially 1.00 mol of $N_2$ and $H_2$ are mixed in a 1.00-L container. At equilibrium it is found that $[NH_3]$ = 0.20.
(a) What is the concentration of $N_2$ and $H_2$ at equilibrium?
(b) What is the value of $K_{eq}$ for the system at this temperature?

* **15-20** At a certain temperature $N_2$, $H_2$, and $NH_3$ are mixed so that the initial concentration of each is 0.50 mol/L. At equilibrium the concentration of $N_2$ is 0.40 mol/L.
(a) Calculate the concentration of $H_2$ and $NH_3$ at equilibrium.
(b) What is the value of $K_{eq}$ at this temperature?

**15-21** Given the following system:

$$4NH_3(g) + 5O_2(g) \rightleftharpoons 4NO(g) + 6H_2O(g)$$

At the start of the reaction, $[NH_3]$ = $[O_2]$ = 1.00. At equilibrium it is found that 0.25 mol/L of $NH_3$ has reacted.
(a) What is the concentration of $O_2$ that reacts?
(b) What is the concentration of all species at equilibrium?
(c) Write the total mass action expression and substitute the proper values for the concentrations of reactants and products.

**15-22** Given the following equilibrium:

$$CO(g) + Cl_2(g) \rightleftharpoons COCl_2(g)$$

At the start of the reaction, [CO] = 0.650 and $[Cl_2]$ = 0.435. At equilibrium it is found that 10.0% of the CO has reacted. What is [CO], $[Cl_2]$, and $[COCl_2]$ at equilibrium? What is the value of $K_{eq}$?

**15-23** For the reaction

$$PCl_3(g) + Cl_2(g) \rightleftharpoons PCl_5(g)$$

$K_{eq}$ = 0.95 at a certain temperature. If $[PCl_3]$ = 0.75 and $[Cl_2]$ = 0.40 at equilibrium, what is the concentration of $PCl_5$ at equilibrium?

**15-24** At a certain temperature, $K_{eq}$ = 46.0 for the reaction

$$4HCl(g) + O_2(g) \rightleftharpoons 2Cl_2(g) + 2H_2O(g)$$

At equilibrium, [HCl] = 0.100, $[O_2]$ =

0.455, and $[H_2O] = 0.675$. What is $[Cl_2]$ at equilibrium?

**15-25** Using the value for $K_{eq}$ calculated in Problem 15-16, what is the concentration of $H_2O$ at equilibrium if the concentrations of the other species present at equilibrium are $[CH_4] = 0.50$, $[CO_2] = 0.24$, and $[H_2] = 0.20$?

* **15-26** For the following equilibrium, $K_{eq} = 56$ at a certain temperature:

$$CH_4(g) + Cl_2(g) \rightleftharpoons CH_3Cl(g) + HCl(g)$$

At equilibrium, $[CH_4] = 0.20$ and $[Cl_2] = 0.40$. What is the equilibrium concentration of $CH_3Cl$ and HCl if they are equal?

* **15-27** Using the equilibrium in Problem 15-14, calculate the equilibrium concentration of $O_3$ if the equilibrium concentration of $O_2$ is 0.80 mol/L.

## Equilibria of Weak Acids and Bases

**15-28** Write the expression for $K_a$ or $K_b$ for each of the following equilibria. Where necessary, complete the equilibrium.

(a) $HOBr + H_2O \rightleftharpoons H_3O^+ + OBr^-$
(b) $NH_3 + H_2O \rightleftharpoons NH_4^+ + OH^-$
(c) $H_2SO_3 + H_2O \rightleftharpoons H_3O^+ + HSO_3^-$
(d) $HSO_3^- + H_2O \rightleftharpoons$ _____ $+ SO_3^{2-}$
(e) $H_3PO_4 + H_2O \rightleftharpoons H_3O^+ +$ _____
(f) $HN(CH_3)_2 + H_2O \rightleftharpoons H_2N(CH_3)_2^+ +$ _____

**15-29** Write the expression for $K_a$ or $K_b$ for each of the following equilibria. Where necessary, complete the equilibrium.

(a) $N_2H_4 + H_2O \rightleftharpoons N_2H_5^+ + OH^-$
(b) $HCN + H_2O \rightleftharpoons H_3O^+ + CN^-$
(c) $H_2SO_3 + H_2O \rightleftharpoons H_3O^+ +$ _____
(d) $H_2PO_4^- + H_2O \rightleftharpoons$ _____ $+ HPO_4^{2-}$

**15-30** A 0.10 *M* solution of a weak acid HX has pH = 5.0. A 0.10 *M* solution of another weak acid HB has pH = 5.8. Which is the weaker acid? Which has the larger value for $K_a$?

**15-31** A hypothetical weak acid HZ has $K_a = 4.5 \times 10^{-6}$. Rank the following 0.10 *M* solutions in order of increasing pH: HZ, $HC_2H_3O_2$, and HOCl.

**15-32** In a 0.20 *M* solution of cyanic acid, HOCN, $[H_3O^+] = [OCN^-] = 6.2 \times 10^{-3}$.

(a) What is [HOCN] at equilibrium?
(b) What is $K_a$?
(c) What is the pH?

**15-33** A 0.58 *M* solution of a weak acid (HX) is 10.0% dissociated.

(a) Write the equilibrium and the mass action expression.
(b) What are $[H_3O^+]$, $[X^-]$, and [HX] at equilibrium?
(c) What is $K_a$?
(d) What is the pH?

**15-34** Nicotine (Nc) is a nitrogen base in water in the same manner as ammonia. Write the equilibrium illustrating this base behavior. In a 0.44 *M* solution of nicotine, $[OH^-] = [NcH^+] = 5.5 \times 10^{-4}$. What is $K_b$ for nicotine? What is the pH of the solution?

**15-35** In a 0.085 *M* solution of carbolic acid, $HC_6H_5O$, the pH = 5.48. What is $K_a$?

**15-36** Novacaine (Nv) is a nitrogen base in water, like ammonia. Write the equilibrium. In a 1.25 *M* solution of novacaine, pH = 11.46. What is $K_b$?

**15-37** In a 0.300 *M* solution of chloroacetic acid, $HOC_2Cl_3$, $[HOC_2Cl_3] = 0.277$ *M* at equilibrium. What is the pH? What is $K_a$?

**15-38** What is the pH of a 0.65 *M* solution of HOBr?

**15-39** What is the pH of a 2.0 *M* solution of HF?

**15-40** What is $[OH^-]$ in a 0.55 *M* solution of $NH_3$?

**15-41** What is $[H_3O^+]$ in a 1.25 *M* solution of $HC_2H_3O_2$?

**15-42** What is the pH of a 1.00 *M* solution of $NH(CH_3)_2$?

**15-43** What is the pH of a buffer made by mixing 0.45 mol of NaCN and 0.45 mol of HCN in 2.50 L of solution?

**15-44** What is the pH of a buffer made by dissolving 1.20 mol of $NH_3$ and 1.20 mol of $NH_4ClO_4$ in 13.5 L of solution?

**15-45** What is the pH of a buffer made by

dissolving 0.20 mol of KOBr and 0.60 mol of HOBr in 850 mL of solution?

**15-46** What is the pH of a buffer that contains 150 g of $HNO_2$ and 150 g of $LiNO_2$?

* **15-47** What is the pH of a buffer made by adding 0.20 mol of NaOH to a 1.00 *M* solution of $HC_2H_3O_2$? Assume no volume change. (*Hint*: First consider the partial neutralization of $HC_2H_3O_2$ by NaOH.)

## Le Châtelier's Principle and Equilibrium

**15-48** The following equilibrium is an important industrial process used to convert $N_2$ to $NH_3$. The ammonia is used mainly for fertilizer. This method for the production of $NH_3$ is called the Haber process.

$$N_2(g) + 3H_2(g) \underset{500\ °C}{\overset{\text{Catalyst}}{\rightleftharpoons}} 2NH_3(g) + \text{heat energy}$$

Determine the direction that the equilibrium will be shifted by the following changes.

(a) Increasing the concentration of $N_2$.
(b) Increasing the concentration of $NH_3$.
(c) Decreasing the concentration of $H_2$.
(d) Decreasing the concentration of $NH_3$.
(e) Compressing the reaction mixture.
(f) Removing the catalyst.
(g) How will the yield of $NH_3$ be affected by raising the temperature to 750 °C?
(h) How will the yield of $NH_3$ be affected by lowering the temperature to 0 °C? How will this affect the rate of formation of $NH_3$?

**15-49** Given the following equilibrium:

$$4NH_3(g) + 5O_2(g) \rightleftharpoons 4NO(g) + 6H_2O(g) + \text{heat energy}$$

How will this system at equilibrium be affected by:

(a) Increasing the concentration of $O_2$
(b) Removing all of the $H_2O$ as it is formed
(c) Increasing the concentration of NO
(d) Compressing the reaction mixture
(e) Increasing the volume of the reaction container
(f) Increasing the temperature
(g) Decreasing the concentration of $O_2$
(h) Addition of a catalyst

**15-50** The following equilibrium takes place at high temperatures.

$$N_2(g) + 2H_2O(g) + \text{heat energy} \rightleftharpoons 2NO(g) + 2H_2(g)$$

How will the yield of NO be affected by:

(a) Increasing $[N_2]$
(b) Decreasing $[H_2]$
(c) Compressing the reaction mixture
(d) Decreasing the volume of the reaction container
(e) Decreasing the temperature
(f) Addition of a catalyst

**15-51** Fortunately for us, the major components of air, $N_2$ and $O_2$, do not react under ordinary conditions. At very high temperatures, however, like those found in an automobile engine, the equilibrium constant for the following reaction becomes larger.

$$N_2(g) + O_2(g) \rightleftharpoons 2NO(g)$$

(a) Does the information given indicate that the formation of NO is an endothermic or an exothermic process?
(b) Since NO is a serious pollutant, it is desirable to minimize its formation in automobile engines. Would a cooler- or hotter-running engine increase the yield of NO?
(c) Would a lower pressure in the engine affect the formation of NO?

Petroleum from a well like this not only supplies us with energy but is the ultimate source of most plastics, synthetic fibers, and other organic based chemicals that we now take for granted.

# Chapter 16
# An Introduction to Organic Chemistry

## Purpose of Chapter 16

In Chapter 16 we introduce some important classes of organic compounds, their unique chemical structure, and their common uses.

## Objectives for Chapter 16

After completion of this chapter, you should be able to:

1. Write the Lewis structures of selected organic compounds. (16-1)
2. Illustrate with Lewis structures the isomers of a specified compound. (16-1)
3. Illustrate why the compound $C_2H_5Cl$ has only two isomers. (16-1)
4. Write condensed and fully condensed formulas for specified hydrocarbons. (16-1)
5. Write the formulas of some simple alkanes from the general formula of the homologous series. (16-2)
6. Determine the longest chain and the names of substituent groups for specified alkanes. (16-2)
7. Describe how petroleum is refined and the uses for the various components that are obtained. (16-2)
8. Write the formulas of some simple alkenes from the general formula. (16-3)
9. Describe how alkenes are used to make polymers and the names and uses of some common polymers. (16-3)
10. Write the formulas of some simple alkynes from the general formula. (16-4)
11. Illustrate the difference in structure and chemical reactivity between the aromatic compound, benzene, and a normal alkene. (16-5)
12. Distinguish between compounds containing a functional group and an alkane. (16-6)
13. Give the unique structural feature, an example, and a common use of the following hydrocarbon derivatives.
    (a) Alcohol (16-7)
    (b) Ether (16-8)
    (c) Amine (16-9)
    (d) Aldehyde (16-10)
    (e) Ketone (16-10)
    (f) Carboxylic acid (16-11)
    (g) Ester (16-11)
    (h) Amide (16-11)

Compounds have long been grouped into two categories: organic and inorganic. **Organic compounds** *are those containing only carbon and hydrogen or carbon, hydrogen, and other elements*. The key element, however, in all organic compounds is carbon, although some carbon compounds such as $CaCO_3$ and LiCN are not considered organic. Before 1828 it was thought that organic compounds could only be synthesized from inorganic compounds (e.g., $H_2O$, $CO_2$, and $NH_3$) by living matter (organisms). Only life had that magic ingredient of nature called the "vital force" that allowed the miracle of organic synthesis. In 1828, urea, an organic waste product of metabolism, was synthesized from the inorganic compound, ammonium cyanate ($NH_4OCN$). The myth of the vital force was immediately laid to rest, and since that time millions of organic compounds have been synthesized from basic materials in chemical laboratories.

Organic compounds are certainly central to our lives. Hydrocarbons (compounds composed only of carbon and hydrogen) are used as fuel to power our cars and to heat our homes. Our bodies are fueled with organic compounds obtained from the food we eat in the form of sugars, carbohydrates, fats, and proteins. This food is made more palatable by organic flavorings, wrapped in organic plastic, and kept from spoiling with organic preservatives. Our clothes are made of organic compounds, whether these compounds come from plant and animal sources (cotton and wool) or are synthetic (nylon and Dacron®). These fabrics are made colorful with organic dyes. When we are ill, we take drugs that may also be organic: aspirin relieves headaches, codeine suppresses coughs, and diazepam (Valium®) calms nerves. These are just a few examples of how we use organic chemicals daily.

The introduction of such a large topic as organic chemistry in one chapter is not an easy assignment. Synthesis, structure, nomenclature, properties, and the reactions of each class of compound are important topics, and they all can't be emphasized. In this text, we will present the following information for each class of compound: the unique feature of the chemical structure, an introduction to the nomenclature, how the compounds are synthesized, and some familiar uses of the compounds.

As background to this chapter, you should review:

1 Lewis structures (Section 6-9)

2 Polarity of covalent bonds and molecules (Sections 6-11 and 6-12)

3 The solubility of polar and nonpolar compounds in solvents (Section 12-2)

## 16-1 Carbon and Its Chemical Bonds

Carbon is a unique element. For one thing, it has an amazing ability to share its electrons with other carbon atoms to form extremely large molecules with molar masses in the millions of grams per mole. No other element rivals carbon in this property. In addition to bonds between carbon atoms, carbon also forms strong bonds to many other elements such as hydrogen, oxygen, and nitrogen. This means that a wide variety of compounds involving carbon are very stable.

Notice that we said that carbon *shares* its electrons with other carbon or other nonmetal atoms. Thus carbon tends to form *covalent* rather than ionic bonds, and this gives rise to many of the characteristic properties of organic molecules. Although some organic compounds are somewhat polar (especially when combined with oxygen), they are usually much less polar than inorganic compounds. Because of this, most organic chemicals are insoluble or only slightly soluble in water. Although most inorganic compounds have high melting and boiling points, most organic compounds melt and boil at much lower temperatures.

Since carbon is in group IVA and has four valence electrons, it must form a total of *four bonds* to achieve an octet of outer electrons. Carbon forms single, double, and triple bonds. Let's review some simple organic compounds to see how these molecules appear when written as Lewis structures.

Single bonds

$$CH_4 \quad \begin{array}{c} \mathrm{H} \\ | \\ \mathrm{H{-}C{-}H} \\ | \\ \mathrm{H} \end{array} \qquad CCl_4 \quad \begin{array}{c} :\ddot{\mathrm{Cl}}: \\ | \\ :\ddot{\underset{..}{\mathrm{Cl}}}{-}\mathrm{C}{-}\ddot{\underset{..}{\mathrm{Cl}}}: \\ | \\ :\underset{..}{\mathrm{Cl}}: \end{array}$$

$$C_2H_6 \quad \begin{array}{c} \mathrm{H}\;\;\mathrm{H} \\ |\;\;\;| \\ \mathrm{H{-}C{-}C{-}H} \\ |\;\;\;| \\ \mathrm{H}\;\;\mathrm{H} \end{array} \qquad C_2H_6O \quad \begin{array}{c} \mathrm{H}\;\;\mathrm{H} \\ |\;\;\;| \\ \mathrm{H{-}C{-}C{-}}\ddot{\underset{..}{\mathrm{O}}}\mathrm{{-}H} \\ |\;\;\;| \\ \mathrm{H}\;\;\mathrm{H} \end{array}$$

Double bonds

$$C_2H_4 \quad \begin{array}{c} \mathrm{H}\diagdown\quad\diagup\mathrm{H} \\ \mathrm{C{=}C} \\ \mathrm{H}\diagup\quad\diagdown\mathrm{H} \end{array} \qquad CH_2O \quad \begin{array}{c} \mathrm{H}\diagdown \\ \mathrm{C{=}}\ddot{\underset{.}{\mathrm{O}}}\!: \\ \mathrm{H}\diagup \end{array}$$

Triple bonds $C_2H_2$ $\quad \mathrm{H{-}C{\equiv}C{-}H}$

**Example 16-1**
Give the Lewis structures for the following compounds.

(a) $C_3H_8$ (b) $C_3H_6$ (c) $H_3C_2N$ (all H's on one C)

**Answers**

$$\begin{array}{lccccccccc} & & & H & & H & & H & & \\ & & & | & & | & & | & & \\ \text{(a)} & H & - & C & - & C & - & C & - & H \\ & & & | & & | & & | & & \\ & & & H & & H & & H & & \end{array}$$

$$\begin{array}{lccccccccc} & & & H & & H & & & & \\ & & & | & & | & & & & \\ \text{(b)} & H & - & C & - & C & = & C & - & H \\ & & & | & & & & | & & \\ & & & H & & & & H & & \end{array} \quad \text{or} \quad \begin{array}{ccccccc} & & H & & & H & \\ & & | & & & | & \\ H & - & C & \text{———} & & C & - H \\ & & & \diagdown & \diagup & & \\ & & & & C & & \\ & & & \diagup & \diagdown & & \\ & & H & & & H & \end{array}$$

$$\begin{array}{lccccc} & & & H & & \\ & & & | & & \\ \text{(c)} & H & - & C & - & C{\equiv}N\colon \\ & & & | & & \\ & & & H & & \end{array}$$

Many times we find that there is more than one Lewis structure for a given formula. Such is the case with butane, $C_4H_{10}$, since two correct structures can be drawn. *Compounds with different structures but the same molecular formula are called* **isomers.**

$$\begin{array}{ccccccccc} & H & & H & & H & & H & \\ & | & & | & & | & & | & \\ H- & C & - & C & - & C & - & C & -H \\ & | & & | & & | & & | & \\ & H & & H & & H & & H & \end{array} \quad \longleftarrow \begin{array}{c} \text{Isomers} \\ \text{of} \\ \text{each other} \end{array} \longrightarrow \quad \begin{array}{ccccccc} & H & & H & & H & \\ & | & & | & & | & \\ H- & C & - & C & - & C & -H \\ & | & & | & & | & \\ & H & & | & & H & \\ & & & | & & & \\ & & H- & C & -H & & \\ & & & | & & & \\ & & & H & & & \end{array}$$

*n*-butane isobutane

We have been representing Lewis structures as if molecules were all flat or two-dimensional. In fact, when carbon is bonded to four other atoms, it exists as a three-dimensional structure where the carbon is located in the center of a tetrahedron bonded to the other atoms at the corners. The two compounds, methane and ethane, are represented in this fashion in Figure 16-1 by means of "ball and stick" models. One important result of the actual geometry is helpful in deciding about the existence of isomers. For example, all hydrogen atoms in the ethane molecule are identical. This means that if a Cl is substituted for one H in ethane no isomers are possible. A ball and stick model of each of the following would be exactly identical and superimposable.

**Figure 16-1** THE GEOMETRY OF $CH_4$ AND $C_2H_6$. In these two molecules the carbon is at the center of a tetrahedron.

$$\begin{array}{ccccc} & \text{Cl} & \text{H} & \\ & | & | & \\ \text{H}- & \text{C} - & \text{C} & -\text{H} \\ & | & | & \\ & \text{H} & \text{H} & \end{array} = \begin{array}{ccccc} & \text{H} & \text{H} & \\ & | & | & \\ \text{H}- & \text{C} - & \text{C} & -\text{H} \\ & | & | & \\ & \text{Cl} & \text{H} & \end{array} = \begin{array}{ccccc} & \text{H} & \text{H} & \\ & | & | & \\ \text{H}- & \text{C} - & \text{C} & -\text{H} \\ & | & | & \\ & \text{H} & \text{Cl} & \end{array} = \begin{array}{ccccc} & \text{H} & \text{H} & \\ & | & | & \\ \text{H}- & \text{C} - & \text{C} & -\text{Cl} \\ & | & | & \\ & \text{H} & \text{H} & \end{array}$$

On the other hand, if a Cl is substituted for an H in $C_3H_8$, to form $_7$Cl, two (and only two) isomers exist, which are illustrated as follows:

$$\begin{array}{cccccc} & \text{H} & \text{H} & \text{H} & \\ & | & | & | & \\ \text{H}- & \text{C} - & \text{C} - & \text{C} & -\text{H} \\ & | & | & | & \\ & \text{Cl} & \text{H} & \text{H} & \end{array} \quad \text{and} \quad \begin{array}{cccccc} & \text{H} & \text{H} & \text{H} & \\ & | & | & | & \\ \text{H}- & \text{C} - & \text{C} - & \text{C} & -\text{H} \\ & | & | & | & \\ & \text{H} & \text{Cl} & \text{H} & \end{array}$$

Although actual geometry is important in the study of organic chemistry, we will continue to represent structures in the simplified manner, but it should be understood that the molecules actually have a three-dimensional nature.

Since the molecular formula by itself does not differentiate among the various isomers, it is necessary to refer to isomers by name and by a detailed structure. When drawing the structure, it can become tedious if all hydrogens and carbons are written out, especially for large molecules. A **condensed formula** in which separate bonds are not written out is helpful. For example,

$$\begin{array}{ccc} & \text{H} & \\ & | & \\ \text{H}- & \text{C} & - \\ & | & \\ & \text{H} & \end{array}$$

is represented as

$$CH_3—$$

and

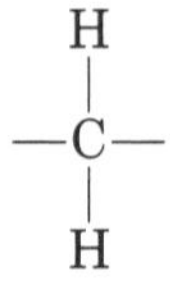

$$\begin{array}{c} H \\ | \\ —C— \\ | \\ H \end{array}$$

is represented as

$$—CH_2—$$

Depending on what we are trying to show, the structure may be partially or fully condensed.

| | Partially condensed | Fully condensed |
|---|---|---|
| *n*-butane* | $CH_3—CH_2—CH_2—CH_3$ | $CH_3(CH_2)_2CH_3$ |
| isobutane* | 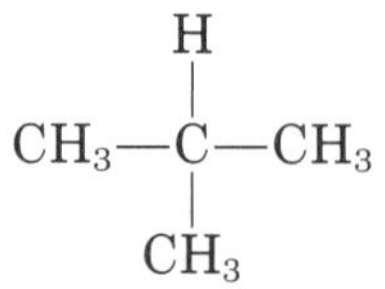 $\begin{array}{c} H \\ \vert \\ CH_3—C—CH_3 \\ \vert \\ CH_3 \end{array}$ | $(CH_3)_3CH$ |

A few other compounds and their isomers are shown in Table 16-1. As you can see, the number of isomers increases as the number of carbons increases, and the addition of a **hetero atom** (*any atom other than carbon or hydrogen*) also increases the number of isomers.

## 16-2 Alkanes

The most basic types of organic compounds are called hydrocarbons. **Hydrocarbons** *contain only carbon and hydrogen.* **Alkanes** *are the simplest of all hydrocarbons, since they contain only carbon and hydrogen joined together by single bonds.* The general formula for open-chained alkanes† is $C_nH_{2n+2}$. The simplest alkane ($n = 1$) is methane, in which the carbon shares a pair of electrons with four different hydrogen atoms. The next

*The *n* is an abbreviation of *normal* and refers to the isomer in which all the C's are bound consecutively in a continuous chain. The *iso* refers to the isomer in which there is a three carbon branch at the end of a continuous chain of carbons.

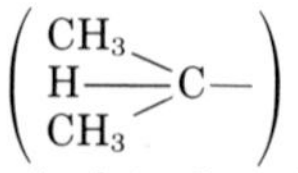

$$\left( \begin{array}{l} CH_3 \\ H—\!\!>C— \\ CH_3 \end{array} \right)$$

†Open-chained hydrocarbons have terminal (end) carbon atoms that are not bonded to each other.

**Table 16-1** Isomers

| Formula | Isomers (Names) | | |
|---|---|---|---|
| $C_5H_{12}$ | $CH_3CH_2CH_2CH_2CH_3$ *n* pentane | $CH_3CH_2CH(CH_3)—CH_3$ isopentane | $CH_3—C(CH_3)_2—CH_3$ neopentane |
| $C_3H_6$ | $H_2C{=}CHCH_3$ propene | $H_2C—CH_2—CH_2$ (ring) cyclopropane | (Notice that the carbons can also be arranged in a ring or *cyclic* structure.) |
| $C_2H_6O$ | $CH_3CH_2OH$ ethanol | $H_3C—O—CH_3$ dimethyl ether | |
| $C_3H_6O$ | $CH_3CH_2CH{=}O$ propanal | $H_3C—C({=}O)—CH_3$ propanone | $H_2C{=}CHCH_2OH$ allyl alcohol and others |

**See Problems 16-1 through 16-6.**

alkane is ethane ($C_2H_6$), and the third alkane is propane ($C_3H_8$). These alkanes are members of a homologous series. *In a* **homologous series,** *the next member differs from the previous one by a constant amount, which in this case is $CH_2$.*

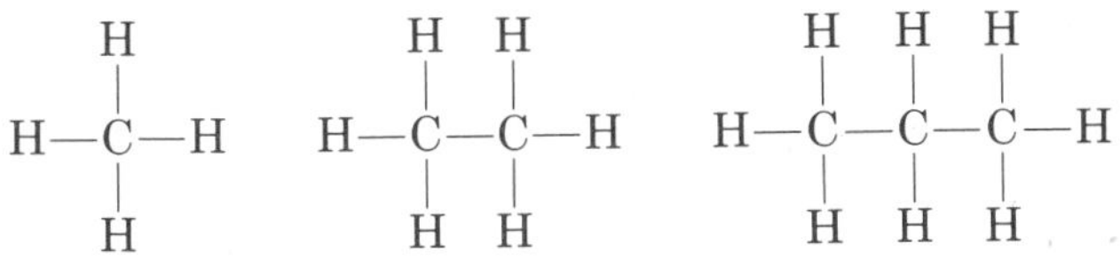

$$\underset{\text{methane}}{H—\overset{H}{\underset{H}{C}}—H} \qquad \underset{\text{ethane}}{H—\overset{H}{\underset{H}{C}}—\overset{H}{\underset{H}{C}}—H} \qquad \underset{\text{propane}}{H—\overset{H}{\underset{H}{C}}—\overset{H}{\underset{H}{C}}—\overset{H}{\underset{H}{C}}—H}$$

The homologous series of alkanes up to 10 carbons, together with the names of the compounds, is given in Table 16-2.

The names in Table 16-2 are the basis of the names of all organic compounds. By altering them slightly, we can name other classes of organic compounds that are discussed later. There are two systems of nomenclature used in organic chemistry. The most systematic is the one devised by the International Union of Pure and Applied Chemistry (the IUPAC system). Although the rules for naming complex molecules can be extensive, we will be concerned with just the basic concept. Compounds are also known by *common* or *trivial* names. Sometimes these names follow a pattern, sometimes they do not. They have been used for so many years that it is hard

**Table 16-2** Alkanes

| Formula | Name | Formula | Name |
|---|---|---|---|
| $CH_4$ | methane | $C_6H_{14}$ | hexane |
| $C_2H_6$ | ethane | $C_7H_{16}$ | heptane |
| $C_3H_6$ | propane | $C_8H_{18}$ | octane |
| $C_4H_{10}$ | butane | $C_9H_{20}$ | nonane |
| $C_5H_{12}$ | pentane | $C_{10}H_{22}$ | decane |

to break the habit of using them. When a chemical that is frequently known by its common name is encountered, that name is given in parentheses.

The name of a simple organic compound has two parts; the *prefix* gives the number of carbons in the longest carbon chain and the *suffix* tells what kind of a compound it is. The underlined portions of the names in Table 16-2 are the prefixes used for compounds containing one through 10 carbons in the longest chain; *meth-* stands for one carbon, *eth-* for two carbons, and so on. The ending used for alkanes is *-ane*. Therefore, the one-carbon alkane is methane, the two-carbon alkane is ethane, etc.

Organic compounds can exist as unbranched compounds (all carbons bound to each other in a continuous chain) or as branched compounds. Previously, we indicated that *n*-butane is an unbranched alkane and that isobutane is a branched alkane. The IUPAC system bases its names on the longest carbon chain in the molecule, whereas the common names frequently include all of the carbons in the name (e.g., isobutane). The longest carbon chain in isobutane is three carbons long and is therefore considered a propane in the IUPAC system. Notice in isobutane that there is a $CH_3$— group attached to the propane chain. In this system of nomenclature, the branches are named separately. *Since the branches can be considered as groups of atoms substituted for a hydrogen, the branches are called* **substituents.** *Substituents that contain one less hydrogen than an alkane are called* **alkyl groups.** Alkyl groups are not compounds by themselves; they must always be attached to some other group or atom. They are named by taking the alkane name, dropping the *-ane* ending, and substituting *-yl.* The most common alkyl groups are given in Table 16-3.

Thus the alkyl substituent in isobutane is a methyl group. the IUPAC name for isobutane is *methyl propane.*

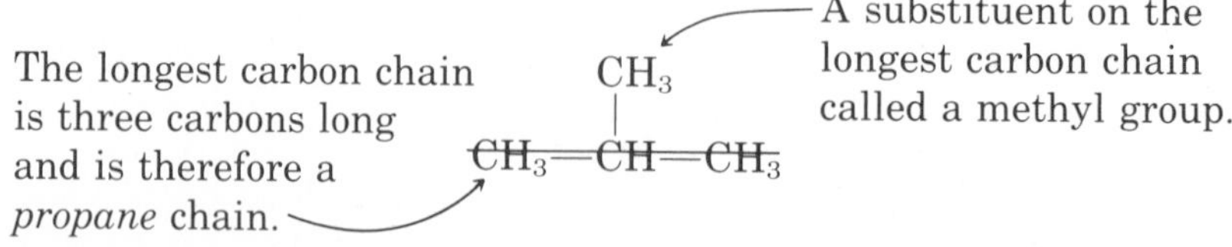
The longest carbon chain is three carbons long and is therefore a *propane* chain.
CH3
CH3—CH—CH3
A substituent on the longest carbon chain called a methyl group.

When the carbons form a ring, the name is prefixed with *cyclo-*. There-

**Table 16-3** Alkyl Groups

| Alkyl Group[a] | Name | Alkyl Group[a] | Name |
|---|---|---|---|
| $CH_3$— | methyl | $(CH_3)_2CH$— | isopropyl |
| $CH_3CH_2$— | ethyl | $CH_3CH_2CH_2CH_2$— | *n*-butyl |
| $CH_3CH_2CH_2$— | *n*-propyl | $(CH_3)_3C$— | *tert*-butyl (or *t*-butyl) |

[a]The dash (—) shows where the alkyl group is attached to a carbon chain or to another atom such as a halogen, oxygen, or a nitrogen.

fore, the three-carbon ring compound

```
     H     H
      \   /
        C
       / \
H—C———C—H
   |      |
   H      H
```

is called cyclopropane. Cycloalkanes have the general formula $C_nH_{2n}$.

There is much more that could be said about naming compounds, but we will limit ourselves to recognizing the number of carbons in the longest chain, the type of compound, and the presence of substituents when given the name of a compound or its structure.

**Example 16-2**

Draw a line through the longest carbon chain in the following compounds and circle the substituents. Name the longest chain.

```
    CH3   CH3  CH3                    CH2—CH2—CH3
    |     |    |                      |
(a) CH2—CH—CH               (c) CH3CH2—CH—CH2—CH2—CH2—CH3
                |
                CH2—CH3

         CH2—CH3                    CH3
         |                          |
(b) CH2—CH—CH2—CH3          (d) H—C—CH2—CH3
    |                               |
    CH3                             CH3
```

**Answers**

```
    CH3  (CH3) (CH3)                   CH2—CH2—CH3
    |     |     |                      |
(a) CH2—CH—CH                (c) (CH3CH2—)CH—CH2—CH2—CH2—CH3
                |
                CH2—CH3                      nonane

          hexane
```

(b) $CH_2$—CH—$CH_2$—$CH_3$ with $CH_2$—$CH_3$ above the CH and $CH_3$ below the first $CH_2$ (circled)
pentane

(d) H—C—$CH_2$—$CH_3$ with $CH_3$ above and $CH_3$ below the C (circled)
butane

Alkanes are nonpolar, since they are made up of only carbon and hydrogen, which have approximately the same electronegativities. Because of this, they are insoluble in water and have low boiling points. Alkanes containing four or fewer carbons are gases, and those with more than 18 carbons are solids (resembling candle wax); the rest are liquids. Alkanes have no odor and are colorless. All are extremely flammable, and the lighter alkanes are also volatile (they evaporate easily).*

There are two major sources of alkanes, natural gas and crude oil (petroleum). Natural gas is mainly methane with smaller amounts of ethane, propane, and butanes. Unlike natural gas, petroleum contains hundreds of compounds, the majority of which are open-chain and cyclic alkanes. Before we can make use of petroleum, it must be separated into groups of compounds with similar properties. Further separation may or may not be carried out depending on the final use of the hydrocarbons.

Crude oil is separated into groups of compounds according to boiling points by distillation in a refinery (see Figure 16-2). In such a distillation, the liquid is boiled and the gases move up a large column that becomes cooler and cooler toward the top. Compounds condense (become liquid) at different places in the column, depending on their boiling points. As the liquids condense, they are drawn off, providing a rough separation of the crude oil. Some of the material is too high boiling to vaporize and remains in the bottom of the column. A drawing showing this process and the various fractions obtained is shown in Figure 16-3. Notice that the fewer the carbon atoms in the alkane, the lower the boiling point (the more volatile it is).

The composition of crude oil itself varies somewhat depending on where it is found. Certain crude oil, such as that found in Nigeria and Libya, is called "light" oil because it is especially rich in the hydrocarbons that are present in gasoline. Otherwise, one fraction can be converted into another by three processes, known as cracking, reforming, and alkylation. **Cracking** *changes large molecules into small molecules.* **Reforming** *removes hydrogens from the carbons and/or changes unbranched hydrocarbons into branched hydrocarbons.* (Branched hydrocarbons perform better in gasoline; that is, they have a higher "octane" rating.) **Alkylation** *takes small molecules and*

*Since all alkanes are nonpolar molecules, the only intermolecular forces present in the liquid state are weak London forces. London forces are roughly proportional to molar mass, which means that the lighter the molecule the more volatile it is (i.e., the lower the boiling point).

**Figure 16-2** AN OIL REFINERY. The crude oil is separated into various fractions in large refineries.

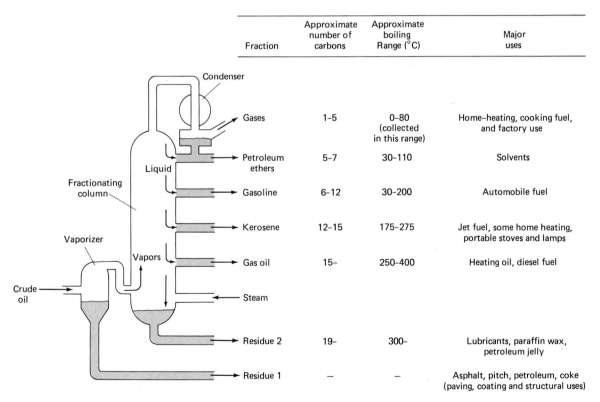

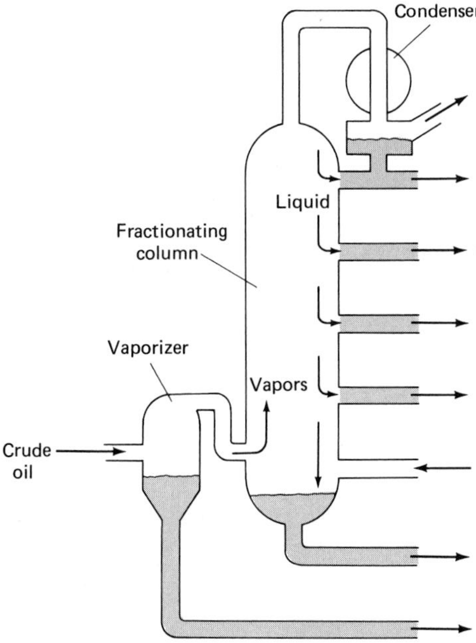

| Fraction | Approximate number of carbons | Approximate boiling Range (°C) | Major uses |
|---|---|---|---|
| Gases | 1–5 | 0–80 (collected in this range) | Home-heating, cooking fuel, and factory use |
| Petroleum ethers | 5–7 | 30–110 | Solvents |
| Gasoline | 6–12 | 30–200 | Automobile fuel |
| Kerosene | 12–15 | 175–275 | Jet fuel, some home heating, portable stoves and lamps |
| Gas oil | 15– | 250–400 | Heating oil, diesel fuel |
| Steam | | | |
| Residue 2 | 19– | 300– | Lubricants, paraffin wax, petroleum jelly |
| Residue 1 | – | – | Asphalt, pitch, petroleum, coke (paving, coating and structural uses) |

**Figure 16-3** THE REFINING OF PETROLEUM. Oil is separated into fractions according to boiling points.

*puts them together to make larger molecules.* In all of these processes, catalysts are used, but there is a different catalyst for each process.

About 96% of all oil and gas is burned as fuel, whereas only 4% is used to make other organic chemicals. As a fuel, hydrocarbons burn to give carbon dioxide, water, and a great deal of heat energy.

$$CH_4 + 2O_2 \xrightarrow{\text{spark}} CO_2 + 2H_2O + \text{heat}$$

$$2C_8H_{18} + 25O_2 \xrightarrow{\text{spark}} 16CO_2 + 18H_2O + \text{heat}$$

Industrially, most synthetic organic chemicals have their ultimate origin in *the alkanes obtained from crude oil.* These **petrochemicals** are put to a wide range of uses in the manufacture of fibers, plastics, coatings, adhesives, synthetic rubber, some flavorings, perfumes, and pharmaceuticals.

## 16-3 Alkenes

**Alkenes** *are hydrocarbons that contain a double bond. Organic compounds with multiple bonds are said to be* **unsaturated.** The general formula for an open-chain alkene with one double bond is $C_nH_{2n}$. The simplest alkene is ethene (common name ethylene), which has the structure

$$\begin{matrix} H & & & & H \\ & \diagdown & & \diagup & \\ & & C{=}C & & \\ & \diagup & & \diagdown & \\ H & & & & H \end{matrix}$$

Alkenes are named by dropping the *-ane* ending of the corresponding alkane name and substituting *-ene*. The double bond is located by numbering the carbon chain in such a way that the double bond is assigned the lowest possible number. Thus, the name of $CH_3CH_2CH_2CH{=}CHCH_3$ is 2-hexene.

Only small amounts of alkenes are found naturally in crude oil; the majority are made from alkanes by the reforming process during the refining of crude oil. When alkanes are heated over a catalyst, hydrogen is lost from the molecule and alkenes together with hydrogen are formed.

$$C_nH_{2n+2} \xrightarrow[\text{heat}]{\text{catalyst}} C_nH_{2n} + H_2$$

$$C_2H_6 \xrightarrow[\text{heat}]{\text{catalyst}} CH_2{=}CH_2 + H_2$$

Most of the alkenes produced industrially are used to make polymers. When certain compounds called initiators are added to an alkene or a mixture of alkenes, the double bond is broken and the alkenes become joined to each other by single bonds. This produces a high-molar mass molecule called a polymer, which has repeating units of the original alkene (called the monomer).

$$\underset{\text{monomer (ethylene)}}{CH_2{=}CH_2} \xrightarrow{\text{initiator}}$$

$$\text{etc.}\,\boxed{-CH_2-CH_2-}\boxed{-CH_2-CH_2-}\boxed{-CH_2-CH_2-}\boxed{-CH_2-CH_2-}\,\text{etc.}$$

Written as $-\!\!(CH_2-CH_2)\!\!-_n$
polymer (polyethylene)

Polymers are named by adding *poly-* to the name of the alkene used to form the polymer. In the example shown above, the polymer was made from ethylene (usually common names are used for polymers), and so the polymer is called polyethylene. Since it would be impossible to write out the structure of a polymer, which may contain thousands of carbons, we abbreviate the structure by giving the repeating unit in parentheses along with a subscript $n$ to indicate that the monomer is repeated many times. Groups attached to the double bond affect the properties of the polymer, and by varying the group, we can vary the uses for which a polymer is suited (see Figure 16-4). Some commonly used polymers and their uses are given in Table 16-4.

Chemicals can add to the double bond to form new single bonds and therefore new compounds. A test based on such a reaction, the addition of bromine to an alkene, is used to show the presence of alkenes. It is a very simple test to perform. If the red color of bromine disappears when it is added to a liquid, this means that multiple bonds are present (the dibromide formed is colorless). We will talk more about this test when we discuss aromatic compounds.

$$CH_2{=}CH_2 + \underset{\text{Red}}{Br_2} \rightarrow \underset{\text{Colorless}}{\underset{Br\quad\;\; Br}{CH_2{-}CH_2}}$$

## 16-4 Alkynes

**Alkynes** *are hydrocarbons that contain a triple bond.* The general formula for an open-chain alkyne with one triple bond is $C_nH_{2n-2}$. The simplest alkyne is ethyne (acetylene), which has the structure

$$H{-}C{\equiv}C{-}H$$

Alkynes are named just like alkenes except that *-yne* is substituted for

**Figure 16-4** POLYMERS. The plastic bottles, the bag, and the covering on the walkway are made of polymers.

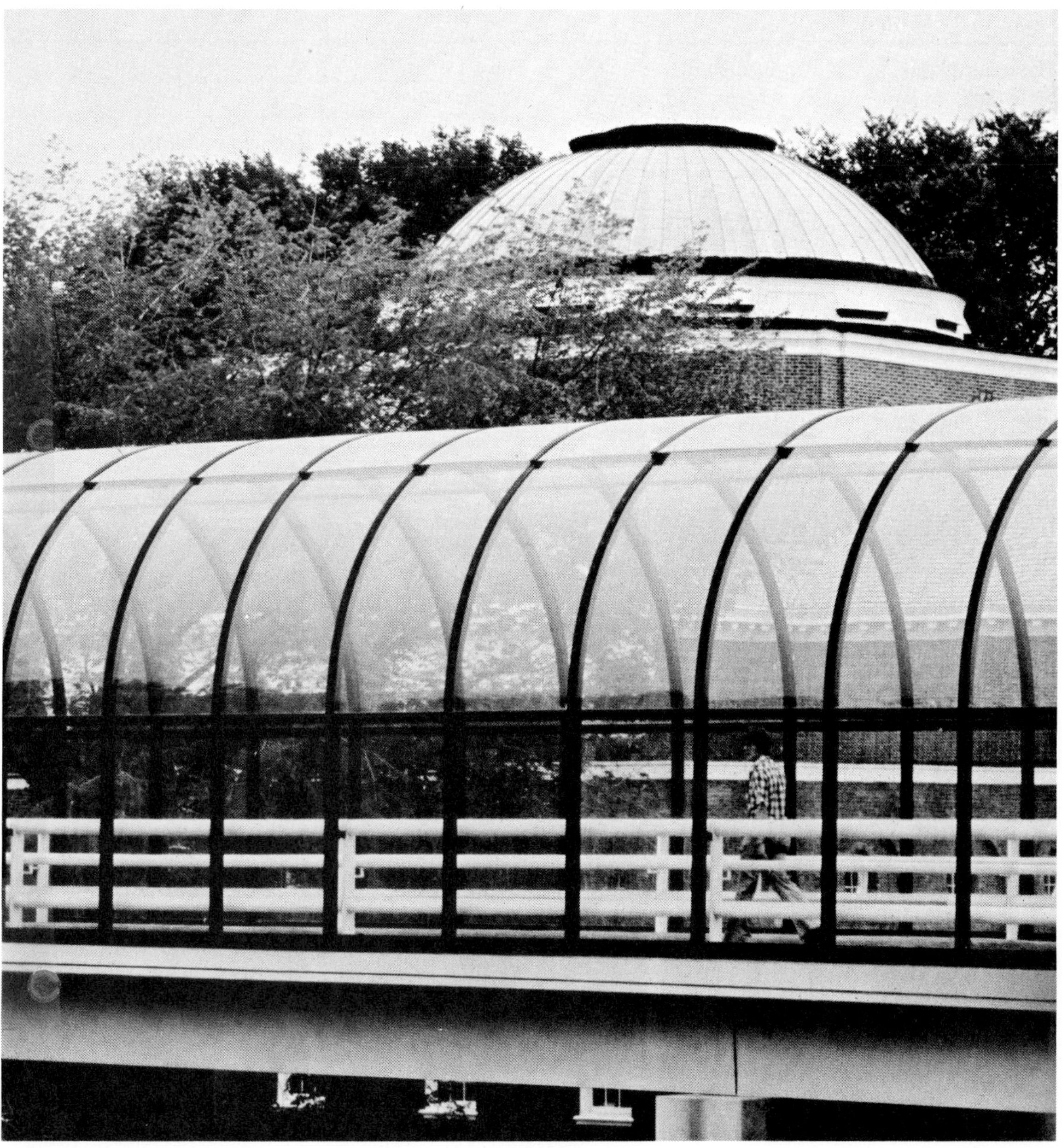

**Table 16-4** Polymers

| Monomer Name, Structure | Polymer Name, Structure | Some Common Trade Names | Uses |
|---|---|---|---|
| ethylene $CH_2{=}CH_2$ | polyethylene $+CH_2-CH_2+_n$ | Polyfilm[a] Marlex[b] | Electrical insulation, packaging (plastic bags), floor covering, plastic bottles, pipes, tubing |
| propylene $CH_2{=}CH(CH_3)$ | polypropylene $(CH_2-CH(CH_3))_n$ | Herculon[c] | Pipes, carpeting, artificial turf, molded auto parts, fibers |
| vinyl chloride $CH_2{=}CH(Cl)$ | polyvinyl chloride (PVC) $(CH_2-CH(Cl))_n$ | Tygon[d] | Wire and cable coverings, pipes, rainwear, shower curtains, tennis court playing surfaces |
| styrene 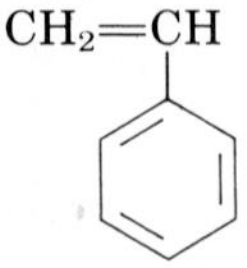 $CH_2{=}CH(C_6H_5)$ | polystyrene $(CH_2-CH(C_6H_5))_n$ 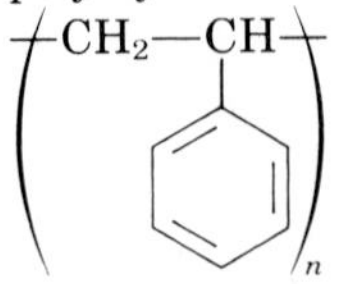 | Styrofoam[a] Styron[a] | Molded objects (combs, toys, brush and pot handles), refrigerator parts, insulating material, phonograph records, clock and radio cabinets |
| tetrafluoroethylene $CF_2{=}CF_2$ | polytetrafluoroethylene $+CF_2-CF_2+_n$ | Teflon[e] Halon[f] | Gaskets, valves, tubing, coatings for cookware |
| methyl methacrylate $CH_2{=}C(CH_3)(CO_2CH_3)$ | polymethyl methacrylate $+CH_2-C(CH_3)(CO_2CH_3)+_n$ | Plexiglas[g] Lucite[e] | Glass substitute, lenses, aircraft glass, dental fillings, artificial eyes, braces |
| acrylonitrile $CH_2{=}CH(CN)$ | polyacrylonitrile $(CH_2-CH(CN))_n$ | Orlon[e] Acrilan[h] | Fibers for clothing, carpeting |

[a]Dow Chemical Co.
[b]Phillips Petroleum Co.
[c]Hercules, Inc.
[d]U.S. Stoneware Co.
[e]E. I. du Pont de Nemours & Co.
[f]Allied Chemical Corp.
[g]Rohm & Haas Co.
[h]Monsanto Industrial Chemicals, Inc.

the alkane ending instead of *-ene*. Thus, $CH_3CH_2CH_2CH_2C{\equiv}CCH_2CH_3$ is named 3-octyne.

Alkynes are not found in nature, but can be prepared synthetically. Acetylene can be made from coal by first reacting the coal with calcium oxide at high temperature and then treating the calcium carbide formed with water:

$$\underset{\text{coal}}{C} + \underset{\text{calcium oxide}}{CaO} \xrightarrow{\Delta} \underset{\text{calcium carbide}}{CaC_2} \xrightarrow{H_2O} \underset{\text{acetylene}}{H{-}C{\equiv}C{-}H} + \underset{\text{calcium hydroxide}}{Ca(OH)_2}$$

A more common method of making acetylene is through the reforming process. Methane is heated in the presence of a catalyst, forming acetylene and hydrogen:

$$2CH_4 \xrightarrow[\Delta]{\text{catalyst}} H{-}C{\equiv}C{-}H + 3H_2$$

Although acetylene has some use in oxyacetylene torches, its biggest use is in the manufacture of other organic compounds used as monomers for polymers. For example, acetylene is used to make vinyl chloride (which is then used to make polyvinyl chloride) and acrylonitrile [used to make Acrilan® and Orlon® (see Table 16-4)].

$$H{-}C{\equiv}C{-}H + HCl \longrightarrow \underset{\text{vinyl chloride}}{(H)(H)C{=}C(H)(Cl)}$$

$$H{-}C{\equiv}C{-}H + HCN \longrightarrow \underset{\text{acrylonitrile}}{(H)(H)C{=}C(H)(CN)}$$

## 16-5 Aromatic Compounds

There is a second class of hydrocarbons known as the aromatics. *An* **aromatic compound** *is a cyclic hydrocarbon containing alternating single and double bonds between adjacent carbon atoms in a six-member ring.**

This discussion will concentrate on one aromatic compound, benzene ($C_6H_6$), and its derivatives. (*A* **derivative** *of a compound is produced by the sub-*

*While there are other types of aromatic compounds, this definition includes the largest class.

*stitution of a group or hetero atom on the molecules of the original compound.)* At first glance, benzene looks like a cyclic alkene with three double bonds. It turns out, however, that the three double bonds do not

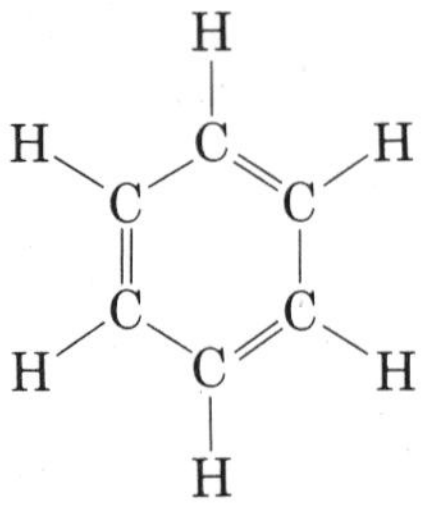

(A) 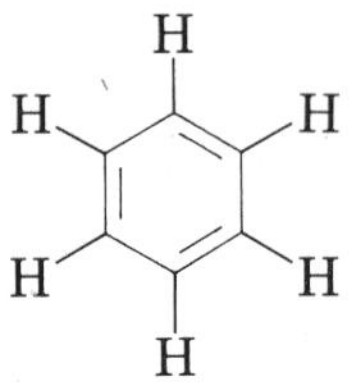 (B) 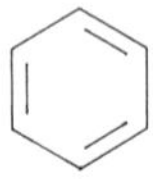 (C)

[Simplified structures can be written by omitting carbons (B) or by omitting carbons and hydrogens (C).]

act like alkene double bonds. For example, benzene does not "decolorize" a solution of bromine as do simple alkenes.

$$CH_2{=}CH_2 + Br_2 \longrightarrow CH_2Br{-}CH_2Br$$

Red Colorless

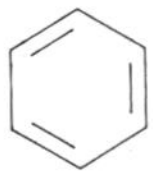

$$C_6H_6 + Br_2 \longrightarrow \text{no reaction}$$

Red Still red

As the preceding reactions illustrate, benzene is less chemically reactive than simple alkenes. It is this property that distinguishes aromatics from alkenes.

Compounds that contain the benzene ring can usually be recognized by name because they are named as derivatives of benzene.

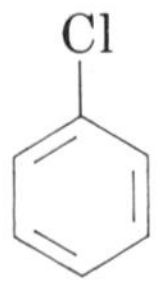

Cl — chlorobenzene

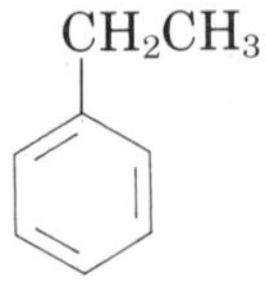

$CH_2CH_3$ — ethylbenzene

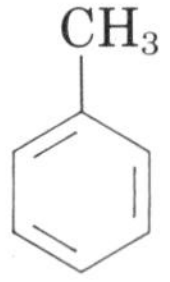

$CH_3$ — methylbenzene (toluene)

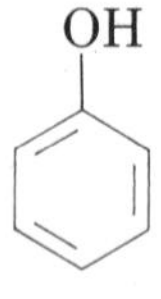

OH — hydroxybenzene (phenol)

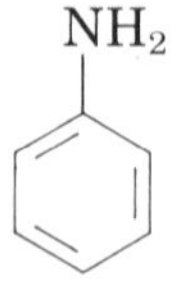

$NH_2$ — aminobenzene (aniline)

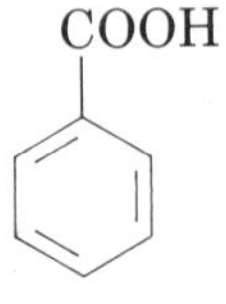

COOH — benzoic acid

Although some benzene and toluene are present in crude oil, additional quantities can be obtained by the reforming process. When cyclohexane and/or hexane are heated with a catalyst, benzene is formed. Coal is another source of benzene. When coal is heated to high temperatures in the absence of air, some benzene is formed.

$$\text{cyclo-}(CH_2)_6 \text{ and/or } CH_3(CH_2)_4CH_3 \xrightarrow{\text{catalyst}} C_6H_6 + 3H_2 \text{ or } 4H_2$$

Benzene and toluene are used mainly as solvents and as starting materials to make other aromatic compounds. Benzene must be used with care, however, because it has been found to be a potent carcinogen (it can cause cancer). Phenol and its derivatives have been used as disinfectants, in the manufacture of dyes, explosives, drugs, and plastics, and as preservatives.

**See Problems 16-7 through 16-20.**

## 16-6 Organic Functional Groups

In contrast to the hydrocarbons, many organic compounds contain alkyl or other hydrocarbon groups attached to other elements, particularly nitrogen and oxygen. This, of course, alters the chemistry of these compounds significantly from that of the basic hydrocarbons. *The part of the molecule that contains the elements other than C is called the* **functional group.** It is called a functional group because that is where most chemical reactions take place. In this respect, *the double bond in alkenes and the triple bond in alkynes are also considered functional groups.* Each functional group has a strong influence on the chemistry of the compounds and thus establishes a new class of compounds. In the following sections we will examine some of these classes of compounds that are established by the presence of a particular functional group.

## 16-7 Alcohols (R—OH)*

The simplest alcohol, methanol (methyl alcohol), has the formula $CH_4O$ and the structure $CH_3$—O—H. *The functional group of alcohol is the OH group* (**hydroxyl group**).

Alcohols are named by taking the alkane name, dropping the *-e* and substituting *-ol*. Common names are obtained by just naming the alkyl group attached to the —OH followed by *alcohol*. Some very useful alcohols (see Figure 16-5) have more than one hydroxyl group. Two of them are shown below.

*The R or R′ represents a hydrocarbon group such as an alkyl or aromatic group.

$$\begin{array}{cccc}
CH_3CH_2OH & CH_3CH_2CH_2OH & \begin{array}{cc} CH_2 - & CH_2 \\ | & | \\ OH & OH \end{array} & \begin{array}{ccc} CH_2 - & CH - & CH_2 \\ | & | & | \\ OH & OH & OH \end{array} \\
\text{ethanol} & \text{propanol} & \text{1,2-ethane}\underline{\text{di}}\text{ol} & \text{1,2,3-propane}\underline{\text{tri}}\text{ol} \\
\text{(ethyl alcohol)} & (n\text{-propyl alcohol}) & \text{(ethylene glycol)} & \text{(glycerol)} \\
 & & [\textit{di} \text{ means two} & [\textit{tri} \text{ means three} \\
 & & \text{hydroxyl groups]} & \text{hydroxyl groups]}
\end{array}$$

Methanol and ethanol can both be obtained from natural sources. Methanol can be prepared by heating wood in the absence of oxygen to about 400 °C; at such high temperatures, methanol, together with other organic compounds, is given off as a gas. Since methanol was once made exclusively by this process, it is often called wood alcohol. Currently, methanol is prepared from synthesis gas, a mixture of CO and $H_2$. When CO and $H_2$ are passed over a catalyst at the right temperature and pressure, methanol is formed:

$$CO + 2H_2 \xrightarrow[\Delta]{\text{catalyst}} CH_3OH$$

**Figure 16-5** ALCOHOLS. A major ingredient in each of these products is an alcohol.

Ethanol (often known simply as alcohol) is formed in the fermentation of various grains (therefore it is also known as grain alcohol) and is the "active ingredient" in all alcoholic beverages. In fermentation, the sugars and carbohydrates in grains are converted to ethanol and carbon dioxide by the enzymes in yeast.

$$\underset{\text{glucose (a sugar)}}{C_6H_{12}O_6} \xrightarrow[\text{enzymes}]{\text{yeast}} 2CH_3CH_2OH + 2CO_2$$

Industrially, ethanol is prepared by reacting ethylene with water in the presence of an acid catalyst.

$$CH_2{=}CH_2 + H_2O \xrightarrow{H^+} CH_3CH_2OH$$

Isopropanol (isopropyl alcohol or rubbing alcohol) can be prepared from propene (propylene) in the same way.

$$CH_3CH{=}CH_2 + H_2O \xrightarrow{H^+} CH_3-\overset{\displaystyle CH_3}{\underset{\displaystyle OH}{C}}-H$$

Methanol has been used as a solvent for shellac, as a denaturant for ethanol (it makes the ethanol undrinkable), and as an antifreeze for automobile radiators. It is very toxic when ingested. In small doses it causes blindness, and in large doses it can cause death.

Ethanol is present in alcoholic beverages such as beer, wine, and liquor. The "proof" of an alcoholic beverage is two times the percent by volume of alcohol. If a certain brand of bourbon is 100 proof, it contains 50% ethanol. Recently, there has been much interest in the use of ethanol as an additive to gasoline, called "gasohol." Ethanol is an excellent solvent and has been used as such in perfumes, medicines, and flavorings. It has also been used as an antiseptic and as a rubbing compound to cleanse the skin and lower a feverish person's temperature. While ethanol is not as toxic as methanol, it can cause coma or death when ingested in large quantities.

Ethylene glycol is the major component of antifreeze and coolant used in automobiles. It is also used to make polymers, the most common of which is Dacron®, a polyester.

Glycerol, which can be obtained from fats, is used in many applications where a lubricant and/or softener is needed. It has been used in pharmaceuticals, cosmetics, foodstuffs, and in some liqueurs. When glycerol reacts with nitric acid, it produces nitroglycerin. Nitroglycerin is a powerful explosive,

$$\underset{\text{glycerol}}{\underset{\displaystyle OH}{CH_2}-\underset{\displaystyle OH}{CH}-\underset{\displaystyle OH}{CH_2}} + 3HNO_3 \rightarrow \underset{\text{nitroglycerin}}{\underset{\displaystyle ONO_2}{CH_2}-\underset{\displaystyle ONO_2}{CH}-\underset{\displaystyle ONO_2}{CH_2}} + 3H_2O$$

but it is also a strong smooth-muscle relaxant and vasodilator and has been used to lower the blood pressure and to treat angina pectoris.

## 16-8 Ethers (R—O—R′)

*An* **ether** *contains an oxygen bonded to two hydrocarbon groups* (rather than one hydrocarbon group and one hydrogen as in alcohols). The simplest ether, dimethyl ether, is an isomer of ethanol and has very different properties.

Ethers are named by giving the alkyl groups on either side of the oxygen and adding *ether.*

| | |
|---|---|
| $H_3O—O—CH_3$<br>dimethyl ether | $CH_3OCH_2CH_3$<br>methyl ethyl ether |
| $CH_3CH_2OCH_2CH_3$<br>diethyl ether | $CH_3CH_2OCH_2CH_2CH_3$<br>ethyl propyl ether |

Diethyl ether is made industrially by reacting ethanol with sulfuric acid. In this reaction, two ethanol molecules are joined together with the loss of an $H_2O$ molecule.

$$2CH_3CH_2OH \xrightarrow{H_2SO_4} CH_3CH_2—O—CH_2CH_3 + H_2O$$

The most commonly known ether, diethyl ether (or ethyl ether or just ether) has, in the past, been used extensively as an anesthetic. It has the advantages of being an excellent muscle relaxant that doesn't affect the blood pressure, pulse rate, or the rate of respiration greatly. On the other hand, ether has an irritating effect on the respiratory passages and often causes nausea. Its flammability is also a drawback due to dangers of fire and explosions. Diethyl ether is rarely used now as an anesthetic. Other anesthetics that do not have its disadvantages have taken its place.

## 16-9 Amines (R—NH₂)

*An* **amine** *contains a nitrogen with single bonds to a hydrocarbon group and a total of two other hydrocarbon groups or hydrogens.* The nitrogen in amines has one pair of unshared electrons similar to ammonia, $NH_3$. As we learned in Chapter 13, $NH_3$ utilizes the unshared pair of electrons to form weakly basic solutions in water. In a similar manner, amines are characterized by their ability to act as bases.

$$\ddot{N}H_3 + H_2O \rightleftharpoons NH_4^+ + OH^-$$

$$CH_3\ddot{N}H_2 + H_2O \rightleftharpoons CH_3NH_3^+ + OH^-$$

Basic solutions

Amines are named by listing the alkyl groups attached to the nitrogen and adding *amine.*

$$\begin{array}{ccc}
\begin{array}{c} CH_3{-}\ddot{N}{-}H \\ | \\ H \end{array} &
\begin{array}{c} CH_3{-}\ddot{N}{-}CH_3 \\ | \\ H \end{array} &
\begin{array}{c} \boxed{CH_3}{-}\ddot{N}{-}\boxed{CH_2CH_2CH_3} \\ | \\ \boxed{CH_2CH_3} \end{array} \\
\text{methyl amine} & \text{dimethyl amine} & \text{methyl} \quad \text{ethyl} \quad \text{propyl amine} \\
 & (di = \text{two methyls}) &
\end{array}$$

Simple amines are prepared by the reaction of ammonia with alkyl halides (e.g., $CH_3Cl$). In the reaction the alkyl group substitutes for the hydrogen on ammonia. The hydrogen and chlorine combine to form HCl.

$$\begin{array}{ccc}
\begin{array}{c} H{-}\ddot{N}{-}\boxed{H + Cl}{-}CH_3 \\ | \\ H \end{array} & \rightarrow &
\begin{array}{c} H{-}\ddot{N}{-}CH_3 + HCl \\ | \\ H \end{array} \\
\text{ammonia} \quad \text{methylchloride} & & \text{methylamine}
\end{array}$$

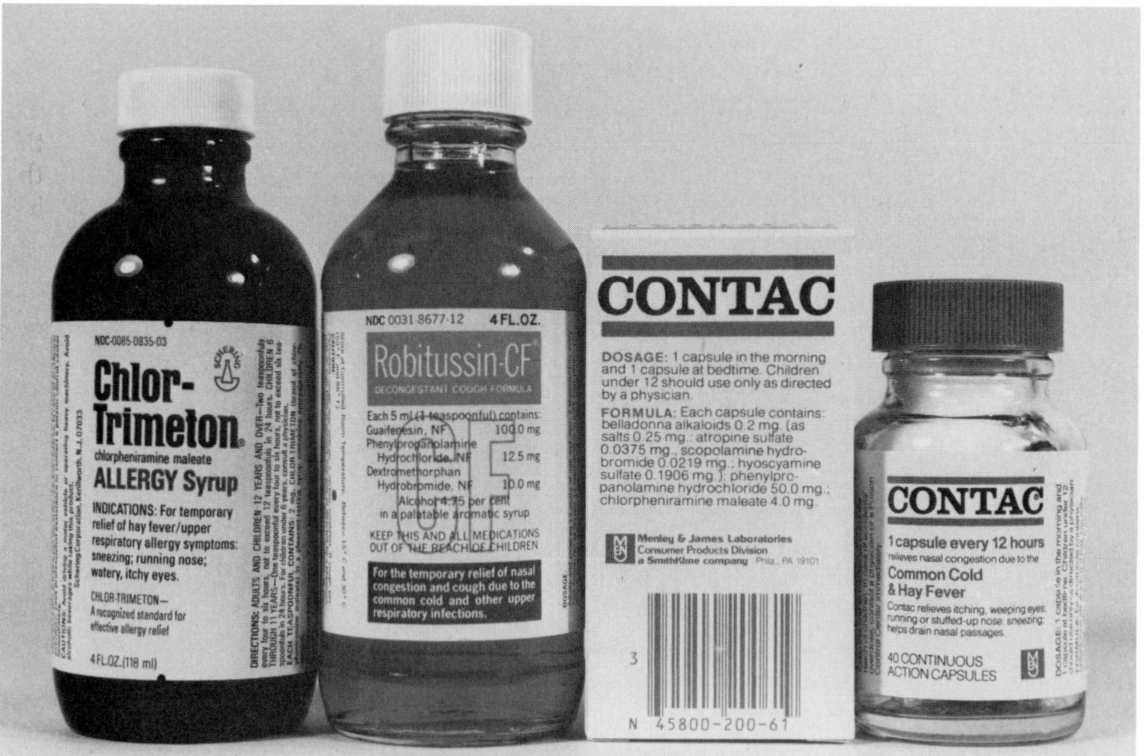


**Figure 16-6** AMINES. The active ingredient in an antihistamine is an organic amine.

Further reaction with the same or different alkyl halides leads to substitution of one or both remaining hydrogens on the methylamine.

Amines are used in the manufacture of dyes, drugs, disinfectants, and insecticides. They also occur naturally in biological systems and are important in many biological processes.

Amine groups are present in many synthetic and naturally occurring drugs (see Figure 16-6). They may be useful as antidepressants, antihistamines, antibiotics, antiobesity preparations, antinauseants, analgesics, antitussives, diuretics, and tranquilizers, among others. Frequently, a drug may have more than one use [codeine is both an analgesic (pain reliever) and an antitussive agent (cough depressant).] Often a drug may be obtained from plants or animals. A class of compounds that falls into this type of drug are the alkaloids, which are amines found in plants. Some common drugs are shown below.

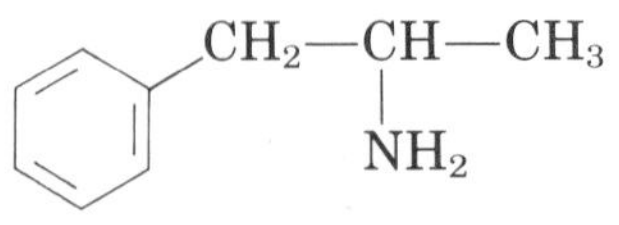

$CH_2—CH(NH_2)—CH_3$

amphetamine
(Benzedrine®;
synthetic
appetite depressant,
stimulant)

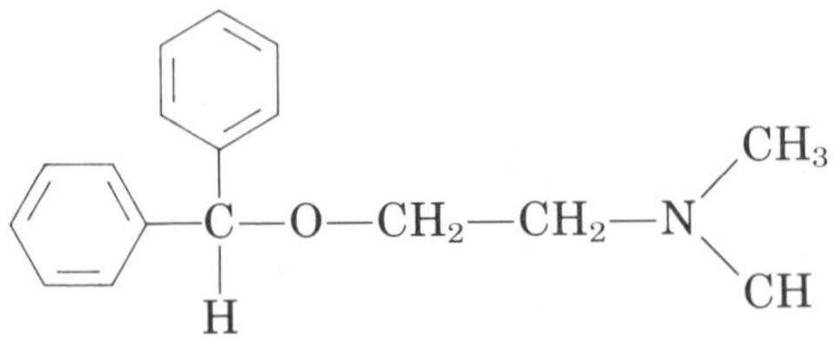

$C(H)—O—CH_2—CH_2—N(CH_3)(CH)$

diphenhydramine
(synthetic
antihistamine)

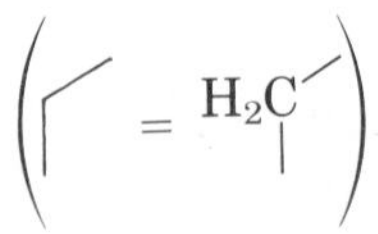

$= H_2C$

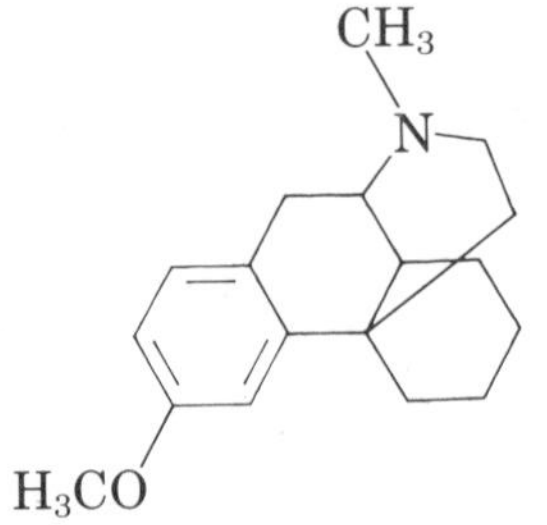

$CH_3$ N $H_3CO$

dextromethorphan
(synthetic
analgesic, antitussive)

$CH_3$ N OH O $H_3CO$

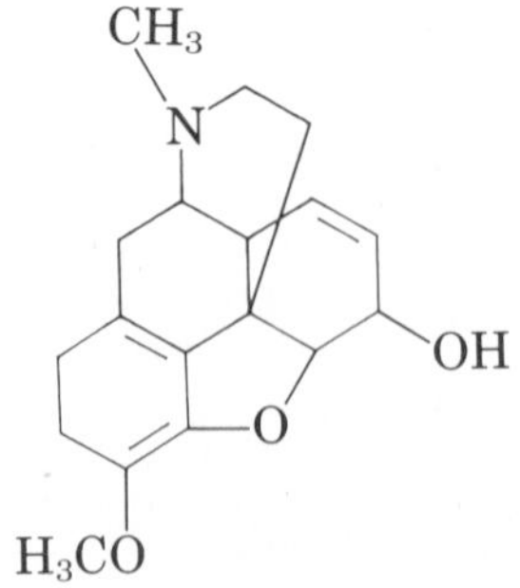

codeine
(from opium;
analgesic, antitussive)

$CH_3$ N HO O OH

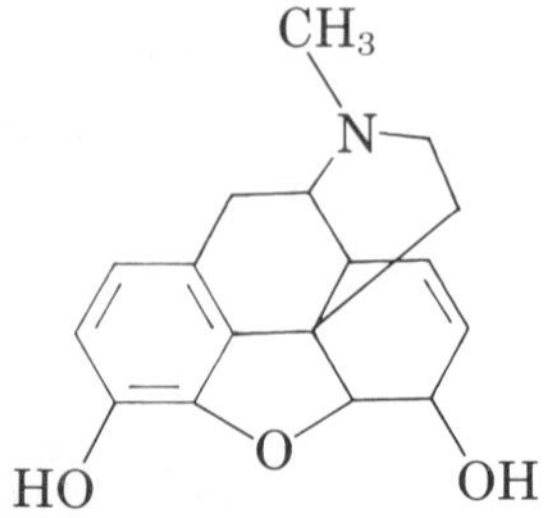

morphine
(from opium;
analgesic)

## 16-10 Aldehydes ($R-\overset{O}{\overset{\|}{C}}-H$) and Ketones ($R-\overset{O}{\overset{\|}{C}}-R'$)

The functional group of aldehydes and ketones is a carbonyl group. A **carbonyl group** *is a carbon with a double bond to an oxygen.* (i.e., $-\overset{O}{\overset{\|}{C}}-$). *In* **aldehydes** *the carbonyl group is bound to one hydrocarbon group and one hydrogen, whereas in* **ketones** *the carbonyl group is bound to two hydrocarbon groups.*

Aldehydes are named by dropping the *-e* of the corresponding alkane name and substituting *-al*. Therefore the two-carbon aldehyde is ethanal. Ketones are named by taking the alkane name, dropping the *-e*, and substituting *-one*. The common names of ketones are obtained by naming the alkyl groups on either side of the C=O and adding *ketone*. Thus, the four-carbon ketone can be called butanone (IUPAC) or methyl ethyl ketone (common).

Aldehydes

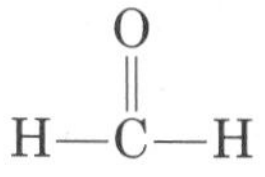

$$H-\overset{O}{\overset{\|}{C}}-H \qquad CH_3-\overset{O}{\overset{\|}{C}}-H$$

methanal (formaldehyde) ethanal (acetaldehyde)

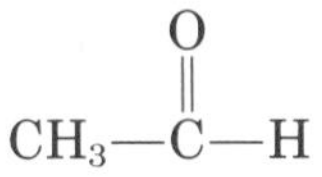

Ketones

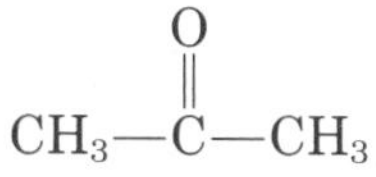

$$CH_3-\overset{O}{\overset{\|}{C}}-CH_3 \qquad CH_3-\overset{O}{\overset{\|}{C}}-CH_2-CH_2-CH_3$$

propanone (acetone or dimethyl ketone) 2-pentanone (methyl propyl ketone)

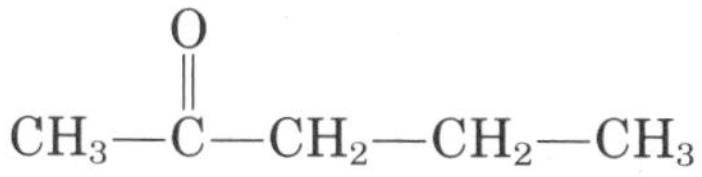

Many aldhydes and ketones are prepared by the oxidation of alcohols. Industrially, the alcohol is oxidized by heating it in the presence of oxygen and a catalyst. About half of the methanol made industrially is used to make formaldehyde by oxidation. Acetone can be prepared in the same fashion by oxidation of isopropanol.

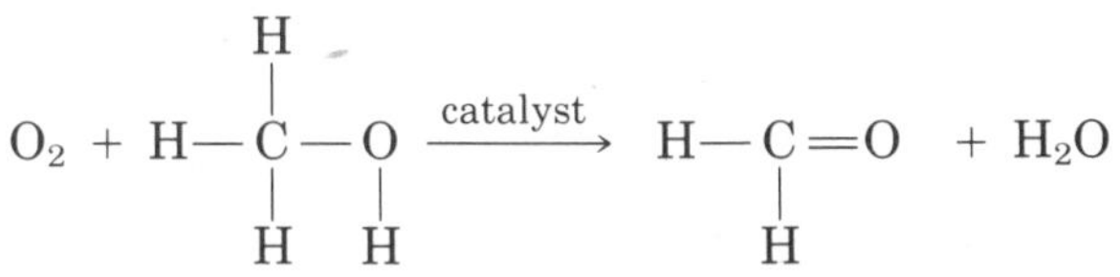

$$O_2 + H-\underset{H}{\overset{H}{\underset{|}{\overset{|}{C}}}}-\underset{H}{\underset{|}{O}} \xrightarrow{\text{catalyst}} H-\underset{H}{\underset{|}{C}}=O + H_2O$$

methanol formaldehyde

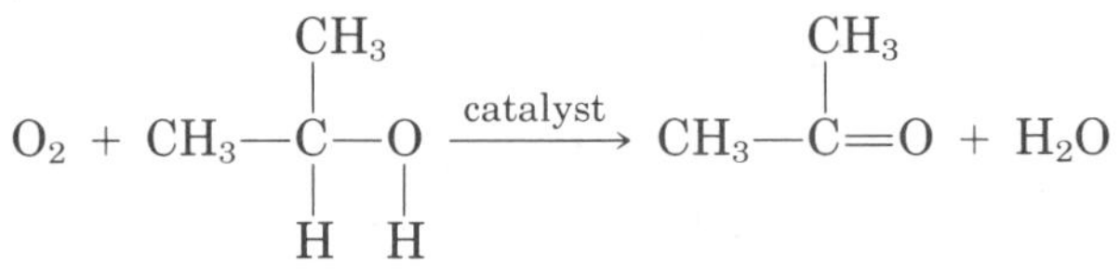

$$O_2 + CH_3\!-\!\underset{\displaystyle H}{\overset{\displaystyle CH_3}{\underset{|}{\overset{|}{C}}}}\!-\!\underset{\displaystyle H}{\underset{|}{O}} \xrightarrow{\text{catalyst}} CH_3\!-\!\overset{\displaystyle CH_3}{\overset{|}{C}}\!=\!O + H_2O$$

isopropanol acetone

Aldehydes and ketones have uses as solvents, in the preparation of polymers, as flavorings, and in perfumes. The simplest aldehyde, formaldehyde, has been used as a disinfectant, antiseptic, germicide, fungicide, and embalming fluid (as a 37%-by-weight water solution). It has also been used in the preparation of polymers such as Bakelite (the first commercial plastic) and Melmac® (used to make dishes). Formaldehyde polymers have also been used as coatings on fabrics to give "permanent press" characteristics.

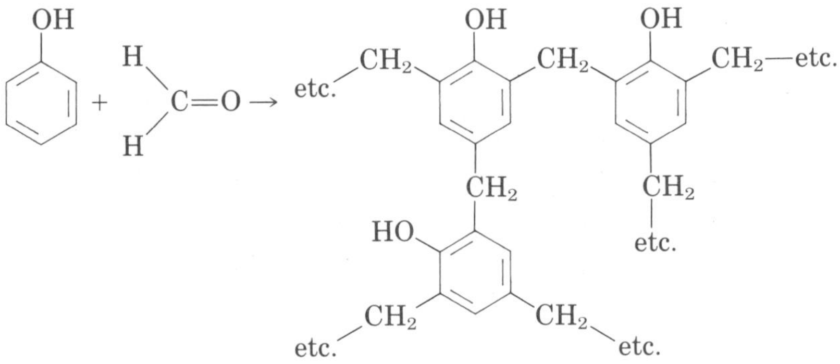

[Structure: phenol + $H_2C=O$ → Bakelite (phenol rings linked by $CH_2$ groups, "etc.")]

Bakelite

The simplest ketone, acetone, has been used mainly as a solvent. It is soluble in water and dissolves relatively polar and nonpolar molecules. It is an excellent solvent for paints and coatings.

More complex aldehydes and ketones, such as the following two examples, are used in flavorings and perfumes.

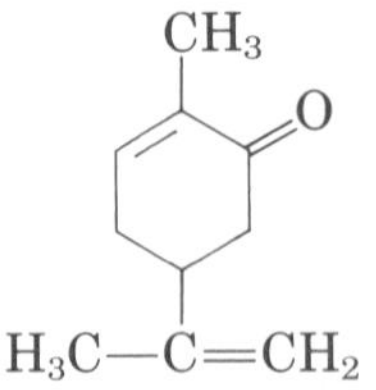

[Structures of carvone and citral]

carvone
(in oil of spearmint,
a flavoring)

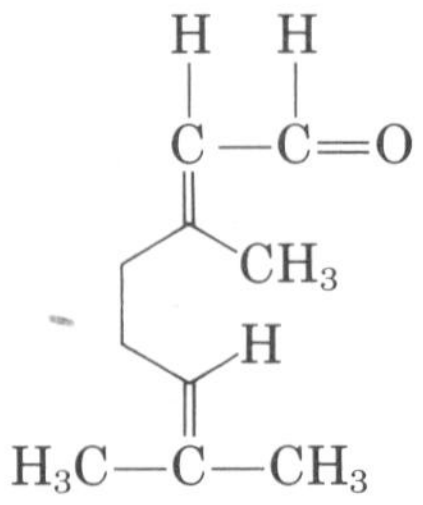

citral
(in oil of lemon grass;
a fragrance)

## 16-11 Carboxylic Acids, ($R-\overset{\displaystyle O}{\overset{\|}{C}}-O-H$), Esters ($R-\overset{\displaystyle O}{\overset{\|}{C}}-O-R'$), and Amides ($R-\overset{\displaystyle O}{\overset{\|}{C}}-NH_2$)

**Carboxylic acids** *contain the functional group* $-\overset{\displaystyle O}{\overset{\|}{C}}-O-H$ *(also written* —(COOH), *which is known as a* **carboxyl group.** They are acidic, organic compounds but not as acidic as inorganic acids such as nitric or sulfuric. Carboxylic acids are easily changed into two derivatives: amides and esters. **Amides** *have an amine group substituted for the hydroxyl group of the acid.* **Esters** *have a hydrocarbon group in place of the hydrogen in the carboxyl group.*

Carboxylic acids are named by dropping the *-e* from the alkane name and substituting *-oic acid.* Esters are named by giving the name of the alkyl group attached to the oxygen, followed by the acid name minus the *-ic* acid ending and substituting *-ate.* Amides are named by dropping the *-oic acid* portion of the carboxylic acid name and substituting *-amide.*

A few examples of carboxylic acids, esters, and amides are as follows.

Acids

| $H-\overset{\displaystyle O}{\overset{\|}{C}}-OH$ | $CH_3-\overset{\displaystyle O}{\overset{\|}{C}}-OH$ | $C_6H_5-\overset{\displaystyle O}{\overset{\|}{C}}-OH$ | $HO-\overset{\displaystyle O}{\overset{\|}{C}}-\overset{\displaystyle O}{\overset{\|}{C}}-OH$ |
|---|---|---|---|
| methanoic acid (formic acid) | ethanoic acid (acetic acid) | benzoic acid | oxalic acid |

Esters

| $H-\overset{\displaystyle O}{\overset{\|}{C}}-O-CH_2CH_3$ | $CH_3-\overset{\displaystyle O}{\overset{\|}{C}}-O-CH_2CH_3$ | $C_6H_5-\overset{\displaystyle O}{\overset{\|}{C}}-O-CH_3$ |
|---|---|---|
| ethyl methanoate (ethyl formate) | ethyl ethanoate (ethyl acetate) | methyl benzoate |

Amides

| $H-\overset{\displaystyle O}{\overset{\|}{C}}-NH_2$ | $CH_3-\overset{\displaystyle O}{\overset{\|}{C}}-NH_2$ | $C_6H_5-\overset{\displaystyle O}{\overset{\|}{C}}-NH_2$ |
|---|---|---|
| methanamide (formamide) | ethanamide (acetamide) | benzamide |

Carboxylic acids are made by oxidizing either alcohols or aldehydes. Formic acid, which was first isolated by distilling red ants, can be made by oxidizing either methanol or formaldehyde.

$$CH_3\text{—}OH \xrightarrow{\text{oxidizing agent}} H\text{—}\overset{\overset{\displaystyle O}{\|}}{C}\text{—}OH \xleftarrow{\text{oxidizing agent}} \underset{H\quad\;\; H}{\overset{\overset{\displaystyle O}{\|}}{C}}$$

Acetic acid can be made from the oxidation of ethanol. In fact, this is what happens when wine becomes "sour." Wine vinegar is produced by the air oxidation of the alcohol in ordinary wine. The sharp, tart flavor of vinegar is due to the acetic acid, which is present in a concentration of about 4 to 5%.

Esters are made from the reaction of alcohols and carboxylic acids. In the reaction, $H_2O$ is split off from the two molecules (OH from the acid and H from the alcohol), leading to the union of the remnants of the two molecules.

$$H\text{—}\overset{\overset{\displaystyle O}{\|}}{C}\text{—}\boxed{OH + H}\text{—}O\text{—}CH_3 \xrightarrow{H^+} H\text{—}\overset{\overset{\displaystyle O}{\|}}{C}\text{—}O\text{—}CH_3 + H_2O$$

Amides are made in a similar fashion, except that the acid is mixed with ammonia or an amine instead of an alcohol.

$$H\text{—}\overset{\overset{\displaystyle O}{\|}}{C}\text{—}\boxed{OH + H}\text{—}NH_2 \xrightarrow{\Delta} H\text{—}\overset{\overset{\displaystyle O}{\|}}{C}\text{—}NH_2 + H_2O$$

$$H\text{—}\overset{\overset{\displaystyle O}{\|}}{C}\text{—}\boxed{OH + H}\text{—}NHCH_3 \xrightarrow{\Delta} H\text{—}\overset{\overset{\displaystyle O}{\|}}{C}\text{—}NHCH_3 + H_2O$$

Carboxylic acids, esters, and amides are frequently present in compounds that have medicinal uses. Salicylic acid has both antipyretic (fever-reducing) and analgesic (pain-relieving) properties. It has the disadvantage, however, of causing severe irritation of the stomach liming. Acetylsalicylic acid (aspirin), which is both an acid and an ester, doesn't irritate the stomach as much. Aspirin is broken up in the small intestine, to form salicylic acid, where it is absorbed. Some people are allergic to aspirin, and must take "aspirin substitutes." The common aspirin substitutes are amides, phenacetin and acetaminophen. Acetaminophen is the active ingredient in Tylenol® and Datril.®

$C_6H_4(OH)COOH$

salicylic acid

$C_6H_4(OCOCH_3)COOH$

acetylsalicylic acid

$H-N(C_6H_4OCH_2CH_3)-C(=O)-CH_3$

phenacetin

$H-N(C_6H_4OH)-C(=O)-CH_3$

acetaminophen

Just as aldehydes and ketones are used as fragance and flavorings, so are esters. Some of these are

$H-\overset{O}{\overset{\|}{C}}OCH_2CH_3$

ethyl formate
(rum)

$CH_3\overset{O}{\overset{\|}{C}}O(CH_2)_4CH_3$

pentyl acetate
(bananas)

$CH_3CH_2CH_2\overset{O}{\overset{\|}{C}}OCH_2CH_3$

ethyl butanoate
(pineapple)

$C_6H_4(OH)-\overset{O}{\overset{\|}{C}}-O-CH_3$

methyl salicylate
(oil of wintergreen)

One of the biggest uses for carboxylic acids, esters, and amides is in the formation of polymers. These are different from the polymers we discussed earlier, in that a small molecule (usually $H_2O$) is given off during the formation of the polymer. Two of the most widely known polymers are Nylon 66 (see Figure 16-7) and Dacron®. The first is a polyamide, made from a diacid and a diamine; the second is a polyester, made from a diacid and a dialcohol (diol). Both of these polymers are used to make fibers.

**See Problems 16-21 through 16-31.**

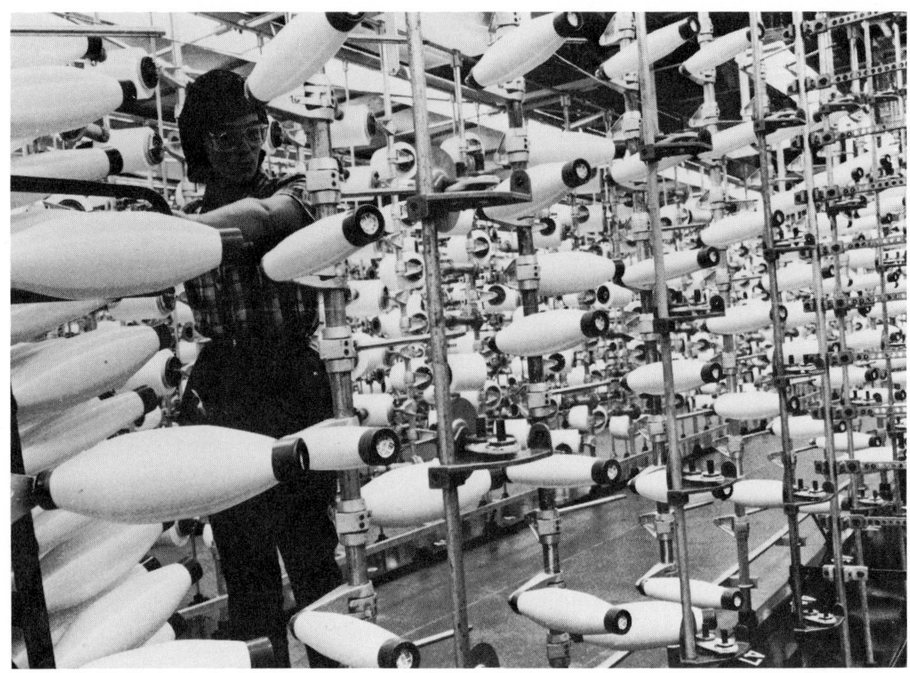

**Figure 16-7** NYLON. Nylon was first synthesized by DuPont chemists. It was one of the first synthetic fibers.

Nylon 66

$$\underset{\text{adipic acid}}{HO\overset{\overset{O}{\|}}{C}(CH_2)_4\overset{\overset{O}{\|}}{C}-}\boxed{OH + H}\underset{\text{1,6-hexanediamine}}{-\overset{\overset{H}{|}}{N}(CH_2)_6NH_2} \xrightarrow{\Delta}$$

$$\underset{\text{Nylon 66}}{\left[-\overset{\overset{O}{\|}}{C}(CH_2)_4\overset{\overset{O}{\|}}{C}NH(CH_2)_6NH-\right]} + H_2O \quad (\text{Amide: } \overset{\overset{O}{\|}}{C}NH)$$

Dacron®

$$\underset{\text{Terephthalic acid}}{HO-\overset{\overset{O}{\|}}{C}-C_6H_4-\overset{\overset{O}{\|}}{C}-}\boxed{OH + H}\underset{\text{ethylene glycol}}{-O-CH_2CH_2-OH} \xrightarrow{H^+}$$

$$\underset{\text{Dacron}}{\left[-\overset{\overset{O}{\|}}{C}-C_6H_4-\overset{\overset{O}{\|}}{C}-O-CH_2CH_2-O-\right]_n} + H_2O \quad (\text{Ester: } \overset{\overset{O}{\|}}{C}-O)$$

## 16-12 Chapter Summary

The only practical way to introduce organic chemistry is to classify various compounds in groups with common properties. Since properties are determined to a large extent by the types, identity, and nature of the chemical bonds, we can use this as a convenient means of classification. The first groupings of organic compounds are as follows: hydrocarbons that contain only carbon and hydrogen, and organic compounds that contain hetero atoms.

Hydrocarbons can be subdivided further into several homologous series. These series are summarized as follows:

| Hydrocarbons | | | | |
|---|---|---|---|---|
| Name of the Homologous Series | General Formula | Example | Name | Comments |
| Alkane | $C_nH_{2n+2}$ | $C_3H_8$ | propane | Alkanes contain only single bonds and are saturated. |
| Alkene | $C_nH_{2n}$ | $C_3H_6$ | propene propylene | Alkenes contain double bonds and are unsaturated. |
| Alkyne | $C_nH_{2n-2}$ | $C_3H_4$ | propyne | Alkynes contain triple bonds and are unsaturated. |
| Aromatic | — | $C_6H_6$ | benzene | Aromatic compounds are cyclic, with alternating double bonds, and are less reactive than alkenes. |

A significant feature of the hydrocarbons is the existence of isomers. Isomers of a compound have the same formula but different structural characteristics. Depending on the difference, isomers may be very different types of compounds (e.g., ethanol and dimethyl ether) or have similar properties (e.g. butane and isobutane). Petroleum and natural gas are the sources of hydrocarbons. By use of the processes of cracking, reforming, and alkylation the hydrocarbons found in natural sources can be used to make many other useful hydrocarbons. The nomenclature of hydrocarbons is somewhat complicated because of the use of formal IUPAC names as well as the more popular common names.

Hydrocarbons all have their important uses: alkanes for fuel, alkenes to make polymers, alkynes to make alkenes for various polymers, and aromatics to make solvents, drugs, and flavors.

The focal point of chemical reactivity in an organic compound is at the functional group. In alkenes and alkynes the double and triple bonds serve

**16-10** Name the longest chain in the compounds in Problem 16-9.

**16-11** $C_nH_{2n}$ can represent an alkane or an alkene. Explain.

**16-12** What is the general formula of a cyclic alkene that contains one double bond?

**16-13** Identify the following as an alkane, alkene, or alkyne. Assume that the compounds are not cyclic.

(a) $C_8H_{16}$
(b) $C_5H_{12}$
(c) $C_4H_8$
(d) $C_{20}H_{38}$
(e) $CH_4$
(f) $C_{10}H_{20}$
(g) $C_{18}H_{38}$

**16-14** Which of the following is the formula of a cyclic alkane?

(a) $C_8H_{18}$
(b) $C_5H_8$
(c) $C_6H_6$
(d) $C_5H_{10}$

**16-15** Which of the following are formulas of alkyl groups?

(a) $C_4H_8$
(b) $C_4H_9$
(c) $C_6H_{10}$
(d) $C_6H_6$
(e) $C_6H_{13}$

**16-16** Give the abbreviated structure of the polymer that would be formed from each of the following alkenes. Name the polymer.

(a) $CH_2{=}CH(CH_3)$ (propylene)
(b) $CH_2{=}CH(F)$ (vinyl fluoride)
(c) $CH_2{=}CF_2$ (difluoroethylene)
(d) $CH_2{=}CH(CO_2CH_3)$ (methylacrylate)

**16-17** A certain polymer has repeating units represented by the formula $\left(\!-CH(F)-CH(CH_3)-\!\right)_n$

Write the formula of the monomer.

**16-18** Naphthalene has the formula $C_{10}H_8$ and is composed of two benzene rings joined together. Write the Lewis structure, including hydrogens, for naphthalene.

* **16-19** Anthracene and phenanthrene are isomers with the formula $C_{14}H_{10}$. Both are composed of three benzene rings that are joined together. Write the Lewis structures for two isomers of this nature.

**16-20** There are three isomers for dichlorobenzene. This is a molecule in which two chlorines are substituted for two hydrogens on a benzene ring. Write the Lewis structures of the three isomers.

## Derivatives of Hydrocarbons

**16-21** Circle and name the functional groups in each of the following compounds:

(a) $CH_3CH{=}CHCH_2\overset{O}{\overset{\|}{C}}H$

(b) $C_6H_4(NH_2)-\overset{O}{\overset{\|}{C}}-CH_3$ (benzene ring with $NH_2$ and $-\overset{O}{\overset{\|}{C}}-CH_3$ in meta positions)

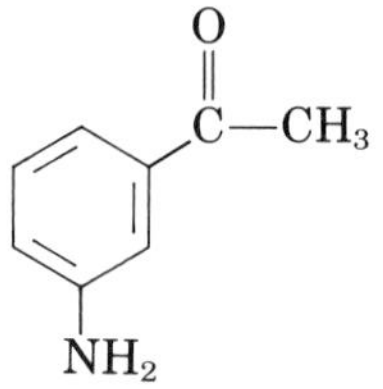

(c) $CH_3OCH_2CH_2\overset{O}{\overset{\|}{C}}NH_2$

(d) $H-C{\equiv}C-CH_2CH_2CH_2OH$

(e) $C_6H_5-CH_2CH_2\overset{O}{\overset{\|}{C}}OCH_3$

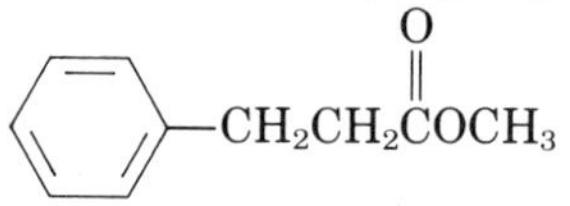

(f) $CH_3CH_2CH_2COOH$

**16-22** Aspartame is an effective sweetener that is used as a low-calorie substitute for sugar. Identify the functional groups present in an aspartame molecule.

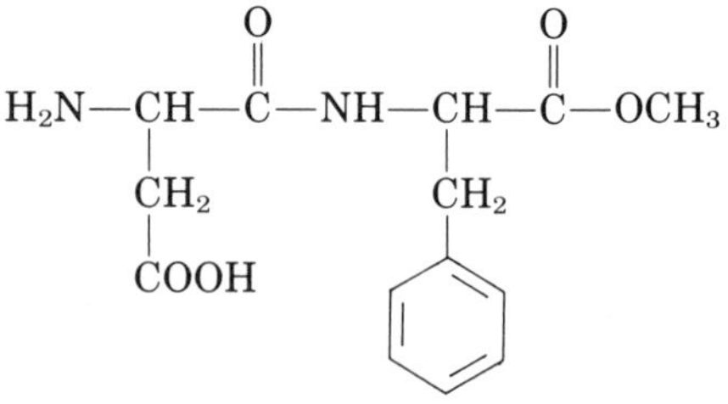

$$H_2N-\underset{\underset{COOH}{|}}{\underset{CH_2}{|}}{CH}-\overset{O}{\overset{\|}{C}}-NH-\underset{\underset{C_6H_5}{|}}{\underset{CH_2}{|}}{CH}-\overset{O}{\overset{\|}{C}}-OCH_3$$

**16-23** Tell how the following differ in structure.

(a) alcohols and ethers
(b) aldehydes and ketones
(c) amines and amides
(d) carboxylic acids and esters

**16-24** Glycine has the formula $C_2H_5O_2N$. It is a member of a class of compounds called amino acids that combine to form proteins. Glycine contains both an amine and a carboxylic acid group. Write the structure for glycine. Can glycine function as an acid, base, neither, or both in water?

**16-25** Which of the eight classes of compounds containing hetero atoms contains a carboxyl group? Which contains a hydroxyl group?

**16-26** What two classes of compounds combine to form esters? What two classes combine to form amides?

**16-27** To what class does each of the following compounds belong?
(a) 3-heptanone
(b) 3-nonene
(c) 2-methylpentane
(d) 2-ethylhexanal
(e) ethylbenzene
(f) propanol

**16-28** To what class does each of the following compounds belong?
(a) 2-octanol
(b) 2-butyne
(c) ethyl pentanoate
(d) 3-ethylhexene
(e) dimethyl ether
(f) aminobenzene

**16-29** Give a general method for making each of the following compounds.
(a) alcohols
(b) ketones
(c) carboxylic acids
(d) esters
(e) amides
(f) aldehydes
(g) amines

**16-30** Tell how the following can be obtained from natural sources.
(a) ethanol
(b) acetic acid
(c) methanol
(d) methane
(e) gasoline

**16-31** Give one possible use for each of the following compounds or class of compounds.
(a) formaldehyde
(b) acetic acid
(c) ethylene glycol
(d) acetylsalicylic acid
(e) amines
(f) esters
(g) alkenes
(h) alkanes

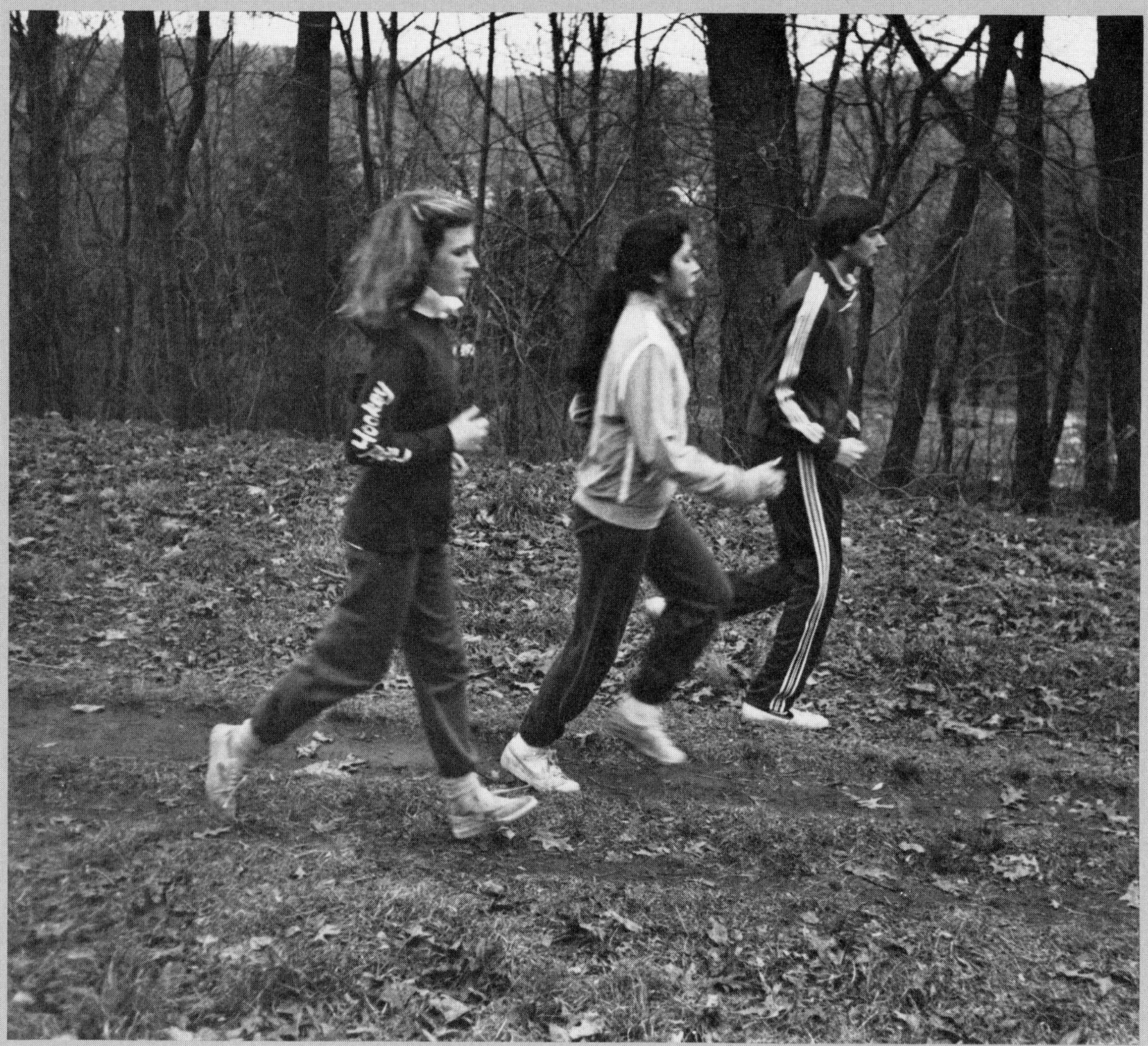


The successful athlete must stay "in shape" physically. The successful chemistry student must stay "in shape" mathematically.

# Foreword to the Appendixes

It is amazing how students who have *confidence* enjoy and do well in chemistry. Confidence, of course, comes from being prepared with the proper background. Just as a professional football player has the proper background of a successful college career, the serious chemistry student needs a proper mathematical background. Many, if not most, students who take this course need to give some early attention to reviewing their math. The reason is not important, regardless of whether it's because of a deficient secondary school background, a change in career intentions, or a lengthy interruption of studies. If the deficiencies are serious, the student should consider taking math *before* this course. More often, a review like that provided by the following appendixes can patch the weak spots so that math is not a hindrance to understanding chemistry.

Appendixes A, B, and C are intended to aid your review of the math and algebra required in the study of college or university chemistry. The first three appendixes begin with a short test, so that you can quickly gauge if the review and exercises that follow are needed.

Appendix D complements and expands on the introduction to problem solving by the unit-factor method in Chapter 2. Students who can become comfortable with the types of problems in this appendix will find themselves in excellent condition for the quantitative aspects of chemistry. Appendix E is an introduction to the use of common logarithms, which are needed to understand pH in Chapter 13. Appendix F is a discussion of the interpretation and construction of graphs, a topic of prime importance in both the social and the natural sciences. Appendix G is a glossary of terms used throughout the text, and finally, Appendix H contains answers to about two-thirds of the problems at the ends of the chapters.

# Appendix A
# Basic Mathematics

This section reviews the following basic mathematical skills.

1 Addition and subtraction

2 Multiplication

3 Roots of numbers

4 Division, fractions, and decimal fractions

5 Multiplication and division of fractions

6 Fractions, percent, and the use of percent

## A Test of Basic Mathematics

Following is a brief test on some basic mathematical skills necessary in chemistry. Work the test as carefully as you can *without a calculator*. Compare your answers with the answers given at the end of the test. If you have 100% of the answers correct, proceed to Appendix B on algebra. If you don't get 100, recheck your calculations. If a mistake or two was due to minor computational errors, then you should move on to Appendix B. If you make more than two mistakes or can't identify the source of your error, work through Appendix A before proceeding. After you have finished, take the test again. Hopefully, you will then get 100. If not, consult with your instructor for further review.

### Math Test

**Addition and Subtraction**

1 Carry out the following calculations:

**(a)** $43.76 + 6.90 + 134.88$ **1(a)** ______

**(b)** $0.668 - 9.232 + 3.445$ **1(b)** ______

**(c)** $5.65 - (-2.34) + (-0.89)$ **1(c)** ______

### Multiplication

**2** Carry out the following calculations. Express the answer to the proper number of significant figures and include units in (c) and (d).

**(a)** $10.7 \times 8.6$ **2(a)** ______

**(b)** $(-0.560) \times 234$ **2(b)** ______

**(c)** $16\ \text{in.} \times 34\ \text{in.}$ **2(c)** ______

**(d)** $13.6\ \text{g/mL} \times 187\ \text{mL}$ **2(d)** ______

### Roots of numbers

**3** Express the following roots in decimal fraction form. Include units and the proper number of significant figures in the answer.

**(a)** $\sqrt{81\ \text{cm}^2}$ **3(a)** ______

**(b)** $\sqrt[3]{512\ \text{in.}^3}$ **3(b)** ______

### Division, Fractions, and Decimal Fractions

**4** Express the following answers in decimal fraction form (e.g., $\frac{3}{5} = 0.6$). Include units and the proper number of significant figures in the answer.

**(a)** $465\ \text{miles} \div 5.7\ \text{hr}$ **4(a)** ______

**(b)** $\dfrac{115\ \text{g}}{7.82\ \text{mL}}$ **4(b)** ______

**(c)** $\dfrac{186\ \text{ft}^3}{4.5\ \text{ft}}$ **4(c)** ______

### Multiplication and Division of Fractions

**5** Express the following answers in decimal fraction form to three significant figures. When a problem has units, include the units in the answer.

**(a)** $(-\frac{9}{4}) \times \frac{24}{35} \times (-\frac{7}{3})$ **5(a)** ______

**(b)** $\frac{3}{4} \div (\frac{6}{7} \times \frac{1}{2})$ **5(b)** ______

**(c)** $\dfrac{187\ \text{torr}/7.45}{760\ \text{torr/atm}}$ **5(c)** ______

**(d)** $(\frac{2}{3}\ \text{cm})^3$ **5(d)** ______

### Fractions, Percent, and the Use of Percent

**6** There are 156 apples in a crate. Seventy apples are Jonathans, 45 are Golden Delicious, and the others are rotten. What is the decimal fraction of rotten apples?

**6** ______

**7** Express the following fractions as percentages to three significant figures.

**(a)** $\frac{3}{8}$ **7(a)** ______

**(b)** 0.0875 **7(b)** ______

**8** Express the following percentages as decimal fractions.

**(a)** 98.9% **8(a)** ______

**(b)** 0.74% **8(b)** ______

**9** **(a)** A young man received a score of 74% on a test with 155 total points. How many points did he receive? (Round off to the nearest integer.)

**(b)** A young woman received 85 points on a different test, which was a score of 74%. How many points were on the test? (Round off to the nearest integer.)

### Answers to Test

**1(a)** 185.54 **1(b)** −5.119 **1(c)** 7.10 **2(a)** 92 **2(b)** −131 **2(c)** 540 in.$^2$ **2(d)** 2540 g **3(a)** 9.0 cm **3(b)** 8.00 in. **4(a)** 82 miles/hr **4(b)** 14.7 g/mL **4(c)** 41 ft$^2$ **5(a)** 3.6 **5(b)** 1.75 **5(c)** 0.0330 atm **5(d)** 0.296 cm$^3$ **6** 0.263 **7(a)** 37.5% **7(b)** 8.75% **8(a)** 0.989 **8(b)** 0.0074 **9(a)** 115 **9(b)** 115

## B A Review of Basic Mathematics

The following is a quick (very quick) refresher of fundamentals of math. This may be sufficient to aid you if you are just a little rusty on some of the basic concepts. For more thorough explanations and practice, however, you are urged to use a more comprehensive math review workbook or consult with your instructor.

### 1 Addition and Subtraction

Since most calculations in this text use numbers expressed in decimal form, we will emphasize the manipulation of these types of numbers. In addition and subtraction, it is important to line up the decimal point carefully before doing the math.

**Example A-1**
Add the following numbers: 16.75, 13.31, and 175.67.

$$\begin{array}{r} \downarrow \\ 16.75 \\ 13.31 \\ \underline{175.67} \\ \underline{\underline{205.73}} \end{array}$$

Subtraction is simply the addition of a negative number. Remember that subtraction of a negative number changes the sign to a plus (two negatives make a positive). For example, $4 - 7 = -3$, but $4 - (-7) = 4 + 7 = 11$.

**Example A-2**
Carry out the following calculations.
(a) $11.8 + 13.1 - 6.1$

$$\begin{array}{r} 11.8 \\ +13.1 \\ \hline 24.9 \end{array} \qquad \begin{array}{r} 24.9 \\ -6.1 \\ \hline 18.8 \\ \hline\hline \end{array}$$

(b) $47.82 - 111.18 - (-12.17)$
This is the same as $47.82 - 111.18 + 12.17$.

$$\begin{array}{r} 47.82 \\ +12.17 \\ \hline 59.99 \end{array} \qquad \begin{array}{r} -111.18 \\ +59.99 \\ \hline -51.19 \\ \hline\hline \end{array}$$

### Exercises

**A-1** Carry out the following calculations.

**(a)** $47 + 1672$
**(b)** $11.15 + 190.25$
**(c)** $114 + 26 - 37$
**(d)** $-97 + 16 - 118$
**(e)** $0.897 + 1.310 - 0.063$
**(f)** $-0.377 - (-0.101) + 0.975$
**(g)** $17.489 - 318.112 - (-0.315) + (-3.330)$

Answers: **(a)** 1719 **(b)** 201.40 **(c)** 103 **(d)** −199 **(e)** 2.144 **(f)** 0.699 **(g)** −303.638

## 2 Multiplication

Multiplication is expressed in various ways as follows:

$$13.7 \times 115.35 = 13.7 \cdot 115.35 = (13.7)(115.35) = 13.7(115.35)$$

If it is necessary to carry out the multiplication in longhand, you must be careful to place the decimal point correctly in the answer. Count the *total* number of digits to the right of the decimal point in both multipliers (three in this example). The answer has that number of digits to the right of the decimal point in the answer.

$$13.\underset{1}{7} \times 2.\underbrace{15}_{2} = 29.\underbrace{455}_{3} = 29.5^{*}$$

(1 + 2 = 3)

*Rounded off to three significant figures. See Section 2-1.

When a number (called a *base*) is multiplied times itself two or more times it is said to be raised to a *power*. The power (called the *exponent*) indicates the number of bases multiplied. For example, the exact values of the following numbers raised to a power are:

$4^2 = 4 \times 4 = 16$ (four "squared")

$2^4 = 2 \times 2 \times 2 \times 2 = 16$ (two to the fourth power)

$4^3 = 4 \times 4 \times 4 = 64$ (four "cubed")

$(14.1)^2 = 14.1 \times 14.1 = 198.81$

In the calculations used in this book, most numbers have specific units. In multiplication, the units are multiplied as well as the numbers. For example,

$$3.7 \text{ cm} \times 4.61 \text{ cm} = 17 \text{ (cm} \times \text{cm)} = 17 \text{ cm}^2$$

$$(4.5 \text{ in.})^3 = 91 \text{ in.}^3$$

In the multiplication of a series of numbers, grouping is possible:

$$(a \times b) \times c = a \times (b \times c)$$

$$3.0 \text{ cm} \times 148 \text{ cm} \times 3.0 \text{ cm} = (3.0 \times 3.0) \times 148 \times (\text{cm} \times \text{cm} \times \text{cm}) = \underline{\underline{1300 \text{ cm}^3}}$$

When multiplying signs, remember:

$$+ \times - = - \qquad + \times + = + \qquad - \times - = +$$

For example, $(-3) \times 2 = -6$; $(-9) \times (-8) = +72$.

## Exercises

**A-2** Carry out the following calculations. For (a) through (d) carry out the multiplication completely. For (e) through (h) round off the answer to the proper number of significant figures and include units.

**(a)** $16.2 \times (-118)$

**(b)** $(4 \times 2) \times 879$

**(c)** $(-8) \times (-2) \times (-37)$

**(d)** $(-47.8) \times (-9.6)$

**(e)** $3.0 \text{ ft} \times 18 \text{ lb}$

**(f)** $17.7 \text{ in.} \times (13.2 \text{ in.} \times 25.0 \text{ in.})$

**(g)** What is the area of a circle where the radius is 2.2 cm? (Area $= \pi r^2$, $\pi = 3.14$.)

**(h)** What is the volume of a cylinder 5.0 in. high with a cross-section radius $= 0.82$ in.? (Volume = area of cross section × height.)

Answers: **(a)** $-1911.6$ **(b)** 7032 **(c)** $-592$ **(d)** 458.88 **(e)** 54 ft. · lb **(f)** 5840 in.$^3$ **(g)** 15 cm$^2$ **(h)** 11 in.$^3$

### 3 Roots of Numbers

A root of a number is a fractional exponent. It is expressed as

$$\sqrt[x]{a} = a^{1/x}$$

If $x$ is not shown (on the left), it is assumed to be 2 and is known as the *square root*. The square root is the number that when multiplied times itself gives the base $a$. For example,

$$\sqrt{4} = 2 \qquad (2 \times 2 = 4)$$

$$\sqrt{9} = 3 \qquad (3 \times 3 = 9)$$

The square root of a number may have either a positive or a negative sign. Generally, however, we are interested only in the positive root in chemistry calculations.

The square roots of numbers that are not even numbers may be computed on a calculator or found in a table. Without these available, an educated approximation can come close to the answer.

**Example A-3**
What is the square root of 54.0 in.$^2$?

**Solution**
We know that

$$\sqrt{49 \text{ in.}^2} = 7.0 \text{ in.}$$

and

$$\sqrt{64 \text{ in.}^2} = 8.0 \text{ in.}$$

(Notice that the square roots of units are also expressed.) You can see that the answer is between 7 in. and 8 in. Try 7.5 in.

$$7.5 \text{ in} \times 7.5 \text{ in.} = 56.2 \text{ in.}^2$$

This is close but a little high. Try 7.4 in.

$$7.4 \text{ in.} \times 7.4 \text{ in.} = 54.8 \text{ in.}^2$$

This is still a little high. Try 7.3 in.

$$7.3 \text{ in.} \times 7.3 \text{ in.} = 53.3 \text{ in.}^2$$

This is a little low, but an approximation of $\underline{\underline{7.35}}$ would be on the money.

The cube root of a number is expressed as

$$\sqrt[3]{b} = b^{1/3}$$

It is the number multiplied by itself three times that gives $b$. For example,

$$\sqrt[3]{27} = 3.0 \qquad (3 \times 3 \times 3 = 27)$$

$$\sqrt[3]{64} = 4.0 \qquad (4 \times 4 \times 4 = 64)$$

**Exercises**

**A-3** Find the following roots. If necessary, approximate the answer and check on a calculator or with a table.

**(a)** $\sqrt{25}$ **(b)** $\sqrt{36\ \text{cm}^2}$ **(c)** $\sqrt{144\ \text{ft}^4}$ **(d)** $\sqrt{40}$

**(e)** $\sqrt{7.0}$ **(f)** $110^{1/2}$ **(g)** $100^{1/3}$ **(h)** $\sqrt[3]{50}$

**(i)** What is the radius of a circle that has an area of 150 $\text{ft}^2$.

**(j)** What is the radius of the cross section of a cylinder if it has a volume of 320 $\text{m}^3$ and a height of 6.0 m?

Answers: **(a)** 5.0 **(b)** 6.0 cm **(c)** 12.0 $\text{ft}^2$ **(d)** 6.3 **(e)** 2.6 **(f)** 10.5 **(g)** 4.64 **(h)** 3.7 **(i)** 6.9 ft **(j)** 4.1 m

### 4 Division, Fractions, and Decimal Fractions

Division problems are expressed in fraction form as follows:

$$\underbrace{88.8 \div 2.44}_{\text{Division form}} = \underbrace{\frac{88.8}{2.44} = 88.8/2.44}_{\text{Common fraction form}} = \underbrace{36.4}_{\text{Decimal fraction form}}$$

In most cases in chemistry, answers are expected in decimal fraction form rather than common fraction form. Therefore, to obtain the answer, the *numerator* (the number on the top) is divided by the *denominator* (the number on the bottom). Before doing the actual calculation, it helps to have a feeling for how the fractional number should look in decimal form. If the numerator is larger than the denominator, the decimal number is greater than 1. If the opposite is true, the decimal number is less than 1.

To carry out the division longhand, it is easier to divide a whole number in the denominator into the numerator. To do this, move the decimal in both numerator and denominator the same number of places. In effect, you are multiplying both numerator and denominator by the same number, which does not change the value of the fraction.

$$\frac{a}{b} = \frac{a \times c}{b \times c}$$

$$\frac{88.8}{2.44} = \frac{88.8 \times 100}{2.44 \times 100} = \frac{8880}{244} = \underline{\underline{36.4}}$$

Many divisions can be simplified by "cancellation." Cancellation is the elimination of common multipliers in the numerator and denominator. This is possible because a number divided by itself is equal to unity (e.g., 25/25 = 1). As in multiplication, all units must also be divided. If identical units appear in both numerator and denominator, they can also be canceled.

$$\frac{a \times \cancel{c}}{b \times \cancel{c}} = \frac{a}{b}$$

$$\frac{\overset{1}{\cancel{190}} \times 4 \cancel{\text{torr}}}{\underset{1}{\cancel{190}} \cancel{\text{torr}}} = \underline{\underline{4}}$$

$$\frac{2500 \text{ cm}^3}{150 \text{ cm}} = \frac{\overset{1}{\cancel{50}} \times 50 \cancel{\text{cm}} \times \text{cm} \times \text{cm}}{\underset{1}{\cancel{50}} \times 3 \cancel{\text{cm}}} = \frac{50 \text{ cm}^2}{3} = \underline{\underline{17 \text{ cm}^2}}$$

$$\frac{2800 \text{ miles}}{45 \text{ hr}} = \frac{\cancel{5} \times 560 \text{ miles}}{\cancel{5} \times 9 \text{ hr}} = \frac{62 \text{ miles}}{1 \text{ hr}} = \underline{\underline{62 \text{ miles/hr}}}$$

This is read as 62 miles "per" one hour or simply 62 miles per hour. The word "per" implies a fraction or a ratio with the unit after "per" in the denominator. If a number is not written or read in the denominator with a unit, it is assumed that the number is unity and is known to as many significant figures as the number in the numerator (i.e., 62 miles per 1.0 hr).

### Exercises

**A-4** Express the following in decimal fraction form.

**(a)** 892 miles ÷ 41 hr

**(b)** 982.6 ÷ 0.250

**(c)** 195 ÷ 2650

**(d)** $\frac{67.5 \text{ g}}{15.2 \text{ mL}}$

**(e)** $\frac{1890 \text{ cm}^3}{66 \text{ cm}}$

**(f)** $\frac{146 \text{ ft} \cdot \text{hr}}{0.68 \text{ ft}}$

**(g)** $\frac{0.8772 \text{ ft}^3}{0.0023 \text{ ft}^2}$

**(h)** $\frac{37.50 \text{ ft}}{0.455 \text{ sec}}$

Answers: **(a)** 22 miles /hr **(b)** 3930 **(c)** 0.0736 **(d)** 4.44 g/mL **(e)** 29 cm$^2$ **(f)** 210 hr **(g)** 380 ft **(h)** 82.4 ft/sec

## 5 Multiplication and Division of Fractions

When two or more fractions are multiplied, all numbers *and units* in both numerator and denominator can be combined into one fraction.

Express the following answers to two significant figures.

**Example A-4**

$$\frac{3}{5} \times \frac{75}{4} \times \frac{16}{7} = \frac{3 \times 75 \times 16}{5 \times 4 \times 7} = \frac{3 \times \overset{15}{\cancel{75}} \times \overset{4}{\cancel{16}}}{\underset{1}{\cancel{5}} \times \underset{1}{\cancel{4}} \times 7} = \frac{180}{7} = \underline{\underline{26}}$$

**Example A-5**

$$\frac{42 \text{ miles}}{\text{hr}} \times \frac{3}{7}\text{hr} \times \frac{5280 \text{ ft}}{\text{miles}} = \frac{\overset{6}{\cancel{42}} \times 3 \times 5280 \cancel{\text{miles}} \times \cancel{\text{hr}} \times \text{ft}}{\underset{1}{\cancel{7}} \cancel{\text{hr}} \times \cancel{\text{miles}}}$$

$$= \underline{\underline{95{,}000 \text{ ft}}}$$

**Example A-6**

$$\frac{3}{4}\text{mol} \times \frac{0.75 \text{ g}}{\text{mol}} \times \frac{1 \text{ mL}}{19.3 \text{ g}} = \frac{3 \times 0.75 \times 1 \times \cancel{\text{mol}} \times \cancel{\text{g}} \times \text{mL}}{4 \times 1 \times 19.3 \times \cancel{\text{mol}} \times \cancel{\text{g}}}$$

$$= \underline{\underline{0.029 \text{ mL}}}$$

The division of one fraction by another is the same as the multiplication of the numerator by the *reciprocal* of the denominator. The reciprocal of a fraction is simply the fraction in an inverted form (e.g., $\frac{3}{5}$ is the reciprocal of $\frac{5}{3}$).

$$\frac{a}{\frac{b}{c}} = a \times \frac{c}{b} \qquad \frac{\frac{a}{b}}{\frac{c}{d}} = \frac{a}{b} \times \frac{d}{c} = \frac{a \times d}{b \times c}$$

**Example A-7**

$$\frac{1650}{\frac{3}{5}} = 1650 \times \frac{5}{3} = \underline{\underline{2750}}$$

**Example A-8**

$$\frac{145 \text{ g}}{\frac{7.50 \text{ g}}{\text{mL}}} = 145 \not{g} \times \frac{1 \text{ mL}}{7.5 \not{g}} = \underline{\underline{19.3 \text{ mL}}}$$

## Exercises

**A-5** Express the following in decimal fraction form. If units are not used, round off the answer to three significant figures. If units are included, round off to the proper number of significant figures and include units in the answer.

**(a)** $\frac{3}{8} \times \frac{4}{7} \times \frac{21}{20}$

**(b)** $\frac{250}{273} \times \frac{175}{300} \times (-6)$

**(c)** $\frac{4}{9} \times (-\frac{5}{8}) \times (-\frac{3}{4})$

**(d)** 195 g/mL × 47.5 mL

**(e)** 0.75 mol × 17.3 g/mol

**(f)** $(3.57 \text{ in.})^2 \times 0.85 \text{ in.} \times \frac{16.4 \text{ cm}^3}{\text{in.}^3}$

**(g)** $\frac{\frac{150}{350}}{\frac{25}{42}}$

**(h)** $\frac{(-\frac{3}{7})}{(-\frac{4}{9})}$

**(i)** $\frac{(-\frac{17}{3})}{\frac{8}{9}}$

**(j)** $\frac{\frac{16}{9} \times \frac{10}{14}}{\frac{5}{6}}$

**(k)** $\frac{75.2 \text{ torr}}{760 \text{ torr/atm}}$

**(l)** $\frac{(55.0 \text{ miles/hr}) \times (5280 \text{ ft/mile}) \times (1 \text{ hr}/60 \text{ min})}{60 \text{ sec/min}}$

**(m)** $\frac{305 \text{ K} \times (62.4 \text{ L} \cdot \text{torr/K} \cdot \text{mol}) \times 0.25 \text{ mol}}{650 \text{ torr}}$

Answers: **(a)** 0.225 **(b)** −3.21 **(c)** 0.208 **(d)** 9260 g **(e)** 13 g **(f)** 180 $\text{cm}^3$ **(g)** 0.720 **(h)** 0.964 **(i)** −6.38 **(j)** 1.52 **(k)** 0.989 atm **(l)** 80.7 ft/sec **(m)** 7.3 L

## 6 Fractions, Percent, and the Use of Percent

In the examples of the fractions thus far we have seen that the units of the numerator can be profoundly different from those of the denominator (e.g., miles/hr, g/mL, etc.). In other problems in chemistry we use fractions to express a component part in the numerator to the total in the denominator. In most cases such fractions are expressed without units and in decimal fraction form.

**Example A-9**

A box of nails contains 985 nails; 415 of these are 6-in. nails, 375 are 3-in. nails, and the rest are roofing nails. What is the fraction of roofing nails?

**Solution**

Roofing nails = total − others = 985 − (415 + 375) = 195

$$\frac{\text{Component}}{\text{Total}} = \frac{195}{375 + 415 + 195} = \underline{\underline{0.198}}$$

**Example A-10**

A mixture contains 4.25 mol of $N_2$, 2.76 mol of $O_2$, and 1.75 mol of $CO_2$. What is the fraction of moles of $O_2$ present in the mixture? (This fraction is known, not surprisingly, as "the mole fraction." The mole is a unit of quantity, like dozen.)

**Solution**

$$\frac{\text{Component}}{\text{Total}} = \frac{2.76}{4.25 + 2.76 + 1.75} = \underline{\underline{0.315}}$$

### Exercises

**A-6** A grocer has 195 dozen boxes of fruit; 74 dozen boxes are apples, 62 dozen boxes are peaches, and the rest are oranges. What is the fraction of the boxes that are oranges?

**A-7** A mixture contains 9.85 mol of gas. A 3.18-mol quantity of the gas is $N_2$, 4.69 mol is $O_2$, and the rest is He. What is the mole fraction of He in the mixture?

**A-8** The total pressure of a mixture of two gases, $N_2$ and $O_2$, is 0.72 atm. The pressure due to $O_2$ is 0.41 atm. What is the fraction of the pressure due to $O_2$?

Answers: **A-6** 0.30 **A-7** 0.201 **A-8** 0.57

The decimal fractions that have just been discussed are frequently expressed as percentages. Percent simply means parts "per" 100. Percent is obtained by multiplying a fraction in decimal form by 100%.

**Example A-11**
If 57 out of 180 people at a party are women, what is the percent women?

**Solution**
The fraction of women in decimal form is

$$\frac{57}{180} = 0.317$$

The percent women is

$$0.317 \times 100\% = 31.7\% \text{ women}$$

The general method used to obtain percent is

$$\frac{\text{Component}}{\text{Total}} \times 100\% = \underline{\qquad\qquad}\% \text{ of component}$$

To change from percent back to a decimal fraction, divide the percent by 100%, which moves the decimal to the left two places.

$$86.2\% = \frac{86.2\%}{100\%} = 0.862 \qquad \text{(fraction in decimal form)}$$

## Exercises

**A-9** Express the following fractions as percent: $\frac{1}{4}$, $\frac{3}{8}$, $\frac{9}{8}$, $\frac{55}{25}$, 0.67, 0.13, 1.75, 0.098.

**A-10** A bushel holds 198 apples, of which 27 are green. What is the percent of green apples?

**A-11** A basket contains 75 pears, 8 apples, 15 oranges, and 51 grapefruit. What is the percent of each?

Answers: **A-9** $\frac{1}{4}$ = 25%, $\frac{3}{8}$ = 37.5%, $\frac{9}{8}$ = 112.5%, $\frac{55}{25}$ = 220%, 0.67 = 67%, 0.13 = 13%, 1.75 = 175%, 0.098 = 9.8% **A-10** 13.6 %, **A-11** 50.3% pears, 5.4% apples, 10.1% oranges, and 34.2% grapefruit

We have seen how the percent is calculated from the total and the component part. We now wish to use percent as a conversion factor to find either the component part as in Example A-12 or the total as in Example

A-13. In so doing, we recall that percent is an expression of a fraction "per 100."

**Example A-12**

A certain crowd at a rock concert was composed of about 87% teenagers. If the crowd totaled 586 people, how many were teenagers?

**Procedure**

The percent can be expressed as a fraction:

$$\frac{87 \text{ teenagers}}{100 \text{ people}}$$

If this fraction is then multiplied by the number of people, the result is the component part or the number of teenagers.

$$586 \cancel{\text{people}} \times \frac{87 \text{ teenagers}}{100 \cancel{\text{people}}} = (586 \times 0.87) = \underline{\underline{510 \text{ teenagers}}}$$

**Example A-13**

A professional baseball player got a hit 28.7% of the times he batted. If he got 246 hits, how many times did he bat?

**Procedure**

The percent can be written in fraction form and then inverted. It thus relates the total at bats to the number of hits:

$$\frac{100 \text{ at bats}}{28.7 \text{ hits}}$$

If this is now multiplied by the number of hits, the result is the total number of at bats.

$$246 \cancel{\text{hits}} \times \frac{100 \text{ at bats}}{28.7 \cancel{\text{hits}}} = \left(\frac{246}{0.287}\right) = \underline{\underline{857 \text{ at bats}}}$$

### Exercises

**A-12** In a certain audience, 45.9% were men. If there were 196 people in the audience, how many women were present?

**A-13** In the alcohol molecule, 34.8% of the weight is due to oxygen. What is the weight of oxygen in 497 g of alcohol?

**A-14** The cost of a hamburger in 1985 is 216% of the cost in 1970. If hamburgers cost $0.75 each in 1970, what do they cost in 1985?

**A-15** In a certain audience, 46.0% are men. If there are 195 men in the audience, how large is the audience?

**A-16** If a solution is 23.3% by weight HCl and it contains 14.8 g of HCl, what is the total weight of the solution?

**A-17** An unstable isotope weighs 131 amu. This is 104% of the weight of a stable isotope. What is the weight of the stable isotope?

Answers: **A-12** 106 women **A-13** 173 g **A-14** $1.62 **A-15** 424 people **A-16** 63.5 g **A-17** 126 amu

# Appendix B
# Basic Algebra

This section reviews the following basic algebra skills.

1 Operations on algebra equations

2 Word problems and algebra equations

3 Direct and inverse proportionalities

## A Test of Basic Algebra

Following is a test on some of the basic algebra skills that are necessary in chemistry. After you take the test, compare your answers to those listed at the end of the test. If you miss more than one question in any one section, you should work through the review following the test. In any case, you will be referred to this review at specific points in the text when certain calculations are first encountered.

### Algebra Test

**Operations on Algebra Equations**

**1** Solve the following equations for $x$. (Isolate all numbers and variables other than $x$ on the right-hand side of the equation.)

**(a)** $4x + 17 = 73$ **1(a)** ______

**(b)** $\frac{y}{3x} + 4 = R$ **1(b)** ______

**(c)** $3x^2 + y = 225 - 2x^2$ **1(c)** ______

**(d)** $\frac{1}{2x + 1} = y$ **1(d)** ______

**Word Problems and Algebra Equations**

**2** Write an algebra equation that expresses each of the following.

**(a)** A number $x$ is six more than $y$.

**(b)** Three times a number $R$ is equal to eight less than $Z$.

**(c)** A number $b$ is 36 less than five times the square of $d$.

**(d)** Four times the product of two numbers $s$ and $t$ is 14% of the sum of two other numbers $p$ and $q$.

**3** Write algebra equations that represent the following (let $x$ = an unknown) and then solve for $x$.

**(a)** The sum of the lengths of two wooden planks is 24 ft. If one plank is three times the length of the other, what is the length of the smaller plank?
3(a) Equation ______ Answer ______

**(b)** Two students received 65% and 80% on a chemistry quiz. The sum of their scores was 180 points. How many points were on the test?
3(b) Equation ______ Answer ______

**(c)** A large brick weighs 1 lb more than twice as much as a smaller brick. The sum of the weights of the two bricks is 28 lb. What is the weight of each brick?
3(c) Equation ______ Answer ______

**(d)** There was $9.87 in a cash register, with one more than twice as many dimes as quarters and three less than four times as many pennies as quarters. How many quarters were in the cash register?
3(d) Equation ______ Answer ______

**4** Answer the following:

**(a)** A small car weighs 100 lb more than two-thirds of the weight of a midsize car. The difference in weight between the two cars is 800 lb. What is the weight of the small car?

**4(a)** ______

**(b)** A major leaguer during his career got 32 home runs, 25 triples, 55 doubles, and the rest singles. The fraction of his hits that were doubles was 0.119. How many singles did he get?

**4(b)** ______

**(c)** An oil refinery held 175 barrels of oil. When refined, each barrel yields 24 gallons of gasoline. If 3120 gallons of gasoline were produced, what percentage of the original barrels of oil were refined?

**4(c)** ______

## Direct and Inverse Proportionalities

**5** Write equations for the following. Use $k$ for a constant of proportionality.

**(a)** A quantity $x$ is directly proportional to $y$.

**5(a)** ______

**(b)** A quantity $z$ is inversely proportional to the square root of $r$.

**5(b)** ______

**(c)** A quantity $q$ is directly proportional to $A$ and inversely proportional to the square of the sum of $B$ and $C$.

**5(c)** ______

**(d)** The cube root of a quantity $t$ is directly proportional to the square of $v$ and inversely proportional to six less than the quantity $w$.

**5(d)** ______

### Answers to Test

**1(a)** $x = 14$ **1(b)** $x = y/(3R - 12)$ **1(c)** $x = \pm\sqrt{45 - y/5}$ **1(d)** $x = (1 - y)/2y$ **2(a)** $x = y + 6$ **2(b)** $3R = Z - 8$ **2(c)** $b = 5d^2 - 36$ **2(d)** $4st = 0.14(p + q)$ **3(a)** $x + 3x = 24$, $x = 6$ ft **3(b)** $0.65x + 0.80x = 180$, $x = 124$ points **3(c)** $x + (2x + 1) = 28$, $x = 9$ lb; 19 lb **3(d)** $0.25x + 0.10(2x + 1) + 0.01(4x - 3) = 9.87$, $x = 20$ quarters **4(a)** 1900 lb **4(b)** 350 **4(c)** 74.3% **5(a)** $x = ky$ **5(b)** $x = k/\sqrt{r}$ **5(c)** $q = kA/(B + C)^2$ **5(d)** $\sqrt[3]{t} = kv^2/(w - 6)$

## B A Review of Basic Algebra

### 1 Operations of Algebra Equations

Many of the problems of chemistry require the use of basic algebra.

Let's examine the following simple algebra equation:

$$a = (b + c)$$

In any algebraic equation the equality remains valid when *identical* operations are performed on both sides of the equation. The following operations illustrate this principle.

1 A quantity $d$ may be added to or subtracted from both sides of the equation.

$$a + d = (b + c) + d$$

$$a - d = (b + c) - d$$

2 Both sides of the equation may be multiplied or divided by the same quantity $d$.

$$a \times d = (b + c) \times d$$

$$\frac{a}{d} = \frac{(b + c)}{d}$$

3 Both sides of the equation may be raised to a power or a root of both sides of an equation may be taken.

$$a^2 = (b + c)^2$$

$$\sqrt{a} = \sqrt{b + c}$$

**4** Both sides of an equation can be inverted

$$\frac{1}{a} = \frac{1}{(b + c)}$$

In addition to operations on both sides of an equation, there are two other points to recall.

**1** As in any fraction, identical multipliers in the numerator and the denominator in an algebraic equation may be cancelled.

$$\frac{a \times \cancel{d}}{\cancel{d}} = a = (b + c)$$

**2** Quantities equal to the same quantity are equal to each other. Thus substitutions for equalities may be made in algebraic equations.

$$a = b + c$$

$$a = 2d + r$$

Therefore, since $a = a$,

$$b + c = 2d + r$$

We can use these basic rules to solve algebraic equations. Usually, we need to isolate one variable on the left-hand side of the equation with all other numbers and variables on the right-hand side of the equation. The following examples illustrate the isolation of a variable on the left-hand side of the equation.

**Example B-1**
Solve for $x$ in $x + 5 = 92$.

**Solution**
Subtract 5 from both sides of the equation.

$$x + 5 \underline{- 5} = 92 \underline{- 5}$$

$$\underline{\underline{x = 87}}$$

In practice, a number or a variable may be moved to the other side of an equation with a change of sign. For example,

$$x \underline{+ 5} = y$$

$$x = y \underline{- 5}$$

**Example B-2**
Solve for $x$ in $x + y + 8 = z + 6$.

**Solution**

$$\begin{aligned} x &= z + 6 \underline{- y - 8} \\ &= z - y + 6 - 8 \\ &= \underline{\underline{z - y - 2}} \end{aligned}$$

**Example B-3**
Solve for $x$ in

$$\frac{x + 8}{y} = z$$

**Solution**
To move the $y$ to the right, multiply both sides of the equation by $y$, This cancels the $y$ on the left and leaves $y$ in the numerator on the right.

$$y \cdot \frac{x + 8}{y} = z \cdot y$$

$$x + 8 = zy$$

$$\underline{\underline{x = zy - 8}}$$

In practice, to move a denominator to the other side of the equation, multiply both sides of the equation by the denominator. To move a numerator, divide both sides by the numerator. For example,

$$\frac{z}{x} = y$$

$$\frac{1}{z} \cdot \frac{z}{x} = y \cdot \frac{1}{z}$$

$$\frac{1}{x} = \frac{y}{z}$$

In these problems, the numerical value of $x$ can be found if values for the other variables are known. For example, if $y = -2$ and $z = 8$ in Example B-3,

$$\begin{aligned} x &= zy - 8 \\ &= (-2)(8) - 8 = -16 - 8 = -24 \end{aligned}$$

**Example B-4**
Solve for $x$ in

$$\frac{4x + 2}{3 + x} = 7$$

**Solution**
Clear of fractions:

$$4x + 2 = 7(3 + x) = 21 + 7x$$

Move $7x$ to the left and 2 to the right:

$$4x - 7x = 21 - 2$$

$$-3x = 19$$

Divide both sides by $-3$:

$$\frac{-3x}{-3} = \frac{19}{-3}$$

$$x = -\frac{19}{3}$$

**Example B-5**
Solve for $T_2$ in

$$\frac{P_1V_1}{T_1} = \frac{P_2V_2}{T_2}$$

**Solution**
Invert both sides of the equation and switch sides (so that $T_2$ is on the left).

$$\frac{T_2}{P_2V_2} = \frac{T_1}{P_1V_1}$$

Multiply both sides by $P_2V_2$ to move $P_2V_2$ to the right-hand side of the equation.

$$\cancel{P_2V_2} \times \frac{T_2}{\cancel{P_2V_2}} = \frac{T_1 \cdot P_2V_2}{P_1V_1}$$

$$T_2 = \frac{T_1P_2V_2}{P_1V_1}$$

**Example B-6**
Solve the following equation for $y$:

$$\frac{2y}{3} + x = 9z + 4$$

**Solution**
Multiply both sides of the equation by 3.

$$2y + 3x = 27z + 12$$

$$2y = 27z + 12 - 3x$$

$$y = \frac{27z + 12 - 3x}{2}$$

**Example B-7**
Solve the following equation for $x$:

$$x^2 - 49 = y$$

$$x^2 = y + 49$$

$$x = \pm \sqrt{y + 49}$$

## Exercises

**B-1** Solve for $x$ in $17x = y - 87$.

**B-2** Solve for $x$ in

$$\frac{y}{x} + 8 = z + 16$$

**B-3** Solve for $T$ in $PV = (\text{wt}/MM)RT$.

**B-4** Solve for $x$ in

$$\frac{7x - 3}{6 + 2x} = 3r$$

**B-5** Solve for $x$ in $18x - 27 = 2x + 4y - 35$. If $y = 3x$, what is the value of $x$?

**B-6** Solve for $x$ in

$$\frac{x}{4y} + 18 = y + 2$$

**B-7** Solve for $x$ in $5x^2 + 12 = x^2 + 37$.

**B-8** Solve for $r$ in

$$\frac{80}{2r} + \frac{y}{r} = 11$$

What is the value of $r$ if $y = 14$?

Answers: **B-1** $x = (y - 87)/17$ **B-2** $x = y/(8 + z)$ **B-3** $T = PV \cdot MM/\text{wt} \cdot R$ **B-4** $x = 3(6r + 1)/(7 - 6r)$ **B-5** $x = (y - 2)/4$. When $y = 3x$, $x = -2$. **B-6** $x = 4y(y - 16)$ **B-7** $x = \pm 2.5$ **B-8** $r = (40 + y)/11$. When $y = 14$, $r = \frac{54}{11}$.

## 2 Word Problems

Eventually, a necessary skill in chemistry is the ability to translate word problems into algebra equations and then solve. The key is to assign a variable (usually $x$) to be equal to a certain quantity and then to treat the variable consistently throughout the equation. Again, examples are the best way to illustrate the problems.

### Example B-8

Translate each of the following to an equation.

(a) A number $x$ is 4 larger than another number $y$.

$$\underline{\underline{x = y + 4}}$$

(b) A number $z$ is three-fourths of $u$.

$$\underline{\underline{z = \tfrac{3}{4}u}}$$

(c) The square of a number $r$ is 16.9% of $w$.

$$\underline{\underline{r^2 = 0.169w}} \qquad \text{(Change percent to a decimal fraction)}$$

(d) A number $t$ is equal to 12 plus the square root of $q$.

$$\underline{\underline{t = 12 + \sqrt{q}}}$$

### Exercises

Write algebraic equations for the following.

**B-9** A number $n$ is 85 smaller than $m$.

**B-10** A number $y$ is one-fourth of $z$.

**B-11** Fifteen percent of a number $k$ is equal to the square of another number $d$.

**B-12** A number $x$ is 14 more than the square root of $v$.

**B-13** Four times the sum of two numbers, $q$ and $w$, is equal to 68.

**B-14** Five times the product of two variables, $s$ and $t$, is equal to 16 less than the square of $s$.

**B-15** Five-ninths of a number $C$ is equal to 32 less than a number $F$.

Answers: **B-9** $n = m - 85$ **B-10** $y = z/4$ **B-11** $0.15k = d^2$
**B-12** $x = \sqrt{v} + 14$ **B-13** $4(q + w) = 68$ **B-14** $5st = s^2 - 16$
**B-15** $\frac{5}{9}C = F - 32$

We now move from the abstract to the real. In the following examples it is necessary to translate the problem into an algebraic expression as in the previous examples. There are two types of examples that we will use. The first you will certainly recognize, but the second type may be unfamiliar, especially if you have just begun the study of chemistry. However, it is *not* important that you understand the units of the chemistry problems at this time. What *is* important is for you to notice that the problems are worked in the same manner regardless of the units.

**Example B-9**
John is two years more than twice as old as Mary. The sum of their ages is 86. How old is each?

**Solution**
Let $x$ = age of Mary. Then $2x + 2$ = age of John.

$$x + (2x + 2) = 86$$

$$3x = 84$$

$$x = \underline{\underline{28}} \quad \text{(age of Mary)}$$

$$[2(28) + 2] = \underline{\underline{58}} \quad \text{(age of John)}$$

**Example B-10**
One mole of $SF_6$ weighs 30.0 g less than four times the weight of 1 mol of $CO_2$. The weight of 1 mol of $SF_6$ plus the weight of 1 mol of $CO_2$ is 190 g. What is the weight of 1 mol of each?

**Solution**
Let $x$ = weight of 1 mol of $CO_2$. Then $4x - 30$ = weight of 1 mol of $SF_6$.

$$x + (4x - 30) = 190$$

$$x = \underline{\underline{44 \text{ g}}} \qquad \text{(weight of 1 mol of } CO_2\text{)}$$

$$[4(44) - 30] = \underline{\underline{146 \text{ g}}} \qquad \text{(weight of 1 mol of } SF_6\text{)}$$

**Example B-11**
Two students took the same test, and their percent scores differed by 10%. If there were 200 points on the test and the total of their point scores was 260 points, what was each student's percent score?

**Procedure**
Set up an equation relating each person's percent scores to their total points (260).

Let $x$ = percent score of higher test

Then $x - 10$ = percent score of lower test

The points that each person scores is the percent in fraction form multiplied by the points on the test.

$$\frac{\%\text{ grade}}{100 \text{ points}} \times (\text{points on test}) = \text{points scored}$$

**Solution**

$$\left[\frac{x}{100}(200 \text{ points})\right] + \left[\frac{x - 10}{100}(200 \text{ points})\right] = 260 \text{ points}$$

$$200x + 200x - 2000 = 26{,}000$$

$$400x = 28{,}000$$

$$x = 70$$

Higher score = $\underline{\underline{70\%}}$ lower score = $70 - 10 = \underline{\underline{60\%}}$

**Example B-12**
If a 8.75-g quantity of sugar represents 65.7% of a mixture, what is the weight of the mixture?

**Solution**
Let $x$ = weight of the mixture. Then

$$\frac{65.7}{100}x = 0.657\,x = 8.75$$

$$x = \frac{8.75}{0.657} = \underline{\underline{13.3 \text{ g}}}$$

**Example B-13**

In a certain year 14.5% of a rancher's herd of 876 cows had two calves each. The rest had none.

(a) How many calves were born?
(b) How many cows did not have a calf?
(c) How large was the herd including calves?

**Solution**

Let $x$ = number of cows having calves. Then $2x$ = number of calves born.

$$x = 0.145 \times 876 = 127 \text{ cows having calves}$$

$$2x = 2 \times 127 = \underline{\underline{254 \text{ calves born}}}$$

$$876 - x = 876 - 127 = \underline{\underline{749 \text{ cows not having calves}}}$$

$$876 + 2x = 876 + 354 = \underline{\underline{1130 \text{ total cattle}}}$$

**Example B-14**

This example is directly analogous to the preceding example. A certain compound, $AX_2$, can break apart (dissociate) into one A and two X's. In a 0.250-mol quantity of $AX_2$, 6.75% of the moles dissociated. (A mole, abbreviated mol, is a unit of quantity, like dozen or gross.)

(a) How many moles of $AX_2$ dissociated?
(b) How many moles of $AX_2$ did not dissociate?
(c) How many total moles of particles ($AX_2$, A, and X) are present?

**Solution**

Let $x$ = number of moles of $AX_2$ that dissociate. Then

$$x = 0.0675 \times 0.250 \text{ mol} = \underline{\underline{0.0169 \text{ mol dissociated}}}$$

The number of moles of $AX_2$ that are not dissociated = $0.250 - x$.

$$0.250 - 0.0169 = \underline{\underline{0.233 \text{ mol undissociated}}}$$

Total moles present

$$= \text{moles } AX_2 \text{ (undissociated)} + \text{moles of A} + \text{mols of X}$$

$$\text{Moles } AX_2 = 0.233$$

$$\text{Moles of A} = x = 0.0169$$

$$\text{Moles of X} = 2x = 0.0338$$

$$\text{Total moles} = 0.233 + 0.0169 + 0.0338 = \underline{\underline{0.284 \text{ mol total}}}$$

**Example B-15**

A used car dealer has Fords, Chevrolets, and Plymouths. There are 120 Fords, 152 Chevrolets, and the rest are Plymouths. If the fraction of Fords is 0.310, how many Plymouths are on the lot?

**Solution**

Let $x$ = number of Plymouths.

$$\text{Fraction of Fords} = \frac{\text{number of Fords}}{\text{total number of cars}} = 0.310$$

$$\frac{120}{120 + 152 + x} = 0.310$$

$$120 = 0.310(272 + x)$$

$$120 = 84.3 + 0.310x$$

$$x = \underline{\underline{115 \text{ Plymouths}}}$$

**Example B-16**

There is a 0.605-mol quantity of $N_2$ present in a mixture of $N_2$ and $O_2$. If the mole fraction of $N_2$ is 0.251, how many moles of $O_2$ are present?

**Solution**

Let $x$ = number of moles of $O_2$. Then

$$\text{Mole fraction } N_2 = \frac{\text{mol of } N_2}{\text{total mol present}} = 0.251$$

$$\frac{0.605}{0.605 + x} = 0.251$$

$$0.605 = 0.251(0.605 + x)$$

$$x = \underline{\underline{1.80 \text{ mol}}}$$

## Exercises

In the following exercises, a problem concerning an everyday situation is followed by one or two closely analogous problems concerning a chemistry situation. In both cases the mechanics of the solution are similar. Only the units differ.

**B-16** The total length of two boards is 18.4 ft. If one board is 4.0 ft longer than the other, what is the length of each board?

**B-17** An isotope of iodine weighs 10 amu less than two-thirds of the weight of an isotope of thallium. The total weight of the two isotopes is 340 amu. What is the weight of each isotope?

**B-18** An isotope of gallium weighs 22 amu more than one-fourth the weight of an isotope of osmium. The difference in the two weights is 122 amu. What is the weight of each?

**B-19** An oil refinery held 175 barrels of oil. When refined, each barrel yields 24 gallons of gasoline. If 3120 gallons of gasoline were produced, what percentage of the original barrels of oil were refined?

**B-20** A solution contained 0.856 mol of a substance $A_2X$. In solution some of the $A_2X$'s break up into A's and X's (notice that each mole of $A_2X$ yields 2 mol of A). If 0.224 mol of A are present in the solution, what percentage of the moles of $A_2X$ dissociated (broke apart)?

**B-21** In Las Vegas, a dealer starts with 264 decks of cards. If 42.8% of the decks were used in an evening, how many jacks (four per deck) were used?

**B-22** A solution originally contains a 1.45-mol quantity of a compound $A_3X_2$. If 31.5% of the $A_3X_2$ dissociates (three A's and two X's per $A_3X_2$), how many moles of A are formed? How many moles of X? How many moles of undissociated $A_3X_2$ remain? How many moles of particles (A's, X's, and $A_3X_2$'s) are present in the solution?

**B-23** The fraction of kerosene that can be recovered from a barrel of crude oil is 0.200. After refining a certain amount of oil, 8.9 gal of kerosene, some gasoline, and 18.6 gal of other products were produced. How many gallons of gasoline were produced?

**B-24** The fraction of moles (mole fraction) of gas A in a mixture is 0.261. If the mixture contains 0.375 mol of gas B and 0.175 mol of gas C as well as gas A, how many moles of gas A are present?

Answers: **B-16** 7.2 ft, 11.2 ft **B-17** thallium, 210 amu; iodine, 130 amu **B-18** gallium, 70 amu; osmium, 192 amu **B-19** 74.3% **B-20** 13.1% **B-21** 452 jacks **B-22** 1.37 mol of A, 0.194 mol of X, 0.99 mol of $A_3X_2$, 3.27 mol total **B-23** 17.0 gallons **B-24** 0.195 mol

### 3 Direct and Inverse Proportionalities

There is one other point that should be included in a review on algebra, and that is direct and inverse proportionalities. We will use these terms often in chemistry.

When a quantity is directly proportional to another, it means that an increase in one variable will cause a corresponding increase of the same percent in the other variable. The direct proportionality is shown as

$$A \propto B \qquad (\propto \text{ proportionality symbol})$$

which reads "$A$ is directly proportional to $B$." A proportionality can be easily converted to an algebraic equation by the introduction of a constant (in our examples designated $k$), called a constant of proportionality. Thus the proportion becomes

$$A = kB$$

or, rearranging,

$$\frac{A}{B} = k$$

Notice that $k$ is not a variable but has a certain numerical value that does not change as do $A$ and $B$ under experimental conditions.

A common, direct proportionality that we will study relates Kelvin temperature $T$ and volume $V$ of a gas at constant pressure. This is written as

$$V \propto T \qquad V = kT \qquad \frac{V}{T} = k$$

(This is known as Charles' law.)

In a hypothetical case, $V = 100$ L and $T = 200$ K. From this information we can calculate the value of the constant $k$.

$$\frac{V}{T} = \frac{100 \text{ L}}{200 \text{ K}} = 0.50 \text{ L/K}$$

A change in volume or temperature requires a corresponding change in the other *in the same direction*. For example, if the temperature of the gas is changed to 300 K, we can see that a corresponding change in volume is required from the following calculation:

$$\frac{V}{T} = k$$

$$V = kT = 0.50 \text{ L/}\cancel{\text{K}} \times 300 \cancel{\text{K}} = \underline{\underline{150 \text{ L}}}$$

When a quantity is inversely proportional to another quantity, an increase in one brings about a corresponding *decrease* in the other. An inverse proportionality between $A$ and $B$ is written as

$$A \propto \frac{1}{B}$$

As before, the proportionality can be written as an equality by introduction of a constant (which has a different value than in the example above).

$$A = \frac{k}{B} \quad \text{or} \quad AB = k$$

A common inverse proportionality that we use relates the volume $V$ of a gas to the pressure $P$ at a constant temperature. This is written as

$$V \propto \frac{1}{P} \qquad V = \frac{k}{P} \qquad PV = k$$

(This is known as Boyle's law.)

In a hypothetical case, $V = 100$ L and $P = 1.50$ atm. From this information we can calculate the value of the constant $k$.

$$PV = k = 1.50 \text{ atm} \times 100 \text{ L} = 150 \text{ atm} \cdot \text{L}$$

A change in volume or pressure requires a corresponding change in the other *in the opposite direction*. For example, if the pressure on the gas is changed to 3.00 atm, we can see that a corresponding change in volume is required.

$$PV = k$$

$$V = \frac{k}{P} = \frac{150 \cancel{\text{atm}} \cdot \text{L}}{3.00 \cancel{\text{atm}}} = \underline{50.0 \text{ L}}$$

Quantities can be directly or inversely proportional to the square, square root, or any other function of another variable or number, as illustrated by the following examples.

---

**Example B-17**

A quantity $C$ is directly proportional to the square of $D$. Write an equality for this statement and explain how a change in $D$ affects the value of $C$.

**Solution**

The equation is

$$\underline{\underline{C = kD^2}}$$

Notice that a change in $D$ will have a significant effect on the value of $C$. For example,

If $D = 1$, then $C = k$.

If $D = 2$, then $C = 4k$.

If $D = 3$, then $C = 9k$.

Notice that when the value of $D$ is doubled, the value of $C$ is increased *fourfold*.

---

**Example B-18**

A variable $X$ is directly proportional to the square of the variable $Y$ and inversely proportional to the square of another variable $Z$.

This can be written as two separate equations if it is assumed that $Y$ is constant when $Z$ varies and vice versa.

$$X = k_1Y^2 \qquad (Z \text{ constant})$$

$$X = k_2/Z^2 \qquad (Y \text{ constant})$$

($k_1$ and $k_2$ are different constants)

This relationship can be combined into one equation when both $Y$ and $Z$ are variables:

$$X = \frac{k_3Y^2}{Z^2}$$

$k_3$ is a third constant that is a combination of $k_1$ and $k_2$.

## Exercises

Write equalities for the following relations.

**B-25** $X$ is inversely proportional to $Y + Z$.

**B-26** $[H_3O^+]$ is inversely proportional to $[OH^-]$.

**B-27** $[H_2]$ is directly proportional to the square root of $r$.

**B-28** $B$ is directly proportional to the square of $y$ and the cube of $z$.

**B-29** The pressure $P$ of a gas is directly proportional to the number of moles $n$ and the temperature $T$, and inversely proportional to the volume $V$.

Answers: **B-25** $X = k/(Y + Z)$ **B-26** $[H_3O^+] = k/[OH^-]$ **B-27** $[H_2] = k\sqrt{r}$ **B-28** $B = ky^2z^3$ **B-29** $P = knT/V$

# Appendix C
# Scientific Notation

This section reviews the following scientific notation skills.

1 Expressing numbers in scientific notation

2 Addition and subtraction

3 Multiplication and division

4 Powers and roots

## A Test of Scientific Notation

Following is a short test on the manipulation of numbers expressed in scientific notation. Since you should be comfortable with the use of such numbers in calculations before beginning Chapter 8, you are urged to study the review if you miss more than one problem in any one section. Compare your answers to those that follow the test. In the following problems, zeros to the left of a decimal point but to the right of a digit (e.g., 92$\underline{00}$) where no decimal point is showing are assumed not to be significant figures (see Section 2-1). Although the numbers used in this appendix do not have units (for simplicity), treat the numbers as measurements so that the answer to a problem shows the proper number of significant figures.

**Expressing Numbers in Scientific Notation**

**1** Express the following numbers in scientific notation in standard form [one digit to the left of the decimal point in the coefficient (e.g., $7.39 \times 10^6$)].

**(a)** 0.00657 **1(a)** ______

**(b)** 14,300 **1(b)** ______

**(c)** 0.00030 **1(c)** ______

**(d)** 3,457,000,000 **1(d)** ______

**2** Change the following numbers expressed in scientific notation to standard form.

**(a)** $0.778 \times 10^5$ **2(a)** ______

**(b)** $145 \times 10^{-8}$ **2(b)** ______

(c) $0.00720 \times 10^{-8}$ 2(c) ______

(d) $4330 \times 10^{-10}$ 2(d) ______

**Addition and Subtraction**

**3** Carry out the following calculations. Express your answer to the proper decimal place.

(a) $0.049 + (3.7 \times 10^{-2}) + (76.6 \times 10^{-3})$ 3(a) ______

(b) $(1.62 \times 10^{6}) + (0.78 \times 10^{4}) - (0.025 \times 10^{7})$ 3(b) ______

(c) $(0.363 \times 10^{-6}) + (71.2 \times 10^{-9}) + (619 \times 10^{-12})$ 3(c) ______

(d) $(481 \times 10^{6}) - (0.113 \times 10^{9}) + (3.5 \times 10^{5})$ 3(d) ______

**Multiplication and Division**

**4** Carry out the following calculations. Express your answer in standard form to the proper number of significant figures.

(a) $(4.8 \times 10^{6}) \times (17.2 \times 10^{-2})$ 4(a) ______

(b) $(186 \times 10^{20}) \div (0.044 \times 10^{5})$ 4(b) ______

(c) $[(55.3 \times 10^{5}) \times (1.5 \times 10^{6})] \div (2.2 \times 10^{-5})$ 4(c) ______

(d) $0.000879 \div 8.22 \times 10^{5}$ 4(d) ______

**Powers and Roots**

**5** Carry out the following calculations. Exprses the answer in standard form.

(a) $(5.0 \times 10^{-5})^{2}$ 5(a) ______

(b) $\sqrt{4.9 \times 10^{5}}$ 5(b) ______

(c) $(1.1 \times 10^{4})^{3}$ 5(c) ______

(d) $\sqrt[3]{1.25 \times 10^{-10}}$ 5(d) ______

Answers: **1(a)** $6.57 \times 10^{-3}$ **1(b)** $1.43 \times 10^{4}$ **1(c)** $3.0 \times 10^{-4}$ **1(d)** $3.457 \times 10^{9}$ **2(a)** $7.78 \times 10^{4}$ **2(b)** $1.45 \times 10^{-6}$ **2(c)** $7.20 \times 10^{-11}$ **2(d)** $4.33 \times 10^{-7}$ **3(a)** 0.163 **3(b)** $1.38 \times 10^{6}$ **3(c)** $4.35 \times 10^{-7}$ **3(d)** $3.68 \times 10^{8}$ **4(a)** $8.3 \times 10^{5}$ **4(b)** $4.2 \times 10^{18}$ **4(c)** $3.8 \times 10^{17}$ **4(d)** $1.07 \times 10^{-9}$ **5(a)** $2.5 \times 10^{-9}$ **5(b)** $7.0 \times 10^{2}$ **5(c)** $1.3 \times 10^{12}$ **5(d)** $5.00 \times 10^{-4}$

## B A Review of Scientific Notation

Although this topic was first introduced in Section 2-2 in this text, we will focus on a review of the mathematical manipulation of numbers expressed in scientific notation in this appendix. Specifically, addition, multiplication, division, and taking the roots of numbers expressed in scientific notation are covered.

As mentioned in Chapter 2, scientific notation makes use of powers of 10 to simplify awkward numbers that employ more than two or three zeros that are not significant figures. The exponent of 10 simply indicates how many times we should multiply or divide a number (called the coefficient) by 10 to produce the actual number. For example, $8.9 \times 10^3 = 8.9$ (the coefficient), multiplied by 10 *three* times, or

$$8.9 \times 10 \times 10 \times 10 = 8900$$

Also, $4.7 \times 10^{-3} = 4.7$ (the coefficient), divided by 10 *three* times, or

$$\frac{4.7}{10 \times 10 \times 10} = 0.0047$$

## 1 Expressing Numbers in Standard Scientific Notation

The method for expressing numbers in scientific notation was explained in Section 2-2. However, in order to simplify a number or to express it in the standard form with one digit to the left of the decimal point in the coefficient, it is often necessary to change a number already expressed in scientific notation. Since this is often done in a hurry, a large number of errors may result. Thus, it is worthwhile to practice moving the decimal point of numbers expressed in scientific notation.

**Example C-1**
Change the following numbers to the standard form:
(a) $489 \times 10^4$ (b) $0.00489 \times 10^8$

**Procedure**
All you need to remember is to raise the power of 10 one unit for each place the decimal point is moved to the left and lower the power of 10 one unit for each place that the decimal point is moved to the right in the coefficient.

**Solution**

$$489 \times 10^4 = (4\overset{+2}{\curvearrowleft}8\,9.) \times 10^4 = 4.89 \times 10^{4+2} = \underline{\underline{4.89 \times 10^6}}$$

$$0.00489 \times 10^8 = (0.0\,0\,4\overset{-3}{\curvearrowright}8\,9) \times 10^8 = 4.89 \times 10^{8-3} = \underline{\underline{4.89 \times 10^5}}$$

As an aid to remembering whether you should raise or lower the exponent as you move the decimal point, it is suggested that you write (or at least imagine) the coefficient on a slant. For each place that you move the decimal point *up,* add one to the exponent. For each place that you move the decimal point *down,* subtract one from the exponent. Notice that the exponent moves up or down with the decimal point.

**Example C-2**

Change the following numbers to the standard form in scientific notation.

(a) $4223 \times 10^{-7}$ (b) $0.00076 \times 10^{18}$

**Solution**

$$4223 \times 10^{-7} = [4223] \times 10^{-7} = 4.223 \times 10^{-7+3} = \underline{\underline{4.223 \times 10^{-4}}}$$

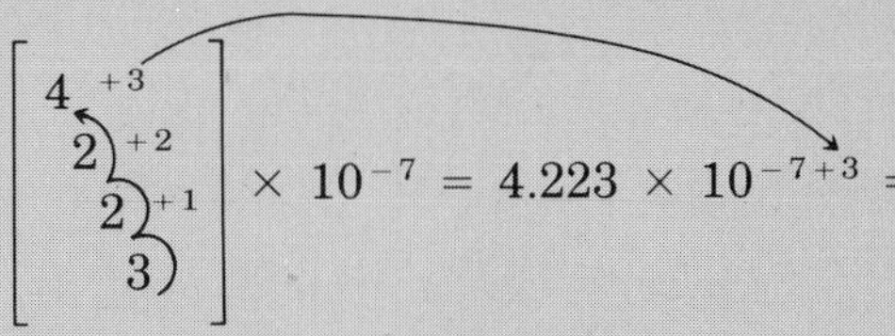

$$0.00076 \times 10^{18} = [0.00076] \times 10^{18} = 7.6 \times 10^{18-4} = \underline{\underline{7.6 \times 10^{14}}}$$

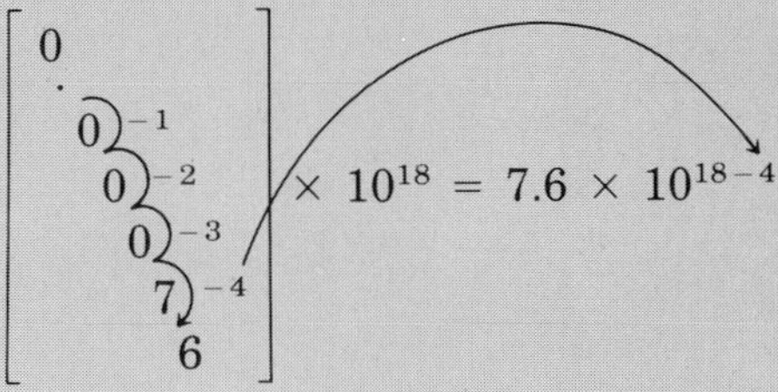

### Exercises

**C-1** Change the following numbers to standard scientific notation with one digit to the left of the decimal point in the coefficient.

**(a)** $787 \times 10^{-6}$ **(d)** $0.0037 \times 10^{9}$

**(b)** $43.8 \times 10^{-1}$ **(e)** $49.3 \times 10^{15}$

**(c)** $0.015 \times 10^{-16}$ **(f)** $6678 \times 10^{-16}$

**C-2** Change the following numbers to a number with two digits to the left of the decimal point in the coefficient.

**(a)** $9554 \times 10^{4}$ **(d)** $116.5 \times 10^{4}$

**(b)** $1.6 \times 10^{-5}$ **(e)** $0.023 \times 10^{-1}$

**(c)** $1 \times 10^{6}$ **(f)** $0.005 \times 10^{23}$

Answers: **C-1(a)** $7.87 \times 10^{-4}$ **C-1(b)** 4.38 **C-1(c)** $1.5 \times 10^{-18}$ **C-1(d)** $3.7 \times 10^{6}$ **C-1(e)** $4.93 \times 10^{16}$ **C-1(f)** $6.678 \times 10^{-13}$ **C-2(a)** $95.54 \times 10^{6}$ **C-2(b)** $16 \times 10^{-6}$ **C-2(c)** $10 \times 10^{5}$ **C-2(d)** $11.65 \times 10^{5}$ **C-2(e)** $23 \times 10^{-4}$ **C-2(f)** $50 \times 10^{19}$

## 2 Addition and Subtraction

Addition or subtraction of numbers in scientific notation can be accomplished only when all coefficients have the same exponent of 10. When all the exponents are the same, the coefficients are added and then multiplied by the power of 10. The correct number of places to the right of the decimal point must be shown as discussed in Section 2-1.

**Example C-3**
Add the following.

$$3.67 \times 10^{-4},\ 4.879 \times 10^{-4},\ \text{and}\ 18.2 \times 10^{-4}$$

**Solution**

$$\begin{array}{l} 3.67 \times 10^{-4} \\ 4.879 \times 10^{-4} \\ \underline{18.2 \times 10^{-4}} \\ 26.749 \times 10^{-4} = \underline{\underline{26.7 \times 10^{-4}}} \end{array}$$

**Example C-4**
Add the following.

$$320.4 \times 10^{3},\ 1.2 \times 10^{5},\ \text{and}\ 0.0615 \times 10^{7}$$

**Solution**
Before adding, change all three numbers to the same exponent of 10.

$$\begin{array}{rcl} 320.4 \times 10^{3} &=& 3.204 \times 10^{5} \\ 1.2 \times 10^{5} &=& 1.2 \times 10^{5} \\ 0.0615 \times 10^{7} &=& \underline{6.15 \times 10^{5}} \\ && 10.554 \times 10^{5} = \underline{\underline{10.6 \times 10^{5}}} = \underline{\underline{1.06 \times 10^{6}}} \end{array}$$

### Exercises

**C-3** Add the following numbers. Express the answer to the proper decimal place.

**(a)** $152 + (8.635 \times 10^{2}) + (0.021 \times 10^{3})$

**(b)** $(10.32 \times 10^{5}) + (1.1 \times 10^{5}) + (0.4 \times 10^{5})$

**(c)** $(1.007 \times 10^{-8}) + (118 \times 10^{-11}) + (0.1141 \times 10^{-6})$

**(d)** $(0.0082) + (2.6 \times 10^{-4}) + (159 \times 10^{-4})$

**C-4** Carry out the following calculations. Express your answer to the proper decimal place.

**(a)** $(18.75 \times 10^{-6}) - (13.8 \times 10^{-8}) + (1.0 \times 10^{-5})$

**(b)** $(1.52 \times 10^{-11}) + (17.7 \times 10^{-12}) - (7.5 \times 10^{-15})$

**(c)** $(481 \times 10^{6}) - (0.113 \times 10^{9}) + (8.5 \times 10^{5})$

**(d)** $(0.363 \times 10^{-6}) + (71.2 \times 10^{-9}) + (519 \times 10^{-12})$

Answers: **C-3(a)** $1.036 \times 10^{3}$ **C-3(b)** $1.18 \times 10^{6}$ **C-3(c)** $1.254 \times 10^{-7}$ **C-3(d)** $2.44 \times 10^{-2}$ **C-4(a)** $2.9 \times 10^{-5}$ **C-4(b)** $3.29 \times 10^{-11}$ **C-4(c)** $3.69 \times 10^{8}$ **C-4(d)** $4.35 \times 10^{-7}$

### 3 Multiplication and Division

When numbers expressed in scientific notation are multiplied, the exponents of 10 are *added*. When the numbers are divided, the exponent of 10 in the denominator (the divisor) is subtracted from the exponent of 10 in the numerator (the dividend).

**Example C-5**
Carry out the following calculation.

$$(4.75 \times 10^{6}) \times (3.2 \times 10^{5})$$

**Solution**
In the first step, group the coefficients and the powers of 10. Carry out each step separately.

$$= (4.75 \times 3.2) \times (10^{6} \times 10^{5})$$

$$= 15.200 \times 10^{6+5} = 15 \times 10^{11} = \underline{\underline{1.5 \times 10^{12}}}$$

**Example C-6**
Carry out the following calculation:

$$(1.62 \times 10^{-8}) \div (8.55 \times 10^{-3})$$

**Solution**

$$= \frac{1.62 \times 10^{-8}}{8.55 \times 10^{-3}} = \frac{1.62}{8.55} \times \frac{10^{-8}}{10^{-3}} = 0.189 \times 10^{-8-(-3)}$$

$$= 0.189 \times 10^{-5} = \underline{\underline{1.89 \times 10^{-6}}}$$

## Exercises

**C-5** Carry out the following calculations. Express your answer to the proper number of significant figures with one digit to the left of the decimal point.

**(a)** $(7.8 \times 10^{-6}) \times (1.12 \times 10^{-2})$

**(b)** $(0.511 \times 10^{-3}) \times (891 \times 10^{-8})$

**(c)** $(156 \times 10^{-12}) \times (0.010 \times 10^{4})$

**(d)** $(16 \times 10^{9}) \times (0.112 \times 10^{-3})$

**(e)** $(2.35 \times 10^{3}) \times (0.3 \times 10^{5}) \times (3.75 \times 10^{2})$

**(f)** $(6.02 \times 10^{23}) \times (0.0100)$

**C-6** Follow the instructions in Problem C-5.

**(a)** $(14.6 \times 10^{8}) \div (2.2 \times 10^{8})$

**(b)** $(6.02 \times 10^{23}) \div (3.01 \times 10^{20})$

**(c)** $(0.885 \times 10^{-7}) \div (16.5 \times 10^{3})$

**(d)** $(0.0221 \times 10^{3}) \div (0.57 \times 10^{18})$

**(e)** $238 \div (6.02 \times 10^{23})$

**C-7** Follow the instructions in Problem C-5.

**(a)** $[(8.70 \times 10^{6}) \times (3.1 \times 10^{8})] \div (5 \times 10^{-3})$

**(b)** $(47.9 \times 10^{-6}) \div [(0.87 \times 10^{6}) \times (1.4 \times 10^{2})]$

**(c)** $1 \div [(3 \times 10^{6}) \times (4 \times 10^{10})]$

**(d)** $1.00 \times 10^{-14} \div [(6.5 \times 10^{5}) \times (0.32 \times 10^{-5})]$

**(e)** $[(147 \times 10^{-6}) \div (154 \times 10^{-6})] \div (3.0 \times 10^{12})$

Answers: **C-5(a)** $8.7 \times 10^{-8}$ **C-5(b)** $4.55 \times 10^{-9}$ **C-5(c)** $1.6 \times 10^{-8}$ **C-5(d)** $1.8 \times 10^{6}$ **C-5(e)** $3 \times 10^{10}$ **C-5(f)** $6.02 \times 10^{21}$ **C-6(a)** 6.6 **C-6(b)** $2.00 \times 10^{3}$ **C-6(c)** $5.36 \times 10^{-12}$ **C-6(d)** $3.9 \times 10^{-17}$ **C-6(e)** $3.95 \times 10^{-22}$ **C-7(a)** $5 \times 10^{17}$ **C-7(b)** $3.9 \times 10^{-13}$ **C-7(c)** $8 \times 10^{-18}$ **C-7(d)** $4.8 \times 10^{-15}$ **C-7(e)** $3.2 \times 10^{-13}$

## 4 Powers and Roots

When a number expressed in scientific notation is raised to a power, the coefficient is raised to the power and the exponent of 10 is *multiplied* by the power.

**Example C-7**
Carry out the following calculation.

$$(3.2 \times 10^3)^2 = (3.2)^2 \times 10^{3\times 2} = 10.24 \times 10^6 = \underline{\underline{1.0 \times 10^7}}$$

$$[(10^3)^2 = 10^3 \times 10^3 = 10 \times 10 \times 10 \times 10 \times 10 \times 10 = 10^6]$$

**Example C-8**
Carry out the following calculation.

$$(1.5 \times 10^{-3})^3 = (1.5)^3 \times 10^{-3\times 3} = \underline{\underline{3.4 \times 10^{-9}}}$$

When the root of a number expressed in scientific notation is taken, take the root of the coefficient and divide the exponent of 10 by the root. Before taking a root, however, adjust the number so that the exponent of 10 divided by the root produces a whole number. Adjust the exponent in the direction that will leave the coefficient as a number greater than 1 (to avoid mistakes).

**Example C-9**
Carry out the following calculation.

$$\sqrt{2.9 \times 10^5}$$

**Solution**
First adjust the number so that the exponent of 10 is divisible by 2.

$$\sqrt{2.9 \times 10^5} = \sqrt{29 \times 10^4} = \sqrt{29} \times \sqrt{10^4} = \sqrt{29} \times 10^{4/2} = 5.4 \times 10^2 = 540$$

**Example C-10**
Carry out the following calculation.

$$\sqrt[3]{6.9 \times 10^{-8}}$$

**Solution**
Adjust the number so that the exponent of 10 is divisible by 3.

$$\sqrt[3]{6.9 \times 10^{-8}} = \sqrt[3]{69 \times 10^{-9}} = \sqrt[3]{69} \times \sqrt[3]{10^{-9}} = 4.1 \times 10^{-9/3} = 4.1 \times 10^{-3}$$

### Exercises

**C-8** Carry out the following operations.

**(a)** $(6.6 \times 10^4)^2$

**(b)** $(0.7 \times 10^6)^3$

**(c)** $(1200 \times 10^{-5})^2$ (It will be easier to square if you change the number to $1.2 \times 10^?$ first.)

**(d)** $(0.035 \times 10^{-3})^3$

**(e)** $(0.7 \times 10^7)^4$

**C-9** Take the following roots. Approximate the answer if necessary.

**(a)** $\sqrt{36 \times 10^4}$

**(b)** $\sqrt[3]{27 \times 10^{12}}$

**(c)** $\sqrt{64 \times 10^9}$

**(d)** $\sqrt[3]{1.6 \times 10^5}$

**(e)** $\sqrt{81 \times 10^{-7}}$

**(f)** $\sqrt{180 \times 10^{10}}$

Answers: **C-8(a)** $4.4 \times 10^9$ **C-8(b)** $3 \times 10^{17}$ **C-8(c)** $1.4 \times 10^{-4}$ **C-8(d)** $4.3 \times 10^{-14}$ **C-8(e)** $2 \times 10^{27}$ **C-9(a)** $6 \times 10^2$ **C-9(b)** $3.0 \times 10^4$ **C-9(c)** $2.5 \times 10^5$ **C-9(d)** 54 **C-9(e)** $2.8 \times 10^{-3}$ **C-9(f)** $1.3 \times 10^6$

# Appendix D
# Problem Solving by the Unit-Factor Method (Dimensional Analysis)

This section reviews the following problem solving skills.

1 The construction of conversion factors
2 The manipulation of units
3 One-step conversions
4 Multiple-step conversions

Two of the biggest stumbling blocks to a good performance in chemistry are the lack of ability to memorize and the lack of problem solving skills. The purpose of this section is to supplement Chapter 2 with a more detailed introduction to problem solving. Here we will take the time to develop the necessary tools from the basics to simple problems and finally to more challenging problems.

A large number of problems in basic chemistry fit into the category of *conversion* problems. Whether it is from grams to milligrams, grams to pounds, moles to number of molecules, volume of a gas to moles, or weight of solute to volume of solution, these problems can all be solved by the same method. Once the method is understood, the actual solution becomes routine. The problems have the same basic feature; a quantity is given in one unit of measurement and you are asked to convert this to another unit of measurement. To solve such problems we use the unit-factor method, which employs one or more conversion factors.

The basic procedure for a one-step conversion is

$$\begin{bmatrix}\text{Given}\\ \text{quantity*}\end{bmatrix} \cancel{\text{old unit}} \times \begin{bmatrix}\text{conversion}\\ \text{factor}\\ \text{quantity}\end{bmatrix} \frac{\text{new unit}}{\cancel{\text{old unit}}} = \begin{bmatrix}\text{requested}\\ \text{quantity}\end{bmatrix} \text{new unit}$$

*This refers to the numerical value of the quantity only.

If you are asked to convert 24 in. to feet, the answer comes easily because of your familiarity with the units and the relationship of inches to feet. To obtain the answer of 2 ft, most of us probably aren't even aware of what we did mathematically. In this appendix, however, we will carefully show what we do with units as well as numbers even for the simple conversions. By getting into the habit of including units in the calculation, the more complex problems can be simplified.

Three types of units are employed in the examples that follow:

**1** English units
- weight—ounces, pounds
- distance—feet, yards, miles
- volume—quarts, gallons
- quantity—dozen, gross
- time—hours, minutes

**2** Alien units
- weight—gerbs
- distance—frims
- volume—moks
- quantity—nums, rims
- time—bleems

**3** Chemical units (metric or SI)
- weight—grams
- distance—meters, kilometers
- volume—milliliters, liters
- quantity—moles
- time—hours, minutes
- and others

There is a reason for this approach. Most of the math you need in general chemistry is quite simple and was reviewed in the previous appendixes. Often, however, the units and the awkward numbers cause confusion. It is the goal of this appendix to end the confusion.

First we will work examples and exercises in familiar English units, where you probably have a feeling for the correct answer before doing the actual calculation. By using the unit-factor method in these problems we are in effect perfecting the use of a "decoder" that can be used with less familiar systems. Next we travel to an alien planet with our "decoder" to work problems with units that have no meaning to us. In these problems we certainly have no feeling for the approximate answer to a calculation, so there is no way to work the problems without using the "decoder." The unit-factor method leads us confidently to the answers in this strange world.

Finally, we return to earth to work examples with the units used in chemistry. To the beginning student these units may also appear alien. However, until these units become familiar, we may rely on the "decoder" to lead us to the correct answers.

In summary, don't be overly concerned at this point if some of the units used in chemistry are unfamiliar. Our calculations in the alien and chemical units is like flying an airplane on instruments. The experienced pilot trusts the instruments to lead safely to the destination when the ground can't be seen. Likewise, you can trust the units to lead safely to the answer when you are not sure whether to multiply or divide. Understandably, pilots prefer to see where they are going with their own eyes. Likewise, you will eventually wish to see where you are going by becoming familiar with the units and calculations.

To be comfortable with the unit-factor method, there are three essential requirements: (1) to understand the origin and function of the conversion factor, (2) to understand the manipulation of units, and, finally, (3) to carry out the calculation itself correctly. We will take up these three points individually.

## 1 The Construction of Conversion Factors

Conversions factors quantitatively relate one unit of measurement to that of another. They may be between similar units [e.g., length (m and km)] or between very different units [e.g., mass (g) and volume (mL)]. The relationships that are used as conversion factors originate in two ways:

1 From an *exact* definition (e.g., $10^3$ m = 1 km, 12 in. = 1 ft, 60 min = 1 hr). These are usually (but not always) relationships within one system of measurement. Exact relationships have unlimited numbers of significant figures.

2 From a measurement (e.g., 1 lb = 454 g, 1.00 mL weighs 4.64 g, 1.00 lb costs $1.75). These are usually between different systems of measurement. The precision of the relationship is indicated by the number of significant figures shown.

Conversion factors are constructed by expressing the equality or equivalency relationship in fraction or *factor* form. For example, 12 in. = 1 ft can be expressed as

$$\frac{12 \text{ in.}}{1 \text{ ft}} \quad \text{or} \quad \frac{1 \text{ ft}}{12 \text{ in.}}$$

(This reads as 12 in. "per" foot or as 1 foot "per" 12 in.) and 1.00 mL weighs 4.64 g as

$$\frac{1.00 \text{ mL}}{4.64 \text{ g}} \quad \text{or} \quad \frac{4.64 \text{ g}}{1.00 \text{ mL}}$$

*These are sometimes called "unity factors" because they relate a quantity in one measurement to "one" of another.* When unity factors from a measurement are shown, it is implied that the "one" is known to as many significant figures as the numerical quantity. Also, in cases where the "one" is in the denominator, it is often neither shown nor read. Thus the relationship between mL and g shown above is simplified in this text as

$$\frac{1\ \text{mL}}{4.64\ \text{g}} \quad \text{or} \quad \frac{4.64\ \text{g}}{\text{mL}} \quad \text{or} \quad 4.64\ \text{g/mL} \quad (4.64\ \text{g "per" mL})$$

Most conversion factors used in calculations are in the form of unity factors. (This is more a matter of convenience than necessity, however.) In the following examples and exercises, we will practice the construction of unity factors from a measurement relating two quantities.

**Example D-1**
If 0.250 miles is found to equal 1320 ft, how many feet are there per mile?

**Solution**
The problem tells us that 1320 ft = 0.250 mile. We put the relationship in factor form and carry out the math.

$$\frac{1320\ \text{ft}}{0.250\ \text{mile}} = \underline{\underline{5280\ \text{ft/mile}}}$$

**Example D-2**
A substance from the alien planet weighs 488 gerbs and has a volume of 14.2 moks. How many gerbs are there per mok?

**Solution**

$$488\ \text{gerbs} = 14.2\ \text{moks}$$

$$\frac{488\ \text{gerbs}}{14.2\ \text{moks}} = \underline{\underline{34.4\ \text{gerbs/mok}}}$$

**Example D-3**
A 0.785-mole (abbreviated mol) quantity of a compound has a weight of 175 g. What is the weight in grams per mole?

**Solution**

$$175\ \text{g} = 0.785\ \text{mol}$$

$$\frac{175\ \text{g}}{0.785\ \text{mol}} = \underline{\underline{223\ \text{g/mol}}}$$

### Exercises

In the following problems, include the units of the answer with the number where appropriate.

**D-1** The unit price of groceries is sometimes listed in cost per ounce. Which of the two has the smallest cost per ounce: 16 oz of baked beans costing \$1.45, or 26 oz costing \$2.10?

**D-2** An 82.3-doz quantity of oranges weighs 247 lb. What is the weight per dozen oranges? How many dozen oranges are there per pound?

**D-3** If two men can dig an 8-ft trench in 2.5 hr, what is the number of feet dug per man per hour? (The units will be ft/man · hr.)

**D-4** There are 1480 years in 3.25 bleems. How many bleems are there per year?

**D-5** There are $8.61 \times 10^5$ objects in 0.574 nums. How many objects are there per num?

**D-6** A gas has a volume of 146 L and a weight of 48.5 g. What is the weight per liter? What is the volume per gram?

**D-7** There are $8.85 \times 10^5$ coulombs in 9.17 faradays. How many coulombs are there per faraday?

**D-8** A quantity of $6.02 \times 10^{23}$ atoms of tungsten weighs 184 g. What is the weight per atom? How many atoms per gram?

Answers: **D-1** The 16-oz can costs \$0.091/oz, the 26-oz can costs \$0.081/oz. **D-2** 3.00 lb/doz, 0.333 doz/lb **D-3** 1.6 ft/man · hr **D-4** $2.20 \times 10^{-3}$ bleems/year **D-5** $1.50 \times 10^6$ objects/num **D-6** 0.332 g/L, 3.01 L/g **D-7** $9.65 \times 10^4$ coulombs/faraday **D-8** $3.06 \times 10^{-22}$ g/atom, $3.27 \times 10^{21}$ atoms/g.

## 2 The Manipulation of Units

The convenience of problem solving by the unit-factor method is that the units lead the way to the correct answer. In this method, the units as well as the numerical quantities are multiplied, divided, or canceled in the calculation. If the units are handled properly to produce the desired unit in the answer, we can be assured that the answer is correct (barring a mathematical error). Because of the importance of units we will review their manipulation in calculations in a little more detail than presented in Appendix A. After this, numerical quantities will be included and we'll do some problems.

**Example D-4**

What are the unit or units of the answers implied by the following calculations?

$$\frac{\text{ft}}{\text{mile}} \times \frac{\text{mile}}{\text{min}} \times \frac{\text{min}}{\text{hr}}$$

Cancel units that occur in both numerator and denominator.

$$\frac{\text{ft}}{\cancel{\text{mile}}} \times \frac{\cancel{\text{mile}}}{\cancel{\text{min}}} \times \frac{\cancel{\text{min}}}{\text{hr}} = \frac{\text{ft}}{\text{hr}}$$

**Example D-5**

$$\frac{\text{frim}^3/\text{mok} \times 1/\text{bleem}}{\text{frim}^2/\text{mok}}$$

Rearrange into one quotient and cancel (see Appendix A for manipulation of fractions).

$$\frac{\text{frim}^{\overset{1}{\cancel{3}}} \times \cancel{\text{mok}}}{\cancel{\text{frim}^2} \times \cancel{\text{mok}} \times \text{bleem}} = \frac{\text{frim}}{\text{bleem}}$$

**Example D-6**

$$\frac{(\text{L} \cdot \text{atm/K} \cdot \text{mol}) \times \text{K}}{\text{atm}}$$

Rearrange into one quotient and cancel.

$$\frac{\text{L} \times \cancel{\text{atm}} \times \cancel{\text{K}}}{\cancel{\text{K}} \times \text{mol} \times \cancel{\text{atm}}} = \frac{\text{L}}{\text{mol}}$$

### Exercises

In the following exercises, write the units of the answers implied from the calculations.

**D-9** $\dfrac{\text{ft}}{\text{mile}} \times \text{lb} \times \text{mile}$

**D-10** $\dfrac{(\text{mile/hr}) \times (\text{ft/mile}) \times (\text{hr/min})}{\text{sec/min}}$

**D-11** $\dfrac{(\text{rim/num}) \times (1/\text{bleem})}{1/\text{num}}$

**D-12** $\dfrac{(\text{gerb/frim}^3) \times \text{mok} \times \text{frim}}{\text{mok/frim}}$

**D-13** $\dfrac{\text{g} \times (\text{L} \cdot \text{torr/K} \cdot \text{mol}) \times \text{K}}{\text{torr} \times \text{L}}$

**D-14** $\dfrac{1/(\text{atm} \cdot \text{L})}{\text{mol} \times (\text{L} \cdot \text{atm/K} \cdot \text{mol})}$

**D-15** $\dfrac{\text{coulomb} \times \text{mol/F}}{(\text{mol/g}) \times \text{coulomb}}$

Answers: **D-9** ft · lb **D-10** ft/sec **D-11** rim/bleem **D-12** gerb/frim **D-13** g/mol **D-14** $\text{K/L}^2 \cdot \text{atm}^2$ **D-15** g/F.

The main lesson to be learned from these examples and exercises is to be sure that the units in the calculation lead to the units desired in the answer. For example, if the answer should be mL, but the units come out to be $\text{g}^2$/mL, the units indicate that you have made a mistake. Be careful in your manipulation of units to avoid *forcing* a cancellation, such as

$$\frac{\text{g}}{\cancel{\text{mL}}} \times \frac{1}{\cancel{\text{mL}}} = \text{g} \qquad \text{(wrong)}$$

$$\frac{\text{g}}{\cancel{\text{mL}}} \times \cancel{\text{mL}} = \text{g} \qquad \text{(right)}$$

### 3 One-Step Conversions

Before we work some examples, a word of warning is in order. In working problems in the English system, this procedure should not become so mechanical that you lose a feeling for the approximate answer. For example, if a car travels 50 miles/hr and you are asked how long it takes to travel 160 miles, you know that the answer is more than 3 hr. If you multiply instead of divide, you should immediately sense that the answer of 8000 hr is ridiculous even without help from the units. *Think about your answer.* It is understood at this point, however, that you may not be able to approximate the answer when working with chemical units. Eventually you will. For example, if the weight of one tiny atom is requested, the answer is very small ($\sim 10^{-23}$ g). An answer of 10 g or even $10^{23}$ g (about the weight of the moon) is out of the money and should be so recognized. Our purpose here is to set you at ease about the calculation itself. When you study the particular topic, you should answer the very important question: "Does this answer make sense?"

In the following problems we will list what is given and requested, a shorthand procedure, the conversion factor properly set up to follow the procedure, and finally the solution. This effort pays off for the more complex problems later.

Remember that, in most calculations, the conversion factor is set up so that the old unit (which is given) is in the denominator and cancels. The new unit (which is requested) is in the numerator and remains in the answer. Obviously, just the opposite would hold if we wished to make a conversion involving the denominator (e.g., miles/hr to miles/min). This all assumes a simple one-step conversion where a conversion factor is available that relates a given measurement to one that is requested. Such conversions are illustrated by the following examples and exercises.

**Example D-7**
If a car travels 55 miles/hr, how many miles does it travel in 8.5 hr?

Given: 8.5 hr

Requested: ___?___ miles

Procedure: hr → miles

Conversion factor: 55 miles/hr

**Solution**

$$8.5\ \cancel{\text{hr}} \times \frac{55\ \text{miles}}{\cancel{\text{hr}}} = \underline{\underline{470\ \text{miles}}}$$

**Example D-8**
If the same car as in Example D-7 travels 785 miles, how many hours does it take?

Given: 785 miles

Requested: ___?___ hr

Procedure: miles → hr

Conversion factor: $\dfrac{1\ \text{hr}}{55\ \text{miles}}$

**Solution**

$$785\ \cancel{\text{miles}} \times \frac{1\ \text{hr}}{55\ \cancel{\text{miles}}} = \underline{\underline{14\ \text{hr}}}$$

Notice in Example D-7 that a small number (8.5) is converted to a large number (470). In such a case, the conversion factor is always larger than unity. The opposite is true in Example D-8, where a large number (785) is converted to a small number (14). Notice that for such a conversion the factor is less than unity. A simple check to make sure that the conversion factor changes the magnitude of the number in the proper direction saves many mistakes.

**Example D-9**

A certain type of chain costs \$0.55/ft. What does it cost per mile? (5280 ft = 1 mile)

Given: \$0.55/ft

Requested: $\frac{\$}{\text{mile}}$

Procedure: $\frac{\$}{\text{ft}} \rightarrow \frac{\$}{\text{mile}}$

Conversion factor: 5280 ft/mile

In this case, the unit of the denominator is to be changed. The factor should therefore have the given unit in the *numerator* and the requested unit in the *denominator*.

**Solution**

$$\frac{\$0.55}{\cancel{\text{ft}}} \times \frac{5280\ \cancel{\text{ft}}}{\text{mile}} = \$2904/\text{mile}$$

One popular exception to the proper use of significant figures is where money is involved. Traditionally, price is expressed to the nearest cent regardless of how it is obtained. An argument to a salesperson in the above example that the price should only be \$2900 because of significant figures would not go very far.

**Example D-10**

A section of cable weighs 18.5 gerbs/frim. How many frims are there in 97.2 gerbs?

Given: 97.2 gerbs

Requested: ___?___ frims

Procedure: gerbs → frims

Conversion factor: $\frac{1 \text{ frim}}{18.5 \text{ gerb}}$

**Solution**

$$97.2 \cancel{\text{gerb}} \times \frac{1 \text{ frim}}{18.5 \cancel{\text{gerb}}} = \underline{\underline{5.25 \text{ frim}}}$$

**Example D-11**
A space ship can travel $1.85 \times 10^6$ frims in 0.240 bleems. How far can it travel in 0.525 bleem?

Given: 0.525 bleem

Requested: ___?___ frims

Procedure: bleems → frims

Conversion factor: $\frac{1.85 \times 10^6 \text{ frims}}{0.240 \text{ bleem}}$

**Solution**

$$0.525 \cancel{\text{bleem}} \times \frac{1.85 \times 10^6 \text{ frims}}{0.240 \cancel{\text{bleem}}} = \underline{\underline{4.05 \times 10^6 \text{ frims}}}$$

**Example D-12**
What is the weight of a 3.25-mol quantity of a substance if it weighs 80.0 g/mol?

Given: 3.25 moles

Requested: ___?___ g

Procedure: mol → g
Conversion factor: 80.0 g/mol

**Solution**

$$3.25 \cancel{\text{mol}} \times \frac{80.0 \text{ g}}{\cancel{\text{mol}}} = \underline{\underline{260 \text{ g}}}$$

**Example D-13**
If there are $6.02 \times 10^{23}$ molecules/mol, how many moles does $5.65 \times 10^{24}$ molecules represent?

Given: $5.65 \times 10^{24}$ molecules

Requested: $\underline{\ ?\ }$ mol

Procedure: molecules → mol

Conversion factor: $\dfrac{1 \text{ mol}}{6.02 \times 10^{23} \text{ molecules}}$

**Solution**

$$5.65 \times 10^{24} \cancel{\text{molecules}} \times \frac{1 \text{ mol}}{6.02 \times 10^{23} \cancel{\text{molecules}}} = \underline{\underline{9.39 \text{ mol}}}$$

**Exercises**

Practice one-step conversions on the following problems.

**D-16** If the unit price of chili is $0.18/oz, what is the cost of a 12-oz can of chili? How many ounces can you buy at this rate for $1.75?

**D-17** If a bullet travels 1860 ft/sec, how far does it travel in 4.50 sec?

**D-18** If there are 4.55 lb of corn per dozen ears, how many dozen ears of corn are in 265 lb?

**D-19** There are $1.50 \times 10^6$ commas per num (remember, a num is a unit of quantity, like dozen, gross, or mole). How many nums is $83 \times 10^8$ commas? How many commas are in 0.420 nums?

**D-20** A resident of an alien planet can lose 0.475 gerbs in 2.15 bleems when on a diet. How many gerbs can be lost in 0.880 bleem? How many bleems would it take to lose 187 gerbs?

**D-21** If 18.0 g of a gas has a volume of 22.4 L under certain conditions, what is the weight of the gas that would have a volume of 1.35 L?

**D-22** A pressure of 1 atm is equal to 760 torr. How many torr are there in 0.66 atm?

**D-23** There are $6.02 \times 10^{23}$ molecules/mol. How many molecules are present in 0.212 mol?

**D-24** Using the relationship from the previous problem, determine how many molecules there are per gram of a particular substance if 1 mol weighs 18.0 g.

**D-25** There are 96,500 coulombs/faraday. How many faradays are equivalent to 4820 coulombs?

Answers: **D-16** $2.16, 9.72 oz **D-17** 8370 ft **D-18** 58.2 doz **D-19** $5.5 \times 10^3$ nums, $6.30 \times 10^5$ commas **D-20** 0.194 gerbs, 846 bleems **D-21** 1.08 g **D-22** 500 torr **D-23** $1.28 \times 10^{23}$ molecules **D-24** $3.34 \times 10^{22}$ molecules/g **D-25** 0.0499 faradays

## 4 Multiple-Step Conversions

Hopefully, the one-step conversions did not cause a major problem despite the unfamiliar units. In any case, the previous exercises included some fairly sophisticated problems encountered in chemistry. We are now ready to advance to some more complex problems. In what follows a direct relationship between the units of what's given and what's requested is not available. In these cases, conversions must be done in two or more steps, with each step requiring a separate conversion factor. The general procedure for a two-step conversion is as follows:

$$\begin{bmatrix}\text{Given}\\ \text{quantity}\end{bmatrix} \cancel{\text{old unit}} \times \begin{bmatrix}\text{conversion factor}\\ \text{quantity (a)}\end{bmatrix} \frac{\cancel{\text{unit X}}}{\cancel{\text{old unit}}}$$

$$\times \begin{bmatrix}\text{conversion factor}\\ \text{quantity (b)}\end{bmatrix} \frac{\text{new unit}}{\cancel{\text{unit X}}} = \begin{bmatrix}\text{requested}\\ \text{quantity}\end{bmatrix} \text{new unit}$$

Notice that the conversion factor for each step leaves a new unit in the numerator while canceling an old unit with the denominator. For example, assume that a procedure is planned as follows for a problem where weight (in lb) is given and the cost (in \$) is requested. If conversion factors are available that relate weight (lb) to quantity (dozen) and quantity (dozen) to cost (\$), the procedure is

$$\text{lb} \xrightarrow{(a)} \text{doz} \xrightarrow{(b)} \$$$

Step (a) converts lb to doz.

$$\cancel{\text{lb}} \times \frac{\text{doz}}{\cancel{\text{lb}}} = \underline{\qquad\qquad} \text{ doz}$$

Step (b) converts doz to \$.

$$\cancel{\text{lbs}} \times \frac{\cancel{\text{doz}}}{\cancel{\text{lbs}}} \times \frac{\$}{\cancel{\text{doz}}} = \$\underline{\qquad\qquad}$$

In the examples and exercises that follow experiment with the units before doing the math. Work with the units of what's given and the units of the conversion factors so that undesired units cancel, leaving only the requested unit or units in the answer. When you are sure of your setup, then do the math. Don't be discouraged if you find the problems with alien and chemical units quite challenging. They are meant to be. The purpose is to present problems where it is almost necessary to use the unit-factor method where the units lead the way to the answer. If you are able to work your way through these exercises successfully, you should be ready for many if not most quantitative problems facing you in general chemistry. Many of the relationships used in the following problems are found in Tables 2-1 and 2-2.

**Example D-14**

If a train travels 75.0 miles/hr, how many miles does it travel in 416 min?

Given: 416 min

Requested: __?__ miles

Procedure: $\text{min} \xrightarrow{(a)} \text{hr} \xrightarrow{(b)} \text{miles}$

Conversion factors: (a) $\dfrac{1\ \text{hr}}{60\ \text{min}}$ (b) $\dfrac{75.0\ \text{miles}}{\text{hr}}$

**Solution**

$$416\ \cancel{\text{min}} \times \frac{1\ \cancel{\text{hr}}}{60\ \cancel{\text{min}}} \times \frac{75.0\ \text{miles}}{\cancel{\text{hr}}} = \underline{\underline{520\ \text{miles}}}$$

**Example D-15**

One gallon of gas costs \$1.22. What is the cost of 48.0 L (the volume unit used in Mexico and Canada?

Given: 48.0 L

Requested: \$__?__

Procedure: $\text{L} \xrightarrow{(a)} \text{qts} \xrightarrow{(b)} \text{gal} \xrightarrow{(c)} \$$

Conversion factors:

(a) $\dfrac{1.06\ \text{qt}}{\text{L}}$ (b) $\dfrac{1\ \text{gal}}{4\ \text{qt}}$ (c) $\dfrac{\$1.22}{\text{gal}}$

**Solution**

$$48.0\ \cancel{\text{L}} \times \frac{1.06\ \cancel{\text{qt}}}{\cancel{\text{L}}} \times \frac{1\ \cancel{\text{gal}}}{4\ \cancel{\text{qt}}} \times \frac{\$1.22}{\cancel{\text{gal}}} = \underline{\underline{\$15.52}}$$

**Example D-16**

If two men can dig an 8-ft trench in 2.5 hr, how long a trench can six men dig in 18 hr if all the men are working at the same rate?

Given: 6 men 18 hr

Requested: __?__ ft

Procedure: men, hr $\xrightarrow{(a)}$ ft

Conversion factor: $\dfrac{8\text{ ft}}{2\text{ men}\cdot 2.5\text{ hr}}$

**Solution**

$$6\,\cancel{\text{men}} \times 18\,\cancel{\text{hr}} \times \frac{8\text{ ft}}{2\,\cancel{\text{men}}\cdot 2.5\,\cancel{\text{hr}}} = \underline{\underline{173\text{ ft}}}$$

---

**Example D-17**

The speed limit of a space ship is 55.0 frims/millibleem ($10^3$ millibleems per bleem). How many frims can a space ship travel in 0.205 bleem?

Given: 0.205 bleem

Requested: __?__ frims

Procedure: bleems $\xrightarrow{(a)}$ millibleems $\xrightarrow{(b)}$ frims

Conversion factors:

$$\text{(a)}\ \frac{10^3\text{ millibleems}}{\text{bleem}} \qquad \text{(b)}\ \frac{55.0\text{ frims}}{\text{millibleem}}$$

**Solution**

$$0.205\,\cancel{\text{bleem}} \times \frac{10^3\,\cancel{\text{millibleems}}}{\cancel{\text{bleem}}} \times \frac{55.0\text{ frims}}{\cancel{\text{millibleem}}} = \underline{\underline{11{,}300\text{ frims}}}$$

---

**Example D-18**

There are $1.50 \times 10^6$ numecules in a num. If the weight of 1 num (the "num mass") of a substance is 24.0 gerbs, what is the weight in gerbs of $7.50 \times 10^5$ numecules of the substance?

Given: $7.50 \times 10^5$ numecules

Requested: __?__ gerbs

Procedure: numecules $\xrightarrow{(a)}$ nums $\xrightarrow{(b)}$ gerbs

Conversion factors:

$$\text{(a)}\ \frac{1\text{ num}}{1.50 \times 10^6\text{ numecule}} \qquad \text{(b)}\ \frac{24.0\text{ gerbs}}{\text{num}}$$

**Solution**

$$7.50 \times 10^5 \text{ \sout{numecule}} \times \frac{1 \text{ \sout{num}}}{1.50 \times 10^6 \text{ \sout{numecule}}} \times \frac{24.0 \text{ gerbs}}{\text{\sout{num}}} = \underline{\underline{12.0 \text{ gerbs}}}$$

**Example D-19**
Using the information given in Example D-18, determine how many numecules there are in 175 gerbs.

Given: 175 gerbs

Requested: ___?___ numecules

Procedure: gerbs $\xrightarrow{(a)}$ nums $\xrightarrow{(b)}$ numecules

Conversion factors:

$$\text{(a) } \frac{1 \text{ num}}{24.0 \text{ gerbs}} \qquad \text{(b) } \frac{1.50 \times 10^6 \text{ numecules}}{\text{num}}$$

**Solution**

$$175 \text{ \sout{gerbs}} \times \frac{1 \text{ \sout{num}}}{24.0 \text{ \sout{gerbs}}} \times \frac{1.50 \times 10^6 \text{ numecules}}{\text{\sout{num}}} = \underline{\underline{1.09 \times 10^7 \text{ numecules}}}$$

**Example D-20**
Gold weighs 19.3 g/mL. What is the volume (in milliliters) of exactly 1 lb of gold?

Given: 1.00 lb

Requested: ___?___ mL

Procedure: lb $\xrightarrow{(a)}$ g $\xrightarrow{(b)}$ mL

Conversion factors:

$$\text{(a) } \frac{454 \text{ g}}{\text{lb}} \qquad \text{(b) } \frac{1 \text{ mL}}{19.3 \text{ g}}$$

**Solution**

$$1.00 \text{ \sout{lb}} \times \frac{454 \text{ \sout{g}}}{\text{\sout{lb}}} \times \frac{1 \text{ mL}}{19.3 \text{ \sout{g}}} = \underline{\underline{23.5 \text{ mL}}}$$

**Example D-21**
There are $6.02 \times 10^{23}$ molecules in 1 mol. If the weight of 1 mol (the "molar mass") of carbon dioxide is 44.0 g, what is the weight of $8.46 \times 10^{22}$ molecules of carbon dioxide?

Given: $8.46 \times 10^{22}$ molecules

Requested: ___?___ g

Procedure: $\text{molecules} \xrightarrow{(a)} \text{mol} \xrightarrow{(b)} \text{g}$

Conversion factors:

$$\text{(a)}\ \frac{1\ \text{mol}}{6.02 \times 10^{23}\ \text{molecules}} \qquad \text{(b)}\ \frac{44.0\ \text{g}}{\text{mol}}$$

**Solution**

$$8.46 \times 10^{22}\ \cancel{\text{molecules}} \times \frac{1\ \cancel{\text{mol}}}{6.02 \times 10^{23}\ \cancel{\text{molecules}}} \times \frac{44.0\ \text{g}}{\cancel{\text{mol}}} = \underline{\underline{6.18\ \text{g}}}$$

**Example D-22**
Using the information given in Example D-21, determine how many molecules are in 890 g of carbon dioxide.

Given: 890 g

Requested: ___?___ molecules

Procedure: $\text{g} \xrightarrow{(a)} \text{mol} \xrightarrow{(b)} \text{molecules}$

Conversion factors:

$$\text{(a)}\ \frac{1\ \text{mol}}{44.0\ \text{g}} \qquad \text{(b)}\ \frac{6.02 \times 10^{23}\ \text{molecules}}{\text{mol}}$$

**Solution**

$$890\ \cancel{\text{g}} \times \frac{1\ \cancel{\text{mol}}}{44.0\ \cancel{\text{g}}} \times \frac{6.02 \times 10^{23}\ \text{molecules}}{\cancel{\text{mol}}} = \underline{\underline{1.22 \times 10^{25}\ \text{molecules}}}$$

## Exercises

The following problems require two or more conversions to obtain the answer.

**D-26** If a train travels at a speed of 85 miles/hr, how many hours does it take to travel 17,000 ft? How many yards can it travel in 37 min? (1760 yd = 1 mile.)

**D-27** A certain size of nail costs \$1.25/lb. What is the cost of 3.25 kg of nails?

**D-28** Two men can push a car 110 yards in 15 min. How many minutes would it take three men to push a car 250 yards at the same rate?

**D-29** If a car travels 55.0 miles/hr, what is its speed in feet per second?

**D-30** Low-tar cigarettes contain 11.0 mg of tar per cigarette. If all the tar gets into the lungs, how many packages of cigarettes (20 cigarettes per package) would have to be smoked to produce 0.500 lb of tar? If a person smoked two packages per day, how many years would it take to accumulate 0.500 lb of tar?

**D-31** If a car gets 24.5 miles to the gallon of gasoline and gasoline costs \$1.22/gal, what would it cost to drive 350 km?

**D-32** A Martian fruit weighs 0.820 gerbs/rim. What is the volume (in moks) of 425 rims if there are 115 gerbs per mok?

**D-33** If there are $1.50 \times 10^6$ numecules/num, how many numecules are in 1 rim if there are 135 rims/num?

**D-34** If numecules weigh 118 gerbs/num, how many gerbs are in $4.50 \times 10^8$ numecules ($1.50 \times 10^6$ numecules/num)? How many numecules are in 45.5 gerbs?

**D-35** A liquid solution from the alien planet contains 0.250 nums of a substance per mok. If the substance weighs 164 gerbs/num, how many gerbs of the substance are in 152 moks of solution?

**D-36** A liquid solution on earth contains 0.250 mol of sulfuric acid per liter. How many grams of sulfuric acid are in 6.75 L of solution if the sulfuric acid weighs 98.0 g/mol? What is the volume (in liters) occupied by 15.5 g of sulfuric acid?

**D-37** If there are $6.02 \times 10^{23}$ molecules/mol, how many molecules of sulfuric acid are in the two liquid solutions in Problem D-36?

**D-38** How many grams of silver are deposited by 46,200 coulombs of electricity if 1 faraday is equal to 96,500 coulombs and 1 faraday deposits 108 g of silver?

**D-39** If 1 mol of a gas occupies 22.4 L under certain conditions, how many liters will 75.0 g of this gas occupy if the gas weighs 60.0 g/mol?

**D-40** One atmosphere pressure is equal to 760.0 torr. What is 355.0 torr expressed in kilopascals (kPa) if 1 atm is equal to 101.4 kPa?

**D-41** There are 3 mol of hydrogen atoms per mole of phosphoric acid. How many moles of hydrogen atoms are in 16.0 g of phosphoric acid if 1 mol of phosphoric acid weighs 98.0 g?

Answers: **D-26** 0.038 hr, $9.2 \times 10^4$ yards **D-27** \$8.94 **D-28** 23 min **D-29** 80.7 ft/sec **D-30** 1030 packages, 1.41 years **D-31** \$10.83 **D-32** 3.03 moks **D-33** $1.11 \times 10^4$ numecules/rim **D-34** $3.54 \times 10^4$ gerbs, $5.78 \times 10^5$ numecules **D-35** 6230 gerbs **D-36** 165 g, 0.633 L **D-37** $1.02 \times 10^{24}$ molecules, $9.52 \times 10^{22}$ molecules **D-38** 51.7 g **D-39** 28.0 L **D-40** 47.36 kPa **D-41** 0.490 mol of H atoms

# Appendix E
# Logarithms

This section reviews the following common logarithm skills.

1 Taking the log of a number between 1 and 10
2 Taking the logs of numbers less than 1 and greater than 10
3 Finding the antilog of a positive log
4 Finding the antilog of a negative log

In Appendix C we learned that scientific notation is particularly useful in expressing very large or very small numbers. In certain areas of chemistry, however, such as in the expression of $H_3O^+$ concentration, even the repeated use of scientific notation becomes tedious. In this case, it is convenient to express the concentration as simply the *exponent of 10. The exponent to which 10 must be raised to give a certain number is called its* **common logarithm**. With common logarithms (or just logs) it is possible to express the coefficient and the exponent of 10 as one number.

Since logarithms are simply exponents of 10, logs of numbers such as 100 can be easily determined. Notice that 100 can be expressed as $10^2$, so that the log of 100 is simply 2. Other examples of simple logs are as follows:

$$
\begin{array}{rlrlrlrl}
1 = 10^0 & & \log 1 = 0 & & 0.1 = 10^{-1} & & \log 0.1 = -1 \\
10 = 10^1 & & \log 10 = 1 & & 0.01 = 10^{-2} & & \log 0.01 = -2 \\
100 = 10^2 & & \log 100 = 2 & & 0.001 = 10^{-3} & & \log 0.001 = -3 \\
1000 = 10^3 & & \log 1000 = 3 & & 0.0001 = 10^{-4} & & \log 0.0001 = -4
\end{array}
$$

The same mathematical operations that apply to exponents also apply to logs. For example, consider the multiplication and division of exponents of 10 and logs.

| | *Exponents* | *Logarithms* |
|---|---|---|
| Multiplication: | $10^4 \times 10^3 = 10^{4+3} = 10^7$ | $\log(10^4 \times 10^3) = \log 10^4 + \log 10^3 = 4 + 3 = 7$ |

Division: $\dfrac{10^{10}}{10^4} = 10^{10-4} = 10^6 \quad \log \dfrac{10^{10}}{10^4} = \log 10^{10} - \log 10^4$

$$= 10 - 4 = 6$$

Taking logs of numbers in scientific notation where the coefficients are exactly 1 produces an integral number. Obviously, since most numbers do not have a coefficent of exactly 1 we will need some help in finding the logs of such numbers. The procedure is outlined in the following two sections.

### 1 Taking the Log of a Number between 1 and 10

Notice that log 1 = 0 and log 10 = 1. The log of numbers between 1 and 10 will therefore be between 0 and 1. The log of these numbers can be conveniently found in a *log table* as found on page 538. This is a four-place log table, which suits our purpose in this text. Two-, three-, and five-place log tables are available that give fewer or more significant figures, respectively. A four-place log table provides the log of a three-significant-figure number directly from the table. In addition, the log of a four-significant figure number can be calculated by interpolation between numbers in the table. The log of a number is found in the body of the table. To find the log of a number between 1 and 10, locate the first two digits in the left-hand column and then move across the table to locate the third at the top of the table. For each significant figure in the original number, include one place to the right of the decimal point.

**Example E-1**

What is the log of (a) 5.82, (b) 3.78, and (c) 1.11?

**Solution**

The answer is between 0 and 1 and is expressed as

0. log

log 5.82 = 0.765 (58 down, 2 across)

log 3.78 = 0.578 (37 down, 8 across)

log 1.11 = 0.045 (11 down, 1 across)

**Exercises**

**E-1** Find logarithms of the following numbers.

**(a)** 9.87 **(b)** 7.65 **(c)** 2.550 **(d)** 1.08

Answers: **(a)** 0.994 **(b)** 0.884 **(c)** 0.4065 **(d)** 0.033

## 2 Taking the Logs of Numbers Less Than 1 and Greater Than 10

To find the log of a number less than 1 or greater than 10, follow these steps:

1 Write the number in standard scientific notation (one digit to the left of the decimal point in the coefficient).

2 Look up the log of the coefficient and write as described in Section 1.

3 The value of the log of the coefficient is added to the value of the exponent of 10. That is a result of the general rule

$$\log (A \times B) = \log A + \log B$$

$$\log (A \times 10^B) = \log A + \log 10^B = (\log A) + B$$

The sum of the two numbers (the exponent plus the log of the coefficient) has two parts. *The number to the right of the decimal point is called the* **mantissa.** As mentioned before, the mantissa is expressed to the same number of decimal places as there are significant figures in the coefficient. *The whole number to the left of the decimal point is called the* **characteristic.** For example,

$$\log (2.65 \times 10^6) = \log 2.65 + \log 10^6 = 0.423 + 6$$
$$= \boxed{6}.\boxed{423}$$

Characteristic Mantissa

**Example E-2**
What is the log of the following numbers?

(a) $4.76 \times 10^3$

$$\log(4.76 \times 10^3) = \log 4.76 + \log 10^3$$
$$= 0.678 + 3 = \underline{\underline{3.678}}$$

(b) 15.8

$$\log 15.8 = \log(1.58 \times 10^1)$$
$$= \log 1.58 + \log 10^1$$
$$= 0.199 + 1 = 1.199$$

(c) 0.0666

$$\log 0.0666 = \log(6.66 \times 10^{-2})$$
$$= \log 6.66 + \log 10^{-2}$$
$$= 0.824 + (-2)$$
$$= 0.824 - 2 = \underline{\underline{-1.176}}$$

Notice that the logs of numbers less than 1 have *negative values.*

(d) $87.1 \times 10^{-8}$

$$\begin{aligned}\log(8.71 \times 10^{-7}) &= \log 8.71 + \log 10^{-7}\\ &= 0.940 + (-7)\\ &= 0.940 - 7 = \underline{\underline{-6.060}}\end{aligned}$$

**Exercises**

**E-2** Find the logs of the following numbers.

**(a)** 7.43 **(b)** 11.8 **(c)** 0.875 **(d)** 0.0878 **(e)** 85.2 **(f)** 1780 **(g)** $7.33 \times 10^4$ **(h)** $7.33 \times 10^{-4}$ **(i)** $4.11 \times 10^{-12}$ **(j)** 4.55 **(k)** $0.0577 \times 10^8$ **(l)** $32.8 \times 10^{-5}$ **(m)** $164 \times 10^6$ **(n)** $0.601 \times 10^{-7}$

**E-3** pH is defined as the negative of the log of a number (e..g, $-\log 10^{-2} = -(-2) = 2$). Find $-\log$ of the following.

**(a)** $8.55 \times 10^{-6}$ **(b)** $3.42 \times 10^{-11}$ **(c)** $71.9 \times 10^{-10}$ **(d)** $0.217 \times 10^{-3}$ **(e)** 1.15

Answers: **E-2(a)** 0.871 **E-2(b)** 1.072 **E-2(c)** −0.058 **E-2(d)** −1.056 **E-2(e)** 1.930 **E-2(f)** 3.250 **E-2(g)** 4.865 **E-2(h)** −3.135 **E-2(i)** −11.386 **E-2(j)** 0.658 **E-2(k)** 6.761 **E-2(l)** −3.484 **E-2(m)** 8.215 **E-2(n)** −7.221 **E-3(a)** 5.068 **E-3(b)** 10.466 **E-3(c)** 8.143 **E-3(d)** 3.664 **E-3(e)** −0.061

## 3 Finding the Antilog of a Positive Log

Not only should we be able to find the log of a number, but we must be able to find *the number whose log has a certain value.* This is called the **antilog** of a log. The procedure is simply the reverse of finding the log. Remember, we are asking the question, "What is the number whose exponent of 10 is a certain value?" For example, what is the antilog ($x$) of 2?

$$\log x = 2$$

$$x = \underline{\underline{10^2}} \qquad \text{since } \log 10^2 = 2$$

When the log is between 0 and 1, the antilog is between 1 and 10. (This is just the reverse of the statement in Section 1.) To find the antilog, locate the given log in the *body* of the log table. Find the two numbers on the left margin and the third on the top which corresponds to this log. Write the three numbers you found with one digit to the left of the decimal point.

**Example E-3**
What is the antilog of:

(a) 0.847

$$\text{Antilog } 0.847 = \underline{\underline{7.03}}$$

(b) 0.021

$$\text{Antilog } 0.021 = \underline{\underline{1.05}}$$

(c) 0.956

$$\text{Antilog } 0.956 = \underline{\underline{9.04}}$$

If the log is a number greater than 1, divide the number into two parts, the mantissa which lies to the right of the decimal and the characteristic which lies to the left. For example,

$$2.567 = 0.567 + 2$$

$$11.118 = 0.118 + 11$$

Take the antilog of the mantissa and express the number as explained above. This becomes the coefficient of the number. The characteristic then becomes the exponent of 10.

**Example E-4**
Find the antilog of (a) 4.655 and (b) 3.897.

(a) $4.655 = 0.655 + 4$

$$\text{Antilog } 0.655 = 4.52$$

$$\text{Antilog } 4 = 10^4$$

$$\text{Answer} = \underline{\underline{4.52 \times 10^4}}$$

(b) $3.897 = 0.897 + 3$

$$\text{Antilog } 0.897 = 7.89$$

$$\text{Antilog } 3 = 10^3$$

$$\text{Answer} = \underline{\underline{7.89 \times 10^3}}$$

## 4 Finding the Antilog of a Negative Log

A negative log must be adjusted so that the mantissa is positive without changing the actual value of the log itself. For example, $-0.182$ is the same number as $(+0.818 - 1)$, and $-3.289$ is the same number as $(+0.711 - 4)$. To do this, first separate and write the mantissa and the characteristic as described before. *Add* 1 to the mantissa and *subtract* 1 from the characteristic. Notice that if you add and subtract the same number from a quantity, you have not actually changed the number (e.g., $15 + 1 - 1 = 15$).

$$-3.289 = \overbrace{-0.289 - 3}^{\text{Separate}} = \overbrace{(-0.289 + 1)}^{\text{Add 1 to mantissa}} \overbrace{- 3 - 1}^{\text{Subtract 1 from characteristic}} = +0.711 - 4$$

$$-0.182 = -0.182 - 0 = (-0.182 + 1) - 0 - 1 = +0.818 - 1$$

Locate the mantissa in the body of the log table as described before and locate the corresponding numbers in the left-hand column and the top. The negative characteristic becomes the negative exponent of 10.

**Example E-5**
Find the following antilogs.

(a) $-0.013$

$$-0.013 = (-0.013 + 1) - 1 = +0.987 - 1$$

$$\text{Antilog } 0.987 = 9.71$$

$$\text{Antilog } (-1) = 10^{-1}$$

$$\text{Answer} = 9.71 \times 10^{-1} = \underline{\underline{0.971}}$$

(b) $-4.548$

$$-4.548 = -0.548 - 4 = (-0.548 + 1) - 4 - 1$$

$$= +0.452 - 5$$

$$\text{Antilog } 0.452 = 2.83$$

$$\text{Antilog } (-5) = 10^{-5}$$

$$\text{Answer} = \underline{\underline{2.83 \times 10^{-5}}}$$

### Exercises

**E-4** Find the antilog of **(a)** 0.813 **(b)** 3.459 **(c)** $-0.699$ **(d)** $-5.783$ **(e)** 10.945 **(f)** $-2.600$ **(g)** $-0.086$ **(h)** 8.401 **(i)** 0.345 **(j)** $-5.996$.

Answers: **E-4(a)** 6.50 **E-4(b)** $2.88 \times 10^{3}$ **E-4(c)** 0.200 **E-4(d)** $1.65 \times 10^{-6}$ **E-4(e)** $8.81 \times 10^{10}$ **E-4(f)** $2.51 \times 10^{-3}$ **E-4(g)** 0.820 **E-4(h)** $2.52 \times 10^{8}$ **E-4(i)** 2.21 **E-4(j)** $1.01 \times 10^{-6}$

Logarithms

| | 0 | 1 | 2 | 3 | 4 | 5 | 6 | 7 | 8 | 9 |
|---|---|---|---|---|---|---|---|---|---|---|
| 10 | 0000 | 0043 | 0086 | 0128 | 0170 | 0212 | 0253 | 0294 | 0334 | 0374 |
| 11 | 0414 | 0492 | 0453 | 0531 | 0569 | 0607 | 0645 | 0682 | 0719 | 0775 |
| 12 | 0792 | 0828 | 0864 | 0899 | 0934 | 0969 | 1004 | 1038 | 1072 | 1106 |
| 13 | 1139 | 1173 | 1206 | 1239 | 1271 | 1303 | 1335 | 1367 | 1399 | 1430 |
| 14 | 1461 | 1492 | 1523 | 1553 | 1584 | 1614 | 1644 | 1673 | 1703 | 1732 |
| 15 | 1761 | 1790 | 1818 | 1847 | 1875 | 1903 | 1931 | 1959 | 1987 | 2014 |
| 16 | 2041 | 2068 | 2095 | 2122 | 2148 | 2175 | 2201 | 2227 | 2253 | 2279 |
| 17 | 2304 | 2330 | 2355 | 2380 | 2405 | 2430 | 2455 | 2480 | 2504 | 2529 |
| 18 | 2553 | 2577 | 2601 | 2625 | 2648 | 2672 | 2695 | 2718 | 2742 | 2765 |
| 19 | 2788 | 2810 | 2833 | 2856 | 2878 | 2900 | 2923 | 2945 | 2967 | 2989 |
| 20 | 3010 | 3032 | 3054 | 3075 | 3096 | 3118 | 3139 | 3160 | 3181 | 3201 |
| 21 | 3222 | 3243 | 3263 | 3284 | 3304 | 3324 | 3345 | 3365 | 3385 | 3404 |
| 22 | 3424 | 3444 | 3464 | 3483 | 3502 | 3522 | 3541 | 3560 | 3579 | 3598 |
| 23 | 3617 | 3636 | 3655 | 3674 | 3692 | 3711 | 3729 | 3747 | 3766 | 3784 |
| 24 | 3802 | 3820 | 3838 | 3856 | 3874 | 3892 | 3909 | 3927 | 3945 | 3962 |
| 25 | 3979 | 3997 | 4014 | 4031 | 4048 | 4065 | 4082 | 4099 | 4116 | 4133 |
| 26 | 4150 | 4166 | 4183 | 4200 | 4216 | 4232 | 4249 | 4265 | 4281 | 4298 |
| 27 | 4314 | 4330 | 4346 | 4362 | 4378 | 4393 | 4409 | 4425 | 4440 | 4456 |
| 28 | 4472 | 4487 | 4502 | 4518 | 4533 | 4548 | 4564 | 4579 | 4594 | 4609 |
| 29 | 4624 | 4639 | 4654 | 4669 | 4683 | 4698 | 4713 | 4728 | 4742 | 4757 |
| 30 | 4771 | 4786 | 4800 | 4814 | 4829 | 4843 | 4857 | 4871 | 4886 | 4900 |
| 31 | 4914 | 4928 | 4942 | 4955 | 4969 | 4983 | 4997 | 5011 | 5024 | 5038 |
| 32 | 5051 | 5065 | 5079 | 5092 | 5105 | 5119 | 5132 | 5145 | 5159 | 5172 |
| 33 | 5185 | 5198 | 5211 | 5224 | 5237 | 5250 | 5263 | 5276 | 5289 | 5302 |
| 34 | 5315 | 5328 | 5340 | 5353 | 5366 | 5378 | 5391 | 5403 | 5416 | 5428 |
| 35 | 5441 | 5453 | 5465 | 5478 | 5490 | 5502 | 5514 | 5527 | 5539 | 5551 |
| 36 | 5563 | 5575 | 5587 | 5599 | 5611 | 5623 | 5635 | 5647 | 5658 | 5670 |
| 37 | 5682 | 5694 | 5705 | 5717 | 5729 | 5740 | 5752 | 5763 | 5775 | 5786 |
| 38 | 5798 | 5809 | 5821 | 5832 | 5843 | 5855 | 5866 | 5877 | 5888 | 5899 |
| 39 | 5911 | 5922 | 5933 | 5944 | 5955 | 5966 | 5977 | 5988 | 5999 | 6010 |
| 40 | 6021 | 6031 | 6042 | 6053 | 6064 | 6075 | 6085 | 6096 | 6107 | 6117 |
| 41 | 6128 | 6138 | 6149 | 6160 | 6170 | 6180 | 6191 | 6201 | 6212 | 6222 |
| 42 | 6232 | 6243 | 6253 | 6263 | 6274 | 6284 | 6294 | 6304 | 6314 | 6325 |
| 43 | 6335 | 6345 | 6355 | 6365 | 6375 | 6385 | 6395 | 6405 | 6415 | 6425 |
| 44 | 6435 | 6444 | 6454 | 6464 | 6474 | 6484 | 6493 | 6503 | 6513 | 6522 |
| 45 | 6532 | 6542 | 6551 | 6561 | 6571 | 6580 | 6590 | 6599 | 6609 | 6618 |
| 46 | 6628 | 6637 | 6646 | 6656 | 6665 | 6675 | 6684 | 6693 | 6702 | 6712 |
| 47 | 6721 | 6730 | 6739 | 6749 | 6758 | 6767 | 6776 | 6785 | 6794 | 6803 |
| 48 | 6812 | 6821 | 6830 | 6839 | 6848 | 6857 | 6866 | 6875 | 6884 | 6893 |
| 49 | 6902 | 6911 | 6920 | 6928 | 6937 | 6946 | 6955 | 6964 | 6972 | 6981 |
| 50 | 6990 | 6998 | 7007 | 7016 | 7024 | 7033 | 7042 | 7050 | 7059 | 7067 |
| 51 | 7076 | 7084 | 7093 | 7101 | 7110 | 7118 | 7126 | 7135 | 7143 | 7152 |
| 52 | 7160 | 7168 | 7177 | 7185 | 7193 | 7202 | 7210 | 7218 | 7226 | 7235 |
| 53 | 7243 | 7251 | 7259 | 7267 | 7275 | 7284 | 7292 | 7300 | 7308 | 7316 |
| 54 | 7324 | 7332 | 7340 | 7348 | 7356 | 7364 | 7372 | 7380 | 7388 | 7396 |

Logarithms

| | 0 | 1 | 2 | 3 | 4 | 5 | 6 | 7 | 8 | 9 |
|---|---|---|---|---|---|---|---|---|---|---|
| 55 | 7404 | 7412 | 7419 | 7427 | 7435 | 7443 | 7451 | 7459 | 7466 | 7474 |
| 56 | 7482 | 7490 | 7497 | 7505 | 7513 | 7520 | 7528 | 7536 | 7543 | 7551 |
| 57 | 7559 | 7566 | 7574 | 7582 | 7589 | 7597 | 7604 | 7612 | 7619 | 7627 |
| 58 | 7634 | 7642 | 7649 | 7657 | 7664 | 7672 | 7679 | 7686 | 7694 | 7701 |
| 59 | 7709 | 7716 | 7723 | 7731 | 7738 | 7745 | 7752 | 7760 | 7767 | 7774 |
| 60 | 7782 | 7789 | 7796 | 7803 | 7810 | 7818 | 7825 | 7832 | 7839 | 7846 |
| 61 | 7853 | 7860 | 7868 | 7875 | 7882 | 7889 | 7896 | 7903 | 7910 | 7917 |
| 62 | 7924 | 7931 | 7938 | 7945 | 7952 | 7959 | 7966 | 7973 | 7980 | 7987 |
| 63 | 7993 | 8000 | 8007 | 8014 | 8021 | 8028 | 8035 | 8041 | 8048 | 8055 |
| 64 | 8062 | 8069 | 8075 | 8082 | 8089 | 8096 | 8102 | 8109 | 8116 | 8122 |
| 65 | 8129 | 8136 | 8142 | 8149 | 8156 | 8162 | 8169 | 8176 | 8182 | 8189 |
| 66 | 8195 | 8202 | 8209 | 8215 | 8222 | 8228 | 8235 | 8241 | 8248 | 8254 |
| 67 | 8261 | 8267 | 8274 | 8280 | 8287 | 8293 | 8299 | 8306 | 8312 | 8319 |
| 68 | 8325 | 8331 | 8338 | 8344 | 8351 | 8357 | 8363 | 8370 | 8376 | 8382 |
| 69 | 8388 | 8395 | 8401 | 8407 | 8414 | 8420 | 8426 | 8432 | 8439 | 8445 |
| 70 | 8451 | 8457 | 8463 | 8470 | 8476 | 8482 | 8488 | 8494 | 8500 | 8506 |
| 71 | 8513 | 8519 | 8525 | 8531 | 8537 | 8543 | 8549 | 8555 | 8561 | 8567 |
| 72 | 8573 | 8579 | 8585 | 8591 | 8597 | 8603 | 8609 | 8615 | 8621 | 8627 |
| 73 | 8633 | 8639 | 8645 | 8651 | 8657 | 8663 | 8669 | 8675 | 8681 | 8686 |
| 74 | 8692 | 8698 | 8704 | 8710 | 8716 | 8722 | 8727 | 8733 | 8739 | 8745 |
| 75 | 8751 | 8756 | 8762 | 8768 | 8774 | 8779 | 8785 | 8791 | 8797 | 8802 |
| 76 | 8808 | 8814 | 8820 | 8825 | 8831 | 8837 | 8842 | 8848 | 8854 | 8859 |
| 77 | 8865 | 8871 | 8876 | 8882 | 8887 | 8893 | 8899 | 8904 | 8910 | 8915 |
| 78 | 8921 | 8927 | 8932 | 8938 | 8943 | 8949 | 8954 | 8960 | 8965 | 8971 |
| 79 | 8976 | 8982 | 8987 | 8993 | 8998 | 9004 | 9009 | 9015 | 9020 | 9025 |
| 80 | 9031 | 9036 | 9042 | 9047 | 9053 | 9058 | 9063 | 9069 | 9074 | 9079 |
| 81 | 9085 | 9090 | 9096 | 9101 | 9106 | 9112 | 9117 | 9122 | 9128 | 9133 |
| 82 | 9138 | 9143 | 9149 | 9154 | 9159 | 9165 | 9170 | 9175 | 9180 | 9186 |
| 83 | 9191 | 9196 | 9201 | 9206 | 9212 | 9217 | 9222 | 9227 | 9232 | 9238 |
| 84 | 9423 | 9248 | 9253 | 9258 | 9263 | 9269 | 9274 | 9279 | 9284 | 9289 |
| 85 | 9294 | 9299 | 9304 | 9309 | 9315 | 9320 | 9325 | 9330 | 9335 | 9340 |
| 86 | 9345 | 9350 | 9355 | 9360 | 9365 | 9370 | 9375 | 9380 | 9385 | 9390 |
| 87 | 9395 | 9400 | 9405 | 9410 | 9415 | 9420 | 9425 | 9430 | 9435 | 9440 |
| 88 | 9445 | 9450 | 9455 | 9460 | 9465 | 9469 | 9474 | 9479 | 9484 | 9489 |
| 89 | 9494 | 9499 | 9504 | 9509 | 9513 | 9518 | 9523 | 9528 | 9533 | 9538 |
| 90 | 9542 | 9547 | 9552 | 9557 | 9562 | 9566 | 9571 | 9576 | 9581 | 9586 |
| 91 | 9590 | 9595 | 9600 | 9605 | 9609 | 9614 | 9619 | 9624 | 9628 | 9633 |
| 92 | 9638 | 9643 | 9647 | 9652 | 9657 | 9661 | 9666 | 9671 | 9675 | 9680 |
| 93 | 9685 | 9689 | 9694 | 9699 | 9703 | 9708 | 9713 | 9717 | 9722 | 9727 |
| 94 | 9731 | 9736 | 9741 | 9745 | 9750 | 9754 | 9759 | 9763 | 9768 | 9773 |
| 95 | 9777 | 9782 | 9786 | 9791 | 9795 | 9800 | 9805 | 9809 | 9814 | 9818 |
| 96 | 9823 | 9827 | 9832 | 9836 | 9841 | 9845 | 9850 | 9854 | 9859 | 9863 |
| 97 | 9868 | 9872 | 9877 | 9881 | 9886 | 9890 | 9894 | 9899 | 9903 | 9908 |
| 98 | 9912 | 9917 | 9921 | 9926 | 9930 | 9934 | 9939 | 9943 | 9948 | 9952 |
| 99 | 9956 | 9961 | 9965 | 9969 | 9974 | 9978 | 9983 | 9987 | 9991 | 9996 |

# Appendix F
# Graphs

This section reviews the following graphing skills.

1 Direct linear relationships

2 Nonlinear relationships

In Appendix B we discussed proportionalities. In a proportionality a variable ($x$) is related to another variable ($y$) by some mathematical operation (i.e., a power, a root, a reciprocal, etc.). A graph translates a proportionality into a line that illustrates the relationship of $x$ to $y$. There are two types of graphs: linear and nonlinear. A linear graph is represented by a straight line and is a result of a direct, first-power proportionality between the variables $x$ and $y$. This is expressed as

$$x \propto y$$

Nonlinear graphs are represented by a curved line and result from any other type of proportionality such as power or an inverse proportion. Examples of such relationships are expressed as

$$x \propto y^2 \quad \text{and} \quad x \propto \frac{1}{y}$$

Graphs are useful in that the results of as few as two experiments (for a linear graph) can be used to give the values of many other measurements.

## 1 Direct Linear Relationships

A real example of a direct linear relationship is between the Kelvin temperature and the volume of a gas measured at constant pressure. This relationship has previously been expressed as

$$V \propto T$$

The following two experiments provide the information necessary to construct a graph:

**Experiment 1** At a temperature of 350 °C, the volume was 12.5 L.

**Experiment 2** At a temperature of 250 °C, the volume was 10.5 L.

Actually, it is always risky to assume that only two experiments provide an accurate graph or give all of the needed information, but for simplicity we will assume that it is valid in this case. We will check the graph with an additional experiment later.

To record this information on a graph, we first need some graph paper with regularly spaced divisions. On the vertical axis (called the ordinate or $y$ axis) we put temperature. Pick divisions on the graph paper that spread out the range of temperatures as much as possible. In our example the range of temperatures is between 250 °C and 350 °C, or 100 degrees. Notice that the two axes do not need to intersect at $x = y = 0$. We then plot the volume on the horizontal axis (called the abscissa or $x$ axis). The range of volumes is between 12.5 L and 10.5 L or 2.0 L.

The graph is shown in Figure F-1. An ⊙ is marked at the location of the two points from the experiments, and a straight line has been drawn between them.

We will now check the accuracy of our graph. A third experiment tells us that at $T = 300$ °C, $V = 11.5$ L. Refer back to the graph. Locate 300 °C on the ordinate and trace that line in to where it intersects the vertical line representing the volume of 11.5 L. Since the point of intersection is right on the line, we have confirmation of the accuracy of the original plot. *Normally, four or five measurements are required to provide enough points on the graph so that a line can be drawn through or near as many points as possible.*

A straight-line graph can be expressed by the linear algebraic equation

$$y = mx + b$$

where

$y =$ a value on the ordinate
$x =$ the corresponding value on the abscissa

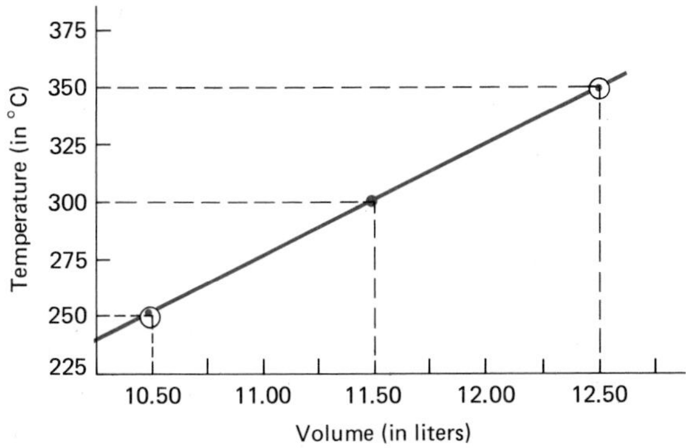


**Figure F-1**

$b$ = the point on the $y$ axis or ordinate where the straight line intersects the axis
$m$ = the slope of the line

The slope of a line is the ratio of the change on the ordinate to the change on the abscissa. It is determined by choosing two widely spaced points on the line ($x_1,y_1$ and $x_2,y_2$). The slope is the difference in $y$ divided by the difference in $x$.

$$m = \frac{y_2 - y_1}{x_2 - x_1}$$

The larger the slope *for a particular graph,* the steeper the straight line. For the graph shown in Figure F-1, if we pick point 1 to be $x_1 = 10.50$, $y_1 = 250$ and point 2 to be $x_2 = 12.50$, $y_2 = 350$, then

$$m = \frac{350 - 250}{12.50 - 10.50} = \frac{100}{2.00} = 50.0$$

The value for $b$ cannot be determined directly from Figure F-1. The value for $b$ is found by determining the point on the $y$ axis where the line intersects when $x = 0$. If we plotted the data on such a graph, we would find that $b = -273$°C. Therefore, the linear equation for this line is

$$y = 50.0x - 273$$

From this equation we can determine the value for $y$ (the temperature) by substitution of any value for $x$ (the volume).

## Exercises

**F-1** A sample of water was heated at a constant rate, and its temperature was recorded. The following results were obtained.

| Experiment | Temp (°C) | Time (min) |
|---|---|---|
| 1 | 10.0 | 0 |
| 2 | 20.0 | 4.5 |
| 3 | 30.0 | 7.5 |
| 4 | 55.0 | 17.8 |
| 5 | 85.0 | 30.0 |

**(a)** Construct a graph that includes all of the above information with temperature on the ordinate and time on the abscissa.

**(b)** Calculate the slope of the line and the linear equation $y = mx +$ b for the line.

**(c)** From the graph find the temperature at 11.0 min and compare it to the value calculated from the equation.

**(d)** From the graph find the time when the temperature is 65.0 °C and compare it to the value calculated from the equation.

Answers: **F-1(a)** Draw the line touching or coming close to as many points as possible. Notice that the straight line does not go through all points. **F-1(b)** $m = 2.5$, $y = 2.5x + 10$ **F-1(c)** From the equation, when $x = 11.0$, $y = 2.5 \cdot 11.0 + 10$, so $y = 37.5$ °C. The graph agrees. **F-1(d)** From the equation, when $y = 65$, $65 = 2.5x + 10$, and $x = 22.0$ min. The graph agrees.

## 2 Nonlinear Relationships

Thus far we have been dealing with direct linear relationships that produce straight lines when graphed. Other relationships, such as an inverse relationship (see Appendix B), produce curves when plotted on a graph. This is illustrated by the following example.

In the following experiments the pressure on a volume of gas was varied and the volume was measured at constant temperature.

| Experiment | Pressure (torr) | Volume (L) |
|---|---|---|
| 1 | 500 | 15.2 |
| 2 | 600 | 12.6 |
| 3 | 760 | 10.0 |
| 4 | 900 | 8.44 |
| 5 | 1200 | 6.40 |
| 6 | 1600 | 4.80 |

This information has been graphed in Figure F-2. Notice that the line has a gradually changing slope, starting with a very steep slope on the left and going to a small slope on the right of the graph. We can interpret this. In the region of the steep slope the increase in pressure causes little decrease in volume compared to the part of the curve with the small slope to the right. Notice that this is a plot of an inverse relationship. That is, the higher the pressure, the lower the volume. If we regraph the information and include much higher pressures and the corresponding volumes as shown in Figure F-3, we see that *the curve approaches the y axis but never actually touches it,* no matter how high the pressure. Such a curve is said to approach the axis **asymptotically.** In our example, the curve would eventually appear to change to a vertical line parallel to the $y$ axis.

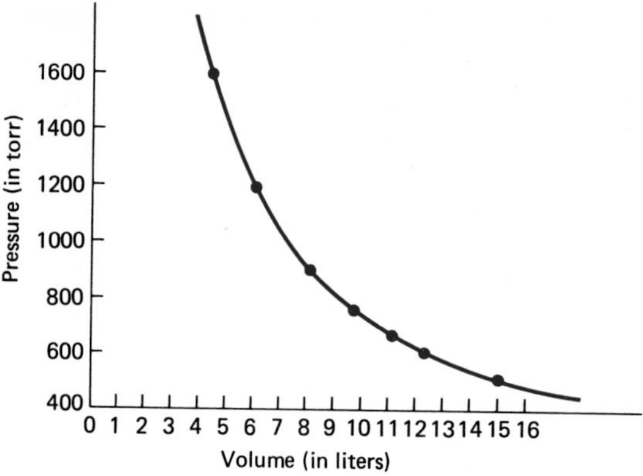


**Figure F-2** An inverse relationship.

## Exercises

**F-2** Construct a graph from the following experimental information. Plot the time on the abscissa and the concentration on the ordinate. Describe what this graph tells you about the change in concentration of HI as a function (how it varies with) the time.

| Experiment | Time (min) | Concentration of HI (mol/L) |
|---|---|---|
| 1 | 0.00 | $2.50 \times 10^{-2}$ |
| 2 | 5.0 | $1.50 \times 10^{-2}$ |
| 3 | 10.0 | $0.90 \times 10^{-2}$ |
| 4 | 15.0 | $0.65 \times 10^{-2}$ |
| 5 | 20.0 | $0.55 \times 10^{-2}$ |
| 6 | 35.0 | $0.50 \times 10^{-2}$ |

**(a)** From the graph, find the concentration of HI at 12.0 min; at 25 min.

**(b)** From the graph, find the time corresponding to when the concentration of HI is $2.00 \times 10^{-2}$; $0.75 \times 10^{-2}$.

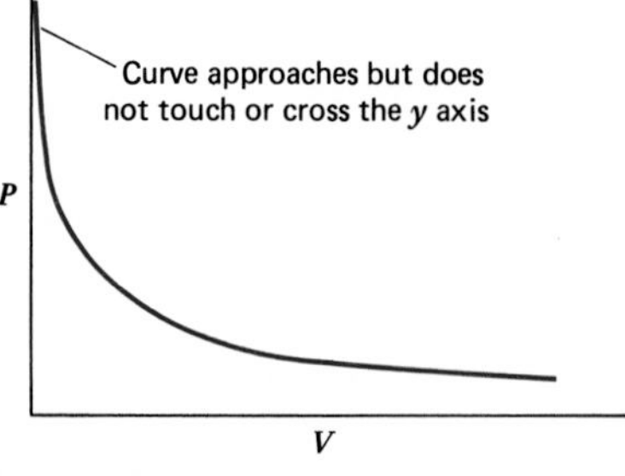


**Figure F-3**

Answers: **F-2** The concentration decreases rapidly at the start (the curve has a large slope) but then changes little after about 25 min. At that point the curve appears to change to a straight line parallel to the $x$ axis. **(a)** At 12.0 min, the concentration is $0.80 \times 10^{-2}$ mol/L; at 25 min, the concentration is about $0.52 \times 10^{-2}$ mol/L. **(b)** When the concentration is $2.00 \times 10^{-2}$ mol/L, the time is 2.4 min; when the concentration is $0.75 \times 10^{-2}$ mol/L, the time is 13.0 min.

# Appendix G
# Glossary*

## A

**Accuracy.** How close the value of a measurement is to the true value. (2)

**Acid.** A substance that increases the $H_3O^+$ ion concentration in water. In the Brønsted definition, an acid is a proton donor. (13)

**Acid anhydride.** A nonmetal oxide that reacts with water to form an oxyacid. (13)

**Acidic solution.** An aqueous solution that has a hydronium ion ($H_3O^+$) concentration greater than $10^{-7}$ *M* at 25 °C (pH less than 7). (13)

**Acid salt.** An ionic compound whose anion contains one or more acidic hydrogens. (13)

**Actinide.** One of 14 elements in which the 5*f* sublevel is filling (Th through Lr). (5)

**Activation energy.** The minimum kinetic energy that colliding reactant molecules must have in order to overcome the potential energy barrier between reactants and products. (15)

**Alcohol.** An organic compound containing at least one hydroxyl group (i.e., R—OH). (16)

**Aldehyde.** An organic compound containing a carbonyl group bound to a hydrogen and a hydrocarbon group or another hydrogen (i.e., $R-\overset{\overset{\displaystyle O}{\|}}{C}-H$). (16)

**Alkali metal.** An element in group IA of the periodic table. (5)

**Alkaline Earth metal.** An element in group IIA of the periodic table. (5)

**Alkane.** A hydrocarbon containing only single bonds (a saturated hydrocarbon). The general formula is $C_nH_{2n+2}$ for open chain and $C_nH_{2n}$ for cyclic alkanes. (16)

**Alkene.** A hydrocarbon containing at least one double bond (an unsaturated hydrocarbon). The general formula for open chain alkenes is $C_nH_{2n}$. (16)

**Alkylation.** The formation of large hydrocarbon molecules from small ones in the refining process. (16)

**Alkyl group.** A substituent that is an alkane minus a hydrogen atom [i.e., $C_2H_6$ (ethane) − H = $C_2H_5$ (ethyl)]. (16)

**Alkyne.** A hydrocarbon containing a triple bond. The general formula for open chain alkynes is $C_nH_{2n-2}$. (16)

**Allotropes.** Different forms of the same element (e.g., graphite and diamond are allotropes of carbon). (5)

**Alpha (α) particle.** A helium nucleus ($^4_2He^{2+}$) that is emitted from a radioactive nucleus. (3)

**Amide.** A derivative of a carboxylic acid in which an amine group replaces the hydroxyl group of the acid (i.e., $R-\overset{\overset{\displaystyle O}{\|}}{C}-NH_2$). (16)

**Amine.** An organic compound containing a nitrogen with single bonds to a hydrocarbon group and a total of two other hydrocarbon groups or hydrogens (i.e., $R-NH_2$). (16)

**Amphoteric.** A property of a substance indicating that it can act as an acid when a base is present or a base when an acid is present. (13)

**Anion.** A negatively charged ion. (3)

**Anode.** The electrode at which oxidation takes place in an electrolytic or voltaic cell. (14)

**Aromatic compounds.** The largest class of aromatic compounds are six-membered cyclic hydrocarbons containing alternating single and double bonds. (16)

**Arrhenius acids and bases.** A definition that defines acids as substances that produce $H^+$ ions and bases as substances that produce $OH^-$ ions in aqueous solution. (13)

* The number in parentheses at the end of each entry refers to the chapter in which the entry is first discussed.

**Atmosphere.** The sea of gases above the solid surface of the earth. Also, the pressure equivalent of 760 torr. (10)

**Atom.** The smallest particle of an element that can exist and enter into chemical reactions. (3)

**Atomic mass unit (amu).** A mass or weight that is exactly $\frac{1}{12}$ of the mass of an atom of $^{12}C$, which is defined as exactly 12 amu. One amu is equal to $1.6605 \times 10^{-24}$ g. (3)

**Atomic number.** The number of protons in the nucleus of the atom of an element. (3)

**Atomic radius.** The distance from the nucleus of an atom to the outermost electron sublevel. (5)

**Atomic theory.** A theory first proposed by John Dalton in 1803 concerning the basic structure of matter as being ultimately composed of atoms. (3)

**Atomic weight.** The weight of a certain isotope compared to $^{12}C$, which is defined as exactly 12 amu. For an element it is the (weighted) average of the atomic weights of all of the natural isotopes present. (3)

**Aufbau principle.** The rule that states that electrons fill the lowest available energy levels (sublevels) first. (4)

**Autoionization.** The partial dissociation of a pure covalent compound to produce a cation and an anion. (13)

**Avogadro's law.** Equal volumes of gases at the same temperature and pressure contain equal numbers of molecules. (10)

**Avogadro's number.** The number of objects or particles in one mole, equal to $6.02 \times 10^{23}$. (8)

## B

**Balanced equation.** A chemical equation that has the same number of atoms of each element and the same total charge on both sides of the equation. (9)

**Barometer.** A closed-end glass tube filled with mercury that is used to measure atmospheric pressure. (10)

**Base.** A substance that increases that $OH^-$ ion concentration in water. In the Brønsted definition, a base is a proton acceptor. (13)

**Basic anhydride.** A metal oxide that reacts with water to form an ionic hydroxide. (13)

**Basic solution.** An aqueous solution that has a hydroxide ion ($OH^-$) concentration greater than $10^{-7}$ *M* at 25 °C (pH greater than 7). (13)

**Battery.** A combination of one or more voltaic cells. (14)

**Beta (*β*) particle.** A high-energy electron emitted from a radioactive nucleus. (3)

**Binary compounds.** Molecules or formula units containing two different elements. (6)

**Bohr model.** A model of the atom in which electrons orbit around a central nucleus in discrete energy levels. (4)

**Boiling point.** The temperature at which bubbles of vapor form in a liquid. This occurs when the vapor pressure of the liquid equals the restraining or atmospheric pressure. (11)

**Boyle's law.** The volume of a quantity of gas is inversely proportional to the pressure at constant temperature. (10)

**Brønsted-Lowry acids and bases.** An acid is a proton ($H^+$) donor and a base is a proton acceptor. (13)

**Buffer solution.** An aqueous solution containing a weak acid and a salt providing its conjugate base (or a weak base and a salt providing its conjugate acid). A buffer solution resists changes in pH. (13)

**Buoyancy.** The ability of a solid or liquid to float in a certain liquid. (2)

## C

**Calorie.** The amount of heat required to raise the temperature of one gram of water 1 °C. The precise definition is 1 cal = 4.184 joule (J). (2)

**Carbonyl group.** A functional group made up of a carbon with a double bond to an oxygen (i.e., $-\overset{\overset{\displaystyle O}{\|}}{C}-$). It is the functional group of aldehydes and ketones. (16)

**Carboyxlic acid.** Organic compounds containing a carboxyl group (i.e., $R-\overset{\overset{\displaystyle O}{\|}}{C}-OH$). (16)

**Catalyst.** A substance that is not consumed in a reaction but whose presence increases the rates of both the forward and the reverse reaction. A catalyst provides an alternate mechanism of lower activation energy but does not shift the point of equilibrium. (15)

**Cathode.** The electrode at which reduction takes place in an electrolytic or voltaic cell. (14)

**Cation.** A positively charged ion. (3)

**Celsius scale (°C).** A temperature scale with 100 equal divisions between the freezing point of water (defined as 0 °C) and the normal boiling point of water (defined as 100 °C). (2)

**Chain reaction.** A self-sustaining reaction. In a nuclear chain reaction one reacting neutron produces an average of three other neutrons that in turn cause reactions. (3)

**Chalcogens.** Elements in group VIA in the periodic table. (5)

**Charles' law.** The volume of a quantity of gas is directly proportional to the Kelvin temperature at constant pressure. (10)

**Chemical bond.** The forces of attraction that hold atoms or ions together in a compound. (3)

**Chemical change.** The change of substances into substances of a different composition. (1)

**Chemical equation.** The representation of a chemical reaction by means of chemical symbols and coefficients. (8)

**Chemical kinetics.** The study of the interrelationships of reaction rates and the mechanism of a reaction. (15)

**Chemical property.** The property of a substance that concerns its ability or tendency to change into, or react with, other substances. (1)

**Chemistry.** The branch of science dealing with the nature, composition, and structure of matter and the changes it undergoes. (1)

**Colligative property.** A property that depends on the relative number of particles of solute and solvent in a solution rather than their identities. Vapor pressure lowering, boiling point elevation, freezing point depresssion, and osmotic pressure are colligative properties of solutions. (12)

**Collision theory.** A theory suggesting that chemical reactions result from collisions of reactant molecules having proper orientation and sufficient energy. (15)

**Combination reactions.** A chemical reaction where one compound results from the union of elements and/or other compounds. (9)

**Combined gas law.** A combination of Boyle's, Gay-Lussac's, and Charles' laws (i.e., $PV/T$ = a constant). (10)

**Combustion reaction.** A chemical reaction in which a compound or element reacts with oxygen. The reaction usually results in the liberation of large amounts of heat and light. (1)

**Compound.** A pure substance composed of two or more different elements combined in fixed proportions. (1)

**Concentrated.** A solution containing a relatively large amount of a specified solute per unit volume. (12)

**Concentration.** The amount of solute present in a specified amount of solvent or solution. (12)

**Condensation.** The formation of a liquid from the vapor or gaseous state. (11)

**Conductor.** A substance such as a metal that allows the flow of electricity. (12)

**Conjugate acid (base).** The conjugate acid of a base is formed by the addition of one $H^+$ to the base. The conjugate base of an acid is formed by the removal of one $H^+$ from the acid. (13)

**Conversion factor.** An equivalency relationship expressed as a ratio. It is used to convert quantities in one unit of measurement into quantities of another. (2)

**Covalent bond.** Two atoms held together in an ion or molecule by sharing one to three pairs of electrons. (6)

**Cracking.** The formation of small hydrocarbon molecules from large ones in the refining process. (16)

**Crystal.** A solid substance made up of ions or molecules in an orderly and regular geometric arrangement. (3)

## D

**Dalton's law of partial pressure.** The total pressure exerted by a mixture of gases is the sum of the partial pressures of the component gases. (10)

**Daniell cell.** A voltaic cell made up of zinc and copper electrodes each immersed in solutions of their ions connected by a salt bridge. (14)

**Decomposition reaction.** A chemical reaction in which one compound is decomposed into elements and/or other compounds. (9)

**Density.** The ratio of the mass of a substance (usually in grams) to its volume (usually in milliliters or cubic centimeters). In other words, it is mass per unit volume. (2)

**Derivative.** A compound produced by the substitution of a group on the molecules of the original or parent compound. (16)

**Dilute.** A solution containing a relatively small amount of a specified solute per unit volume. (12)

**Dilution.** The preparation of a dilute solution from a concentrated solution by addition of more solvent. (12)

**Dipole.** The existence of a partial charge separation in a polar covalent bond. (6)

**Discrete spectra.** The definite wavelengths of light (spectral lines) that are emitted by the hot gaseous atoms of an element. (4)

**Dissociation.** The breakup of one substance into simpler substances. This includes the formation of ions when certain compounds dissolve in water. (13)

**Distillation.** The process of separation of a solution into its components by vaporization (boiling) followed by condensation of the more volatile component. (1)

**Double bond.** A covalent bond in which two pairs of electrons are shared between two atoms. (6)

**Double displacement reaction.** A reaction involving an exchange of ions in a solution that leads to the formation of a precipitate or a covalent compound. (9)

**Dry cell.** A voltaic cell composed of zinc and graphite electrodes immersed in an aqueous paste of $NH_4Cl$ and $MnO_2$. (14)

**Dynamic equilibrium.** The state at which opposing changes occur at equal rates. In a chemical reaction it is the point at which the rate of the forward reaction equals the rate of the reverse reaction. (13)

## E

**Electricity.** A flow of electrons through a conductor. (12)

**Electrode.** A conducting surface at which reactions take place in a voltaic or electrolytic cell. (14)

**Electrolysis.** The process of forcing electrical energy through an electrolytic cell, thereby causing a chemical reaction. (14)

**Electrolyte.** A compound whose aqueous solution or pure melt conducts electricity. (12)

**Electrolytic cell.** An electrochemical cell that converts electrical energy into chemical energy. (14)

**Electron.** A subatomic particle with a charge of negative one ($-1$). In an atom, it exists outside of the nucleus and its mass is very small compared with that of the entire atom. (3)

**Electron affinity.** The energy released when a gaseous atom adds an electron to form a gaseous ion. (5)

**Electron configuration.** The arrangement of electrons in an atom according to shell and sublevel. (4)

**Electronegativity.** A measure of the tendency of a certain atom to attract electrons to itself in a chemical bond. (6)

**Electrostatic forces.** Forces of attraction between centers of opposite charge and forces of repulsion between centers of the same charge. (3)

**Element.** A pure substance that is one of the 108 basic forms of nature. Elements cannot be separated into simpler substances by ordinary chemical means. (1)

**Empirical formula.** A formula showing the simplest whole-number ratio of the atoms of the elements that make up the compound. (8)

**Endothermic reaction.** A chemical reaction that absorbs heat from the surroundings. (1)

**Energy.** The ability to do work. It can take the form of light, heat, mechanical, chemical, electrical, or nuclear. It can be of two types: kinetic or potential. (1)

**Equilibrium.** See *dynamic equilibrium.*

**Equilibrium constant.** A number that defines the position of equilibrium for a particular reaction at a specified temperature. (15)

**Ester.** A derivative of a carboxylic acid in which a hydrocarbon group (R′) replaces the hydrogen on the oxygen in the parent acid (i.e.,

$$R-\overset{\overset{\displaystyle O}{\|}}{C}-O-R'$$

). (16)

**Ether.** An organic compound containing an oxygen bonded to two hydrocarbon groups (i.e., R — O — R′). (16)

**Evaporation.** The vaporization of a liquid below its boiling point. (11)

**Exact number.** A number that results from a definition or an actual count. (2)

**Excited state.** The presence of an electron in an atom in an energy state higher than the lowest available state. (4)

**Exothermic reaction.** A chemical reaction that releases heat energy to the surroundings. (1)

**Extrapolate.** To extend a line on a graph beyond the experimental points. (10)

## F

**Fahrenheit scale (°F).** A temperature scale with 180 equal divisions between the freezing point of water (32 °F) and the normal boiling point of water (212 °F). (2)

**Family.** See *group.*

**Filtration.** A laboratory procedure whereby a solid phase is separated from a liquid phase in a heterogeneous mixture. (1)

**Fission.** The splitting of a large, unstable nucleus into two smaller nuclei of similar size with a production of energy. (3)

**Formula.** The symbols of the elements and the number of atoms of each element that make up a compound. (3)

**Formula unit.** The formula of an ionic compound. It represents the ratio of ions present. (3)

**Formula weight.** The total weight of all of the atoms in a molecule (or formula unit for an ionic compound) expressed in atomic mass units (amu). (8)

**Freezing point.** See *melting point.*

**Functional group.** The part of an organic molecule where most of the chemical reactions occur. (16)

**Fusion (nuclear).** The combination of two small nuclei to form a larger nucleus with the production of energy. (3)

## G

**Gamma ray.** A high-energy form of light emitted from a radioactive nucleus. (3)

**Gas.** One of the three physical states of matter. A gas does not have a fixed shape or volume and thus fills any container uniformly. (1)

**Gas laws.** The various laws governing the behavior of gases. Gas laws are consistent with the kinetic theory of gases. (10)

**Gay-Lussac's law.** The pressure of a gas is directly proportional to the Kelvin temperature at constant volume. (10)

**Graham's law of diffusion.** The rate of diffusion of a gas is inversely proportional to the square root of its formula weight. (10)

**Ground state.** The lowest possible energy state available to an electron in an atom. (4)

**Group.** All of the elements in a vertical column in the periodic table. (4)

## H

**Half-life.** The time required for one-half of a given sample of an isotope to undergo radioactive decay. (3)

**Half-reaction.** A balanced equation for an oxidation or a reduction process. (14)

**Halogen.** An element from group VIIA of the periodic table. (5)

**Heat of fusion.** The heat required to melt a given mass of a substance at its melting point. (11)

**Heat of vaporization.** The heat required to vaporize a given mass of a substance at its boiling point. (11)

**Heterogeneous matter.** A nonuniform mixture of substances consisting of two or more phases with definite boundaries between the phases. (1)

**Homogeneous matter.** Matter existing in one phase or physical state with the same properties and composition throughout. Homogeneous matter may be either a solution or a pure substance. (1)

**Homologous series.** A series of related compounds in which one member differs from the next by a constant number of atoms. (16)

**Hund's rule.** Electrons occupy separate orbitals of the same energy level with their spins parallel. (4)

**Hydrate.** A compound that contains one or more bonded water molecules that can generally be removed by heating. (11)

**Hydration.** The formation of attractive interactions between polar water molecules and solute ions or molecules. (12)

**Hydrocarbon.** An organic compound containing only carbon and hydrogen. (16)

**Hydrogen bond.** A relatively strong dipole–dipole interaction between a hydrogen bonded to a highly electronegative atom (N, O, or F) and a highly electronegative atom on another molecule. (11)

**Hydrolysis.** The reaction of an anion with water to produce a basic solution or the reaction of a cation with water to produce an acidic solution. (13)

**Hydrometer.** A device used to directly measure the density or specific gravity of a liquid. (2)

**Hydronium ion.** The $H_3O^+$ ion, which represents the $H^+$ ion in aqueous solution. (13)

**Hydroxyl group.** The —OH group as found in an organic compound. It is the functional group of alcohols. (16)

## I

**Ideal gas.** A hypothetical gas whose molecules are considered to have no volume or interactions with each other. An ideal gas obeys the ideal gas law at all temperatures and pressures. (10)

**Ideal gas law.** $PV = nRT$. (10)

**Infrared light.** Light with wavelengths somewhat longer than red light. (4)

**Inner transition element.** Elements in which the *f* sublevel is filling (the lanthanides and actinides). (5)

**Ion.** An atom or a group of atoms that have acquired an electrical charge. (3)

**Ion product of water.** The equilibrium product of the $H_3O^+$ and $OH^-$ molarities in water. At 25 °C the product, $K_w$, is a constant that is equal to $1.0 \times 10^{-14}$. (13)

**Ionic bond.** The ion–ion electrostatic attractions between oppositely charged ions. (6)

**Ionization.** The process of forming an ion or ions from a molecule or atom. (6)

**Ionization constant.** The equilibrium constant for the ionization (dissociation) of a weak acid ($K_a$) or a weak base ($K_b$). (15)

**Ionization energy.** The energy required to remove an electron from a gaseous atom to form a gaseous ion. (5)

**Isomers.** Compounds with the same molecular formula but different molecular structures. (16)

**Isotopes.** Atoms with different mass numbers but the same atomic number. (3)

## K

**Kelvin scale (K).** The temperature scale in which zero Kelvin is known as absolute zero, which is the lowest possible temperature. The Kelvin scale ($T$) relates to the Celsius scale [$t$(C)] by $T = [t(\mathrm{C}) + 273]$ K. (2)

**Ketone.** An organic compound containing a carbonyl group bonded to two hydrocarbon groups (i.e., $\mathrm{R{-}\overset{\overset{\displaystyle O}{\|}}{C}{-}R'}$). (16)

**Kinetic energy.** Energy due to motion. K.E. = $\frac{1}{2}$[mass × (velocity)$^2$]. (1)

**Kinetic theory of gases.** The assumptions of the nature of a gas that explain the gas laws and other observations. (10)

## L

**Lanthanide.** One of the 14 elements in which the 4*f* sublevel is filling (Ce through Lu). (5)

**Lattice.** The regular three-dimensional arrangement of ions or molecules in a solid crystal. (6)

**Law of mass action.** The relationship that governs the relative distribution of reactants and products at equilibrium. It is composed of an equilibrium constant and a ratio called the mass action expression. (15)

**Lead-acid battery.** A rechargeable voltaic battery composed of lead and lead dioxide electrodes in a sulfuric acid solution. (14)

**Le Châtelier's principle.** When stress is applied to a system at equilibrium, the system reacts in such a way so as to counteract the stress. (15)

**Lewis structure.** The representation of an atom, molecule, or ion showing all outer *s* and *p* electrons as dots and/or dashes. (6)

**Limiting reactant.** The reactant that would produce the least amount of products if completely consumed. (9)

**Liquid.** One of the three physical states of matter. Liquids have a fixed volume but not a fixed shape. Liquids flow and thus take the shape of the bottom of the container. (1)

## M

**Mass.** The quantity of matter (usually in grams or kilograms) that a substance contains. (2)

**Mass action expression.** The fraction containing the concentrations of products in the numerator and reactants in the denominator, each concentration raised to the power corresponding to its coefficient in the balanced equation. (15)

**Mass number** The number of nucleons in an atom. (3)

**Matter.** Anything that has mass and occupies space. (1)

**Measurement.** The quantity, dimensions, or extent of something, usually in comparison to a standard. (2)

**Mechanism.** The means by which reactant molecules collide or otherwise behave so as to eventually form products. (15)

**Melting point.** The temperature at which a solid changes to a liquid. (11)

**Metal.** An element that can lose one or more electrons easily to form a cation. Physically, metals are solids (except Hg) and they conduct heat and electricity. (5)

**Metalloid.** An element with properties intermediate between metals and nonmetals. (5)

**Metallurgy.** The conversion of metal ores into metals. (1)

**Metathesis reaction.** A double displacement reaction in which a solid compound forms from soluble compounds. (9)

**Metric system.** A system of measurement based on multiples of 10. (2)

**Molality.** A unit of concentration equal to the moles of solute divided by the kilograms of solvent. (12)

**Molar mass.** The mass in grams of one mole of an element or compound. (8)

**Molar number.** See *Avogadro's number.*

**Molar volume.** The volume of one mole of gas at STP, which is 22.4 L. (10)

**Molarity.** A unit of concentration equal to the moles of solute divided by the liters of solution. (12)

**Mole.** The SI unit of quantity. It is equal to $6.02 \times 10^{23}$ particles. (8)

**Molecular formula.** The formula of a compound showing the exact number of atoms of each element in one molecule of a compound. (8)

**Molecular weight.** See *formula weight.*

**Molecule.** A group of two or more atoms held together by covalent bonds. (3)

**Monomer.** Single molecules that can be joined together in a chain to form a repeating structure called a polymer. (16)

## N

**Net ionic equation.** The balanced ionic equation written without spectator ions. (12)

**Neutral solution.** An aqueous solution in which $[H_3O^+] = [OH^-] = 1.0 \times 10^{-7}$ *M*. In a neutral solution pH = 7.0 at 25 °C. (13)

**Neutralization.** The reaction of an acid and a base to form a salt and water. (13)

**Neutron.** A neutral subatomic particle that exists in the nucleus of the atom. It has a mass of approximately 1 amu. (3)

**Noble gas.** An element in group **0** in the periodic table. (5)

**Nonconductor.** A substance that does not allow the passage of an electric current under normal conditions. (12)

**Nonelectrolyte.** A covalent compound that dissolves in water without formation of ions. A solution of a nonelectrolyte does not conduct electricity. (12)

**Nonmetals.** Elements in the upper right-hand corner of the periodic table. They are characterized by high ionization energies. (5)

**Nonpolar bond.** A covalent bond where the electrons are shared equally between the two atoms. (6)

**Normal boiling point.** The temperature at which the vapor pressure of a liquid is equal to 1 atm pressure. (11)

**Nuclear equation.** A symbolic representation of the changes of a nucleus or nuclei into other nuclei and particles. (3)

**Nuclear reactor.** A device that generates electrical energy from a controlled nuclear fission reaction. (3)

**Nucleons.** The protons and neutrons that make up the nucleus of the atom. (3)

**Nucleus.** The small, dense interior of the atom containing the nucleons. (3)

## O

**Octet rule.** The generalization that states that the atoms of the representative elements tend to form bonds in such a manner as to have access to eight valence electrons. (6)

**Orbital.** A region of space in which there is the highest probability of finding an electron. Each orbital of a certain sublevel can hold up to two electrons. (4)

**Organic chemistry.** The branch of chemistry that deal with most of the compounds of carbon. (5)

**Osmosis.** The process whereby there is a net movement of solvent molecules through a semipermeable membrane from a dilute to a concentrated solution. (12)

**Osmotic pressure.** The pressure on one side of a membrane that is needed to prevent the net movement of solvent molecules from a dilute to a concentrated solution. (12)

**Oxidation.** The loss of electrons that results in an increase in oxidation state. (14)

**Oxidation–reduction (redox) reactions.** A chemical reaction involving an exchange of electrons between two substances. (14)

**Oxidation state.** The charge that the atoms of an element would have if the electrons in each bond are arbitrarily assigned to the more electronegative element. (7)

**Oxidizing agent.** A substance that accepts the electrons from the substance oxidized (the reducing agent). (14)

**Oxyacid.** An acid derived from hydrogen plus an oxyanion. (7)

**Oxyanion.** An anion containing oxygen and at least one other element. (7)

## P

**Partial pressure.** The pressure due to one component in a mixture of gases. (10)

**Pauli exclusion principle.** A principle that states that no two electrons in an atom can reside in the same orbital with the same spin. (4)

**Percent by weight.** A unit of concentration where the weight of solute is expressed as a percent of the weight of the solution. (12)

**Percent composition.** The weight of a component substance (i.e., an element) expressed as a percent of the total weight (i.e., the formula weight). (8)

**Percent yield.** The ratio of the actual yield to the theoretical yield times 100%. (9)

**Period.** The elements between noble gases in the periodic table. (5)

**Periodic Table.** An arrangement of elements in order of increasing atomic number. Elements with the same number of outer electrons are arranged in vertical columns. (4)

**pH.** An expression of the $H_3O^+$ concentration in an aqueous solution as the negative of the logarithm of the $H_3O^+$ molarity. (13)

**Phase.** A physically distinct state, either solid, liquid, or gas. (1)

**Physical change.** A change that does not alter the chemical composition of a substance. Usually a change in physical state or shape. (1)

**Physical property.** A property of a substance that can be observed or measured without changing the substance into some other substance. (1)

**Physical science.** A science concerned with the natural laws of matter other than those concerned with life or the earth. (2)

**pOH.** An expression of the $OH^-$ concentration in an aqueous solution as the negative of the logarithm of the $OH^-$ molarity. (13)

**Polar covalent bond.** A covalent bond in which the pair of electrons resides closer to one atom than the other, thereby establishing a separation of charge (a dipole). (6)

**Polyatomic ion.** An ion containing more than one atom. (6)

**Polymer.** A high-molecular-weight substance made from repeating units of an alkene or from combinations of other molecules. (16)

**Potential energy.** Energy that is available because of position or composition. (2)

**Precipitate.** A solid compound that is formed in solution and settles to the bottom of the container. (12)

**Precision.** How close repeated measurements come to each other. The more precise a measurement, the more significant figures that are expressed. (2)

**Pressure.** Force per unit area. (10)

**Principal quantum number (*n*).** A number that

corresponds to a particular shell or orbit that is occupied by the electrons in an atom. (4)

**Products.** The compounds and/or elements that result from a chemical reaction. (9)

**Property.** A description of a unique, observable characteristic or trait. (1)

**Proton.** A positively charged subatomic particle that exists in the nucleus of an atom. The mass of a proton is approximately 1 amu. (3)

**Pure substance.** Matter that has definite and unchanging composition and properties. Pure substances are either elements or compounds. (1)

## R

**Radiation.** Particles or high-energy light rays that are emitted by an atom or a nucleus of an atom. (3)

**Radioactive decay series.** A series of elements formed from the successive emission of alpha and beta particles starting from a long-lived isotope and ending with a stable isotope. (3)

**Radioactivity.** The emission of energy or particles from an unstable nucleus. (3)

**Radionuclide.** A radioactive isotope. (3)

**Rate law.** A relationship between the rate of a reaction at a certain temperature and the concentrations of any reactants that affect the rate of the reaction. (15)

**Reactants.** The compounds or elements that undergo a chemical reaction. (9)

**Redox reaction.** See *oxidation–reduction reaction.*

**Reducing agent.** A substance that donates the electrons to the substance reduced (the oxidizing agent). (14)

**Reduction.** The gain of electrons, resulting in a decrease in oxidation state. (14)

**Reforming.** The rearrangement of long-chain hydrocarbons to branched hydrocarbons and/or the removal of hydrogen from hydrocarbons. (16)

**Representative elements.** Groups IA through VIIA in the periodic table. (5)

**Resonance structures.** The representation of all possible Lewis structures of a molecule or ion having the same number of electrons and arrangement of atoms. (6)

## S

**Salt.** An ionic compound formally composed of a positive ion from a base and a negative ion from an acid. (13)

**Saturated hydrocarbon.** A hydrocarbon containing only single covalent bonds (an alkane). (16)

**Saturated solution.** A solution containing the maximum amount of a specified solute that can dissolve at that temperature. (12)

**Scientific notation.** A number where nonsignificant zeros in a very large or small number are expressed as power of 10. (2)

**Semipermeable membrane.** A thin porous membrane that allows passage of small solvent molecules but blocks passage of solute molecules or ions. (12)

**Shell (orbit).** The principal energy state for electrons in an atom. Its number relates to the most probable distance of the electrons from the nucleus. (4)

**Shielding.** The effect that one electron has on another as to the nuclear charge it feels. (5)

**Significant figure.** A valid digit (i.e., 0 through 9) in a measurement. The number of significant figures in a measurement refers to its degree of precision. (2)

**SI units.** The international system of units of measurement. (2)

**Single replacement reaction.** A chemical reaction in which an element is substituted for another element in a compound (9)

**Solid.** One of the three physical states of matter. A solid is characterized by a fixed volume and a fixed shape. (1)

**Solubility.** The maximum amount of a solute that dissolves in a given amount of solvent at a certain temperature. (12)

**Solute.** A substance that dissolves in a solvent to form a homogeneous solution. (12)

**Solution.** A homogeneous mixture of two or more substances with uniform properties and composition. (1)

**Solvent.** A medium that disperses another substance called a solute to form a homogeneous mixture. The solvent and the solution exist in the same physical state. (12)

**Specific gravity.** The ratio of the density of a

substance to the density of water at the same temperature. (2)

**Specific heat.** The heat required to raise the temperature of one gram of a substance one degree Celsius (1 °C). (2)

**Spectator ion.** An ion that is not directly involved in an aqueous reaction and does not appear in the net ionic equation. (12)

**Spectrum.** The separate color components of a beam of light. (4)

**Stock method.** The naming of metal ions in a compound by writing the name of the metal followed by its oxidation state in Roman numerals. (7)

**Stoichiometry.** The quantitative relationships among reactants and products in a chemical reaction. (9)

**STP.** Standard temperature and pressure: 1 atm pressure (760 torr) and 0 ° (273 K). (10)

**Strong electrolyte.** A substance that is essentially dissociated completely into ions in aqueous solution. These solutions are good conductors of electricity. (12)

**Structural formula.** The formula of a compound written in such a way that illustrates the sequence of bonds. (3)

**Sublevel.** One of the subdivisions (*s*, *p*, *d*, or *f*) of a shell of electrons in an atom. The orbitals of each sublevel have their characteristic shape. (4)

**Sublimation.** The vaporization of a solid. (11)

**Substance.** A pure form of matter (i.e., an element or a compound). (1)

**Substituent.** An atom or group of atoms that can substitute for a hydrogen atom in an organic compound. (16)

**Symbol.** The letter or letters used to represent an element. (1)

## T

**Temperature.** A measure of the intensity of heat (the average kinetic energy) within any form of matter. (2)

**Theoretical yield.** The maximum amount of a product that can form when at least one of the reactants is completely consumed. (9)

**Thermometer.** A device used to measure temperature. (2)

**Torr.** A unit of gas pressure equivalent to one millimeter of mercury (1 mm Hg). (10)

**Total ionic equation.** A reaction in aqueous solution involving ionic compounds written so that all ions are shown independently in the equation. (12)

**Transition element.** Elements where the *d* sublevel is filling. These elements are designated as the B elements in the periodic table. (5)

**Transmutation.** The changing of one element into another by radioactive decay or a nuclear reaction. (3)

**Triple bond.** A covalent bond where three pairs of electrons are being shared between two atoms. (6)

## U

**Ultraviolet light.** Light with wavelengths somewhat shorter than violet light. (4)

**Unit-factor.** A problem solving method whereby units and conversion factors are used as an aid in obtaining the answer. (2)

**Unity factor.** A conversion factor that relates a quantity of a certain unit to a quantity of "one" of another unit. (2)

**Unsaturated hydrocarbon.** A hydrocarbon that contains at least one double or triple bond. (16)

**Unsaturated solution.** A solution containing less than the maximum amount of a specified solute that can dissolve at that temperature. (12)

## V

**Valence electrons.** The outer *s* and *p* electrons of a representative element. (6)

**Vapor pressure.** The pressure exerted by the vapor in equilibrium with its liquid or solid form at a specified temperature. (11)

**Volatile.** A compound having a significant vapor pressure. (11)

**Voltaic cell.** A device that converts the chemical energy from a spontaneous redox reaction directly to electrical energy. (14)

W

**Wavelength ($\lambda$).** The distance between two peaks on a wave. (4)

**Weak electrolyte.** A molecular compound that dissolves in water but is only partially dissociated to produce a limited concentration of ions. These solutions are weak conductors of electricity. (12)

**Weight.** The measure of the attraction of gravity for a substance. (2)

# Appendix H
# Answers to Problems

## Chapter 1

**1-1** A liquid (c) has a fixed volume but not a fixed shape.

**1-2** (a) liquid phase only

**1-3** (a) homogeneous (b) heterogeneous (c) heterogeneous (d) homogeneous (e) homogeneous (f) homogeneous (g) heterogeneous (h) heterogeneous (i) homogeneous

**1-5** (a) liquid (b) various solid phases (c) gas and liquid (d) liquid (e) solid (f) liquid (g) solid and liquid (h) liquid and gas (i) gas

**1-7** Carbon dioxide is a compound that is composed of carbon and oxygen. It can be *made* from a mixture of carbon and oxygen, but carbon dioxide is no longer a mixture after the elements have chemically combined.

**1-8** (a) compound (b) element (c) element (d) compound (e) compound (f) compound

**1-10** (a) solution (b) heterogeneous mixture (c) element (d) compound (e) solution

**1-13** (a) Ba (b) Ne (c) Cs (d) Pt (e) Mn (f) W

**1-15** (a) boron (b) bismuth (c) germanium (d) uranium (e) cobalt (f) mercury (g) beryllium (h) arsenic

**1-16** (a) physical (b) chemical (c) physical (d) chemical (e) chemical (f) physical (g) physical (h) chemical (i) physical

**1-18** (a) chemical (b) physical (c) physical (d) chemical (e) physical

**1-20** When a log burns, most of the compounds formed in the combustion are gases and dissipate into the atmosphere. Only some solid residue (ashes) is left. When zinc and sulfur (both solids) combine, the only product is a solid so there is no weight change. When iron burns, however, its only product is a solid. It weighs more than the original iron because it has combined with oxygen gas in the air.

**1-21** (a) exothermic (b) endothermic (c) endothermic (d) exothermic (e) exothermic

**1-22** Gasoline is converted into *heat* energy when it burns. The heat energy moves the pistons, which is *mechanical* energy. The mechanical energy turns the alternator, which generates *electrical* energy. The electrical energy is converted into *chemical* energy in the battery.

**1-24** (a) potential (b) kinetic (c) potential (It is stored energy because of composition.) (d) kinetic (e) kinetic

## Chapter 2

**2-1** (b) 74.212 gal (the most significant figures)

**2-2** A device used to produce a measurement may provide a reproducible answer to several significant figures, but if the device itself is inaccurate (such as a ruler with the tip broken off) the measurement is inaccurate.

**2-4** (a) three (b) two (c) three (d) one (e) four (f) two (g) two (h) three

**2-6** (a) 16.0 (b) 1.01 (c) 0.665 (d) 4890 (e) 87,600 (f) 0.0272 (g) 301

**2-8** (a) 188 (b) 12.90 (c) 2300 (d) 48 (e) 0.84

**2-10** (a) 120 $cm^2$ (b) 394 ft · lb (c) 2 cm (d) 2.3 in.

**2-12** (a) $1.57 \times 10^2$ (b) $1.57 \times 10^{-1}$ (c) $3.00 \times 10^{-2}$ (d) $4.0 \times 10^7$ (e) $3.49 \times 10^{-2}$ (f) $3.2 \times 10^4$ (g) $3.2 \times 10^{10}$ (h) $7.71 \times 10^{-4}$ (i) $2.34 \times 10^3$

**2-13** (a) 0.000476 (b) 6550 (c) 0.00788 (d) 48,900 (e) 4.75 (f) 0.0000034

**2-14** (a) $4.89 \times 10^{-4}$ (b) $4.56 \times 10^{-5}$ (c) $7.8 \times 10^3$ (d) $5.71 \times 10^{-2}$ (e) $4.975 \times 10^8$ (f) $3.0 \times 10^{-4}$

**2-16** (a) $1.597 \times 10^{-3}$ (b) $2.30 \times 10^7$ (c) $3.5 \times 10^{-5}$ (d) $2.0 \times 10^{14}$

**2-18** (a) $3.1 \times 10^{10}$ (b) $2 \times 10^9$ (c) $4 \times 10^{13}$ (d) 14 (e) $2.56 \times 10^{-14}$

**2-20** (a) $9 \times 10^7$ (b) $8.7 \times 10^7$ (c) $8.70 \times 10^7$

**2-22** (a) 720 cm, 7.2 m, $7.2 \times 10^{-3}$ km (b) $5.64 \times 10^4$ mm, 5640 cm, 0.0564 km (c) $2.50 \times 10^5$ mm, $2.50 \times 10^4$ cm, 250 m

**2-23** (a) 8.9 g, $8.9 \times 10^{-3}$ kg (b) $2.57 \times 10^4$ mg, 0.0257 kg (c) $1.25 \times 10^6$ mg, 1250 g

*Note:* Your answer may vary from those below in the last significant figure depending on the conversion factor used.

**2-25** (a) 12 = 1 doz (c) 3 ft = 1 yd (e) $10^3$ m = 1 km

**2-26** (a) 4.8 miles, $2.6 \times 10^4$ ft, 7.8 km (b) 2380 ft, 725 m, 0.725 km (c) 1.70 miles, 2740 m, 2.74 km (d) 4.21 miles, $2.22 \times 10^4$ ft, 6780 m

**2-27** (a) 27.1 qt, 25.6 L (b) 170 gal, 630 L (c) $2.08 \times 10^3$ gal, $8.33 \times 10^3$ qt

**2-29** 55.4 kg

**2-30** $28.0\ \cancel{m} \times \frac{100\ \cancel{cm}}{\cancel{m}} \times \frac{1\ \cancel{in.}}{2.54\ \cancel{cm}} \times \frac{1\ \cancel{ft}}{12\ \cancel{in.}} \times \frac{1\ yd}{3\ \cancel{ft}} = 30.6\ yd$
The team should look for a new punter.

**2-31** 0.354 L

**2-33** 14.6 gal

**2-35** $1/5\ \cancel{\text{gal}} \times \dfrac{4\ \text{qt}}{\cancel{\text{gal}}} = 0.800\ \text{qt} \qquad 0.800\ \cancel{\text{qt}} \times \dfrac{1\ \cancel{\text{L}}}{1.06\ \cancel{\text{qt}}} \times \dfrac{1\ \text{mL}}{10^{-3}\ \cancel{\text{L}}} = \underline{\underline{755\ \text{mL}}}$

There is slightly more in a "fifth" than in 750 mL.

**2-36** 89 km/hr

**2-37** $17,797 or $17,776 depending on which weight factor used. When it comes to money, the price is usually not rounded off.

**2-39** $\dfrac{\$1.12}{\cancel{\text{gal}}} \times \dfrac{1\ \cancel{\text{gal}}}{4\ \cancel{\text{qt}}} \times \dfrac{1.06\ \cancel{\text{qt}}}{\text{L}} = \underline{\underline{\$0.297/\text{L}}}$ : 80.0 L costs $23.76

**2-40** 551 miles costs $29.39.

$482\ \cancel{\text{km}} \times \dfrac{0.621\ \cancel{\text{mi}}}{\cancel{\text{km}}} \times \dfrac{1\ \cancel{\text{gal}}}{21.0\ \cancel{\text{mi}}} \times \dfrac{\$1.12}{\cancel{\text{gal}}} = \$15.96$

**2-41** 1360 km

**2-43** 382 nails

**2-44** $28.20

**2-46** 2100 sec or 0.58 hr

**2-48** 2.60 g/mL

**2-50** $285\ \cancel{\text{g}} \times \dfrac{1\ \text{mL}}{13.6\ \cancel{\text{g}}} = 21.0\ \text{mL}$

**2-51** 1450 g

**2-52** 670 g

**2-53** 625 mL

**2-54** 0.951 g/mL; 4790 mL; pumice floats on water but sinks in ethyl alcohol.

**2-56** 2080 g

**2-58** One liter of water weighs 1000 g, 1 L of gasoline weighs 670 g.

**2-60** $\left(\dfrac{2.54\ \text{cm}}{\text{in.}}\right)^3 = \dfrac{16.4\ \text{cm}^3}{\text{in.}^3} = \dfrac{16.4\ \text{mL}}{\text{in.}^3} \qquad \left(\dfrac{12\ \text{in.}}{\text{ft}}\right)^3 = \dfrac{1728\ \text{in.}^3}{\text{ft}^3}$

$\dfrac{1.00\ \cancel{\text{g}}}{\cancel{\text{mL}}} \times \dfrac{1.00\ \text{lb}}{454\ \cancel{\text{g}}} \times \dfrac{16.4\ \cancel{\text{mL}}}{\cancel{\text{in.}^3}} \times \dfrac{1728\ \cancel{\text{in.}^3}}{\text{ft}^3} = 62.4\ \text{lb/ft}^3$

**2-61** $2.0 \times 10^5$ lb (100 tons)

**2-63** 572 °F

**2-64** 24 °C

**2-66** −38 °F

**2-68** 95 °F

**2-69** (a) 320 K (b) 296 K (c) 200 K (d) 261 K (e) 291 K (f) 244 K

**2-70** (a) −98 °C (b) 22 °C (c) 27 °C (d) −48 °C (e) 600 °C

**2-71** Since $t(\text{C}) = t(\text{F})$ sustitute $t(\text{C})$ for $t(\text{F})$ and set the two equations equal.

$$[t(\text{C}) \times 1.8] + 32 = \frac{[t(\text{C}) - 32]}{1.8} \qquad t(\text{C}) = \underline{\underline{-40\ °\text{C}}}$$

**2-72** 0.204 cal/g · °C

**2-74** $\dfrac{150\ \cancel{\text{cal}}}{50.0\ \cancel{\text{g}}} \times \dfrac{1\ \cancel{\text{g}}\cdot°\text{C}}{0.092\ \cancel{\text{cal}}} = 33\ °\text{C rise} \qquad 25 + 33 = \underline{\underline{58\ °\text{C}}}$ (final temperature)

This compares to a 3 °C rise in temperature for 50.0 g of water.

**2-75** 121 cal

**2-76** 58 − 25 = 33 °C rise in temperature $\dfrac{1\ \text{g}\cdot\cancel{°\text{C}}}{0.106\ \cancel{\text{cal}}} \times \dfrac{16.0\ \cancel{\text{cal}}}{33\ \cancel{°\text{C}}} = \underline{4.6\ \text{g}}$

**2-78** 45 °C

**2-79** 2910 joules

**2-80** 140 g of water

**2-82** Heat lost by metal = heat gained by water

$$100.0\ \text{g} \times 68.7\ °\text{C} \times \text{specific heat} = 100.0\ \cancel{\text{g}} \times 6.3\ \cancel{°\text{C}} \times \frac{1.00\ \text{cal}}{\cancel{\text{g}}\cdot\cancel{°\text{C}}}$$

specific heat = 0.092 cal/g · °C (The metal is copper.)

## Chapter 3

**3-1** (b) and (e)

**3-2** (c)

**3-4** ${}^{16}_{8}\text{O}$, ${}^{14}_{7}\text{N}$, ${}^{28}_{14}\text{Si}$, ${}^{40}_{20}\text{Ca}$

**3-6** (a) 21 *p*, 21 *e*, 24 *n* (b) 90 *p*, 90 *e*, 142 *n* (c) 87 *p*, 87 *e*, 136 *n*
(d) 38 *p*, 38 *e*, 52 *n*

**3-8** (a) ${}^{96}_{42}\text{Mo}$ (b) ${}^{108}_{47}\text{Ag}$ (c) ${}^{28}_{14}\text{Si}$ (d) ${}^{39}_{19}\text{K}$ (e) ${}^{190}_{76}\text{Os}$

**3-10** 5.81 × 12.00 = 69.7 amu The element is Ga.

**3-12** (a) H: $\frac{1}{12}$ × 8.00 = 0.667 (b) N: $\frac{14}{12}$ × 8.00 = 9.33 (c) Na: 15.3
(d) Ca: 26.7

**3-14** 79.9 amu

**3-15** 28.09 amu

**3-17** Let $X$ = decimal fraction of ${}^{35}\text{Cl}$ and $Y$ = decimal fraction of ${}^{37}\text{Cl}$. Since there are only two isotopes present, $X + Y = 1$, $Y = 1 - X$.

$$(X \times 35) + (Y \times 37) = 35.5$$

$$(X \times 35) + [(1 - X) \times 37] = 35.5$$

$$X = 0.75 \quad (75\%\ ^{35}Cl)$$

$$Y = 0.25 \quad (25\%\ ^{37}Cl)$$

**3-19** (a) Re: at. no. 75, at. wt. 186.2 (b) Co: at. no. 27, at. wt. 58.9332 (c) Br: at. no. 35, at. wt. 79.904 (d) Si: at. no. 14, at. wt. 28.086

**3-20** (e) $N_2$, diatomic element (b) CO, diatomic compound

**3-21** (a) six carbons, four hydrogens, and two chlorines (b) two carbons, six hydrogens, and one oxygen (c) one copper, one sulfur, 18 hydrogens, and 13 oxygens (d) nine carbons, eight hydrogens, and four oxygens (e) two aluminums, three sulfurs, and twelve oxygens (f) two nitrogens, eight hydrogens, one carbon, and three oxygens

**3-22** (a) 12 (b) 9 (c) 33 (d) 21 (e) 17 (f) 14

**3-23** (a) 8 (b) 7 (c) 4 (d) 3

**3-24** (a) $SO_2$ (b) $CO_2$ (c) $H_2SO_4$ (d) $Ca(ClO_4)_2$ (e) $(NH_4)_3PO_4$ (f) $C_2H_2$

**3-26** Neon would appear as individual spheres that are widely spaced. Oxygen would appear as molecules with two spheres joined together.

**3-28** (a) 19 *p*, 18 *e* (b) 35 *p*, 36 *e* (c) 16 *p*, 18 *e* (d) 7 + 16 = 23 *p*, 24 *e* (e) 13 *p*, 10 *e* (f) 11 *p*, 10 *e*

**3-30** (a) ${}^{90}_{38}Sr^{2+}$ (b) ${}^{52}_{24}Cr^{3+}$ (c) ${}^{79}_{34}Se^{2-}$ (d) ${}^{14}_{7}N^{3-}$ (e) ${}^{139}_{57}La^{3+}$

**3-32** (a) ${}^{0}_{-1}e$ (b) ${}^{90}_{38}Sr$ (c) ${}^{231}_{90}Th$ (d) ${}^{41}_{20}Ca$ (e) ${}^{210}_{81}Tl$

**3-34** (a) ${}^{230}_{90}Th \rightarrow {}^{226}_{88}Ra + {}^{4}_{2}He$
(b) ${}^{214}_{84}Po \rightarrow {}^{210}_{82}Pb + {}^{4}_{2}He$
(c) ${}^{210}_{84}Po \rightarrow {}^{210}_{85}At + {}^{0}_{-1}e$
(d) ${}^{239}_{94}Pu \rightarrow {}^{235}_{92}U + {}^{4}_{2}He$
(e) ${}^{14}_{6}C \rightarrow {}^{14}_{7}N + {}^{0}_{-1}e$

**3-35** ${}^{234}_{90}Th$, ${}^{234}_{91}Pa$, ${}^{234}_{92}U$, ${}^{230}_{90}Th$, ${}^{226}_{88}Ra$, ${}^{222}_{86}Rn$, ${}^{218}_{84}Po$, ${}^{214}_{82}Pb$, ${}^{214}_{83}Bi$, ${}^{210}_{81}Tl$, ${}^{210}_{82}Pb$, ${}^{210}_{83}Bi$, ${}^{210}_{84}Po$, ${}^{206}_{82}Pb$

**3-36** 2.5 g

**3-38** (a) ${}^{90}_{39}Y$ (b) 50 yr

**3-40** about 11,500 yr old

**3-42** (a) ${}^{87}_{34}Se$ (b) $4{}^{1}_{0}n$ (c) ${}^{97}_{38}Sr$

**3-43** (a) ${}^{35}_{16}S$ (b) ${}^{2}_{1}H$ (c) ${}^{30}_{15}P$ (d) $8\ {}^{0}_{-1}e$ (e) ${}^{254}_{102}No$ (f) ${}^{237}_{92}U$ (g) $4{}^{1}_{0}n$

**3-45** Fission occurs when a large nucleus is split into two smaller nuclei and neutrons. Fusion occurs when two small nuclei join to form a larger nucleus. In both processes the mass of the products is less than the mass of the original nuclei. This mass has been converted to energy according to Einstein's relationship, $E = mc^2$.

**3-46** $^{239}_{93}\text{Np}$, $_{-1}^{0}e$

## Review Test for Chapters 1–3

### Multiple Choice

**1.** (a) **2.** (d) **3.** (b) **4.** (c) **5.** (c) **6.** (d) **7.** (a) **8.** (a) **9.** (b) **10.** (c) **11.** (c) **12.** (a) **13.** (d) **14.** (a) **15.** (b) **16.** (c) **17.** (d) **18.** (d) **19.** (a) **20.** (a) **21.** (c) **22.** (d) **23.** (a) **24.** (a) **25.** (b)

### Problems

**1.** (a) 174 (b) 0.00232 (c) 18,400 **2.** 17.17 cm **3.** 69 in.$^2$ **4.** 13 g/mL **5.** $2.0 \times 10^7$ cm$^2$ **6.** 0.014 mL **7.** $1.2 \times 10^{12}$ **8.** $2 \times 10^{-9}$ **9.** $1.6 \times 10^{13}$ **10.** 74.4 in. **11.** 41 L **12.** 317 km/hr **13.** $1.00 \times 10^2$ g/cm **14.** 4.48 g/cm$^3$ = 4.48 g/mL **15.** 60.0 mL **16.** 44 °C **17.** an increase of 49 Fahrenheit degrees **18.** 0.0688 cal/g · °C **19.** 121.8 amu, Sb, antimony **20.** 114.6 amu

### Chapter 4

**4-1** (b) Sr and (d) Mg

**4-3** Ultraviolet light has shorter wavelengths but higher energy than visible light. Ultraviolet light can damage tissue, thus causing a burn.

**4-4** Since these two shells are close in energy, transitions of electrons from these two levels to the $n = 1$ shell have similar energy. Thus the wavelengths of light from the two transitions are very close together.

**4-5** Since these two shells are comparatively far apart in energy, transitions from these two levels to the $n = 1$ shell have comparatively different energies. Thus the wavelengths of light from the two transitions are quite different. [The $n = 3$ to $n = 1$ transition has a shorter wavelength (more energy) than the $n = 2$ to $n = 1$ transition.]

**4-6** 32 electrons

**4-8** The ground state for a lithium atom is the $n = 2$ shell. Thus the $n = 3$ shell is an excited state, and light can be emitted when the electron falls to the ground state.

**4-9** $2d$, $1p$, $3f$

**4-10** $s-2$, $p - 6$, $d - 10$, $f - 14$

**4-13** The $n = 5$ shell holds $2n^2$ or 50 electrons. The $s$, $p$, $d$, and $f$ sublevels hold a total of 32 electrons. Therefore, the $g$ sublevel holds 18 electrons.

**4-14** (a) 6*s* (b) 5*p* (c) 4*p* (d) 4*d*

**4-16** 4*s*, 4*p*, 5*s*, 4*d*, 5*p*, 6*s*, 4*f*

**4-17** (a) Mg: $1s^22s^22p^63s^2$
(b) Ge: $1s^22s^22p^63s^23p^64s^23d^{10}4p^2$
(c) Pd: $1s^22s^22p^63s^23p^64s^23d^{10}4p^65s^24d^8$
(d) Si: $1s^22s^22p^63s^23p^2$

**4-19** (a) S: $[Ne]3s^23p^4$
(b) Zn: $[Ar]4s^23d^{10}$
(c) Pu: $[Rn]7s^26d^15f^5$
(d) I: $[Kr]5s^24d^{10}5p^5$

**4-21** (a) F (b) Ga (c) Ba (d) Gd (e) Cu

**4-23** (a) VIIA (b) IIIA (c) IIA (e) IB

**4-24** (a) IVA (b) O (noble gases) (c) IB (d) Pr and Pa

**4-25** (a) B (b) Ar (c) K, Cr, and Cu (d) Ga (e) Hf

**4-27** (a) $ns^2$ (b) $ns^2(n-1)d^{10}$ (c) $ns^2np^4$ (d) $ns^2(n-1)d^2$

**4-28** He does not have a filled *p* sublevel. (There is no 1*p* sublevel.)

**4-30** The theoretical order of filling is 6*d*, 7*p*, 8*s*, 5*g*. The 6*d* is completed at element no. 112. The 7*p* and 8*s* fill at element no. 120. Thus element no. 121 would theoretically begin the filling of the 5*g* sublevel. This assumes filling in the order indicated by Figure 4-11.

**4-32** (a) three (b) five (c) none (There is no 2*d* sublevel.) (d) one
(e) seven

**4-33** 3*s*, one; 3*p*, three; 3*d*, five; total = nine

**4-34** If there are 18 electrons in the 5*g* sublevel, there would be nine orbitals.

**4-36** A 4*p* orbital is shaped roughly like a two-sided baseball bat with two "lobes" lying along one of the three axes. This shape represents the region of highest probability of finding the electrons.

**4-38** (a) This is excluded by Hund's rule since electrons are not shown in separate orbitals of the same sublevel with parallel spins.
(b) This is correct.
(c) This is excluded by the Aufbau principle because the 2*s* orbital should fill before the 2*p*.
(d) This is excluded by the Pauli exclusion principle since the two electrons in the 2*s* orbital are shown with the same spin.

**4-39**

3*s* 3*p*

(a) S [↑↓] [↑↓ | ↑ | ↑]

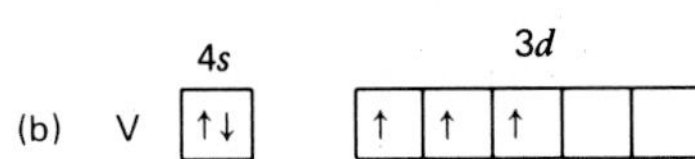

**6-2** (a) group IIIA (b) group VA (c) group IIA

**6-3** The electrons from filled inner sublevels are not involved in chemical bonding.

**6-5** (b) Sr and S (c) H and K (d) Al and F

**6-7** (b) $S^-$ (c) $Cr^{2+}$ (e) $In^+$ (f) $Pb^{2+}$ (h) $Tl^{3+}$

**6-8** They have the noble gas configuration of He, which requires only two electrons.

**6-9** (a) $K^+$ (b) $\cdot\ddot{O}:^-$ (c) $:\ddot{I}:^-$ (d) $:\ddot{P}:^{3-}$ (e) $\dot{B}a^+$ (f) $\cdot\ddot{X}e:^+$ (g) $Sc^{3+}$

**6-11** (b) $0^-$ (e) $Ba^+$ (f) $Xe^+$

**6-12** (a) $Mg^{2+}$ (b) $Ga^{3+}$ (pseudo-noble gas) (c) $Br^-$ (d) $S^{2-}$ (e) $P^{3-}$

**6-15** $Se^{2-}$, $Br^-$, $Rb^+$, $Sr^{2+}$, $Y^{3+}$

**6-16** $Tl^{3+}$ has a full 5*d* sublevel. This is a pseudo-noble gas configuration.

**6-17** $Cs_2S$, $Cs_3N$, $BaBr_2$, BaS, $Ba_3N_2$, $InBr_3$, $In_2S_3$, InN

**6-18** (a) $CaI_2$ (b) CaO (c) $Ca_3N_2$ (d) CaTe (e) $CaF_2$

**6-20** (a) $Cr^{3+}$ (b) $Fe^{3+}$ (c) $Mn^{2+}$ (d) $Co^{2+}$ (e) $Ni^{2+}$ (f) $V^{3+}$

**6-22** (a) $H_2Se$ (b) $GeH_4$ (c) ClF (d) $Cl_2O$ (e) $NCl_3$ (f) $CBr_4$

**6-24**

(a)
```
    H   H
    |   |
H — C — C — H
    |   |
    H   H
```

(b) H—Ö̤—Ö̤—H

(c)
```
:F̤—N̈—F̤:
     |
    :F̤:
```

(d) :C̤̈l—S̤̈—C̤̈l:

(e)
```
     H         H
     |         |
H — C — Ö̤ — C — H
     |         |
     H         H
```

```
     H   H
     |   |
H — C — C — Ö̤ — H
     |   |
     H   H
```

**6-25** (a) $SO_4^{2-}$ (b) $ClO_3^-$ (c) $SO_3^{2-}$ (d) $PO_4^{3-}$ (e) $C_2^{2-}$

**6-27** (a) :C≡O: (b)
```
.Ö:    Ö:
   \\  /
     S
     |
    :Ö:
```
(c) $K^+$[:C≡N:]$^-$

(d)
```
 H—Ö:
    |
   :S—Ö:
    |
   :Ö—H
```

**6-28** (a) .N̈=O=N̈. (b) $Ca^{2+}$ $\left[\cdot\ddot{O}\diagup\ddot{N}\diagdown\ddot{O}:\right]^-_2$

(c) 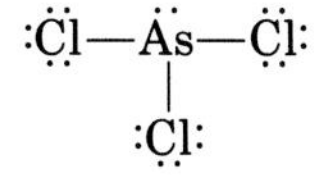 :Cl—As—Cl: with :Cl: below As  (d) 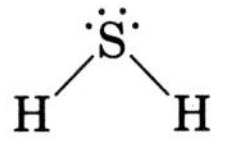 H—S—H (bent, S with two lone pairs)  (e) 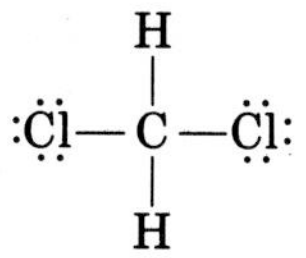 :Cl—C—Cl: with H above and H below C

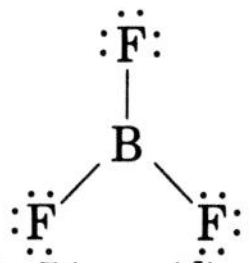

(f) 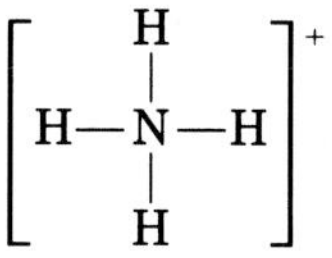 $[H{-}N{-}H$ with H above and H below N$]^{+}$

**6-29** (a) 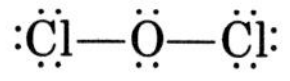 :Cl—O—Cl:  (b) 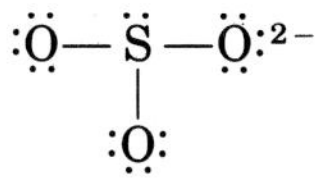 $:O{-}S{-}O:^{2-}$ with :O: below S  (c) $H_2C{=}CH_2$ (H, H on each C)

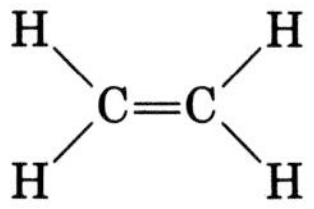

(d) 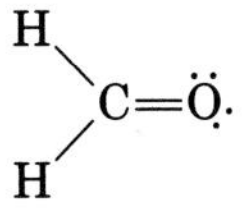 $H_2C{=}O$ (H, H on C; O with two lone pairs)  (e) :F: —B— with :F: and F: (B bonded to three F)  (f) $:N{\equiv}O:^{+}$

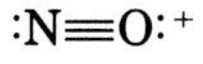

**6-32** (1)(a) 17  (b) 31  (c) $51_3$  (d) 97  (e) 7(13)  (f) $10(13)_2$  (g) $67_2$
(h) $3_26$
(2) On Zerk, six electrons fill the outer *s* and *p* orbitals to make a noble gas structure. Therefore we have a "sextet" rule on Zerk.
(a) 1—7:  (b) $3^{+}1{:}^{-}$ (ionic)  (c) 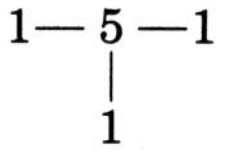 1—5—1 with 1 below 5  (d) $9^{+}{:}7{:}^{-}$(ionic)

(e) :7—13:  (f) $10^{2+}(:13:^{-})_2$(ionic)

(g) :7—6—7:  (h) $(3^{+})_2{:}6{:}^{2-}$(ionic)

**6-33** (b) $K_2SO_4$

**6-34** (a) 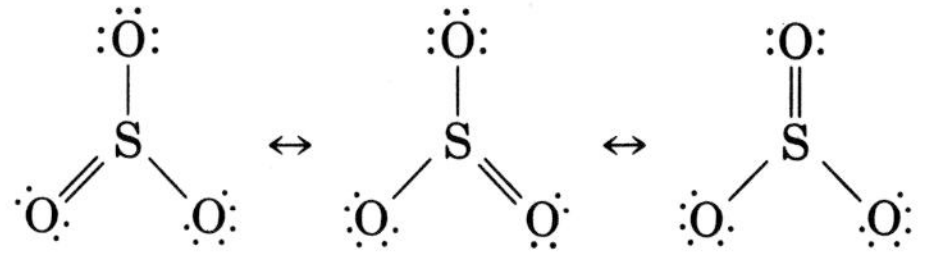 $SO_3$: three resonance structures, each with one S=O double bond and two S—O single bonds ↔ ↔

(b) 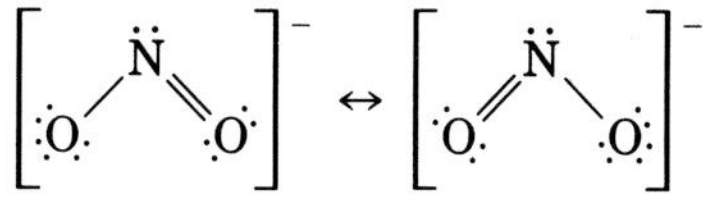 $[O{-}N{=}O]^{-} \leftrightarrow [O{=}N{-}O]^{-}$

(c) 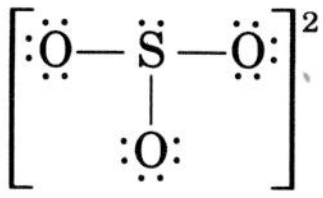 $[:O{-}S{-}O:]^{2-}$ with :O: below S (only one structure)

**6-36** $[H_3B{-}C(=O)O]^{2-} \leftrightarrow [H_3B{-}C(O)=O]^{2-}$ (H—B—C with H above and below B; C bonded to two O)

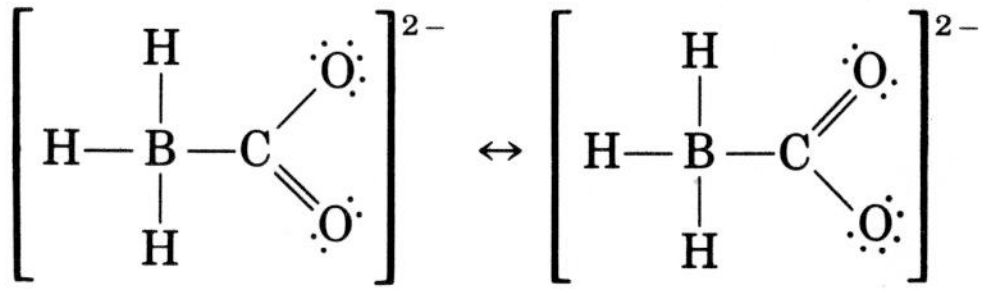

**6-37** A resonance hybrid is the actual structure of a molecule or ion that is implied by the various resonance structures. For example, the two structures shown in Problem 6-36 imply that both C—O bonds have properties that are halfway between those of a single bond and a double bond.

**6-38** :Ö—C≡O: ↔ :O≡C—Ö: The two resonance structures imply that both C—O bonds are halfway between a single and a triple bond, which is a double bond. This is the same as the one structure shown below.

:Ö=C=Ö:

**6-39** Cs, Ba, Be, B, C, Cl, O, F

**6-40** (a) $\overset{\delta-}{N}—\overset{\delta+}{H}$ (b) $\overset{\delta+}{B}—\overset{\delta-}{H}$ (c) $\overset{\delta+}{Li}—\overset{\delta-}{H}$ (d) $\overset{\delta-}{F}—\overset{\delta+}{O}$ (e) $\overset{\delta-}{O}—\overset{\delta+}{Cl}$ (f) $\overset{\delta-}{S}—\overset{\delta+}{Se}$
(g) $\overset{\delta-}{C}—\overset{\delta+}{B}$ (h) $\overset{\delta+}{Cs}—\overset{\delta-}{N}$ (i) C—S (nonpolar)

**6-41** (i) nonpolar, (b) = (f), (d) = (e) = (g), (a), (c), (h)

**6-42** Only (d) is predicted to be ionic on this basis. The electronegativity difference is 2.5.

**6-44** (b) I—I and (d) N—Cl

**6-46** (a) nonpolar (Equal dipoles cancel.) (b) polar (unequal dipoles) (c) polar (Equal dipoles do not cancel.) (d) nonpolar (Equal dipoles cancel.) (e) polar (unequal dipoles)

**6-48** Since the H—S bond is much less polar than the H—O bond, the resultant molecular dipole is much less. The $H_2S$ molecule is less polar than the $H_2O$ molecule.

**6-50** The $CHF_3$ molecule is more polar than the $CHCl_3$ molecule. The C—F bond is more polar, which means that the resultant molecular dipole is larger for $CHF_3$.

## Chapter 7

**7-1** (a) Pb +4, O −2 (b) P +5, O −2 (c) C −1, H +1 (d) N −2, H +1 (e) Li +1, H −1 (f) B +3, Cl −1 (g) Rb +1, Se −2 (h) Bi +3, S −2

**7-3** (a) Li (e) K (g) Rb

**7-5** +3

**7-6** (a) P +5 (b) C +3 (c) Cl +7 (d) Cr +6 (e) S +6 (f) N +5 (g) Mn +7

**7-8** $K_3N$ (−3), $N_2H_4$(−2), $NH_2OH$ (−1), $N_2$(O), $N_2O$ (+1), NO (+2), $N_2O_3$ (+3), $N_2O_4$ (+4), $Ca(NO_3)_2$ (+5)

**7-9** (a) lithium fluoride (b) barium telluride (c) strontium nitride (d) barium hydride (e) aluminum chloride

**7-11** (a) $Rb_2Se$ (b) $SrH_2$ (c) RaO (d) $Al_4C_3$ (e) $BeF_2$

**7-13** (a) bismuth(V) oxide (b) tin(II) sulfide (c) tin(IV) sulfide (d) copper(I) telluride (e) titanium(IV) carbide

**7-15** (a) $Cu_2S$ (b) $V_2O_3$ (c) AgBr (d) $Ni_2C$ (e) $CrO_3$

**7-17** (a) $Bi_2O_5$ (c) $SnS_2$ (e) TiC (e) $CrO_3$

**7-19** (c) $ClO_3^-$

**7-21** ammonium, $NH_4^+$

**7-22** (b) permanganate ($MnO_4^-$) (c) perchlorate ($ClO_4^-$) (e) phosphate ($PO_4^{3-}$) (f) oxalate ($C_2O_4^{2-}$)

**7-23** (a) chromium(II) sulfate (b) aluminum sulfite (c) iron(II) cyanide (d) rubidium hydrogen carbonate (e) ammonium carbonate (f) ammonium nitrate (g) bismuth(III) hydroxide

**7-25** (a) $Mg(MnO_4)_2$ (b) $Co(CN)_2$ (c) $Sr(OH)_2$ (d) $Tl_2SO_3$ (e) $In(HSO_4)_3$ (f) $Fe_2(C_2O_4)_3$ (g) $(NH_4)_2Cr_2O_7$ (h) $Hg_2(C_2H_3O_2)_2$

**7-27** (a) sodium chloride (b) sodium hydrogen carbonate (c) calcium carbonate (d) sodium hydroxide (e) sodium nitrate (f) ammonium chloride (g) aluminum oxide (h) calcium hydroxide (i) potassium hydroxide

**7-28** (a) $Ca_2XeO_6$ (b) $K_4XeO_6$ (c) $Al_4(XeO_6)_3$

**7-29** (a) phosphonium fluoride (b) potassium hypobromite (c) cobalt(III) periodate (d) calcium silicate (e) aluminum phosphite (f) chromium(II) molybdate

**7-30** (a) Si (b) I (c) B (d) Kr (e) H

**7-32** (a) carbon disulfide (b) boron trifluoride (c) tetraphosphorus decoxide (d) dibromine trioxide (e) sulfur trioxide (f) dichlorine monoxide (g) phosphorus pentachloride (h) sulfur hexafluoride

**7-34** (a) $P_4O_6$ (b) $CCl_4$ (c) $IF_3$ (d) $Cl_2O_7$ (e) $SF_6$ (f) $XeO_2$

**7-36** (a) hydrochloric acid (b) nitric acid (c) hypochlorous acid (d) permanganic acid (e) periodic acid (f) hydrobromic acid

**7-37** (a) HCN (b) $H_2Se$ (c) $HClO_2$ (d) $H_2CO_3$ (e) HI (f) $HC_2H_3O_2$

**7-39** (a) hypobromous acid (b) periodic acid (c) phosphorous acid (d) molybdic acid (e) perxenic acid

**7-40** $H_3AsO_3$, arsenous acid; $H_3AsO_4$, arsenic acid

## Review Test on Chapters 4–7

### Multiple Choice

**1.** (c) **2.** (c) **3.** (b) **4.** (c) **5.** (a) **6.** (a) **7.** (a) **8.** (b) **9.** (d) **10.** (d) **11.** (b) **12.** (a) **13.** (b) **14.** (b) **15.** (b) **16.** (a) **17.** (c) **18.** (d)

**19.** (c) **20.** (c) **21.** (a) **22.** (c) **23.** (d) **24.** (b) **25.** (c) **26.** (d)
**27.** (b) **28.** (e) **29.** (d) **30.** (d)

## Problems

1. aluminum (Al) (a) IIIA, representative element (b) [Ne]$3s^23p^1$ (c) solid (d) metal (e) +3 (f) neon (g) $Al_2S_3$, $AlBr_3$, AlN (h) aluminum sulfide, aluminum bromide, aluminum nitride
2. nitrogen (N) (a) VA, representative element (b) $1s^22s^22p^3$ (c) three (d) nonmetal (e) gas (f) $N_2$ (g) :N≡N: (h) −3 (i) neon (j) oxygen, positive; boron, negative (k) $Mg_3N_2$, $Li_3N$, $NF_3$ (l) magnesium nitride, lithium nitride, nitrogen trifluoride (m) :F̤—N̈—F̤: (with :F̤: bonded below N)
3. (a) O (b) He (c) Cu (d) Te (e) Y (f) Sb (g) F (h) Pt (i) P (j) Ba
4. (a) $Mg^{2+}SO_4^{2-}$ magnesium sulfate $Mg^{2+}$ [:Ö:, :Ö—S—Ö:, :Ö:]$^{2-}$

   (b) no ions nitrous acid H—Ö—N̈=Ö:

   (c) $Li^+NO_3^-$ lithium nitrate $Li^+$ [Ö=N, :Ö, Ö:]$^-$

   (d) $(Co^{3+})_2(CO_3^{2-})_3$ cobalt(III) carbonate $2(Co^{3+})$ 3[Ö=C, :Ö, Ö:]$^{2-}$

5. (a) $N_2O_3$ no ions (Ö=)(:Ö—)N—N̈=Ö.

   (b) $Cr_2(SO_3)_3$ $(Cr^{3+})_2(SO_3^{2-})_3$ $2Cr^{3+}$ 3[:Ö—S̈—Ö:, :Ö:]$^{2-}$

   (c) $Fe(OH)_2$ $Fe^{2+}(OH^-)_2$ $Fe^{2+}$ 2[:Ö—H]$^-$

   (d) $SrC_2O_4$ $Sr^{2+}(C_2O_4^{2-})$ $Sr^{2+}$ [:Ö—C(=Ö)—C(=Ö)—Ö:]$^{2-}$

   (e) HI no ions H—Ï:

## Chapter 8

**8-1** 79.4 g

**8-3** $16.0\ \cancel{\text{g Cu}} \times \dfrac{?\text{g X}}{63.5\ \cancel{\text{g Cu}}} = 49.1\ \text{g X}; \quad ? = 195\ \text{g}$ The element is platinum.

**8-4** 71.5 g of Cu

**8-6** Atomic weight is 16.0 g. The element is oxygen. The formula is $CO$.

**8-7** 302 g of Br

**8-9** (a) 0.468 mol of P, $2.82 \times 10^{23}$ atoms (b) 150 g of Rb, $1.05 \times 10^{24}$ atoms (c) 27.0 g, 1.00 mol (d) 5.00 mol, Ge (e) $7.96 \times 10^{-23}$ g, $1.66 \times 10^{-24}$ mol

**8-11** (a) 63.5 g (b) 16 g (c) 40.1 g

**8-13** (a) $1.93 \times 10^{25}$ (b) $6.02 \times 10^{23}$ (c) $1.20 \times 10^{24}$

**8-15** $50.0\ \cancel{\text{g Al}} \times \dfrac{1.00\ \text{mol}}{27.0\ \cancel{\text{g Al}}} = 1.85\ \text{mol Al}$

$50.0\ \cancel{\text{g Fe}} \times \dfrac{1.00\ \text{mol}}{55.8\ \cancel{\text{g Fe}}} = 0.896\ \text{mol Fe}$

There are more moles of atoms (or more atoms) in 50.0 g of Al.

**8-16** $20.0\ \cancel{\text{g Ni}} \times \dfrac{1.00\ \text{mol}}{58.7\ \cancel{\text{g Ni}}} = 0.341\ \text{mol Ni}$

$2.85 \times 10^{23}\ \cancel{\text{atoms}} \times \dfrac{1.00\ \text{mol Ni}}{6.02 \times 10^{23}\ \cancel{\text{atoms}}} = 0.473\ \text{mol Ni}$

The $2.85 \times 10^{23}$ atoms contain more Ni.

**8-18** (a) 107 amu (b) 80.1 amu (c) 108 amu (d) 98.1 amu (e) 106 amu (f) 60.0 amu (g) 460 amu

**8-20** (a) 189 g, $6.32 \times 10^{24}$ molecules (b) 0.339 g, $5.00 \times 10^{-3}$ mol (c) 0.219 mol, $1.32 \times 10^{23}$ molecules (d) 0.0209 g, $7.22 \times 10^{19}$ molecules (e) 598 g, 7.48 mol (f) $7.62 \times 10^{-3}$ mol, $4.59 \times 10^{21}$ molecules

**8-22** C 5.10 mol, H 15.3 mol, O 2.55 mol, 23.0 mol total; C 61.2 g, H 15.4 g, O 40.8 g, 117.4 g total

**8-23** 0.135 mol of $Ca(ClO_3)_2$, 0.135 mol of Ca, 0.270 mol of Cl, 0.810 mol of O, 1.215 mol of atoms.

**8-25** $1.50\ \cancel{\text{mol H}_2\text{SO}_4} \times \dfrac{2\ \cancel{\text{mol H}}}{\cancel{\text{mol H}_2\text{SO}_4}} \times \dfrac{1.01\ \text{g H}}{\cancel{\text{mol H}}} = 3.03\ \text{g H}$

48.2 g of S, 72.0 g of O

**8-27** $1.20 \times 10^{22}\ \cancel{\text{molecules}} \times \dfrac{1.00\ \text{mol O}_2}{6.02 \times 10^{23}\ \cancel{\text{molecules}}} = 0.0199\ \text{mol O}_2\ \text{molecules}$

$0.0199\ \cancel{\text{mol O}_2} \times \dfrac{2\ \text{mol O atoms}}{\cancel{\text{mol O}_2}} = 0.0398\ \text{mol O atoms}$

$$0.0199 \text{ \cancel{mol O_2}} \times \frac{32.0 \text{ g}}{\cancel{\text{mol O}_2}} = 0.637 \text{ g O}_2$$

Same weight as O atoms

**8-29** N, 25.89%; O, 74.11%

**8-30** Si, 46.7%; O, 53.3%

**8-32** (a) $C_2H_6O$: C, 52.1%; H, 13.1%; O, 34.8% (b) $C_3H_6$: C, 85.5%; H, 14.5% (c) $C_9H_{18}$: C, 85.7%; H, 14.3% (essentially the same as $C_3H_6$) (d) $Na_2SO_4$: Na, 32.4%; S, 22.6%; O, 45.0% (e) $(NH_4)_2CO_3$: N, 29.1%; C, 12.5%; H, 8.41%; O, 49.9%

**8-34** Na, 12.1%; B, 11.3%, O, 71.4%; H, 5.27%

**8-36** $125 \cancel{\text{ g Na}_2\text{C}_2\text{O}_4} \times \frac{24.0 \text{ g C}}{134 \cancel{\text{ g Na}_2\text{C}_2\text{O}_4}} = \underline{\underline{22.4 \text{ g C}}}$

**8-37** 4.73 lb

**8-39** 1398 lb

**8-40** (a) and (d)

**8-41** (a) FeS (b) $SrI_2$ (c) $KClO_3$ (d) $I_2O_5$ (e) $Fe_3O_4$ (f) $C_3H_5Cl_3$

**8-43** $N_2O_3$

**8-45** $KO_2$

**8-47** $MgC_2O_4$

**8-48** $CH_2Cl$

**8-50** $N_2SH_8O_3$

**8-51** $1.20 \cancel{\text{ g CO}_2} \times \frac{1.00 \cancel{\text{ mol CO}_2}}{44.0 \cancel{\text{ g CO}_2}} \times \frac{1 \text{ mol C}}{1 \cancel{\text{ mol CO}_2}} = 0.0273 \text{ mol C}$

$$0.489 \cancel{\text{ g H}_2\text{O}} \times \frac{1.00 \cancel{\text{ mol H}_2\text{O}}}{18.0 \cancel{\text{ g H}_2\text{O}}} \times \frac{2 \text{ mol H}}{1 \cancel{\text{ mol H}_2\text{O}}} = 0.0543 \text{ mol H}$$

$$\frac{0.0273}{0.0273} = 1 \qquad \frac{0.0543}{0.0273} = 2 \qquad \underline{\underline{CH_2}}$$

**8-52** $C_9H_{12}Cl_{12}$

**8-54** $B_2C_2H_6O_4$

**8-56** $I_6C_6$

## Chapter 9

**9-1** (a) $Cl_2(g)$ (b) $C(s)$ (c) $K_2SO_4(s)$ (d) $H_2O(l)$ (e) $P_4(s)$ (f) $H_2(g)$ (g) $Br_2(l)$ (h) $NaBr(s)$ (i) $S_8(s)$ (j) $Na(s)$ (k) $Hg(l)$ (l) $CO_2(g)$

**9-2** (a) $CaCO_3 \rightarrow CaO + CO_2$
(b) $4Na + O_2 \rightarrow 2Na_2O$
(c) $H_2SO_4 + 2NaOH \rightarrow Na_2SO_4 + 2H_2O$
(d) $2H_2O_2 \rightarrow 2H_2O + O_2$
(e) $Si_2H_6 + 8H_2O \rightarrow 2Si(OH)_4 + 7H_2$
(f) $2Al + 2H_3PO_4 \rightarrow 2AlPO_4 + 3H_2$
(g) $Ca(OH)_2 + 2HCl \rightarrow CaCl_2 + 2H_2O$
(h) $Na_2NH + 2H_2O \rightarrow NH_3 + 2NaOH$
(i) $3Mg + N_2 \rightarrow Mg_3N_2$
(j) $CaC_2 + 2H_2O \rightarrow C_2H_2 + Ca(OH)_2$
(k) $2C_2H_6 + 7O_2 \rightarrow 4CO_2 + 6H_2O$

**9-4** (a) $2Na(s) + 2H_2O \rightarrow H_2(g) + 2NaOH(aq)$
(b) $2KClO_3(s) \rightarrow 2KCl(s) + 3O_2(g)$
(c) $NaCl(aq) + AgNO_3(aq) \rightarrow AgCl(s) + NaNO_3(aq)$
(d) $2H_3PO_4(aq) + 3Ca(OH)_2(aq) \rightarrow Ca_3(PO_4)_2(s) + 6H_2O$

**9-6** (For Problem 9-2) (à) decomposition (b) combustion, combination (c) double displacement (d) decomposition (e) double displacement (OH from $H_2O$ replaces H on $Si_2H_6$) (f) single replacement (g) double displacement (h) double displacement (i) combination (j) double displacement (k) combustion (For Problem 9-4) (a) single replacement (b) decomposition (c) double displacement (d) double displacement

**9-8** (a) $2K(s) + Cl_2(g) \rightarrow 2KCl(s)$
(b) $Ca(s) + 2H_2O(l) \rightarrow Ca(OH)_2(aq) + H_2(g)$
(c) $2C_6H_6(l) + 15O_2(g) \rightarrow 12CO_2(g) + 6H_2O(l)$
(d) $Na_2S(aq) + Cu(NO_3)_2(aq) \rightarrow CuS(s) + 2NaNO_3(aq)$
(e) $2Au_2O_3(s) \rightarrow 4Au(s) + 3O_2(g)$

**9-10** (a) $\frac{1 \text{ mol Mg}}{2 \text{ mol HCl}}, \frac{1 \text{ mol Mg}}{1 \text{ mol MgCl}_2}, \frac{1 \text{ mol Mg}}{1 \text{ mol H}_2}, \frac{2 \text{ mol HCl}}{1 \text{ mol H}_2}, \frac{2 \text{ mol HCl}}{1 \text{ mol MgCl}_2}$

(b) $\frac{2 \text{ mol C}_4\text{H}_{10}}{13 \text{ mol O}_2}, \frac{2 \text{ mol C}_4\text{H}_{10}}{8 \text{ mol CO}_2}, \frac{2 \text{ mol C}_4\text{H}_{10}}{10 \text{ mol H}_2\text{O}}, \frac{13 \text{ mol O}_2}{8 \text{ mol CO}_2}, \frac{13 \text{ mol O}_2}{10 \text{ mol H}_2\text{O}}$

**9-12** (a) mol $H_2O \rightarrow$ mol $H_2$

$$0.400 \cancel{\text{mol H}_2\text{O}} \times \frac{2 \text{ mol H}_2}{2 \cancel{\text{mol H}_2\text{O}}} = \underline{\underline{0.400 \text{ mol H}_2}}$$

(b) g $O_2 \rightarrow$ mol $O_2 \rightarrow$ mol $H_2O$

$$0.640 \cancel{\text{g O}_2} \times \frac{1 \cancel{\text{mol O}_2}}{32.0 \cancel{\text{g O}_2}} \times \frac{2 \text{ mol H}_2\text{O}}{1 \cancel{\text{mol O}_2}} = \underline{\underline{0.0400 \text{ mol H}_2\text{O}}}$$

(c) g $O_2 \rightarrow$ mol $O_2 \rightarrow$ mol $H_2$

$$0.032 \cancel{\text{g O}_2} \times \frac{1 \cancel{\text{mol O}_2}}{32.0 \cancel{\text{g O}_2}} \times \frac{2 \text{ mol H}_2}{1 \cancel{\text{mol O}_2}} = \underline{\underline{0.0020 \text{ mol H}_2}}$$

(d) g $H_2 \rightarrow$ mol $H_2 \rightarrow$ mol $H_2O \rightarrow$ g $H_2O$

$$0.400 \cancel{\text{g H}_2} \times \frac{1 \cancel{\text{mol H}_2}}{2.02 \cancel{\text{g H}_2}} \times \frac{2 \cancel{\text{mol H}_2\text{O}}}{2 \cancel{\text{mol H}_2}} \times \frac{18.0 \text{ g H}_2\text{O}}{\cancel{\text{mol H}_2\text{O}}} = \underline{\underline{3.56 \text{ g H}_2\text{O}}}$$

**9-13** (a) 1.35 mol of $CO_2$, 1.80 mol of $H_2O$, 2.25 mol of $O_2$ (b) 4.80 g of $H_2O$
(c) 1.10 g of $C_3H_8$ (d) 44.0 g of $C_3H_8$
(e) molecules $O_2 \rightarrow$ mol $O_2 \rightarrow$ mol $CO_2 \rightarrow$ g $CO_2$

$$1.20 \times 10^{23}\ \cancel{\text{molecules}} \times \frac{1\ \cancel{\text{mol O}_2}}{6.02 \times 10^{23}\ \cancel{\text{molecules}}} \times \frac{3\ \cancel{\text{mol CO}_2}}{5\ \cancel{\text{mol O}_2}} \times \frac{44.0\ \text{g CO}_2}{\cancel{\text{mol CO}_2}} = \underline{\underline{5.26\ \text{g CO}_2}}$$

(f) 0.0997 mol of $H_2O$

**9-15** 47.2 g of $N_2$

**9-17**
$$125\ \cancel{\text{g Fe}_2\text{O}_3} \times \frac{1.00\ \cancel{\text{mol Fe}_2\text{O}_3}}{160\ \cancel{\text{g Fe}_2\text{O}_3}} \times \frac{2\ \cancel{\text{mol Fe}_3\text{O}_4}}{3\ \cancel{\text{mol Fe}_2\text{O}_3}} \times \frac{3\ \cancel{\text{mol FeO}}}{1\ \cancel{\text{mol Fe}_3\text{O}_4}} \times \frac{1\ \cancel{\text{mol Fe}}}{1\ \cancel{\text{mol FeO}}} \times \frac{55.85\ \text{g Fe}}{\cancel{\text{mol Fe}}} = \underline{\underline{87.3\ \text{g Fe}}}$$

**9-18** 0.730 g of HCl

**9-20** 16.9 kg of $HNO_3$

**9-22** $8.15 \times 10^3$ g of CO

**9-23** 30.0 g of $SO_3$ (theoretical yield); 70.7% yield

**9-26** Theoretical yield $\times \frac{\%\ \text{yield}}{100\%}$ = actual yield; theoretical yield $= \dfrac{\text{actual yield}}{\dfrac{\%\ \text{yield}}{100\%}}$

$$\text{Theoretical yield} = \frac{250\ \text{g}}{0.700} = 357\ \text{g N}_2$$

g $N_2 \rightarrow$ mol $N_2 \rightarrow$ mol $H_2 \rightarrow$ g $H_2$

$$357\ \cancel{\text{g N}_2} \times \frac{1\ \cancel{\text{mol N}_2}}{28.0\ \cancel{\text{g N}_2}} \times \frac{4\ \cancel{\text{mol H}_2}}{1\ \cancel{\text{mol N}_2}} \times \frac{2.02\ \text{g H}_2}{\cancel{\text{mol H}_2}} = \underline{\underline{103\ \text{g H}_2}}$$

**9-27** 86.4 %

**9-28** If 86.4% is converted to $CO_2$, 13.6% is converted to CO. Thus 0.136 × 57.0 g = 7.75 g of $C_8H_{18}$ is converted to CO. Notice that 1 mol of $C_8H_{18}$ would form 8 mol of CO (because of the eight carbons in $C_8H_{18}$). Thus,

$$7.75\ \cancel{\text{g C}_8\text{H}_{18}} \times \frac{1.00\ \cancel{\text{mol C}_8\text{H}_{18}}}{114\ \cancel{\text{g C}_8\text{H}_{18}}} \times \frac{8\ \cancel{\text{mol CO}}}{1\ \cancel{\text{mol C}_8\text{H}_{18}}} \times \frac{28.0\ \text{g CO}}{\cancel{\text{mol CO}}} = \underline{\underline{15.2\ \text{g CO}}}$$

**9-31** $2KClO_3 \rightarrow 2KCl + 3O_2$

Find the weight of $KClO_3$ needed to produce 12.0 g $O_2$.

g $O_2 \rightarrow$ mol $O_2 \rightarrow$ mol $KClO_3 \rightarrow$ g $KClO_3$

30.6 g $KClO_3$ needed to produce 12.0 g $O_2$.

$$\text{Percent purity} = \frac{\text{weight of compound}}{\text{weight of sample}} \times 100\% = \frac{30.6\ \text{g}}{50.0\ \text{g}} \times 100\% = \underline{\underline{61.2\%}}$$

**9-32** 4.49% $FeS_2$

**9-34** 1.00 mol of $H_2$ ($H_2SO_4$ is the limiting reactant.)

$$1.00 \text{ mol } H_2SO_4 \times \frac{2 \text{ mol Al}}{3 \text{ mol } H_2SO_4} = 0.667 \text{ mol of Al used;}$$

$$0.800 - 0.667 = \underline{\underline{0.133 \text{ mol Al remain}}}$$

**9-36** 40.0 g of $O_2$ forms 0.833 mol of $N_2$, 1.50 mol of $NH_3$ forms 0.750 mol of $N_2$. Therefore, $NH_3$ is the limiting reactant and the yield of $N_2$ is 0.750 mol.

**9-37** $$20.0 \text{ g } AgNO_3 \times \frac{1.00 \text{ mol } AgNO_3}{170 \text{ g } AgNO_3} \times \frac{2 \text{ mol AgCl}}{2 \text{ mol } AgNO_3} = 0.118 \text{ mol AgCl}$$

$$10.0 \text{ g } CaCl_2 \times \frac{1.00 \text{ mol } CaCl_2}{111 \text{ g } CaCl_2} \times \frac{2 \text{ mol AgCl}}{1 \text{ mol } CaCl_2} = 0.180 \text{ mol AgCl}$$

Since $AgNO_3$ produces the smallest yield, it is the limiting reactant.

$$0.118 \text{ mol AgCl} \times \frac{143 \text{ g AgCl}}{\text{mol AgCl}} = 16.9 \text{ g AgCl}$$

Convert moles of AgCl (from limiting reactant) to grams of $CaCl_2$ used.

$$0.118 \text{ mol AgCl} \times \frac{1 \text{ mol } CaCl_2}{2 \text{ mol AgCl}} \times \frac{111 \text{ g } CaCl_2}{\text{mol } CaCl_2} = 6.5 \text{ g } CaCl_2 \text{ used}$$

$$10.0 - 6.5 = \underline{\underline{3.5 \text{ g } CaCl_2 \text{ remaining}}}$$

**9-39** products: 3.53 g of $H_2O$, 2.93 g of NO, 4.71 g of S
reactants remaining: 3.9 g of $HNO_3$

**9-40** $NH_3$ is the limiting reactant. The theoretical yield based on $NH_3$ is 141 g of NO. The percent yield is 28.4%.

## Chapter 10

**10-1** (a) 2.17 atm (b) 0.0266 torr (c) 9560 torr (d) 0.0558 atm
(e) 3.68 lb/in.$^2$ (f) 11 kPa

**10-3** force = 2450 g, $P$ = 2450 g/12.0 cm$^2$ = 204 g/cm$^2$

$$1 \text{ atm} = 76.0 \text{ cm Hg} = 1030 \text{ g/cm}^2; \; 204 \text{ g/cm}^2 \times \frac{1 \text{ atm}}{1030 \text{ g/cm}^2} = \underline{\underline{0.198 \text{ atm}}}$$

**10-4** 0.0102 atm

**10-6** For a column of Hg that is 1 cm$^2$ in area and 76.0 cm high, weight = 76.0 cm × 1 cm$^2$ × 13.6 g/cm$^3$ = 1030 g. For water, volume × density = weight, volume = height × area, height × 1 cm$^2$ × 1.00 g/cm$^3$ = 1030 g, $h$ = 1030 cm.

$$1030 \text{ cm} \times \frac{1 \text{ in.}}{2.54 \text{ cm}} \times \frac{1 \text{ ft}}{12 \text{ in.}} = 33.8 \text{ ft}$$

**10-8** The molecules in the air are comparatively very far apart, meaning that air is mostly empty space. The molecules of water are close together so that more molecules are encountered when you move your arms.

**10-10** A gas can be compressed. A tire is never "filled" with a gas since more can be added.

**10-12** When the pressure is high, the gas molecules are forced closer together. In a highly compressed gas, the molecules can occupy an appreciable part of the total volume. When the temperature is low, molecules have a lower average velocity. If there is some attraction, they can momentarily stick together when moving slowly.

**10-13** 10.2 L

**10-15** 67.9 torr

**10-16** $\frac{V_{\text{final}}}{V_{\text{initial}}} = \frac{1}{15} = \frac{P_{\text{initial}}}{P_{\text{final}}}$; $P_{\text{final}} = 14.2$ atm

**10-18**

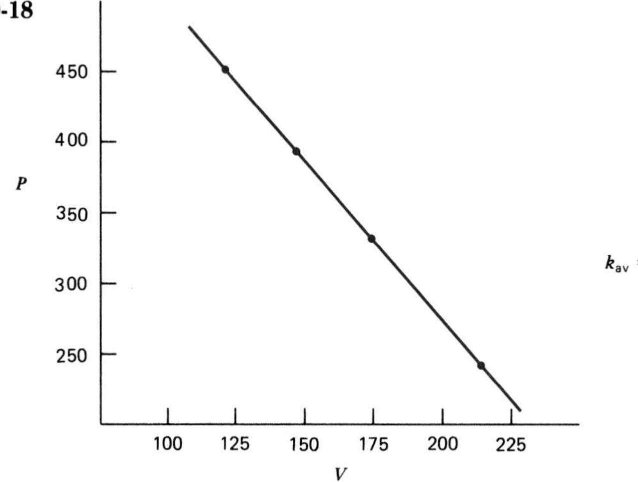


$k_{av} = 5.61 \times 10^4$ mL · torr

**10-20** 1.94 L

**10-22** 77 °C

**10-23** $T_2 = 341$ K (68 °C)

**10-24** $2.60 \times 10^4$ L

**10-27** 2.94 atm

**10-28** 191 K (−82 °C)

**10-29** 596 K (323 °C)

**10-33** (a) and (d)

**10-35** 1.24 atm

**10-36** 76.8 L

**10-37** 88.5 K (−185 °C)

**10-40** 258 K (−15 °C)

**10-42** 1270 miles/hr

**10-43** $SF_6$, $SO_2$, $N_2O$ = $CO_2$, $N_2$, $H_2$

**10-44** rate $N_2$ = 1.19 rate Ar

**10-47** 127 g/mol

**10-49** $\frac{r_{235_U}}{r_{238_U}} = \sqrt{\frac{352}{349}} = 1.004$; $r_{235_U} = 1.004 r_{238_U}$

**10-50** 762 torr

**10-52** 6.8 torr

**10-54** 730 torr

**10-56** $N_2$, 756 torr; $O_2$, 84 torr; $SO_2$, 210 torr

**10-57** $N_2$, 1.02 atm; $CO_2$, 1.73 atm

**10-58** $P_{N_2}$ = 300 torr, $P_{O_2}$ = 170 torr (when compressed from 4.00 L to 2.00 L) $P_{CO_2}$ = 225 torr, $P_{total}$ = 695 torr

**10-60** 1.70 L

**10-62** $\frac{V_1}{V_2} = \frac{n_1}{n_2}$; $n_2 = 5.47 \times 10^{-3}$ mol (original gas + added $N_2$); $n_{N_2} = n_{total} - n_{original} = 5.47 \times 10^{-3} - 2.50 \times 10^{-3} = 2.97 \times 10^{-3}$ mol $N_2$;
$2.97 \times 10^{-3} \cancel{\text{mol N}_2} \times \frac{28.0 \text{ g N}_2}{\cancel{\text{mol N}_2}} = \underline{\underline{0.0832 \text{ g N}_2}}$

**10-64** 98 °C

**10-66** 13.0 g $NH_3$

**10-67** 589 torr

**10-69** 0.0148 g $H_2$

**10-71** Weight of He ($4.2 \times 10^6$ g) $9.2 \times 10^3$ lb, weight of air ($3.0 \times 10^7$ g) $6.6 \times 10^4$ lb, lifting power = 57,000 lb; lifting power with $H_2$ = 61,000 lb. $H_2$ is combustible whereas He is not. In 1937 the German airship *Hindenberg* (which used $H_2$) was destroyed by fire with great loss of life.

**10-72** Molar mass is 41.9 g; empirical formula is $CH_2$; molecular formula is $C_3H_6$.

**10-74** 7.64 L

**10-76** 113 L

**10-78** $1.39 \times 10^{-3}$ g

**10-79** 1.24 g/L

**10-81** 34.0 g/mol

**10-83** $1.00\text{ L} \times \frac{273\,\cancel{K}}{298\,\cancel{K}} \times \frac{1.20\,\cancel{\text{atm}}}{1.00\,\cancel{\text{atm}}} = 1.10\text{ L (STP)} \quad \frac{3.60\text{ g}}{1.10\text{ L}} = 3.27\text{ g/L (STP)}$

**10-84** Find moles of $N_2$ in 1 L at 500 torr and 22 °C. $n = 0.0272$ mol $N_2$ = 0.762 g; density = 0.762 g/L (500 torr, 22 °C).

**10-86** 25.8 L $CO_2$

**10-88** 3.47 L

**10-90** (a) g $C_4H_{10}$ → mol $C_4H_{10}$ → mol $CO_2$ → vol $CO_2$ (STP)

$$85.0\,\cancel{\text{g } C_4H_{10}} \times \frac{1.00\,\cancel{\text{mol } C_4H_{10}}}{58.0\,\cancel{\text{g } C_4H_{10}}} \times \frac{8\,\cancel{\text{mol } CO_2}}{2\,\cancel{\text{mol } C_4H_{10}}} \times \frac{22.4\text{ L}}{\cancel{\text{mol } CO_2}} = 131\text{ L}$$

(b) 96.3 L $O_2$
(c) 124 L $CO_2$

**10-91** 2080 kg Zr (2.29 tons)

**10-93** $n_{O_2} = 1.45$ mol $O_2$; $1.45\,\cancel{\text{mol } O_2} \times \frac{1\text{ mol } CO_2}{2\,\cancel{\text{mol } O_2}} = 0.725$ mol $CO_2$

$V_{CO_2} = \underline{\underline{11.9\text{ L}}}$

## Review Test on Chapters 8–10

### Multiple Choice

**1.** (c) **2.** (a) **3.** (d) **4.** (a) **5.** (b) **6.** (c) **7.** (a) **8.** (c) **9.** (d) **10.** (b)
**11.** (e) **12.** (c) **13.** (c) **14.** (d) **15.** (e) **16.** (d) **17.** (c) **18.** (c) **19.** (c)
**20.** (b) **21.** (a) **22.** (c) **23.** (e) **24.** (e) **25.** (c) **26.** (c) **27.** (c) **28.** (d)
**29.** (b) **30.** (b)

### Problems

**1.** 1.50 mol of $K_2SO_4$ weighs 261 g, contains $9.03 \times 10^{23}$ formula units, contains 6.00 mol of O atoms, contains 117 g K, and contains $1.81 \times 10^{24}$ K atoms.
**2.** $C_4H_{10}N_2$ has a molar mass of 86.1 g/mol, is 32.5% by weight N, and has an empirical formula of $C_2H_5N$.
**3.** $Fe_3O_4$
**4.** $C_2H_{14}B_{10}$
**5.** (a) $6K(s) + Al_2Cl_6 \rightarrow 6KCl(s) + 2\,Al$
(b) $2C_6H_6 + 15\,O_2(g) \rightarrow 12\,CO_2(g) + 6\,H_2O$
(c) $Cl_2O_3 + H_2O \rightarrow 2HClO_2$
(d) $3K_2S + 2H_3PO_4 \rightarrow 3H_2S(g) + 2K_3PO_4$
(e) $B_4H_{10} + 12\,H_2O \rightarrow 4B(OH)_3 + 11\,H_2(g)$
**6.** (a) 0.0233 mol of $SbF_3$ (b) 0.826 mol of $SbCl_3$ (c) $1.10 \times 10^3$ g of $CCl_4$ (d) $4.75 \times 10^{25}$ molecules of $CCl_4$ (e) 3.12 L of freon
**7.** 10.0 L of $O_2$ at STP contains 0.446 mol of $O_2$, contains $2.69 \times 10^{23}$ molecules

of $O_2$, has a volume of 8.06 L at −53 °C, has a volume of 13.1 L at 580 torr, has a pressure of 1.10 atm at 27 °C, and has a volume of 21.2 L if 0.500 mol of a gas is added.

**8.** molar mass = 146 g/mol (from the ideal gas law); empirical formula = $C_3ClH_2$; molecular formula = $C_6Cl_2H_4$

## Chapter 11

**11-1** The two equal O—H bond dipoles would cancel if they were at an angle of 180° (linear).

**11-3** 

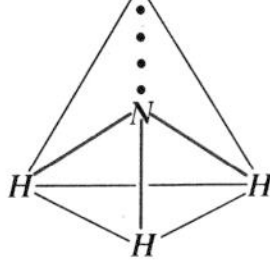


NH$_3$ is a polar molecule because the three equal N—H bond dipoles do not cancel.

**11-4** $PH_3$ is polar like $NH_3$. However, since the P—H bond is much less polar than the N—H bond, the $PH_3$ molecule is less polar than $NH_3$.

**11-5** (a), (b), and (f)

**11-7** (a), (c), (d), and (f)

**11-9** Molecules in both gases and liquids are undergoing frequent collisions. Since the molecules in a gas are much farther apart than in the liquid, they move much farther between collisions, which means they mix faster.

**11-12** OCS has a higher melting point because it is a polar molecule.

**11-14** $H_2O$ has a higher melting point. The H—O bond is more polar than the F—O bond because of the greater difference in electronegativity between the atoms. Thus the molecular dipole is greater in $H_2O$, which indicates a higher melting point.

**11-16** The average kinetic energy of the molecules is the same at the same temperature regardless of the physical state.

**11-18** Because $H_2O$ molecules have an attraction for each other, moving them apart increases the potential energy. Since the molecules in a gas at 100 °C are farther apart than in a liquid at 100 °C, the potential energy of the gas molecules is greater.

**11-19** (a), (b), and (c)

**11-20** $H_2O$, 6.89 g; NaCl, 4.44 g; benzene, 18.1 g

**11-21** Ethyl ether, 555 cal; $H_2O$, 2000 cal. Water is more effective than ether in holding heat. $1.30 \times 10^4$ J melt 125 g of ethyl alcohol.

**11-23** 1690 cal

**11-25** $NH_3$, $1.46 \times 10^5$ cal; freon, $1.74 \times 10^4$ cal. On the basis of weight, $NH_3$ is more effective than freon.

**11-27** Condensation releases $6.22 \times 10^5$ J and cooling releases $8.6 \times 10^4$ J, for a total of $7.08 \times 10^5$ J.

**11-29**

| | | |
|---|---|---|
| Heat ice: | $132\ \cancel{g} \times 20\ \cancel{°C} \times \dfrac{0.492\ \text{cal}}{\cancel{g}\cdot\cancel{°C}}$ | $= 1{,}300$ cal |
| Melt ice at 0 °C | $132\ \cancel{g} \times 79.8\ \text{cal}/\cancel{g}$ | $= 10{,}500$ cal |
| Heat $H_2O$ | $132\ \cancel{g} \times 100\ \cancel{°C} \times \dfrac{1.00\ \text{cal}}{\cancel{g}\cdot\cancel{°C}}$ | $= 13{,}200$ cal |
| Vaporize $H_2O$ at 100 °C | $132\ \cancel{g} \times 540\ \text{cal}/\cancel{g}$ | $= \underline{\underline{71{,}300\ \text{cal}}}$ |
| | | $= 96{,}300$ cal total |

**11-31** Let $Y$ = the weight of the sample in grams. Then

$$\frac{540\ \text{cal}}{\text{g}} Y + \left(25\ °\text{C} \times Y \times \frac{1.00\ \text{cal}}{\text{g}\cdot °\text{C}}\right) = 6780\ \text{cal}$$

$$Y = \underline{\underline{12.0\ \text{g}}}$$

**11-33** $2.83 \times 10^4$ cal

**11-34** 17 °C

**11-36** This is a comparatively high heat of fusion, so the melting point should also be comparatively high.

**11-38** This is a comparatively high boiling point, so the compound probably also has a comparatively high melting point.

**11-40** Equilibrium refers to a state where opposing forces are balanced. In the case of a liquid in "equilibrium" with its vapor, it means that a molecule escaping to the vapor is replaced by one condensing to the liquid.

**11-42** gas

**11-43** Ethyl alcohol boils at about 52 °C at that altitude. At 10 °C ethyl ether is a gas at that altitude.

**11-45** No, at that temperature the pressure would have to be at least 250 torr for water not to boil.

**11-47** $V = 100$ L, $T = 34 + 273 = 307$ K, $P = 0.70 \times 39.0$ torr $= 27.3$ torr. Using the ideal gas law, $n = 0.143$ mol $H_2O$ $= \underline{\underline{2.57\ \text{g}\ H_2O}}$

**11-49** Alcohol is more volatile (i.e., has a higher vapor pressure) than water. It thus evaporates more rapidly and produces a pronounced cooling effect.

**11-51** The liquid other than water. The higher the vapor pressure, the faster the liquid evaporates and the faster the liquid cools.

**11-53** (a) $2K(s) + 2H_2O(l) \rightarrow 2KOH(aq) + H_2(g)$
(b) $Br_2(l) + H_2O(l) \rightarrow HBr(aq) + HOBr(aq)$

(c) $K_2O(s) + H_2O(l) \rightarrow 2KOH(aq)$
(d) $CO_2(g) + H_2O(l) \rightarrow H_2CO_3(aq)$
(e) $C_2H_4(g) + 3O_2(g) \rightarrow 2CO_2(g) + 2H_2O(l)$

**11-55** (a) magnesium carbonate pentahydrate (b) sodium sulfate decahydrate
(c) iron(III) bromide hexahydrate

**11-57** 54.2% $H_2O$

**11-59** Let $x$ = number of $H_2O$ molecules. Then the decimal fraction of $H_2O$ equals

$$\frac{\text{g } H_2O}{\text{g of compound}}$$

Thus
$$\frac{18x}{120.3 + 18x} = 0.511 \qquad x = 7$$

Formula is $MgSO_4 \cdot 7H_2O$.

## Chapter 12

**12-1** An aqueous solution of AB is a good conductor of electricity. AB is dissociated into ions such as $A^+$ and $B^-$. A solution of AC is a weak conductor of electricity. AC is only partially dissociated into ions (i.e., $AC \rightleftharpoons A^+ + C^-$). A solution of AD is a nonconductor of electricity. AD is present as undissociated molecules in solution.

**12-3** (a) $K_2Cr_2O_7 \rightarrow 2K^+(aq) + Cr_2O_7^{2-}(aq)$
(b) $Li_2SO_4 \rightarrow 2Li^+(aq) + SO_4^{2-}(aq)$
(c) $Cs_2SO_3 \rightarrow 2Cs^+(aq) + SO_3^{2-}(aq)$
(d) $Ca(ClO_3)_2 \rightarrow Ca^{2+}(aq) + 2ClO_3^-(aq)$
(e) $2(NH_4)_2S \rightarrow 4NH_4^+(aq) + 2S^{2-}(aq)$
(f) $4Ba(OH)_2 \rightarrow 4Ba^{2+}(aq) + 8OH^-(aq)$

**12-4** $PbSO_4$, $MgSO_3$, $Ag_2O$

**12-6** (a) $2KI(aq) + Pb(C_2H_3O_2)_2(aq) \rightarrow PbI_2(s) + 2KC_2H_3O_2(aq)$
(b) no reaction
(c) no reaction
(d) $BaS(aq) + Hg_2(NO_3)_2(aq) \rightarrow Hg_2S(s) + Ba(NO_3)_2(aq)$
(e) $FeCl_3(aq) + 3KOH(aq) \rightarrow Fe(OH)_3(s) + 3KCl(aq)$

**12-8** (a) $2K^+(aq) + S^{2-}(aq) + Pb^{2+}(aq) + 2NO_3^-(aq) \rightarrow$
$PbS(s) + 2K^+(aq) + 2NO_3^-(aq)$

$S^{2-}(aq) + Pb^{2+}(aq) \rightarrow PbS(s)$

(b) $2NH_4^+(aq) + CO_3^{2-}(aq) + Ca^{2+}(aq) + 2Cl^-(aq) \rightarrow$
$CaCO_3(s) + 2NH_4^+(aq) + 2Cl^-(aq)$

$CO_3^{2-}(aq) + Ca^{2+}(aq) \rightarrow CaCO_3(s)$

(c) $2Ag^+(aq) + 2ClO_4^-(aq) + 2Na^+(aq) + CrO_4^{2-}(aq) \rightarrow$
$Ag_2CrO_4(s) + 2Na^+(aq) + 2ClO_4^-(aq)$

$$2Ag^+(aq) + CrO_4^{2-}(aq) \rightarrow Ag_2CrO_4(s)$$

**12-9** (a) $2K^+(aq) + 2I^-(aq) + Pb^{2+}(aq) + 2C_2H_3O_2^-(aq) \rightarrow PbI_2(s) + 2K^+(aq) + 2I^-(aq)$

(d) $Ba^{2+}(aq) + S^{2-}(aq) + Hg_2^{2+}(aq) + 2NO_3^-(aq) \rightarrow Hg_2S(s) + Ba^{2+}(aq) + 2NO_3^-(aq)$

(e) $Fe^{3+}(aq) + 3Cl^-(aq) + 3K^+(aq) + 3OH^-(aq) \rightarrow Fe(OH)_3(s) + 3K^+(aq) + 3Cl^-(aq)$

**12-11** (a) $Pb^{2+}(aq) + 2I^-(aq) \rightarrow PbI_2(s)$

(d) $Hg_2^{2+}(aq) + S^{2-}(aq) \rightarrow Hg_2S(s)$

(e) $Fe^{3+}(aq) + 3OH^-(aq) \rightarrow Fe(OH)_3(s)$

**12-13** (a) $CuCl_2(aq) + Na_2CO_3(aq) \rightarrow CuCO_3(s) + 2NaCl(aq)$
Filter the solid $CuCO_3$.

(b) $(NH_4)_2SO_3(aq) + Pb(NO_3)_2(aq) \rightarrow PbSO_3(s) + 2NH_4NO_3(aq)$
Filter the solid $PbSO_3$.

(c) $2KI(aq) + Hg_2(NO_3)_2(aq) \rightarrow Hg_2I_2(s) + 2KNO_3(aq)$
Filter the solid $Hg_2I_2$.

(d) $NH_4Cl(aq) + AgNO_3(aq) \rightarrow AgCl(s) + NH_4NO_3(aq)$
Filter off the solid AgCl; the desired product remains after the water is removed by boiling.

(e) $Ca(C_2H_3O_2)_2(aq) + K_2CO_3(aq) \rightarrow CaCO_3(s) + 2KC_2H_3O_2(aq)$
Filter the solid $CaCO_3$; the desired product remains after the water is removed by boiling.

**12-14** 1.49%

**12-16** 0.375 mol of NaOH

**12-19** 8.81%

**12-20** 0.542 *M*

**12-24** (a) 0.873 *M* (b) 1.40 *M* (c) 12.4 L (d) 41.3 g (e) 0.294 *M*
(f) 0.024 mol (g) 310 mL (h) 49.0 g (i) 2.00 mL

**12-25** $1.17 \times 10^{-3}$ *M*

**12-26** $[Ba^{2+}] = 0.166$, $[OH^-] = 0.332$

**12-27** 1.84 *M*

**12-29** 1.32 g/mL

**12-30** 0.833 L

**12-32** Volume of 0.800 *M* NaOH needed is 313 mL. Slowly add the concentrated NaOH to about 500 mL of water in a 1-L volumetric flask. Dilute to the 1-L mark with water.

**12-34** 0.140 *M*

**12-35** 0.72 L

**12-36**

$$0.250\,\cancel{L} \times \frac{0.200\,\cancel{mol}}{\cancel{L}} \times \frac{60.0\,\cancel{g}}{\cancel{mol}} \times \frac{1.00\text{ mL}}{1.05\,\cancel{g}} = 2.86\text{ mL}$$

**12-38** $n = (0.150\,\cancel{L} \times 0.250\text{ mol}/\cancel{L}) + (0.450\,\cancel{L} \times 0.375\text{ mol}/\cancel{L}) = 0.2062\text{ mol}$

$$V = 0.150\text{ L} + 0.450\text{ L} = 0.600\text{ L}$$

$$n/V = 0.344\ M$$

**12-39** $\left(0.500\,\cancel{L} \times 0.250\dfrac{\cancel{mol}}{\cancel{L}}\right) \times \dfrac{1\ \cancel{mol\ Cr(OH)_3}}{3\ \cancel{mol\ KOH}} \times \dfrac{103\text{ g Cr(OH)}_3}{\cancel{mol\ Cr(OH)_3}}$
$= 4.29\text{ g Cr(OH)}_3$

**12-41** 145 g $BaSO_4$

**12-42** 4.08 L

**12-44** 0.653 L

**12-46** NaOH is the limiting reactant and produces 1.75 g of $Mg(OH)_2$.

**12-48** The solute is 2.50 *m* in both solvents.

**12-50** 1.00 *m*

**12-51** 15.8 g

**12-53** Let $x$ = grams of KBr; then the grams of $H_2O = 1000 - x$.

$$\frac{\text{mol solute}}{\text{kg solvent}} = m = 1.00$$

$$\frac{x\ \cancel{g}/119\ \cancel{g}/\text{mol}}{\left(\dfrac{1000 - x}{1000}\right)\text{kg}} = 1.00$$

$x = 106$ g; wt of $H_2O$ = $\underline{\underline{894\text{ g}}}$

**12-54** −0.37 °C

**12-56** 834 g

**12-57** 101.4 °C

**12-59** 2.93 *m*

**12-60** 80.5 °C

**12-62** −14.5 °C

**12-64** (a) −5.8 °C (b) −6.4 °C (c) −5.0 °C

**12-65** The salty water removes water from the cells of the skin by osmosis. After a prolonged period one would dehydrate and become thirsty.

**12-68** One can concentrate a dilute solution by boiling away some solvent if the solute is not volatile. If the solution is separated from pure solvent by a semipermeable membrane, reverse osmosis can be applied. Pressure must be exerted on the solution that is greater than its osmotic pressure to cause solvent molecules to migrate to the pure solvent. As the solution becomes more concentrated, the osmotic pressure becomes greater and the corresponding pressure that is applied must be increased.

## Chapter 13

**13-1** (a) $HNO_3$, nitric acid (b) $HNO_2$, nitrous acid (c) $HClO_3$, chloric acid (d) $H_2SO_3$, sulfurous acid

**13-2** (a) CsOH, cesium hydroxide (b) $Sr(OH)_2$, strontium hydroxide (c) $Al(OH)_3$, aluminum hydroxide (d) $Mn(OH)_3$, manganese(III) hydroxide

**13-4** (a) $HNO_3 + H_2O \rightarrow H_3O^+ + NO_3^-$
(b) $HNO_2 + H_2O \rightarrow H_3O^+ + NO_2^-$
(c) $HClO_3 + H_2O \rightarrow H_3O^+ + ClO_3^-$
(d) $H_2SO_3 + H_2O \rightarrow H_3O^+ + HSO_3^-$

**13-6** (a) acid (b) salt (c) acid (d) acid salt (e) salt (f) base

**13-7** (a) $HC_2H_3O_2 + KOH \rightarrow KC_2H_3O_2 + H_2O$
(b) $2HI + Ca(OH)_2 \rightarrow CaI_2 + 2H_2O$
(c) $H_2SO_4 + Ca(OH)_2 \rightarrow CaSO_4 + 2H_2O$

**13-9** (b) $2H^+(aq) + 2I^-(aq) + Ca^{2+}(aq) + 2OH^-(aq) \rightarrow Ca^{2+}(aq) + 2I^-(aq) + 2H_2O$

$2H^+ + 2OH^- \rightarrow 2H_2O$

(c) $2H^+(aq) + SO_4^{2-}(aq) + Ca^{2+}(aq) + 2OH^-(aq) \rightarrow CaSO_4(s) + 2H_2O$
(net ionic equation the same as the total ionic equation)

**13-10** (a) $2HBr + Ca(OH)_2 \rightarrow CaBr_2 + 2H_2O$
(b) $2HClO_2 + Sr(OH)_2 \rightarrow Sr(ClO_2)_2 + 2H_2O$
(c) $2H_2S + Ba(OH)_2 \rightarrow Ba(HS)_2 + 2H_2O$
(d) $H_2S + 2LiOH \rightarrow Li_2S + 2H_2O$

**13-12** $LiOH + H_2S \rightarrow LiHS + H_2O$
$LiOH + LiHS \rightarrow Li_2S + H_2O$

**13-14** $H_2S + NaOH \rightarrow NaHS + H_2O$

**13-15** $Ca(OH)_2(aq) + 2H_3PO_4(aq) \rightarrow Ca(H_2PO_4)_2(aq) + 2H_2O$

**13-17** $NH(CH_3)_2 + H_2O \rightleftharpoons HNH(CH_3)_2^+ + OH^-$

**13-18** HX is a weak acid. The percent ionization is $\frac{0.010}{0.10} \times 100\% = 10\%$.

**13-20** $[H_3O^+] = 0.55$

**13-21** $[H_3O^+] = 0.030 \times 0.55 = 0.016$

**13-23** From the first ionization, $[H_3O^+] = 0.354$. From the second ionization $[H_3O^+] = 0.25 \times 0.354 = 0.088$. Therefore, the total $[H_3O^+] = 0.354 + 0.088 = 0.442$.

**13-25** The system would not be at equilibrium if $[H_3O^+] = [OH^-] = 10^{-2}\ M$. Therefore, $H_3O^+$ reacts with $OH^-$ until the concentration of each is reduced to $10^{-7}\ M$ (i.e., a neutralization reaction occurs).

$$H_3O^+ + OH^- \rightarrow 2H_2O$$

**13-26** (a) $[H_3O^+] = K_w/[OH^-] = 10^{-14}/10^{-12} = 10^{-2}$

(b) $[H_3O^+] = 10^{-15}$
(c) $[OH^-] = 5.0 \times 10^{-10}$

**13-28** $[H_3O^+] = 0.0250$; $[OH^-] = 4.00 \times 10^{-13}$

**13-29** In lye $[H_3O^+] = 3.92 \times 10^{-15}$; in ammonia, $[H_3O^+] = 2.5 \times 10^{-12}$.

**13-30** (a) acidic (b) basic (c) acidic

**13-32** (a) acidic (b) basic (c) acidic (d) basic

**13-34** (a) 6.00 (b) 9.00 (c) 12.00 (d) 9.40 (e) 10.19

**13-36** (a) $1.0 \times 10^{-3}$ (b) $2.9 \times 10^{-4}$ (c) $1.0 \times 10^{-6}$ (d) $2.4 \times 10^{-8}$ (e) $2.0 \times 10^{-13}$

**13-38** For Problem 13-34: (a) acidic (b) basic (c) basic (d) basic (e) basic
For Problem 13-36: (a) acidic (b) acidic (c) acidic (d) basic (e) basic

**13-40** pH = 1.12

**13-42** pH = 12.56

**13-43** (a) strongly acidic (b) strongly basic (c) weakly acidic (d) strongly basic (e) neutral (f) weakly basic (g) weakly acidic (h) strongly acidic

**13-45** (a) $NO_3^-$ (b) $HSO_4^-$ (c) $PO_4^{3-}$ (d) $CH_3^-$ (e) $OH^-$ (f) $NH_2^-$

**13-48** (a) $NH_3 + H_2O \rightarrow NH_4^+ + OH^-$
$NH_3$ = $B_1$, $NH_4^+$ = $A_1$; $H_2O$ = $A_2$, $OH^-$ = $B_2$

(b) $N_2H_4 + H_2O \rightarrow N_2H_5^+ + OH^-$
$N_2H_4$ = $B_1$, $N_2H_5^+$ = $A_1$; $H_2O$ = $A_2$, $OH^-$ = $B_2$

(c) $HS^- + H_2O \rightarrow H_2S + OH^-$
$HS^-$ = $B_1$, $H_2S$ = $A_1$; $H_2O$ = $A_2$, $OH^-$ = $B_2$

(d) $H^- + H_2O \rightarrow H_2 + OH^-$
$H^-$ = $B_1$, $H_2$ = $A_1$; $H_2O$ = $A_2$, $OH^-$ = $B_2$

(e) $F^- + H_2O \rightarrow HF + OH^-$
$F^-$ = $B_1$, $HF$ = $A_1$; $H_2O$ = $A_2$, $OH^-$ = $B_2$

**13-49** Brønsted-Lowry base: $HS^- + H_3O^+ \rightleftharpoons H_2S + H_2O$
Brønsted-Lowry acid: $HS^- + OH^- \rightleftharpoons S^{2-} + H_2O$

**13-51** (a) $HNO_3 + H_2O \rightarrow H_3O^+ + NO_3^-$
(b) $NO_2^- + H_3O^+ \rightleftharpoons HNO_2 + H_2O$
(c) N.R.
(d) $HOCl + H_2O \rightleftharpoons H_3O^+ + OCl^-$
(e) N.R.
(f) $HClO_4 + H_2O \rightarrow H_3O^+ + ClO_4^-$

**13-53** (a) $HS^-$ (b) $H_3O^+$ (c) $OH^-$ (d) $NH(CH_3)_2$

**13-55** (a) $F^- + H_2O \rightleftharpoons HF + OH^-$
(b) $SO_3^{2-} + H_2O \rightleftharpoons HSO_3^- + OH^-$
(c) $H_2N(CH_3)_2^+ + H_2O \rightleftharpoons HN(CH_3)_2 + H_3O^+$
(d) $HPO_4^{2-} + H_2O \rightleftharpoons H_2PO_4^- + OH^-$
(e) $CN^- + H_2O \rightleftharpoons HCN + OH^-$
(f) $Li^+$ does not hydrolyze.

**13-57** $CaC_2(s) + 2H_2O \rightarrow C_2H_2(g) + 2OH^-(aq) + Ca^{2+}(aq)$

**13-60** (a) neutral (neither ion hydrolyzes) (b) acidic (cation hydrolysis) (c) basic (anion hydrolysis) (d) neutral (neither ion hydrolyzes) (e) acidic (cation hydrolysis) (f) basic (anion hydrolysis)

**13-62** Cation: $NH_4^+ + H_2O \rightleftharpoons NH_3 + H_3O^+$
Anion: $CN^- + H_2O \rightleftharpoons HCN + OH^-$
Since the solution is basic, the anion hydrolysis reaction must take place to a greater extent than the cation hydrolysis.

**13-64** (a), (b), (e), (h), and (i)

**13-65** There is no equilibrium involved when HCl dissolves in water. A reservoir of un-ionized acid must be present to replace any $H_3O^+$ lost when a strong base is added. Likewise, the $Cl^-$ ion does not exhibit base behavior in water, so it cannot react with any $H_3O^+$ added to the solution.

**13-68** (a) $Sr(OH)_2$ (b) $H_2SeO_4$ (c) $H_3PO_4$ (d) CsOH (e) $HNO_2$ (f) $HClO_3$

**13-70** $LiOH(aq) + CO_2(g) \rightarrow LiHCO_3(aq)$

**13-71** $2LiNO_3$

**13-73** (a) acidic: $H_2S + H_2O \rightleftharpoons HS^- + H_3O^+$
(b) basic: $ClO^- + H_2O \rightleftharpoons HClO + OH^-$
(c) neutral
(d) basic: $NH_3 + H_2O \rightleftharpoons NH_4^+ + OH^-$
(e) acidic: $HN_2H_4^+ + H_2O \rightleftharpoons N_2H_4 + H_3O^+$
(f) basic: $Ba(OH)_2 \rightarrow Ba^{2+} + 2OH^-$
(g) neutral
(h) basic: $NO_2^- + H_2O \rightleftharpoons HNO_2 + OH^-$
(i) acidic: $H_2SO_3 + H_2O \rightleftharpoons H_3O^+ + HSO_3^-$
(j) acidic: $Cl_2O_3 + H_2O \rightarrow 2HClO_2$; $HClO_2 + H_2O \rightleftharpoons H_3O^+ + ClO_2^-$

**13-75** LiOH, strongly basic, pH = 13.0; $SrBr_2$, neutral, pH = 7.0; KOCl, weakly basic, pH = 10.2; $NH_4Cl$, weakly acidic, pH = 5.2; HI, strongly acidic, pH = 1.0. The concentration is 0.10 *M*.

## Chapter 14

**14-1** (a) S (+6) (b) Co (+3) (c) U (+6) (d) N (+5) (e) Cr (+6)
(f) Mn (+6)

**14-3** (a), (c), and (f)

**14-4** (a) oxidation (b) reduction (c) oxidation (d) reduction
(e) reduction

**14-6** (a) $Br^-$; $Br_2$; $MnO_2$; $Mn^{2+}$; $MnO_2$; $Br^-$
(b) $CH_4$; $CO_2$; $O_2$; $CO_2$ and $H_2O$; $O_2$; $CH_4$
(c) $Fe^{2+}$; $Fe^{3+}$; $MnO_4^-$; $Mn^{2+}$; $MnO_4^-$; $Fe^{2+}$

**14-8**

$-5e \times (4) = -20e^-$

$-3 \qquad +2$

(a) $4NH_3 + 5O_2 \rightarrow 4NO + 6H_2O$

$0 \qquad -2 \qquad -2$

$+4e^- \times (5) = +20e^-$

$-4e^- \times (1) = -4e^-$

$0 \qquad +4$

(b) $Sn + 4HNO_3 \rightarrow SnO_2 + 4NO_2 + 2H_2O$

$+5 \qquad +4$

$+1e^- \times (4) = +4e^-$

(c) Before the number of electrons is calculated, notice that the $Na_2CrO_4$ in the products must first be multiplied by 2 since there are two Cr's in $Cr_2O_3$ in the reactants.

$+2e^- \times (3) = +6e^-$

$+5 \qquad +3$

$Cr_2O_3 + 2Na_2CO_3 + 3KNO_3 \rightarrow 2CO_2 + 2Na_2CrO_4 + 3KNO_2$

$+6 \qquad +12$

$-6e^- \times (1) = -6e^-$

$-4e^- \times (3) = -12e^-$

(d) $0 \qquad +4$

$3Se + 2BrO_3^- + 3H_2O \rightarrow 3H_2SeO_3 + 2Br^-$

$+5 \qquad -1$

$+6e^- \times (2) = +12e^-$

**14-10** (a) $2H_2O + Sn^{2+} \rightarrow SnO_2 + 4H^+ + 2e^-$
(b) $2H_2O + CH_4 \rightarrow CO_2 + 8H^+ + 8e^-$
(c) $e^- + Fe^{3+} \rightarrow Fe^{2+}$

(d) $6H_2O + I_2 \rightarrow 2IO_3^- + 12H^+ + 10e^-$
(e) $e^- + 2H^+ + NO_3^- \rightarrow NO_2 + H_2O$

**14-12** (a)

$$\begin{array}{rl|l} & S^{2-} \rightarrow S + 2e^- & \times 3 \\ & 3e^- + 4H^+ + NO_3^- \rightarrow NO + 2H_2O & \times 2 \\ \hline & 3S^{2-} + 8H^+ + 2NO_3^- \rightarrow 3S + 2NO + 4H_2O & \end{array}$$

(b) $I_2 + 2S_2O_3^{2-} \rightarrow 2I^- + S_4O_6^{2-}$

(c)

$$\begin{array}{rl|l} & H_2O + SO_3^{2-} \rightarrow SO_4^{2-} + 2H^+ + 2e^- & \times 3 \\ & 6e^- + 6H^+ + ClO_3^- \rightarrow Cl^- + 3H_2O & \times 1 \\ \hline & 3SO_3^{2-} + ClO_3^- \rightarrow Cl^- + 3SO_4^{2-} & \end{array}$$

(d) $2H^+ + 2Fe^{2+} + H_2O_2 \rightarrow 2Fe^{3+} + 2H_2O$
(e) $AsO_4^{3-} + 2I^- + 2H^+ \rightarrow I_2 + AsO_3^{3-} + H_2O$
(f) $4Zn + 10H^+ + NO_3^- \rightarrow 4Zn^{2+} + NH_4^+ + 3H_2O$

**14-14** (a) $2OH^- + SnO_2^{2-} \rightarrow SnO_3^{2-} + H_2O + 2e^-$
(b) $6e^- + 4H_2O + 2ClO_2^- \rightarrow Cl_2 + 8OH^-$
(c) $6OH^- + Si \rightarrow SiO_3^{2-} + 3H_2O + 4e^-$
(d) $8e^- + 6H_2O + NO_3^- \rightarrow NH_3 + 9OH^-$

**14-16** (a)

$$\begin{array}{rl|l} & 8OH^- + S^{2-} \rightarrow SO_4^{2-} + 4H_2O + 8e^- & \times 1 \\ & I_2 + 2e^- \rightarrow 2I^- & \times 4 \\ \hline & S^{2-} + 8OH^- + 4I_2 \rightarrow SO_4^{2-} + 8I^- + 4H_2O & \end{array}$$

(b) $8MnO_4^- + 8OH^- + I^- \rightarrow 8MnO_4^{2-} + IO_4^- + 4H_2O$
(c) $2H_2O + BiO_3^- + SnO_2^{2-} \rightarrow SnO_3^{2-} + Bi(OH)_3 + OH^-$

(d)

$$\begin{array}{rl|l} & 32OH^- + CrI_3 \rightarrow CrO_4^{2-} + 3IO_4^- + 16H_2O + 27e^- & \times 2 \\ & 2e^- + Cl_2 \rightarrow 2Cl^- & \times 27 \\ \hline & 2CrI_3 + 64OH^- + 27Cl_2 \rightarrow 2CrO_4^{2-} + 6IO_4^- + 32H_2O + 54Cl^- & \end{array}$$

**14-19** Zn reacts at the anode and $MnO_2$ reacts at the cathode.

$$Zn + 2MnO_2 + 2H_2O \rightarrow Zn(OH)_2 + 2MnO(OH)$$

**14-21** $H_2$ is oxidized at the anode and $O_2$ is reduced at the cathode.

$$2H_2 + O_2 \rightarrow 2H_2O$$

**14-23** $Ni^{2+} + Cd \rightarrow Cd^{2+} + Ni$
$Cd^{2+} + Zn \rightarrow Zn^{2+} + Cd$
$Cd^{2+}$ is a stronger oxidizing agent than $Zn^{2+}$ but is weaker than $Ni^{2+}$. It appears to have strength comparable to $Fe^{2+}$.

**14-25** $3F_2(g) + 2Au(s) \rightarrow 2AuF_3(aq)$
$Au^{3+} + 3e^- \rightarrow Au$ belongs between the $F_2$ and $Cl_2$ half-reactions in Table 14-2.

**14-26** (a) yes (b) N.R. (c) yes (d) N.R. (e) yes

**14-28** (a) $CuCl_2(aq) + Fe(s) \rightarrow FeCl_2(aq) + Cu(s)$
(A coating of Cu forms on the nails.)
(b) N.R.
(c) N.R.
(d) $Ni(s) + Pb(NO_3)_2(aq) \rightarrow Ni(NO_3)_2(aq) + Pb(s)$

(e) $2Na(s) + 2H_2O \rightarrow 2NaOH(aq) + H_2(g)$
(f) N.R.

**14-30** $Fe + 2H_2O \rightarrow$ N.R.
$Fe + 2H^+ \rightarrow Fe^{2+} + H_2$
Acid rain has a higher $H^+(aq)$ concentration, thus making the second reaction possible.

**14-32** The spontaneous reaction is:

$$Pb(NO_3)_2(aq) + Fe(s) \rightarrow Fe(NO_3)_2(aq) + Pb(s)$$

$$\text{Anode: } Fe(s) \rightarrow Fe^{2+}(aq) + 2e^-$$

$$\text{Cathode: } Pb^{2+}(aq) + 2e^- \rightarrow Pb(s)$$

**14-33** The Daniell cell is more powerful. The greater the separation between oxidizing and reducing agent in the table, the more powerful the cell.

**14-35** The strongest oxidizing agent is reduced the easiest. Thus the reduction of $Ag^+$ (i.e., $Ag^+ + e^- \rightarrow Ag$) occurs first. This procedure can be used to purify silver.

## Review Test on Chapters 11–14

### Multiple Choice

**1.** (d) **2.** (d) **3.** (b) **4.** (a) **5.** (a) **6.** (c) **7.** (e) **8.** (d) **9.** (b) **10.** (c)
**11.** (d) **12.** (e) **13.** (a) **14.** (c) **15.** (c) **16.** (e) **17.** (d) **18.** (b)
**19.** (b) **20.** (b) **21.** (e) **22.** (c) **23.** (c) **24.** (b) **25.** (c) **26.** (b)
**27.** (c) **28.** (a) **29.** (d) **30.** (e)

### Problems

**1.** $1.82 \times 10^4$ cal
**2.** (a) 0.480 $M$ (b) 0.0873 $M$ (c) 625 mL
**3.** (a) $Pb(NO_3)_2(aq) + 2NaI(aq) \rightarrow PbI_2(s) + 2NaNO_3(aq)$
(b) 1.7 L
**4.** (a) $H_2SO_4(aq) + 2LiOH(aq) \rightarrow Li_2SO_4(aq) + 2H_2O$
(b) 150 mL
**5.** (a) $Sn + 4HNO_3 \rightarrow SnO_2 + 4NO_2 + 2H_2O$
(b) 3.2 g $NO_2$
**6.** (a) $Pb^{2+} + 2NO_3^- + 2Na^+ + 2I^- \rightarrow PbI_2 + 2Na^+ + 2NO_3^-$
$Pb^{2+} + 2I^- \rightarrow PbI_2$
(b) $2Li^+ + 2OH^- + 2H^+ + SO_4^{2-} \rightarrow 2H_2O + 2Li^+ + SO_4^{2-}$
$2OH^- + 2H^+ \rightarrow 2H_2O$
**7.** $-0.809$ °C
**8.** (a) Sn (b) $HNO_3$ (c) $SnO_2$ (d) $NO_2$ (e) $HNO_3$ (f) Sn
**9.** (a) $HOCl + H_2O \rightleftharpoons H_3O^+ + OCl^-$
(b) $NH_3 + H_2O \rightleftharpoons NH_4^+ + OH^-$

(c) $KOH + HNO_2 \rightarrow KNO_2 + H_2O$
(d) $H_3AsO_3 + 2NaOH \rightarrow Na_2HAsO_3 + 2H_2O$

**10.** (a) N.R. (b) $CaO + H_2O \rightarrow Ca(OH)_2$ (c) $NH_4^+ + H_2O \rightleftharpoons NH_3 + H_3O^+$
(d) $CO_2 + H_2O \rightarrow H_2CO_3$ (e) $ClO^- + H_2O \rightleftharpoons HClO + OH^-$ (f) N.R.

**11.** (a) $1.5 \times 10^{-11}$ (b) pH = 10.82 (c) pOH = 3.18

**12.** (a) nonspontaneous
(b) $2Cr^{3+} + 3Mn \rightarrow 2Cr + 3Mn^{2+}$
(c) yes: $3I_2 + 2Cr \rightarrow 2Cr^{3+} + 6I^-$

## Chapter 15

**15-1** The molecules must have the proper orientation relative to each other and must have the minimum amount of energy.

**15-3** The rate of the forward reaction was at a maximum at the beginning of the reaction, the rate of the reverse reaction was at a maximum at the point of equilibrium.

**15-5** (a) As the temperature increases, the frequency of collisions between molecules increases as well as the average energy of the collisions. Both contribute to the increased rate of reaction.
(b) The cooking of eggs initiates a chemical reaction that occurs more slowly at lower temperatures.
(c) The average energy of colliding molecules at room temperature is not sufficient to initiate a reaction between $H_2$ and $O_2$.
(d) A higher concentration of $O_2$ increases the rate of combustion.
(e) When a solid is finely divided, a greater surface area is available for collisions with oxygen molecules. Thus it burns faster.
(f) The souring of milk is a chemical reaction that slows as the temperature drops. It takes several days in a refrigerator.

**15-7** Products are easier to form because the activation energy for the forward reaction is less than for the reverse reaction. This is true of all exothermic reactions. The system should come to equilibrium faster starting with pure reactants.

**15-9** (a) $K_{eq} = \dfrac{[COCl_2]}{[CO][Cl_2]}$ (b) $K_{eq} = \dfrac{[CO_2][H_2]^4}{[CH_4][H_2O]^2}$

(c) $K_{eq} = \dfrac{[Cl_2]^2[H_2O]^2}{[HCl]^4[O_2]}$ (d) $K_{eq} = \dfrac{[CH_3Cl][HCl]}{[CH_4][Cl_2]}$

**15-11** products

**15-13** There will be an appreciable concentration of both reactants and products present at equilibrium.

**15-14** $K_{eq} = 0.34$

**15-16** $K_{eq} = \dfrac{[CO_2][H_2]^4}{[CH_4][H_2O]^2} = \dfrac{[2.20/30][4.00/30]^4}{[6.20/30][3.00/30]^2} = 0.0112$

**15-18** (a) $[H_2] = [I_2] = 0.30$

(b) $[H_2] = 0.30$; $[I_2] = 0.50$
(c) $[HI] = 0.60 - 0.20 = 0.40$; $[H_2] = 0.10$; $[I_2] = 0.10$
(d) $K_{eq} = 6.3 \times 10^{-2}$
(e) For the reverse reaction, $K_{eq} = 16$. This is a smaller value than that used in Table 15-1. This indicates that the equilibrium in this problem was established at a different temperature than that in Table 15-1.

**15-20** (a) If $[N_2]_{eq} = 0.40$, then 0.10 mol/L of $N_2$ reacted. This means that 0.30 mol/L $H_2$ reacted forming an additional 0.20 mol/L of $NH_3$.
$[H_2] = 0.50 - 0.30 = 0.20$ mol/L
$[NH_3] = 0.50 + 0.20 = 0.70$ mol/L
(b) $K_{eq} = 153$

**15-21** (a) The concentration of $O_2$ that reacts is

$$0.25 \cancel{\text{mol NH}_3} \times \frac{5 \text{ mol } O_2}{4 \cancel{\text{mol NH}_3}} = 0.31 \text{ mol } O_2$$

(b) $[NH_3] = 0.75$; $[O_2] = 0.69$; $[NO] = 0.25$; $[H_2O] = 0.38$
(c) $K_{eq} = \dfrac{[NO]^4[H_2O]^6}{[NH_3]^4[O_2]^5} = \dfrac{[0.25]^4[0.38]^6}{[0.75]^4[0.69]^5}$

**15-23** $[PCl_5] = 0.28$

**15-25** $[H_2O] = 0.26$

**15-26** Let $x = [HCl] = [CH_3Cl]$, then $\dfrac{x^2}{[CH_4][Cl_2]} = K_{eq} = 56.$

$$\frac{x^2}{(0.20)(0.40)} = 56 \qquad x^2 = 4.48 \qquad x = \underline{\underline{2.1 \text{ mol/L}}}$$

**15-28** (a) $K_a = \dfrac{[H_3O^+][OBr^-]}{[HOBr]}$ (d) $K_a = \dfrac{[H_3O^+][SO_3^{2-}]}{[HSO_3^-]}$

(b) $K_b = \dfrac{[NH_4^+][OH^-]}{[NH_3]}$ (e) $K_a = \dfrac{[H_3O^+][H_2PO_4^-]}{[H_3PO_4]}$

(c) $K_a = \dfrac{[H_3O^+][HSO_3^-]}{[H_2SO_3]}$ (f) $K_b = \dfrac{[H_2N(CH_3)_2^+][OH^-]}{[HN(CH_3)_2]}$

**15-30** The acid HB is weaker because it produces a smaller $H_3O^+$ concentration (higher pH). The stronger acid HX has the larger value of $K_a$.

**15-32** (a) $[HOCN] = 0.20 - 0.0062 = 0.19$
(b) $K_a = \dfrac{[H_3O^+][OCN^-]}{[HOCN]} = \dfrac{(6.2 \times 10^{-3})(6.2 \times 10^{-3})}{(0.19)} = 2.0 \times 10^{-4}$
(c) pH = 2.21

**15-33** (a) $HX + H_2O \rightleftharpoons H_3O^+ + X^-$; $K_a = \dfrac{[H_3O^+][X^-]}{[HX]}$
(b) $[H_3O^+] = 0.100 \times 0.58 = 0.058 = [X^-]$
$[HX] = 0.58 - 0.058 = 0.52$
(c) $K_a = 6.5 \times 10^{-3}$
(d) pH = 1.24

**15-36** $Nv + H_2O \rightleftharpoons NvH^+ + OH^-$
$K_b = 6.6 \times 10^{-6}$

**15-37** $[H_3O^+] = 0.300 - 0.277 = 0.023$; pH $= 1.64$; $K_a = 1.9 \times 10^{-3}$

**15-38** pH $= 4.43$

**15-40** $[OH^-] = 3.2 \times 10^{-3}$

**15-42** pH $= 12.43$

**15-43** pH $= 9.40$

**15-45** pH $= 8.20$

**15-47** pH $= 4.14$

**15-48** (a) right (b) left (c) left (d) right (e) right (f) no effect (g) decrease (h) increase the yield but greatly decrease the rate of formation of $NH_3$

**15-50** (a) increase (b) increase (c) increase (d) decrease (e) decrease (f) no effect

**15-51** (a) endothermic (b) cooler (c) no effect, since there are the same number of moles of gas on both sides of the equation

## Chapter 16

**16-1**

```
          H                     H             H
          |                      \           /
(a)   H—C—Br:          (b)        C=C=C
          |                      /           \
          H                     H             H

       H           H   H              H   H   H   H
        \          |   |              |   |   |   |
(c)      C=C—C—C—H          H—C—C=C—C—H
        /     |   |   |               |               |
       H      H   H   H               H               H

        H   H                H       H
        |   |                 \     /
    H—C—C—H                   C         H
        |   |                 /     \    /
    H—C—C—H           H—C———C—C—H
        |   |                 |       |    \
        H   H                 H       H     H

        H
        |   ..
(d) H—C—N—H
        |   |
        H   H

        H   H                  H         H
        |   |      ..          |    ..   |
(e) H—C—C—N—H,          H—C—N—C—H
        |   |   |              |    |    |
        H   H   H              H    H    H
```

**(f)** 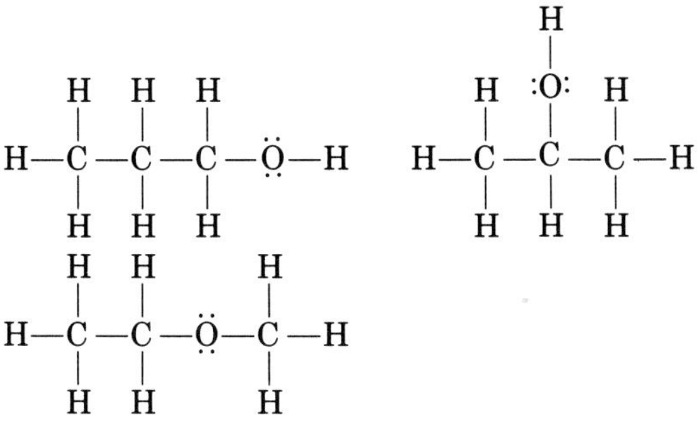
$$\begin{array}{ccccccccc} & & H & & H & & H & & \\ & & | & & | & & | & & \\ H & — & C & — & C & — & C & — & \ddot{\underset{..}{O}} & — & H \\ & & | & & | & & | & & \\ & & H & & H & & H & & \end{array} \qquad \begin{array}{ccccccc} & & & & H & & \\ & & & & | & & \\ & & H & & :\!O\!: & & H \\ & & | & & | & & | \\ H & — & C & — & C & — & C & — & H \\ & & | & & | & & | \\ & & H & & H & & H \end{array}$$

$$\begin{array}{ccccccccc} & & H & & H & & & & H \\ & & | & & | & & & & | \\ H & — & C & — & C & — & \ddot{\underset{..}{O}} & — & C & — & H \\ & & | & & | & & & & | \\ & & H & & H & & & & H \end{array}$$

**16-3** $CH_3—CH_2—CH_2—CH_2—CH_2—CH_3$ $\quad$ $\begin{array}{c} CH_3—CH_2—CH_2—CH—CH_3 \\ \qquad\qquad\qquad\qquad | \\ \qquad\qquad\qquad\qquad CH_3 \end{array}$

$$\begin{array}{c} CH_3—CH_2—CH—CH_2—CH_3 \\ | \\ CH_3 \end{array} \qquad \begin{array}{c} \qquad\qquad CH_3 \\ \qquad\qquad | \\ CH_3—CH_2—C—CH_3 \\ \qquad\qquad | \\ \qquad\qquad CH_3 \end{array} \qquad \begin{array}{c} CH_3—CH—CH—CH_3 \\ \quad | \qquad | \\ \quad CH_3 \; CH_3 \end{array}$$

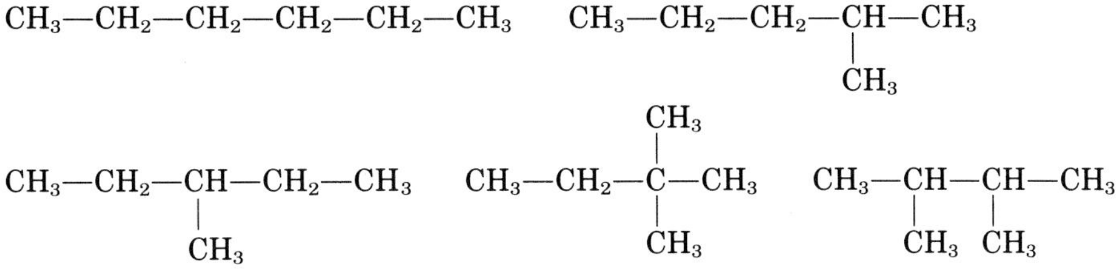

**16-5** (a) no (no oxygen in second compound) (b) no (unequal number of carbons) (c) yes (d) no (unequal number of carbons) (e) These two structures represent the same compound.

**16-7** (a) six (*hex-*) (b) eight (*oct-*) (c) nine (*non-*) (d) seven (*hept-*)

**16-9** (a) $CH_3$—, methyl; $CH_3CH_2$—, ethyl (b) $(CH_3)_2CH$—, isopropyl (c) $CH_3CH_2$—, ethyl; $(CH_3)_3C$—, *t*-butyl (d) $CH_3CH_2CH_2$—, *n*-butyl; $CH_3CH_2CH_2CH_2$—, *n*-propyl

**16-11** This general formula is that of a straight-chain alkene with one double bond or it can also apply to a cyclic alkane.

**16-13** (a) alkene (b) alkane (c) alkene (d) alkyne (e) alkane (f) alkene (g) alkane

**16-14** (c)

**16-16** **(a)** 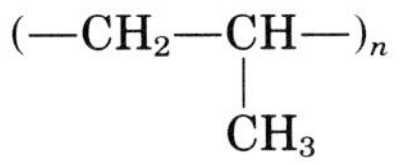 $\begin{array}{c} (—CH_2—CH—)_n \\ \qquad | \\ \qquad CH_3 \end{array}$ polypropylene

**(b)** 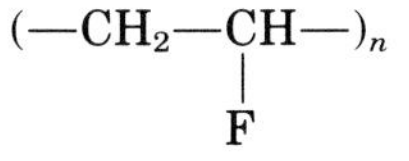 $\begin{array}{c} (—CH_2—CH—)_n \\ \qquad | \\ \qquad F \end{array}$ polyvinylfluoride

**(c)** $(—CH_2—CF_2—)_n$ polydifluoroethylene

**(d)** 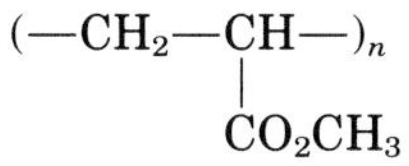 $\begin{array}{c} (—CH_2—CH—)_n \\ \qquad | \\ \qquad CO_2CH_3 \end{array}$ polymethyacrylate

**16-18**

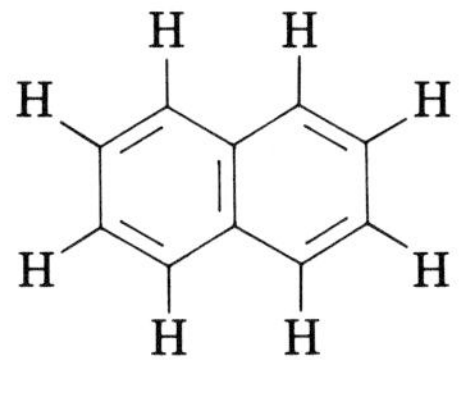

(Naphthalene structure: two fused benzene rings with eight H atoms.)

H H

H H

H H

H H

**16-20** Cl, Cl (1,2-positions); Cl, Cl (1,3-positions); Cl, Cl (1,4-positions) on benzene rings

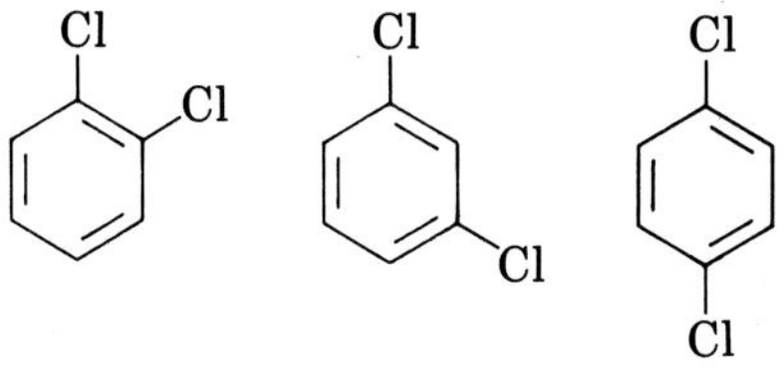

**16-21** **(a)** —C=C—, alkene; and $\overset{\overset{\displaystyle O}{\|}}{C}$—H, aldehyde
**(b)** —$NH_2$, amine; and $-\underset{\underset{\displaystyle O}{\|}}{C}-$, ketone
**(c)** —O—, ether; and $-\underset{\underset{\displaystyle O}{\|}}{C}-NH_2$, amide
(d) —C≡C—, alkene; and —OH alcohol
(e) —$COOCH_3$, ester
(f) —COOH, carboxylic acid

**16-23** (a) In an ether the O is between two C's. In an alcohol the O is between a C and a H.
(b) In an aldehyde the carbonyl (i.e., C=O) is between a C and a H. In a ketone the carbonyl is between two C's.
(c) In an amine the $NH_2$ group is attached to a hydrocarbon group. In an amide the $NH_2$ group is attached to a carbonyl group.
(d) Carboxylic acids have a hydrogen attached to an oxygen; in esters the hydrogen is replaced by a hydrocarbon group.

**16-25** Carboxylic acids contain a carboxyl group, and alcohols contain a hydroxyl group.

**16-27** (a) ketone (b) alkene (c) alkane (d) aldehyde (e) aromatic (f) alcohol

# Photo Credits

**Chapter 1:**
Opener: Dr. William J. Jahoda/Photo Researchers. Page 4: Figure 1.1: Bruce Anspach/EPA Newsphoto. Page 7: Figure 1.2(a): Courtesy The American Museum of Natural History. Figure 1.2(b): Gordon S. Smith/Photo Researchers. Figure 1.2(c): Courtesy Good Year. Page 13: Figure 1.10(a): Kathy Bendo. Figure 1.10(b): Peter M. Lerman. Figure 1.10(c): Peter M. Lerman. Page 15: Figure 1.13: Peter M. Lerman.

**Chapter 2:**
Opener: Frank Siteman/Jeroboam. Page 30: Figure 2.2(a): Courtesy Mt. Wilson and Palomar Observatories. Figure 2.2(c): P. Kolberg/Taurus Photos. Page 32: Figure 2.3(a): Peter M. Lerman. Figure 2.3(b): Kathy Bendo and Paul Cope. Figure 2.3(c): Courtesy Mettler Instrument Corp. Page 35: Figure 2.5(a): Courtesy Ohio Department of Transportation. Figure 2.5(b): Peter M. Lerman. Figure 2.5(c): Kathy Bendo. Page 46: Figure 2.9: Lorinda Morris.

**Chapter 3:**
Opener: A.V. Crewe. Page 73: Figure 3.12: Courtesy University of Texas at Houston. Page 75: Figure 3.14: Courtesy Fermilab. Page 80: Figure 3.18: Eric Knoll/Taurus Photos.

**Chapter 4:**
Opener: Pat Benic/UPI/Bettmann Archive. Page 93: Figure 4.1(a): Peter M. Lerman. Figure 4.1(b): Peter M. Lerman.

**Chapter 5:**
Opener: Mimi Forsyth/Monkmeyer Press Photo. Page 120: Figure 5.3: Peter M. Lerman. Page 121: Figure 5.4: Kathy Bendo. Figure 5.5: Kathy Bendo. Page 122: Figure 5.6 left: Peter M. Lerman. Figure 5.6 center: Courtesy of the Diamond Institute. Figure 5.6 right: Kathy Bendo. Figure 5.7 left: Kathy Bendo. Figure 5.7 right: Peter M. Lerman. Page 123: Figure 5.8 left: Courtesy Linde Division, Union Carbide. Figure 5.8 right: Joel E. Arem, PhD. Figure 5.9 left: Life Science Library. Figure 5.9 right: Peter M. Lerman.

**Chapter 6:**
Opener: Kathy Bendo.

**Chapter 7:**
Opener: Peter M. Lerman. Page 177: Figure 7.1(a): Peter M. Lerman. Figure 7.1(b): Peter M. Lerman.

**Chapter 8:**
Opener: Florence M. Harrison/Taurus Photos. Page 189: Figure 8.1: Peter M. Lerman. Page 195: Figure 8.2: Peter M. Lerman. Page 204: Figure 8.3: Peter M. Lerman.

**Chapter 9:**
Opener: Bethlehem Steel-Erie Mining Co. Page 216: Figure 9.2: Doug Magee/Art Resource. Page 217: Figure 9.3: Peter M. Lerman. Page 218: Figure 9.4: Peter M. Lerman. Page 219: Figure 9.5(a): Peter M. Lerman. Figure 9.5(b): Peter M. Lerman.

**Chapter 10:**
Opener: NASA.

**Chapter 11:**
Opener: Paolo Koch/Rapho-Photo Researchers. Page 294: Figure 11.10(a): Peter M. Lerman. Figure 11.10(b): Peter M. Lerman.

**Chapter 12:**
Opener: Menschen Freund/Taurus Photos. Page 326: Figure 12.13(a): Peter M. Lerman. Figure 12.13(b): EPA/Art Resource.

**Chapter 13:**
Opener: Peter M. Lerman. Page 335: Figure 13.1: Kathy Bendo. Page 347: Figure 13.8: Peter M. Lerman. Page 361: Figure 13.11(a): Westfälisches Amt Fur Denkmalpflege. Figure 13.11(b): Westfälisches Amt Fur Denkmalpflege.

**Chapter 14:**
Opener: Daniel S. Brody/Art Resource. Page 373: Figure 14.1(a): Peter M. Lerman. Figure 14.1(b): Life Science Library. Figure 14.1(c): Kathy Bendo. Page 389: Figure 14.8: Peter M. Lerman.

**Chapter 15:**
Opener: Ken Karp. Page 426: Figure 15.11: The Sulfer Institute. Page 429: Figure 15.12: Visione/Art Resource.

**Chapter 16:**
Opener: Alan Pitcairn/Grant Heilman. Page 441: Figure 16.1: Peter M. Lerman. Page 447: Figure 16.2: Courtesy of Standard Oil Co. of California. Page 450: Figure 16.4(a): Exxon Chemical Company, USA. Figure 16.4(b): Courtesy Owen-Illinois. Page 451: Figure 16.4(c): Courtesy E.I. du Pont de Nemours & Company. Page 456: Figure 16.5: Kathy Bendo. Page 459: Figure 16.6: Kathy Bendo. Page 466: Figure 16.7: Courtesy of Dupont Company.

**APPENDIX:**
Opener: Harvey Stein.

# Index

# The Modern Periodic Table of the Elements

| 1 (atomic number) |
|---|
| H |
| 1.0079 (atomic mass) |

| PERIODS | IA | IIA | IIIB | IVB | VB | VIB | VIIB | VIII | VIII | VIII | IB | IIB | IIIA | IVA | VA | VIA | VIIA | NOBLE GASES O |
|---|---|---|---|---|---|---|---|---|---|---|---|---|---|---|---|---|---|---|
| 1 | 1<br>H<br>1.00797 | | | | | | | | | | | | | | | | | 2<br>He<br>4.00260 |
| 2 | 3<br>Li<br>6.941 | 4<br>Be<br>9.01218 | | | | | | | | | | | 5<br>B<br>10.81 | 6<br>C<br>12.01115 | 7<br>N<br>14.0067 | 8<br>O<br>15.9994 | 9<br>F<br>18.99840 | 10<br>Ne<br>20.179 |
| 3 | 11<br>Na<br>22.98977 | 12<br>Mg<br>24.305 | | | | | | | | | | | 13<br>Al<br>26.98154 | 14<br>Si<br>28.086 | 15<br>P<br>30.97376 | 16<br>S<br>32.06 | 17<br>Cl<br>35.453 | 18<br>Ar<br>39.948 |
| 4 | 19<br>K<br>39.098 | 20<br>Ca<br>40.08 | 21<br>Sc<br>44.9559 | 22<br>Ti<br>47.90 | 23<br>V<br>50.9414 | 24<br>Cr<br>51.996 | 25<br>Mn<br>54.9380 | 26<br>Fe<br>55.847 | 27<br>Co<br>58.9332 | 28<br>Ni<br>58.71 | 29<br>Cu<br>63.546 | 30<br>Zn<br>65.38 | 31<br>Ga<br>69.72 | 32<br>Ge<br>72.59 | 33<br>As<br>74.9216 | 34<br>Se<br>78.96 | 35<br>Br<br>79.904 | 36<br>Kr<br>83.80 |
| 5 | 37<br>Rb<br>85.4678 | 38<br>Sr<br>87.62 | 39<br>Y<br>88.9059 | 40<br>Zr<br>91.22 | 41<br>Nb<br>92.9064 | 42<br>Mo<br>95.94 | 43<br>Tc<br>98.9062 | 44<br>Ru<br>101.07 | 45<br>Rh<br>102.9055 | 46<br>Pd<br>106.4 | 47<br>Ag<br>107.868 | 48<br>Cd<br>112.40 | 49<br>In<br>114.82 | 50<br>Sn<br>118.69 | 51<br>Sb<br>121.75 | 52<br>Te<br>127.60 | 53<br>I<br>126.9045 | 54<br>Xe<br>131.30 |
| 6 | 55<br>Cs<br>132.9054 | 56<br>Ba<br>137.34 | 57<br>*La<br>138.9055 | 72<br>Hf<br>178.49 | 73<br>Ta<br>180.9479 | 74<br>W<br>183.85 | 75<br>Re<br>186.2 | 76<br>Os<br>190.2 | 77<br>Ir<br>192.22 | 78<br>Pt<br>195.09 | 79<br>Au<br>196.9665 | 80<br>Hg<br>200.59 | 81<br>Tl<br>204.37 | 82<br>Pb<br>207.19 | 83<br>Bi<br>208.9804 | 84<br>Po<br>(210) | 85<br>At<br>(210) | 86<br>Rn<br>(222) |
| 7 | 87<br>Fr<br>(223) | 88<br>Ra<br>226.0254 | 89<br>†Ac<br>(227) | 104<br>Ku<br>(261) | 105<br>Ha<br>(260) | 106 | 107 | | 109 | | | | | | | | | |

| | | | | | | | | | | | | | | |
|---|---|---|---|---|---|---|---|---|---|---|---|---|---|---|
| * | 58<br>Ce<br>140.32 | 59<br>Pr<br>140.9077 | 60<br>Nd<br>144.24 | 61<br>Pm<br>(147) | 62<br>Sm<br>150.4 | 63<br>Eu<br>151.96 | 64<br>Gd<br>157.25 | 65<br>Tb<br>158.9254 | 66<br>Dy<br>162.50 | 67<br>Ho<br>164.9304 | 68<br>Er<br>166.264 | 69<br>Tm<br>168.9342 | 70<br>Yb<br>174.06 | 71<br>Lu<br>174.97 |
| † | 90<br>Th<br>232.0381 | 91<br>Pa<br>231.0359 | 92<br>U<br>238.029 | 93<br>Np<br>237.0482 | 94<br>Pu<br>(244) | 95<br>Am<br>(243) | 96<br>Cm<br>(247) | 97<br>Bk<br>(247) | 98<br>Cf<br>(251) | 99<br>Es<br>(254) | 100<br>Fm<br>(257) | 101<br>Md<br>(258) | 102<br>No<br>(255) | 103<br>Lr<br>(256) |